Advances in Biochemistry in Health and Disease

Series Editor: Naranjan S. Dhalla
Winnipeg, Manitoba, Canada

Volume 1: S. K. Cheema (ed), *Biochemistry of Atherosclerosis*

Volume 2: S. W. Schaffer and M-Saadeh Suleiman (eds), *Mitochondria: The Dynamic Organelle*

Volume 3: A. K. Srivastava and M. B. Anand-Srivastava (eds), *Signal Transduction in the Cardiovascular System in Health and Disease*

Volume 4: B. Ostadal and N. S. Dhalla (eds), C*ardiac Adaptations-Molecular Mechanisms*

Volume 5: B. I. Jugdutt and N. S. Dhalla (eds), *Cardiac Remodeling-Molecular Mechanisms*

Volume 6: J. L. Mehta and N. S. Dhalla (eds), *Biochemical Basis and Therapeutic Implications of Angiogenesis*

Volume 7: S. Chakraborti and N. S. Dhalla (eds), *Proteases in Health and Disease*

Volume 8: N. S. Dhalla and S. Chakraborti (eds), *Role of Proteases in Cellular Dysfunction*

Volume 9: B. Turan and N. S. Dhalla (eds), *Diabetic Cardiomyopathy: Biochemical and Molecular Mechanisms*

Volume 10: P. S. Tappia and N. S. Dhalla (eds), *Phospholipases in Heath and Disease*

Volume 11: G. D. Lopaschuk and N. S. Dhalla (eds), *Cardiac Energy Metabolism in Health and Disease*

Volume 12: K. Ahmed, O-G. Issinger, and R. Szyszka (eds), *Protein Kinase CK2 Cellular Function in Normal and Disease States*

More information about this series at http://www.springer.com/series/7064

Khalil Ahmed • Olaf-Georg Issinger
Ryszard Szyszka
Editors

Protein Kinase CK2 Cellular Function in Normal and Disease States

Springer

Editors
Khalil Ahmed
Department of Laboratory Medicine
and Pathology
University of Minnesota Medical School
and Minneapolis VA Health Care System
Minneapolis, MN, USA

Olaf-Georg Issinger
Department of Biochemistry
and Molecular Biology
University of Southern Denmark
Odense, Denmark

Ryszard Szyszka
Department of Molecular Biology
The John Paul II Catholic University
of Lublin Institute of Biotechnology
Lublin, Poland

Advances in Biochemistry in Health and Disease
ISBN 978-3-319-14543-3 ISBN 978-3-319-14544-0 (eBook)
DOI 10.1007/978-3-319-14544-0

Library of Congress Control Number: 2015933162

Springer Cham Heidelberg New York Dordrecht London

Printed on acid-free paper

Springer International Publishing AG Switzerland is part of Springer Science+Business Media (www.springer.com)

Preface

The selection of "Protein Kinase CK2 Cellular Function in Normal and Disease State," as part of the Springer series "Advances in Biochemistry in Health and Disease" (Volume 12) under the leadership of Professor Naranjan S. Dhalla, represents an important mark of the recognition of the emerging importance of Protein Kinase CK2 as a "master regulator" of cell function.

The field of protein kinases represents a particularly important area of biochemistry with over 500 corresponding genes (representing about 1.7 % of the genome). In the human kinome, most of the protein kinases form a "network" of relationships; however, protein kinase CK2 appears to stand on its own. Protein kinase CK2 (the acronym being derived from the misnomer casein kinase 2) is primarily a protein Ser/Thr kinase, although a small activity to certain Tyr residues has been detected. Its structure consists of a tetrameric complex with two catalytic subunits (α and α′) that are linked via the regulatory subunit (β); the abundance and composition of the subunits vary with tissues. As far as we know, CK2 does not require "activation" by chemical modifications, which suggests that it is a constitutively active protein kinase. This raises the question as to how its diverse functions are regulated. A number of hypotheses exist to address this paradox and include shuttling to various loci in response to specific signals, aggregation/disaggregation of the holoenzyme as a modulator of its activity, and interaction with a number of other proteins that allow for the signal transduction.

Function of CK2 in cells in normal and disease states has been the subject of much investigation for over four decades. Given the very large number of its potential substrates localized in different parts of the cell, numerous investigations have contributed to its role in, for example, the regulation of protein and nucleic acid synthesis, signals associated with cell growth, proliferation and cell death, diverse signal transduction pathways, viral activity, inflammation, angiogenesis, neuronal function, and organogenesis. These aspects highlight the importance of CK2 in different areas of cell biology in normal and disease states; however, as mentioned below, a preponderant number of investigations on CK2 focus on its function in cancer biology.

A number of important features of CK2 are well recognized—this is one of the most highly conserved and ubiquitous eukaryotic enzymes and it is essential for cell survival. Originally, it was thought that CK2 was predominantly involved in cell growth and proliferation in normal and cancer cells. However, later it became apparent that besides its role in cell proliferation, CK2 is also able to regulate cell death, and this aspect of CK2 function has provided a key link of the kinase to the cancer cell phenotype. Indeed, an important observation on CK2 in cancer is that it has been found to be elevated in all cancers examined compared with its level in corresponding normal cells. Importantly, this elevation is observed consistently at the protein level. Equally of note is the observation that its dysregulation in cancer is not due to any detected mutations or isoforms. Further, it appears that even a moderate downregulation of CK2 in cancer cells (compared with the normal cells) induces potent death activity in cancer cells. Thus, currently there is a large interest in CK2 in the cancer field, with much effort being directed toward developing strategies toward its targeting for cancer therapy. The theoretical concept is that targeting CK2 for cancer therapy would serve as a double-edged sword since both proliferation and evasion from death would be impacted, and further since cells cannot survive without CK2, the result would be elimination of the cancer cells.

It may be worth noting that working on protein kinase CK2 has distinct challenges largely due to the fact that it is not a trivial task to determine the activity of CK2 relating to specific functions. This attributes to the fact that it is not easy to identify precise modulations in the activity of CK2 in cells because of its localization in various cellular compartments. Further, unlike protein kinases such as MAP kinase where phospho-specific antibodies can be used to follow their "activation," no such approach is possible for CK2. Thus, one has to largely rely on measuring the changes in activity by following the alterations in phosphorylation levels of downstream targets of CK2. This in its own right can be a challenge due to the existence of a large number of CK2 substrates localized at different subcellular locations. Thus, one has to be careful in selecting these markers that can be used for following the effects on CK2 activity in the cell; a few examples include tracking changes in phosphorylation of molecules such as NFκB p65, Cdc37, and AKT, as well as changes in protein B23. This issue becomes particularly important in developing cancer therapy approaches targeting CK2 where it may be important to follow several marker substrates to determine the activity response in cells. Several contributions in this book provide an insight into these issues.

The book is divided into several sections to provide a cohesive set of topics and associated articles. The various headings in the order of presentation are as follows: Regulation and Structure–Function Studies of CK2; CK2 Control of Organismal and Cellular Functions; Function of CK2 in Cancer and its Therapeutic Targeting; and Studies Involving Small Molecule Inhibitors.

The first section (Part I) entitled "Regulation and Structure–Function Studies of CK2" includes four chapters. The first chapter is by Baier et al. and presents a review of the general aspects of yeast CK2 biology discussing aspects of structure and interacting partners of CK2, with a detailed discussion of the mechanism of regulation of yeast CK2 by the anti-silencing protein 1 (Asf1). The second chapter is by Hochscherf et al. and focuses on their pioneering studies on crystallographic structure of CK2 subunits. The authors review the extensive data they have generated on the structural dynamics of CK2 and the insights gained from these studies on the activity of the enzyme. The third chapter is by Lolli and Battistutta and is also devoted to the structural aspects of CK2 describing the structural nature of the interaction of the CK2 subunits which contribute to its enzymic activity. The fourth chapter in this section by Vélez-Bermúdez and Riera deals with maize CK2 (as a model of plant CK2) and discusses the unique characteristics of the regulation of this enzyme dealing with the functional aspects of CK2β1, resulting in the identification of ZmTGH as a novel partner of this subunit with a possible role in splicing process in the plant enzyme.

The second section (Part II) entitled "CK2 Control of Organismal and Cellular Functions" covers nine ensuing chapters. The first chapter in this section is by Ortega et al. and deals with the function of CK2 in organ development. The authors discuss their pioneering studies on the involvement of CK2 in embryonic development in different models as well as the role of CK2α in organ formation, homeostasis, and physiology. The chapter by Majot et al. continues along similar lines and presents an elaborate discussion of the function of CK2 in the developmental aspects of *Drosophila* through interaction of CK2 with numerous E(spl) proteins, thereby regulating different signaling pathways. The chapter by Welker et al. focuses on tissue-specific functions of CK2. These authors discuss their studies that have contributed to the investigation of novel substrates/partners and unique features of CK2 interactions with them in a tissue-specific manner. The subsequent contributions focus largely on the cellular functions of CK2. In this context, the chapter by Girardi and Ruzzene presents a detailed account of the function of CK2 as it pertains to the phosphorylation of Akt and effect on the Akt pathway. The chapter by Nimmanon and Taylor describes the knowledge that has emerged on the function of CK2 in zinc signaling, identifying ZIP channels as novel substrates of CK2 and the function of these interactions on cell biology and pathology. The chapter by Trembley et al. presents the original conceptual development of the function of CK2 as a suppressor of apoptosis and attempts to link the functional activity of CK2 to diverse pathways in the cell death machinery. The chapter points to a novel role of CK2 inhibition in triggering Ca^{2+} signaling that rapidly targets mitochondria to initiate the earliest events in the induction of apoptosis. The chapter by Litchfield and Gyenis continues on the theme of CK2 interaction with other protein substrates and provides a detailed discussion of the novel approaches employed to study the involvement of CK2-mediated phosphorylation of substrates and regulation of other cellular pathways that impact on the cellular biology in normal and disease states. The chapter by Homma et al. describes the identification of CK2 substrate targets in the nucleus during cell cycle progression, identifying the involvement of CK2 in the dynamics of gene regulation. The final chapter in this section is by Tibaldi et al. and addresses the issue of CK2's distinction from the genuine casein kinases. Their contribution to the identification of G-CK/Fam20C protein kinase and aspects of its similarity to CK2 in terms of pleiotropy and involvement of certain pathological states is discussed.

The next series of chapters under Part III fall under the heading "Function of CK2 in Cancer and Its Therapeutic Targeting." This section covers several articles that discuss the extensive work currently in progress for the function of CK2 in cancer and its targeting for cancer therapy. The chapter by Roelants et al. details their novel studies on the role of CK2 in the pathogenesis of renal cell carcinoma (RCC) and discusses the observations on the expression of various subunits of CK2 in a cohort of RCC samples and the potential of CK2 targeting for therapy of RCC. The contribution by Benveniste et al. discusses the involvement of CK2 in the oncogenic inflammatory signaling pathways with a focus on two important signaling pathways (NFκB and JAK/STAT) in glioma, breast cancer, pancreatic cancer, multiple myeloma, leukemia, head and neck squamous cell carcinoma, and non-small cell lung cancer. The chapter by Perea et al. describes the progress in their originally described therapeutic agent CIGB-300 for targeting CK2 phosphorylation sites in target proteins. They present the current status of their investigations on the pharmacokinetic studies of this therapeutic drug and the latest information on their observations of clinical evaluation in Phase II trial in women with cervical cancer. The final chapter in this series by Ahmed et al. provides the account of their ongoing approach to molecular targeting of CK2 in cancer therapy. Based on their original postulate that targeting of CK2 in cancer therapy may be most usefully employed by a cancer cell directed approach, they provide the latest information on the use of a nanoencapsulated RNAi approach to specifically target CK2 catalytic subunits in prostate cancer and head and neck squamous cell carcinoma.

The final section (Part IV) of the book focuses on "Studies Involving CK2 Small Molecule Inhibitors." The various articles in this section describe the ongoing effort at identifying new small molecule inhibitors that may be useful for therapeutic targeting of CK2 and for other biochemical studies of the enzyme function. The first chapter under this section is by Rasmussen et al. who have described the screening of the Drug Therapeutic Program (DTP) library of the NIH/NCI against CK2 and the identification of several small molecule compounds that exert varying degrees of inhibition of CK2 in vitro and in cell culture models, thus indicating the potential development of novel inhibitors of diverse natural background. The chapter by Miyata discusses in detail the inhibitors of CK2 and DYRK family protein kinases describing the overlap in their activity and the biological and clinical implications of the two kinases. The final chapter in this section by Baier and Szyszka discusses the underlying mechanisms that may influence the activity of CK2 inhibitors, pointing to the role of the nature of the substrate/enzyme complex which might be influenced by the composition of CK2 tetramer.

The various chapters in this book are authored by individuals who have maintained a keen dedication to the field of protein kinase CK2. An international committee on protein kinase CK2 holds an International Conference on CK2 every 3 years, and the latest meeting was held at The John Paul II Catholic University of Lublin in 2013 sponsored by IUBMB. The authors in this book also participated in that conference and their contributions reflect the latest developments in the field. These contributions also provide reviews on the development of the various concepts on the functions of CK2. Thus, this book should provide an important starting point to investigators interested in cancer biology, protein kinases, and development of targeted therapies. The diversity of the topics should be of interest to graduate students who may wish to enhance their understanding of protein kinase CK2 and potentially undertake investigations in this area in the future.

The editors would like to express their deep appreciation of the continued support by Professor Naranjan S. Dhalla, Editor-in-Chief of the Springer series entitled "Advances in Biochemistry in Health and Disease." We appreciate the help of Ms. Susan Safren as well as the staff at the Springer Media, New York, in the preparation of this book. Thanks are also due to Dr. Janeen Trembley, Dr. Betsy Kren, and Dr. Andrea Baier for their assistance during the preparation of this work.

Minneapolis, USA — Khalil Ahmed
Odense, Denmark — Olaf-Georg Issinger
Lublin, Poland — Ryszard Szyszka

Contents

About the Editors

Khalil Ahmed is a Professor at the University of Minnesota and a Senior Research Career Scientist at Minneapolis VA Health Care System, Minneapolis, Minnesota, USA. He has a long history of studies on functional biology of protein kinase CK2 in normal and neoplastic cells, originally described the signal-mediated dynamic shuttling of CK2 in the cell, and he discovered the role of CK2 as a suppressor of apoptosis. His current research focuses on studies of mechanism of CK2 regulation of cell death and on developing molecular therapeutic strategies using a nanomedicine approach for the treatment of prostate and other cancers.

Olaf-Georg Issinger is a Professor at the University of Southern Denmark, Odense, Denmark. He has been involved in the initial cloning and characterization of protein kinase CK2, a prerequisite for elucidation of its structure in the absence and presence of specific interaction molecules. His current research focuses on the exploration of cellular signaling pathways in cell lines with respect to the role of various protein kinases using newly characterized kinase inhibitors identified by screening small chemical compound libraries.

Ryszard Szyszka is a Professor at The John Paul II Catholic University of Lublin, Poland. He is head of the Department of Molecular Biology and Dean of the Faculty of Biotechnology and Environmental Sciences. His research focuses on the identification and characterization of new substrates of protein kinase CK2 from *Saccharomyces cerevisiae*. Further areas of interest include the structure and regulation of yeast CK2 and discovery of novel CK2 inhibitors.

Contributors

Khalil Ahmed Department of Laboratory Medicine and Pathology, University of Minnesota Medical School and Minneapolis VA Health Care System, Minneapolis, MN, USA

Department of Laboratory Medicine and Pathology, University of Minnesota, Minneapolis, MN, USA

Masonic Cancer Center, University of Minnesota, Minneapolis, MN, USA

Cellular and Molecular Biochemistry Research Laboratory (151), Veterans Affairs Medical Center, One Veterans Drive, Minneapolis, MN, USA

Ewa Alikowska Department of Molecular Biology, Faculty of Biotechnology and Environmental Sciences, Institute of Biotechnology, The John Paul II Catholic University of Lublin, Lublin, Poland

Daniel F. Alonso Molecular Oncology Laboratory, National University of Quilmes, Buenos Aires, Argentina

Valentin Arnoux Centre hospitalo-universitaire Albert Michallon, service d'urologie, Grenoble, France

Giorgio Arrigoni Department of Biomedical Sciences and CNR, Institute of Neuroscience, University of Padova, Padova, Italy

Proteomic Center of Padova University, Padova, Italy

Cédric Aubert Département d'Anatomie et Cytologie Pathologiques et Plateforme de Génétique Moléculaire des Tumeurs, Pôle de Biologie et Pathologie, CHU Grenoble, La Tronche, France

Andrea Baier Department of Molecular Biology, Faculty of Biotechnology and Environmental Sciences, Institute of Biotechnology, The John Paul II Catholic University of Lublin, Lublin, Poland

Idania Baladrón Clinical Research Area, Center for Genetic Engineering and Biotechnology, Havana, Cuba

Mohna Bandyopadhyay Department of Biology, West Virginia University, Morgantown, WV, USA

Roberto Battistutta Department of Chemical Sciences, University of Padua, Padua, Italy

Venetian Institute for Molecular Medicine (VIMM), Padua, Italy

Fernando Benavent Molecular Oncology Laboratory, National University of Quilmes, Buenos Aires, Argentina

Etty N. Benveniste Department of Cell, Developmental and Integrative Biology, University of Alabama at Birmingham, Birmingham, AL, USA

Benoit Bestgen INSERM, Unité 1036, Biology of Cancer and Infection, Grenoble, France

UJF-Grenoble 1, Biology of Cancer and Infection, Grenoble, France

CEA, DSV/iRTSV, Biology of Cancer and Infection, Grenoble, France

Ashok P. Bidwai Department of Biology, West Virginia University, Morgantown, WV, USA

Jean Boutonnat Département d'Anatomie et Cytologie Pathologiques et Plateforme de Génétique Moléculaire des Tumeurs, Pôle de Biologie et Pathologie, CHU Grenoble, La Tronche, France

Luca Cesaro Department of Biomedical Sciences and CNR, Institute of Neuroscience, University of Padova, Padova, Italy

Pierre Champelovier Département d'Anatomie et Cytologie Pathologiques et Plateforme de Génétique Moléculaire des Tumeurs, Pôle de Biologie et Pathologie, CHU Grenoble, La Tronche, France

Claude Cochet INSERM, Unité 1036, Biology of Cancer and Infection, Grenoble, France

UJF-Grenoble 1, Biology of Cancer and Infection, Grenoble, France

CEA, DSV/iRTSV, Biology of Cancer and Infection, Grenoble, France

Giorgio Cozza Department of Biomedical Sciences and CNR, Institute of Neuroscience, University of Padova, Padova, Italy

Laurence David-Boudet Département d'Anatomie et Cytologie Pathologiques et Plateforme de Génétique Moléculaire des Tumeurs, Pôle de Biologie et Pathologie, CHU Grenoble, La Tronche, France

Jean Luc Descotes Centre hospitalo-universitaire Albert Michallon, service d'urologie, Grenoble, France

Isabel Dominguez Department of Medicine, Boston University School of Medicine, Boston, MA, USA

Eve Duchemin-Pelletier INSERM, Unité 1036, Biology of Cancer and Infection, Grenoble, France

UJF-Grenoble 1, Biology of Cancer and Infection, Grenoble, France

CEA, DSV/iRTSV, Biology of Cancer and Infection, Grenoble, France

Hernán G. Fariña Molecular Oncology Laboratory, National University of Quilmes, Buenos Aires, Argentina

Odile Filhol INSERM, Unité 1036, Biology of Cancer and Infection, Grenoble, France

UJF-Grenoble 1, Biology of Cancer and Infection, Grenoble, France

CEA, DSV/iRTSV, Biology of Cancer and Infection, Grenoble, France

Roberto Gómez ELEA laboratories, Buenos Aires, Argentina

Claudia Götz Medizinische Biochemie und Molekularbiologie, Universität des Saarlandes, Homburg, Germany

Idrián García Clinical Research Area, Center for Genetic Engineering and Biotechnology, Havana, Cuba

Yanelda García Clinical Research Area, Center for Genetic Engineering and Biotechnology, Havana, Cuba

Sofia Giacosa INSERM, Unité 1036, Biology of Cancer and Infection, Grenoble, France

UJF-Grenoble 1, Biology of Cancer and Infection, Grenoble, France

CEA, DSV/iRTSV, Biology of Cancer and Infection, Grenoble, France

Cristina Girardi Department of Biomedical Sciences, University of Padova, Padova, Italy

Lidia González Clinical Research Area, Center for Genetic Engineering and Biotechnology, Havana, Cuba

G. Kenneth Gray PhD Program in Biological and Biomedical Sciences, Harvard University, Boston, MA, USA

Barbara Guerra BMB, SDU, Odense, Denmark

Laszlo Gyenis Department of Biochemistry, Schulich School of Medicine & Dentistry, University of Western Ontario, London, ON, Canada

Department of Oncology, Schulich School of Medicine & Dentistry, University of Western Ontario, London, ON, Canada

Jennifer Hochscherf KinaseDetect ApS, Kruså, Denmark

Miwako Kato Homma Department Biomolecular Science, Fukushima Medical University School of Medicine, Fukushima, Japan

Yoshimi Homma Department Biomolecular Science, Fukushima Medical University School of Medicine, Fukushima, Japan

Olaf-Georg Issinger KinaseDetect ApS, Kruså, Denmark

Lucas M. Jozwick Department of Biology, West Virginia University, Morgantown, WV, USA

Maysoun Kassem Département d'Anatomie et Cytologie Pathologiques et Plateforme de Génétique Moléculaire des Tumeurs, Pôle de Biologie et Pathologie, CHU Grenoble, La Tronche, France

Hiroko Kozuka-Hata Medical Proteomics Laboratory, The Institute of Medical Sciences, The University of Tokyo, Tokyo, Japan

Betsy T. Kren Cellular and Molecular Biochemistry Research Laboratory (151), Minneapolis Veterans Affairs Health Care System, Minneapolis, MN, USA

Department of Laboratory Medicine and Pathology, University of Minnesota, Minneapolis, MN, USA

Masonic Cancer Center, University of Minnesota, Minneapolis, MN, USA

Mathieu Laramas Centre hospitalo-universitaire Albert Michallon, service d'oncologie médicale, Grenoble, France

David W. Litchfield Department of Biochemistry, Schulich School of Medicine & Dentistry, University of Western Ontario, London, ON, Canada

Department of Oncology, Schulich School of Medicine & Dentistry, University of Western Ontario, London, ON, Canada

Graziano Lolli Department of Chemical Sciences, University of Padua, Padua, Italy

Venetian Institute for Molecular Medicine (VIMM), Padua, Italy

Jean-Alexandre Long Centre hospitalo-universitaire Albert Michallon, service d'urologie, Grenoble, France

Adam T. Majot Department of Biology, West Virginia University, Morgantown, WV, USA

Braden C. McFarland Department of Cell, Developmental and Integrative Biology, University of Alabama at Birmingham, Birmingham, AL, USA

Anne McLeer-Florin INSERM U 823, Institut A Bonniot, Université J Fourier, Grenoble, France

Département d'Anatomie et Cytologie Pathologiques et Plateforme de Génétique Moléculaire des Tumeurs, Pôle de Biologie et Pathologie, CHU Grenoble, La Tronche, France

Yoshihiko Miyata Department of Cell and Developmental Biology, Graduate School of Biostudies, Kyoto University, Kyoto, Japan

Mathias Montenarh Medizinische Biochemie und Molekularbiologie, Universität des Saarlandes, Homburg, Germany

Karsten Niefind Department of Chemistry, Institute of Biochemistry, University of Cologne, Cologne, Germany

Thirayost Nimmanon Cardiff School of Pharmacy and Pharmaceutical Sciences, Cardiff University, Cardiff, UK

Masato Ogura Department Biomolecular Science, Fukushima Medical University School of Medicine, Fukushima, Japan

Charina E. Ortega Department of Medicine, Boston University School of Medicine, Boston, MA, USA

Masaaki Oyama Medical Proteomics Laboratory, The Institute of Medical Sciences, The University of Tokyo, Tokyo, Japan

Dominique Pasquier Département d'Anatomie et Cytologie Pathologiques et Plateforme de Génétique Moléculaire des Tumeurs, Pôle de Biologie et Pathologie, CHU Grenoble, La Tronche, France

Silvio E. Perea Molecular Oncology Laboratory, Biomedical Research Area, Center for Genetic Engineering and Biotechnology, Havana, Cuba

Yasser Perera Molecular Oncology Laboratory, Biomedical Research Area, Center for Genetic Engineering and Biotechnology, Havana, Cuba

Catherine Pillet CEA, DSV/iRTSV, Biologie à Grande Echelle, Grenoble, France

Lorenzo A. Pinna Department of Biomedical Sciences and CNR, Institute of Neuroscience, University of Padova, Padova, Italy

Lawrence Prince-Wright Department of Medicine, Boston University School of Medicine, Boston, MA, USA

Fatima Qaiser Cellular and Molecular Biochemistry Research Laboratory (151), Minneapolis Veterans Affairs Health Care System, Minneapolis, MN, USA

Department of Laboratory Medicine and Pathology, University of Minnesota, Minneapolis, MN, USA

Jean-Jacques Rambeaud Centre hospitalo-universitaire Albert Michallon, service d'urologie, GrenobleFrance

Tine D. Rasmussen Clinical Pathology, OUH, Odense C, Denmark

Vilcy Reyes Molecular Oncology Laboratory, Biomedical Research Area, Center for Genetic Engineering and Biotechnology, Havana, Cuba

Marta Riera Center for Research in Agricultural Genomics (CRAG), CSIC-IRTA-UAB-UB Consortium, Barcelona, Spain

Arielis Rodríguez Proteomic Department, Biomedical Research Area, Center for Genetic Engineering and Biotechnology, Havana, Cuba

Caroline Roelants INSERM, Unité 1036, Biology of Cancer and Infection, Grenoble, France

UJF-Grenoble 1, Biology of Cancer and Infection, Grenoble, France

CEA, DSV/iRTSV, Biology of Cancer and Infection, Grenoble, France

Meng Rui Medizinische Biochemie und Molekularbiologie, Universität des Saarlandes, Homburg, Germany

Maria Ruzzene Department of Biomedical Sciences, University of Padova, Padova, Italy

Alexander Schnitzler Department of Chemistry, Institute of Biochemistry, University of Cologne, Cologne, Germany

Laure Schoutteten INSERM, Unité 1036, Biology of Cancer and Infection, Grenoble, France

UJF-Grenoble 1, Biology of Cancer and Infection, Grenoble, France

CEA, DSV/iRTSV, Biology of Cancer and Infection, Grenoble, France

Christina Servas Medizinische Biochemie und Molekularbiologie, Universität des Saarlandes, Homburg, Germany

Takeshi Shibata ABSciex. K.K., Tokyo, Japan

Tyler R. Sizemore Department of Biology, West Virginia University, Morgantown, WV, USA

Toshiyuki Suzuki Department Biomolecular Science, Fukushima Medical University School of Medicine, Fukushima, Japan

Ryszard Szyszka Department of Molecular Biology, The John Paul II Catholic University of Lublin Institute of Biotechnology, Lublin, Poland

Kathryn M. Taylor Cardiff School of Pharmacy and Pharmaceutical Sciences, Cardiff University, Cardiff, UK

Elena Tibaldi Department of Molecular Medicine, University of Padova, Padova, Italy

Céline Tisseyre INSERM, Unité 1036, Biology of Cancer and Infection, Grenoble, France

UJF-Grenoble 1, Biology of Cancer and Infection, Grenoble, France

CEA, DSV/iRTSV, Biology of Cancer and Infection, Grenoble, France

Janeen H. Trembley Cellular and Molecular Biochemistry Research Laboratory (151), Minneapolis Veterans Affairs Health Care System, Minneapolis, MN, USA

Department of Laboratory Medicine and Pathology, University of Minnesota, Minneapolis, MN, USA

Masonic Cancer Center, University of Minnesota, Minneapolis, MN, USA

Gretchen Unger GeneSegues, Inc., Chaska, MN, USA

Isabel Cristina Vélez-Bermúdez Center for Research in Agricultural Genomics (CRAG), CSIC-IRTA-UAB-UB Consortium, Barcelona, Spain

Carmen Valenzuela Clinical Research Area, Center for Genetic Engineering and Biotechnology, Havana, Cuba

Sabrina Welker Medizinische Biochemie und Molekularbiologie, Universität des Saarlandes, Homburg, Germany

Part I
Regulation and Structure–Function Studies of CK2

Yeast Asf1 Protein as Modulator of Protein Kinase CK2 Activity

Andrea Baier, Ewa Alikowska, and Ryszard Szyszka

Abstract Natural modulators of protein kinase CK2 activity might be divided into two groups. Substances of the first one, polycations, like polyamines and polylysine are able to stimulate enzyme activity. On the other hand, compounds like heparin (polyanions) have inhibitory properties towards CK2 activity. The sequence of Asf1 possesses five potential phosphorylation sites for CK2, but it seems that it does not underlie phosphorylation. Yeast Asf1 amino acid sequence contains a characteristic acidic fragment at its C-terminus. Such a sequence, called pseudosubstrate region, of regulatory subunits is present in several protein kinases, like PKA and PKG, and can be also found in the regulatory subunit of CK2. Experimental data suppose a modulating effect of Asf1 towards protein kinase CK2 in a different manner when comparing the influence on each catalytic subunits itself as well as the corresponding holoenzymes.

Keywords Protein kinase CK2 • Autoinhibitory region • Regulation • Asf1 • Yeast

1 Introduction

CK2 has traditionally been classified as a protein serine/threonine kinase ubiquitously distributed in all eukaryotes. The human enzyme mostly appears as a hetero-tetrameric protein complex consisting of a dimer of regulatory β subunits (28 kDa) and two catalytic α (42 kDa) and/or α' (38 kDa) subunits with a stoichiometry of either $\alpha_2\beta_2$, $\alpha'_2\beta_2$ or $\alpha\alpha'\beta_2$ as a holoenzyme [1, 2]. In the case of the baker's yeast, *Saccharomyces cerevisiae*, the holoenzyme requires the presence of both regulatory subunits β and β'. Genetic studies in *S. cerevisiae* [3] and in mice [4, 5] demonstrate

A. Baier (✉) • E. Alikowska
Department of Molecular Biology, Faculty of Biotechnology and Environmental Sciences, Institute of Biotechnology, The John Paul II Catholic University of Lublin, Lublin, Poland
e-mail: baier@kul.pl

R. Szyszka
Department of Molecular Biology, The John Paul II Catholic University of Lublin Institute of Biotechnology, Lublin, Poland

K. Ahmed et al. (eds.), *Protein Kinase CK2 Cellular Function in Normal and Disease States*, Advances in Biochemistry in Health and Disease 12, DOI 10.1007/978-3-319-14544-0_1

that CK2 is essential for cell viability and animal embryonic development. Biochemical and functional analyses of this enzyme reveal that it can phosphorylate over 300 proteins implicated in the regulation of many cellular functions, notably gene expression, signal transduction and DNA/RNA and protein synthesis [6]. It is expectable that the unique pleiotropy of CK2 justifies at least in part two properties of this enzyme: the minimum consensus sequence for phosphorylation defined as S/T-X-X-D/E/pS/pY [6] and its constitutive activity [1, 2].

2 CK2 Interaction Partners

The CK2 activity is detected in cell or tissue extracts even in the absence of any cofactors (e.g. second messengers) or when it is expressed in bacteria. These facts led to the general conclusion that CK2 is unregulated or constitutively active. However, there are some mechanisms that contribute to the regulation of CK2 in the cell. They include the regulated expression and assembly of the enzyme [7, 8], regulation by reversible phosphorylation of regulatory β subunits [9, 10] as well as the catalytic α subunit [11, 12] and regulatory interactions [2, 13].

Studies in yeast and human cells demonstrate that different forms of CK2 interact with a large number of cellular proteins [13, 14]. The identification of several CK2-interacting proteins is consistent with the presumption that CK2 may be directly or indirectly regulated by interacting proteins. CK2-interacting proteins can be categorised by the way how they affect the kinase activity. The first category represented by nucleolin [15] or Nopp140 [16] most likely reflects enzyme-substrate interaction. Second, there are proteins that may directly affect the CK2 catalytic activity, such as FGF-1 and FGF-2 [17], heat shock protein 90 (Hsp90) and Cdc37 [18]. Fibroblast growth factor-1 (FGF-1) interacts with CK2α and CK2β in vitro and in vivo. In vitro, FGF-1 and FGF-2 are phosphorylated by CK2, and the presence of FGF-1 or FGF-2 was found to enhance CK2β autophosphorylation. A correlation between the mitogenic potential of FGF-1 mutants and their ability to bind to CK2α was observed and the possible involvement of CK2 in the FGF-induced stimulation of DNA synthesis [17]. Cdc37 is a protein kinase-targeting molecular chaperone, and its function in cooperation with Hsp90 is required for various signalling kinases. The phosphorylation of Cdc37 by CK2 at the conserved N-terminal Ser-13 is prerequisite for the efficient Cdc37 activity binding to protein kinases including Akt, Cdk4, MOK and Raf1. In addition, the phosphorylation of Cdc37 by CK2 was shown to be crucial for the recruitment of Hsp90 to the protein kinase-Cdc37 complexes. These findings indicated that a subpopulation of CK2 forms complexes with Hsp90 and Cdc37 in the cytoplasm and phosphorylates Cdc37 and thus regulates the molecular chaperone activity of Cdc37. Since CK2 activity depends on Cdc37, CK2 and Cdc37 constitute a positive feedback machinery to control multiple Cdc37-dependent signalling protein kinases [18]. Third, there is a set of proteins that probably play a role in the indirect regulation of CK2. Those proteins work as an adaptor/scaffold/targeting proteins such as FAF1 [19], tubulin [20] and CKIP-1 [21]. First of these proteins—the Fas-associated factor FAF1—is an in vitro and in vivo substrate of protein kinase CK2, and the putative physiological function of FAF1 phosphorylation is connected with the influence on nuclear import ability of FAF1 polypeptide [19]. In the case of tubulin, a direct association between the CK2α and the CK2α' subunit and tubulin was observed. There was no binding of the CK2β subunit to tubulin. Thus, tubulin was identified as a new binding partner of CK2 catalytic subunits [20]. CKIP-1 is an interaction partner of CK2 with a number of protein-protein interaction motifs, including an N-terminal pleckstrin homology domain (PH). The CKIP-1 protein can recruit CK2 to the plasma membrane. Furthermore, the PH domain of CKIP-1 was found to be required for interactions with CK2 and for the recruitment of CK2 to the plasma membrane. The examination of CK2α for a region that mediates interactions with CKIP-1 revealed a putative HIKE domain, a complex motif found exclusively in proteins that bind pleckstrin homology domains. However, mutations within this motif were not able to abolish CKIP-1-CK2 interactions suggesting that this motif by itself may not be sufficient to mediate interactions. Overall, these results provide novel insights into how CK2, a predominantly nuclear enzyme, is targeted to the plasma membrane and perhaps more importantly how it might be regulated [21].

Finally, there are a number of protein interaction partners that either disrupt or enhance the ability of CK2 to phosphorylate certain protein substrates. As an example, the peptidyl-prolyl isomerase (Pin1) interacts in a phosphorylation-dependent manner with CK2 and with several proteins involved in cell cycle events. Pin1 can interact with CK2 complexes that contain CK2α. Furthermore, Pin1 can interact directly with the C-terminal domain of CK2α that contains residues that are phosphorylated in vitro by $p34^{Cdc2}$ and in mitotic cells. Substitution of the CK2α phosphorylation sites with alanines resulted in decreased interactions between Pin1 and CK2. Pin1 inhibits the Thr^{1342} phosphorylation on human topoisomerase II alpha by CK2. Topoisomerase II alpha also interacts with Pin1 suggesting that the effect of Pin1 on the phosphorylation of Thr^{1342} could result from its interactions with both CK2 and topoisomerase II alpha [22].

Analysis of protein-protein interaction in mammalian cells provides also evidences for functional specialisation of CK2α and CK2α'. There are proteins specifically interacting either with the catalytic subunit CK2α or CK2α'. Examples for such proteins are the isomerase Pin1 [11, 22] and protein phosphatase PP2A [23, 24] interacting only with CK2α. Further specific partners for CK2α are protein CKIP-1 [25], cyclin H [26], transcription factor Egr1 [27] and heat shock proteins (Hsp) [28] interacting with α subunits, but not reacting with CK2α'.

Based on literature data analysis, we are able to distinguish among CK2 substrates three main groups: (1) substrates preferentially phosphorylated by free catalytic subunit, (2) substrates preferentially phosphorylated by CK2 holoenzyme and (3) substrates phosphorylated by both molecular forms of CK2. The driving force for binding the protein substrate seems to be the charge and shape complementarity between CK2 and substrate apart from the suitable conformation and accessibility of the phosphorylation site. Most CK2 substrates present large acidic charges on their surfaces. However, in the case of smaller peptides, the presence of the CK2 consensus sequence in right conformation and accessibility is the main requirement for binding as the problem of steric hindrance due to its small size is expected to be not high [29]. Using different peptide substrates, it was estimated that the activity of the CK2 holoenzyme is somewhat higher than that of the free catalytic subunit. However notable, there are protein substrates which either require the presence of the regulatory subunit or conversely are phosphorylated by the catalytic subunits but not by the holoenzyme (e.g. calmodulin [30]). The phosphorylation of the latter group by the holoenzyme is triggered and dramatically enhanced by polybasic peptides (but not polyamines) such as polylysine. The effect of the β subunit on these particular substrates is probably mediated by specific interactions with the protein substrates, perhaps in combination with polycationic stimulators [1].

3 CK2 Structure

CK2 is a Ser/Thr protein kinase having heterotetrameric $\alpha_2\beta_2$ structure with some unusual features among the eukaryotic protein kinases: (1) several acidic determinants are deciding about CK2 substrate specificity; (2) both ATP and GTP can be utilised as phosphoryl donors; (3) it is insensitive to any known second messengers and the regulatory properties of CK2 are poorly understood; and (4) it displays high constitutive activity with over 350 cellular protein substrates known to date [1, 2]. The precise function of two non-catalytic β subunits is still poorly understood; however, it has been shown that their functions are (1) altering the CK2α/α' substrate specificity, (2) protecting the catalytic subunit against denaturing agents and conditions and (3) modulating CK2 catalytic activity.

The catalytic domains of human CK2α and CK2α' are products of different genes located on different chromosomes [31]. Their similar enzymatic properties in vitro are the result of high identity between both sequences with about 90 % [32]. In the case of yeast homologues, the amino acid sequence of catalytic subunits shows only about 60 % identity [33]. The middle part of the catalytic domain contains the high structural similar sequences, whereas in the C-terminal sequences, there exist differences between CK2α and CK2α'. The C-terminus of the CK2α subunit possesses Pro-rich phosphoacceptor sites whose phosphorylation generates a binding site of isomerase Pin1 [11].

Alignments of the predicted amino acid sequence of CK2α subunits of several organisms led to the conclusion that the primary structure of CK2α exhibits conserved features. The conserved sequences $D^{173}WG^{175}$ and $G^{197}PE^{199}$, which are responsible for the formation of CK2α activation loop, are situated between the β-sheet-rich region at the N-terminus and the α-helix-rich C-terminus. The centre of the activation loop is formed by residues R^{155}, D^{156} and H^{160}. Experimental data show that always only one catalytic subunit is active [34]. During catalysis the phosphate group interacts with the nitrogen atoms within the Rossmann fold. This sequence is present in all protein kinases responsible for the binding of nucleotide triphosphates and comprises amino acid sequence $G^{46}XGXXS^{51}$ in human CK2α [35]. Arg191, Arg195 and Lys198 are conserved residues in the C-terminal loop and are responsible for the recognition of acidic residue at position n+1 in the phosphor-acceptor sequence [36].

Human α subunit contains a basic Lys-rich region comprising amino acids $K^{74}KKKKIR^{80}$ within the αC segment that is responsible for the interaction with the acidic region (aa 55–64) of the regulatory subunit. It was shown that the presence of the β subunit may affect the phosphorylation of some substrates such as calmodulin [37]. Furthermore, this basic region also determines the sensitivity to heparin. The deletion of aa 74–77 abolished the sensitivity to polyanionic compounds or in the case of some substrates causes an increase of kinase activity [38]. The basic cluster is also responsible for interaction with heat shock protein Hsp90 [39], nucleolin [40] and recognition of acidic determinants at n+3 in the consensus substrate sequence [36].

There are evidences showing the functional differences between the subunits CK2α and CK2α'. In mammalian cells, CK2α is phosphorylated at Thr^{344}, Thr^{360}, Ser^{362} and Ser^{370} within its unique C-domain by $p34^{cdc2}$ kinase, during the cell cycle procession, while CK2α' is not phosphorylated in this way [11]. Furthermore, the c-Fgr kinase belonging to the Scr family covalently modifies CK2α at Tyr^{255} [41]. In comparison, the effects of induced expression of kinase-inactive CK2α differed significantly from the effects of induced expression of kinase-inactive CK2α'. Of particular interest is the dramatic attenuation of proliferation that is observed following induction of CK2α', but not following induction of CK2α. These results provide evidence for functional specialisation of CK2 isoforms in mammalian cells [42].

Besides the catalytic subunit, the primary structure of CK2β subunit also contains typical conserved features: (1) two main phosphorylation sites (S^2S^3SEE) [43], (2) the Nopp140 interaction site (aa 5–20), (3) two highly conserved amino acids (E^{20} to K^{33}) which are used for the export of CK2 as an ectokinase, (4) one destruction box (aa 46–54), (5) an acidic loop that has autoinhibitory and downregulatory functions of CK2 activity (aa 55–64), (6) residues involved in interaction with cell cycle regulators p21 and p53 (aa 106–116, aa 124, 134, 141, 145–149, 152), (7) dimer interface residues (aa 143–148), (8) one region containing CK2α binding residues (aa 187–192), (9) residues responsible for binding A-Raf and Mos (aa 187–205), (10) four cysteine residues responsible for zinc finger formation (C^{109}, C^{114}, C^{137} and C^{140}) involved in β subunit dimerization and (11) serine residues near the C-terminus phosphorylated by $p34^{cdc2}$ in vitro.

The CK2β subunit possesses a stretch of acidic amino acids encompassing residues $D^{55}LEPDEELED^{64}$ [44, 45]. Based on its similarity to the clusters of amino acids that are typically observed in CK2 substrates, this region is a reminiscent of autoinhibitory sequences that have been identified in a large number of other protein kinases [46, 47]. Indeed, this acidic loop has autoinhibitory and downregulatory function towards CK2 activity [2]. Polyamines, known as CK2 activators, bind to this region of CK2β.

The crystal structure of tetrameric CK2 demonstrates that this acidic stretch is located in a considerable distance from the active site of the catalytic subunit within the same tetramer [34]. Nevertheless, as may be the case for autophosphorylation, it is conceivable that interactions between aa residues 55–64 of CK2β and aa residues 74–80 of CK2α occur through higher-order interactions between CK2 tetramers [2].

ScAsf1	170	DEEEEDDEEE DDDEDDEDDE	190
ScCK2β	72	DLEAMSDEEE DEDDVVEEDE	92
HsCK2β	55	DLEPDEELED NPNQQSDLIE	75

Fig. 1 Sequence alignment of Asf1 aa 170–190, human CK2β aa 55–75 and yeast CK2β aa 72–92

Mutations of acidic residues at positions Asp55, Glu57 and Glu59-Asp64 of the CK2β subunit to Ala led to an up to fourfold more active holoenzyme than the wild-type holoenzyme. At the same time, these mutants were no longer sensitive towards the stimulatory effect of polylysine. This finding supported the autoinhibitory character of this acidic N-terminal cluster, especially that Asp55 and Glu57 are involved in an intrinsic downregulation of CK2 basal activity and are implicated in the responsiveness to various effectors [45, 48, 49].

Another data demonstrated that Pro58 located in the middle of this sequence also possesses an important structural feature that affects downregulatory function of CK2β towards the catalytic subunits. Mutation of this Pro to Ala resulted in a similar effect to that of mutations of the acidic residues alone, namely, it produces hyperactive β subunits that stimulate the CK2α activity to a greater extent than the wild-type regulatory subunit CK2β [50].

The acidic sequences of CK2β were used to search for potential CK2-interacting protein partners. From those results of the database search, yeast Asf1 protein was selected. The acidic C-terminal stretch of yeast Asf1 protein (i.e., **D^{170}DEEEEDDEEEDDDEDDEDDEDDD^{193}**) possesses high similarity to the acidic fragment of regulatory CK2β subunit (i.e., **D^{55}LEPDEELED^{64}**) and the acidic surroundings of the substrate recognition site of potential CK2 protein substrate candidates (Fig. 1).

4 Anti-Silencing Function 1 Protein

Amino acid sequences of CK2β autoinhibitory domains were used to query *Saccharomyces cerevisiae* databases for proteins which may have significant role in regulation of apoptosis. From few candidates Asf1 protein was chosen as a potential CK2-modulating protein. Yeast Asf1 protein differs from its animal homologues and contains (1) N-terminal 155 amino acids long sequence with high homology (65–90 %) to other eukaryotic species and (2) long poly-acidic stretch present in fungi (in baker's yeast over 20 amino acid long cluster of aspartic and glutamic acids) (Fig. 2). This pseudosubstrate acidic cluster may influence CK2 antiapoptotic activity in the cell, what is observed as induction of apoptosis in case of overexpression of yeast Asf1 [51, 52].

In *S. cerevisiae* Asf1 contains 279 amino acids divided into two different regions. The N-terminal part comprising aa 1–155 is highly conserved between eukaryotes (65–90 % identity) and contains mainly β-sheet structures. The C-terminal region

1 MSIVSLLGIK VLNNPAKFTD P**Y**EFEITFEC LESLKHDLEW KLTYVGSSRS
51 LDHDQELDSI LVGPVPVGVN KFVFSADPPS AELIPASELV SVTVILLSCS
101 YDGREFVRVG YYVNNE**Y**DEE ELRENPPAKV QVDHIVRNIL AEKPRVTRFN
151 IVWDNENEGD LYPPEQPGVD DEEEEDDEEE DDDEDDEDDE DDDQEDGEGE
201 AEEAAEEEEE EEEKTEDNE**T** NLEEEEEDIE N**S**DGDEEEGE EEVGSVDKNE
251 DGNDKKRRKI EGGSTDIES**T** PKDAARSTN

Fig. 2 Amino acid sequence of yeast Asf1 protein. The acidic fragment is *underlined*; potential CK2 phosphorylation sites are *bolded*

(<15 % identity) is less conserved and possesses a disturbed molecular structure. In fungi and yeast, the C-terminal sequence contains a long acid fragment which is not conserved in other eukaryotic organisms (Fig. 3). The human homologue of yeast Asf1 protein was cloned as Asf1A and is called CIA (CCGI interacting factor A). Analysing the amino acid sequence 1–153, the identity between CIA and yeast Asf1 is about 58 %, whereas the C-terminal is completely different [53]. In later studies a second homologue, called Asf1B, was cloned having 71 % identity with the amino acid sequence of Asf1A [54]. Alignment of different Asf1 homologues (Fig. 3) revealed the domain structures of CIA: the N-terminal evolutionarily conserved domain (residues 1–169 in Asf1) and the C-terminal divergent region rich in acidic residues in cia1/Asf1, while in fungal CIA this part is rich in serine and threonine [55].

Yeast Asf1 protein amino acid sequence analysis allowed the distinction of several sites designated as the minimum consensus sequence for CK2 substrates (Fig. 2). However, no phosphorylation of Asf1 by different forms of yeast CK2 in vitro could be detected.

5 CK2-Asf1 Interaction

Analysing different studies towards modulators of CK2 activity, we can estimate typical compounds having similar influence towards CK2 isoforms from several organisms [30, 33, 56–58]. Within this group of compounds are polybasic peptides (polylysine), polyanions (heparin, spermine) and CK2 inhibitors like benzotriazoles and benzimidazoles.

The amino acid sequence of the Asf1 protein possesses five potential phosphorylation sites for protein kinase CK2: 2 Tyr sites, 1 Ser site and 2 Thr sites. In enzymatic studies with different CK2 forms, no phosphorylation could be detected.

Then, a possible effect of Asf1 protein was examined using the recombinant Asf1 protein expressed in bacteria cells as well as the recombinant human catalytic subunits CK2α and CK2α' as well as the corresponding holoenzymes. Asf1 exhibits inhibitory effect towards both subunits and holoenzymes. Interestingly, as in the case of other modulators, like heparin, polylysine, NaCl or some inhibitors, the influence differed between them (Fig. 4). The α subunit is activated in the presence

```
Dm   -MAKVHITNVVVLDNPSSFFNPFQFELTFECIEELKEDLEWKMIYVGSAESEEHDQVLDT 59
Hs   -MAKVSVLNVAVLENPSPFHSPFRFEISFECSEALADDLEWKIIYVGSAESEEFDQILDS 59
Af   -MSVVSLLGVKIQNNPAPFLAPYQFEITFECLEQLQKDLEWKLTYVGSATSSEYDQELDS 59
Pr   -MSVVSLLGVKNLNNPAPFTASYQFEITFECLEQLQKDLEWKLTYVGSATSAEYDQELDS 59
Sc   -MSIVSLLGIKVLNNPAKFTDPYEFEITFECLESLKHDLEWKLTYVGSSRSLDHDQELDS 59
Ce   MASRVNIVQVQILDNPAMFVDKFKLEITFEVFEHLPHDLEWELVYVGSGTSRDFDQVLDS 60
       : * :  :    :**: *    :.:*::**  * * .****:: ****. * :.** **:

Dm   IYVGPVPEGRHIFVFQADPPDVSKIPEPDAVGVTIVLLTCSYRGQEFVRVGYYVNNDYAD 119
Hs   VLVGPVPAGRHMFVFQADAPNPSLIPETDAVGVTVVLITCTYHGQEFIRVGYYVNNEYLN 119
Af   LFVGPIPVGVNKFIFEAEAPDLKRIPTSEILGVTVILLTCSYDGREFVRVGYYVNNEYDS 119
Pr   LLVGPIPVGVNKFIFEADAPDVKRIPTSEMLGVTVILLTCSYDGREFVRVGYYVNNEYDS 119
Sc   ILVGPVPVGVNKFVFSADPPSAELIPASELVSVTVILLSCSYDGREFVRVGYYVNNEYDE 119
Ce   ALVGPIPEGRHKFVFDADHPDISKIPVDDIVGVSVLLLRCKYNDQEFINMGWFVANEYTE 120
       ***:* * : *:*.*: *. . **  : :.*:::*: *.* .:**:.:*::* *:* .

Dm   PEMRENPPTKPLFEKLTRNILASKPRVTRFKINWDYGHINGNGNGVENGHQDEMATDGPS 179
Hs   PELRENPPMKPDFSQLQRNILASNPRVTRFHINWDN----------NMDRLEAIETQDPS 169
Af   EDLSAEPPAKPIIERIRRNILAEKPRVTRFAIKWDSEESAPAEYPPDQPEADILEDDSAA 179
Pr   EELAAEPPAKPVIERIRRNVLAEKPRVTRFAIKWDSDDSAPAEYPPDQPEADGLDDDSGA 179
Sc   EELRENPPAKVQVDHIVRNILAEKPRVTRFNIVWDNENEG-DLYPPEQPGVDDEEEEDDE 178
Ce   EELKENPPSQPLIEKLSRKVETEDLRITTFPIRWTDEDPVAEPVEDEANRVFAEDDLMPL 180
      ::  :** :  ..:: *:: :.. *:* * * *                :

Dm   TSEA-----ASAVIHPEDDNSLAMPMENGI--------------KALNENSNSLAMEC-- 218
Hs   LGCG-----LPLNCTPIKGLGLPGCIPG-----------------LLPENS----MDCI- 202
Af   YGAEEAELEAALVRELADAERDVKSEDHEMEGAEP-AIKEEEEEEDISDAESEDIEDESDD 238
Pr   YGAEERELEAALLKELEDSNKPAEGEDHEMEGAEAPAGKEDDEEEISDAESEDLEAESDD 239
Sc   EEDDDEDDEDDEDDDQEDGEGEAEEAAEEEEEEEE---KTEDNETNLEEEEEDIENSDGD 235
Ce   NDDGQ---EDDDEEEEDDDEMEANAEEVDLNESFN-----ERLANALDGAEQKGADEKME 232
                         .                              .     .

Dm   -----------------------------------------------
Hs   -----------------------------------------------
Af   DEEDLDEEEAGDGDEDVEMGDDSEQKDDGPKADSTNQHSHQPEVMVH 285
Pr   DEDELDEEEGGDADEDVEMGDDAEHKDDAAKPT----HQPQPELMVH 282
Sc   EEE--GEEEVGSVDKNEDGNDKKRRKIEGGSTDIES-TPKDAARSTN 279
Ce   DDGANEDVDMADDEPGVQINTDTKVPESMAEPLSDKTNNEMVQ---- 275
```

Fig. 3 Alignment of Asf1 proteins from different organisms: Dm, *Drosophila melanogaster*; Ce, *Caenorhabditis elegans*; Af, *Aspergillus flavus*; Pr, *Penicillium roqueforti*; Sc, *Saccharomyces cerevisiae*; Hs, *Homo sapiens*. Identical amino acids are indicated by an *asterisk*, strongly conserved by *colon* and weakly conserved by a *dot*

of low concentration of Asf1; at higher concentration, it is inhibited. The α' subunit is stronger inhibited than the α subunit. Changing the conditions in the phosphorylation reaction had different results for the α and α' subunits. The inhibition of both catalytic subunits in the presence of 10 or 20 μM ATP revealed another mode of action between the subunits (Fig. 4a). There was no change in the inhibitory effect estimated in the case of α', whereas the α subunit is more inhibited in the presence of the lower ATP concentration, suggesting an ATP-dependent inhibition mode. On the other side, changing the protein substrate concentration (yeast acidic ribosomal

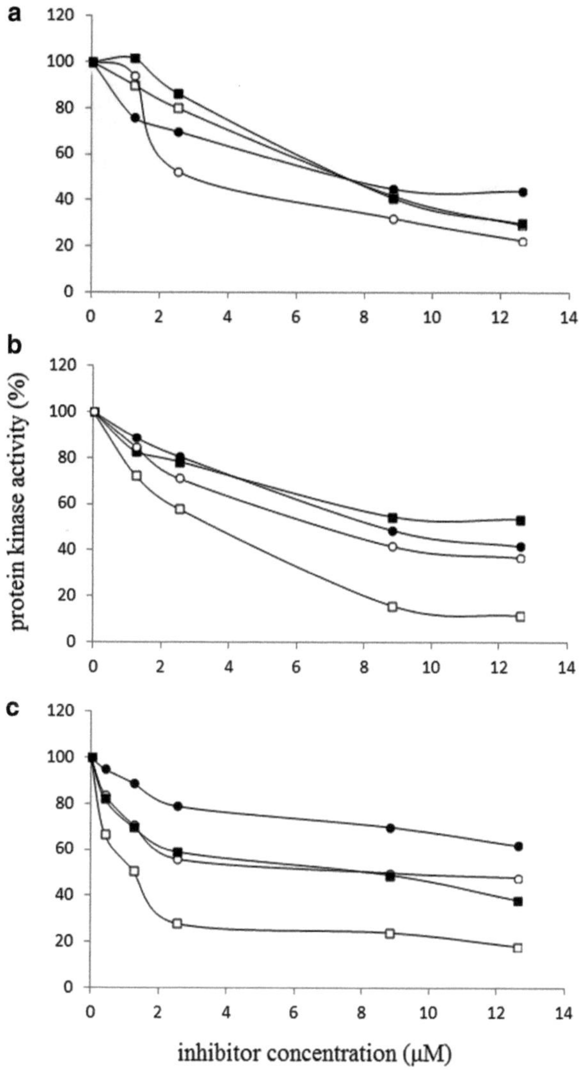

Fig. 4 Studies on the influence of Asf1 protein towards CK2 activity. (**a**) Effect of different ATP concentrations on the activity of the catalytic subunits: CK2α/20 µM ATP (*filled circle*), CK2α/10 µM ATP (*open circle*), CK2α'/20 µM ATP (*filled square*), CK2α'/10 µM ATP (*open square*). (**b**) Effect of different P2B concentrations on the activity of the catalytic subunits: CK2α/7.8 µg P2B (*filled circle*), CK2α/1.3 µg P2B (*open circle*), CK2α'/6.5 µg P2B (*filled square*), CK2α'/1.3 µg P2B (*open square*). (**c**) Effect of different Asf1 constructs on the activity of the catalytic subunits: CK2α/Asf1 aa 169–279 (*filled circle*), CK2α/Asf1 aa 1–279 (*open circle*), CK2α'/Asf1 aa 169–279 (*filled square*), CK2α'/Asf1 aa 1–279 (*open square*)

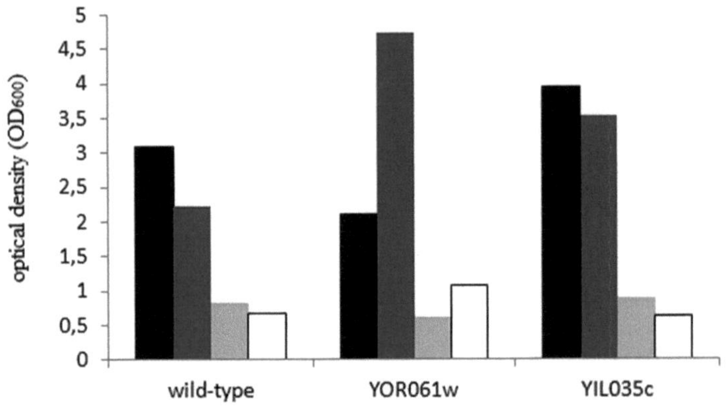

Fig. 5 Studies on the influence of Asf1 protein towards yeast cells growth rate. Optical density was determined after 6 h in the presence of low amount of Asf1 (medium containing glucose) and overexpressed Asf1 (medium containing galactose): *black*, glucose/control; *dark grey*, glucose/+Asf1; *light grey*, galactose/control; *white*, galactose/+Asf1

protein P2B), the extent of inhibition is changed only in the case of the α' subunit (Fig. 4b). The phosphorylating activity is higher in case of higher substrate concentration, suggesting a competitive inhibition mode towards the protein substrate. Regarding the question if the whole Asf1 molecule is necessary for the enzyme inhibition, two Asf1 constructs were tested comprising either amino acids 1–169 or amino acids 170–279. As shown in Fig. 4c, the C-terminal part (aa 170–279) is responsible for the decrease of CK2 activity. The Asf1 molecule lacking the C-terminal acidic stretch has no CK2 inhibiting properties. Using holoenzymes the modulating effect by Asf1 is lower.

First data propose that the effect of Asf1 protein towards protein kinase CK2 is also detectable in vivo. In experiments with *S. cerevisiae*, Asf1 transformed cells were analysed for their CK2 activity. Cell lysates containing the CK2-Asf1 complex possesses lower phosphorylating activity than lysates from non-transformed cells [59]. These in vivo results confirm the results from enzymatic activity studies. In further experiments the growth rate of yeast cells was estimated to measure the influence of Asf1 overexpression (medium with galactose) on cell division. Therefore, *S. cerevisiae* mutant (deletion of one catalytic subunit) cells were grown either in the presence or absence of Asf1 protein and compared with the wild-type strain (Fig. 5). Interestingly, the cell division of the α mutant (YOR061w) is strongly inhibited compared to the wild type. Surprisingly, yeast cells lacking the α' subunit (YIL035c) is growing faster when the Asf1 protein is overexpressed. In case of basal expression (medium with glucose) of Asf1 protein, the mutant only expressing the α subunit exhibits more than twice faster growth than the control strain.

6 Conclusions

Present results suggest a comparable interaction of CK2α with Asf1 protein to that of CK2α and CK2β. Yeast Asf1 contains an acidic stretch possessing the minimal consensus sequence for CK2 substrates. It is similar to the acidic loop, also called pseudosubstrate sequence, of the regulatory subunit CK2β from yeast and human.

A possible mechanism of Asf1 is the blocking of the active site of the catalytic subunit. In this case it would be possible to inhibit the phosphorylation of proteins catalysed by free catalytic subunits, but not the holoenzyme. Asf1 protein may act like the CK2β subunit and decrease or even abolish the kinase activity as in case of calmodulin.

The different effects of Asf1 towards CKα and CK2α' correlate with observations using benzotriazole and benzimidazole derivatives. There are compounds possessing different inhibitory activities on CK2 subunits. Such compounds may help to understand different functions of the α and α' subunits.

References

1. Pinna LA (2002) Protein kinase CK2: a challenge to canons. J Cell Sci 115:3873–3878
2. Litchfield DW (2003) Protein kinase CK2: structure, regulation and role in cellular decisions of life and death. Biochem J 369:1–15
3. Glover CV (1998) On the physiological role of the casein kinase II in *Saccharomyces cerevisiae*. Prog Nucleic Acid Res Mol Biol 59:95–133
4. Xu X, Toselli PA, Russel LD, Seldin DC (1999) Globozoospermia in mice lacking the casein kinase II alpha' catalytic subunit. Nat Genet 23:118–123
5. Lou DY, Dominguez I, Toselli P, Landesman-Bollag E, O'Brien C, Seldin DC (2008) The alpha catalytic subunit of protein kinase CK2 is required for mouse embryonic development. Mol Cell Biol 28:131–139
6. Meggio F, Pinna LA (2003) One-thousand-and-one substrates of protein kinase CK2? FASEB J 17:349–368
7. Bosc DG, Luscher B, Litchfield DW (1999) Expression and regulation of protein kinase CK2 during the cell cycle. Mol Cell Biochem 191:213–222
8. Martel V, Filhol O, Nueda A, Gerber D, Beintez MJ, Cochet C (2001) Visualization and molecular analysis of nuclear import of protein kinase CK2 subunits in living cells. Mol Cell Biochem 227:81–90
9. Bosc DG, Slonimski E, Sichler C, Litchfield DW (1995) Phosphorylation of casein kinase II by $p34^{cdc2}$: identification of phosphorylation sites using phosphorylation site mutants *in vitro*. J Biol Chem 270:25872–25878
10. Paleń E, Traugh J (1991) Phosphorylation of casein kinase II. Biochemistry 30:5586–6690
11. Messenger MM, Saulnier RB, Gilchrist AD, Diamond P, Gorbsky GJ, Litchfield DW (2002) Interactions between protein kinase CK2 and Pin-1: evidence for phosphorylation-dependent interactions. J Biol Chem 275:14295–14306
12. Heriche JK, Chambaz EM (1999) Protein kinase CK2alpha is a target is a target for Abl and Brc-Abl tyrosine kinases. Oncogene 17:13–18
13. Laszlo G, Litchfield DW (2008) The emerging CK2 interactome: insights into the regulation and functions of CK2. Mol Cell Biochem 316:8–14

14. Olsten M, Litchfield DW (2004) Order or chaos? An evaluation of the regulation of protein kinase CK2. Biochem Cell Biol 82:681–693
15. Li D, Dobrowolska G, Krebs E (1996) The physical association of casein kinase 2 with nucleolin. J Biol Chem 271:15662–15668
16. Li D, Meier T, Dobrowolska G, Krebs E (1997) Specific interaction between casein kinase 2 and the nucleolar protein Nopp140. J Biol Chem 272:3373–3379
17. Skjerpen CS, Nilsen T, Wesche J, Olsens S (2002) Binding of FGF-1 variants to protein kinase CK2 correlates with mitogenicity. EMBO J 21:4058–4069
18. Miyata Y, Nishida E (2005) CK2 binds, phosphorylates, and regulates its pivotal substrate Cdc37, an Hsp90-cochaperone. Mol Cell Biochem 274:171–179
19. Olsen BB, Jessen V, Hojrup P, Issinger O-G, Boldyreff B (2003) Protein kinase CK2 phosphorylates the Fas-associated factor FAF1 *in vivo* and influences its transport into the nucleus. FEBS Lett 546:218–222
20. Faust M, Schuster N, Montenarch M (1999) Specific binding of protein kinase CK2 catalytic subunits to tubulin. FEBS Lett 462:51–62
21. Olsten ME, Canton DA, Zhang C, Walton PA, Litchfiels DW (2004) The Pleckstrin homology domain of CK2 interacting protein-1 is required for interactions and recruitment of protein kinase CK2 to the plasma membrane. J Biol Chem 279:42114–42127
22. Zhou XZ, Lu PJ, Wulf G, Lu KP (1999) Phosphorylation-dependent prolyl isomerization: a novel signaling regulatory mechanism. Cell Mol Life Sci 56:788–806
23. Heriche JK, Lebrin F, Rabilloud T, Leroy D, Chambaz EM, Goldberg Y (1997) Regulation of protein phosphatase 2A by direct interaction with casein kinase 2α. Science 276:952–955
24. Szyszka R. (2005) Kinaza białkowa CK2—oddziaływania białko-białko i ich konsekwencje. Na pograniczu chemii i biologii. Tom XII, 143–168.
25. Bosc DG, Graham KC, Saulnier RB, Zhang C, Prober D, Gietz RD, Litchfield DW (2000) Identification and characterization of CKIP-1, a novel pleckstrin homology domain-containing protein that interacts with protein kinase CK2. J Biol Chem 275:14295–14306
26. Faust M, Montenarch M (2002) Subcellular localization of protein kinase CK2. A key to its function? Cell Tissue Res 301:329–340
27. Jain N, Mahendran R, Philip R, Guy GR, Tan YH, Cao X (1996) Casein kinase II associates with Egr-1 and acts as a negative modulator of its DNA binding and transcription activities in NIH 3 T3 cells. J Biol Chem 271:13530–13536
28. Roher N, Miro F, Boldyreff B, Llorens F, Plana M, Issinger IG, Itarte E (2001) The C-terminal domain of human grp94 protects the catalytic subunit of protein kinase CK2 (CK2α) against thermal aggregation. Eur J Biochem 268:429–436
29. Rekha N, Srinivasan N (2003) Structural basis of regulation and substrate specificity of protein kinase CK2 deduced from the modeling of protein-protein interactions. BMC Struct Biol 3:4
30. Guerra B, Boldyreff B, Sarno S, Cesaro L, Issinger O-G, Pinna LA (1999) CK2: a protein kinase in need of control. Pharmacol Ther 82:303–313
31. Wirkner U, Voss H, Lichter P, Pyerin W (1994) Human protein kinase CK2 genes. Cell Mol Biol Res 40:489–499
32. Bodenbach L, Fauss J, Robitzki A, Krehan A, Lorenz P, Lozeman FJ, Pyerin W (1994) Recombinant human casein kinase II. A study with the complete set of subunits (alpha, alpha' and beta), site-directed autophosphorylation mutants and a bicistronically expressed holoenzyme. Eur J Biochem 220:263–273
33. Sajnaga E, Kubiński K, Szyszka R (2008) Catalytic activity of mutants of yeast protein kinase CK2α. Acta Biochim Pol 55:767–776
34. Niefind K, Guerra B, Ermakowa I, Issinger OG (2001) Crystal structure of human protein kinase CK2: insights into basic properties of the CK2 holoenzyme. EMBO J 20:5320–5331
35. Rao ST, Rossmann MG (1973) Comparison of super-secondary structures in proteins. J Mol Biol 76:241–256
36. Sarno S, Vaglio P, Meggio F, Issinger O-G, Pinna LA (1996) Protein kinase CK2 mutants defective in substance recognition. Purification and kinetic analysis. J Biol Chem 271: 10595–10601

37. Sarno S, Marin O, Ghisellini P, Meggio F, Pinna LA (1998) Biochemical evidence that the N-terminal segments of the alpha subunit and the beta subunit play interchangeable roles in the activation of protein kinase CK2. FEBS Lett 44:29–33
38. Vaglio P, Sarno S, Marin O, Meggio F, Issinger O-G, Pinna LA (1996) Mapping the residues of protein kinase CK2 alpha subunit responsible for responsiveness to polyanionic inhibitors. FEBS Lett 380:25–28
39. Miyata Y, Yahara I (1995) Interaction between casein kinase II and the 90-kDa stress protein, HSP90. Biochemistry 34:8123–8129
40. Li D, Dobrowolska G, Krebs EG (1999) Identification of proteins that associate with protein kinase CK2. Mol Cell Biochem 191:223–228
41. Donella-Deana A, Cesaro L, Sarno S, Ruzzene M, Brunati AM, Marin O, Vilk G, Dohery-Kirby A, Lajoie G, Litchfield DW, Pinna LA (2003) Tyrosine phosphorylation of protein kinase CK2 by Src-related tyrosine kinases correlates with increased catalytic activity. Biochem J 372:841–849
42. Vilk G, Saulnier RB, St PR, Litchfield DW (1999) Inducible expression of protein kinase CK2 in mammalian cells. Evidence for functional specialization of CK2 isoforms. J Biol Chem 274:14406–14414
43. Boldyreff B, James P, Staudenmann W, Issinger O-G (1993) Ser2 is the autophosphorylation site in the beta subunit from bicistronically expressed human casein kinase-2 and from native rat liver casein kinase-2 beta. Eur J Biochem 218:515–521
44. Allende JE, Allende CC (1995) Protein kinase CK2: an enzyme with multiple substrates and a puzzling regulation. FASEB J 9:313–323
45. Leroy D, Heriche JK, Filhol O, Chambaz EM, Cochet C (1997) Binding of polyamines to an autonomous domain of the regulatory subunit of protein kinase CK2 induces a conformational change in holoenzyme. A proposed role for kinase stimulation. J Biol Chem 272: 20820–20827
46. Soderling TR (1990) Protein kinases. Regulation by autoinhibitory domains. J Biol Chem 265:1823–1826
47. Meggio F, Boldyreff B, Issinger O-G, Pinna LA (1994) Casein kinase 2 down-regulation and activation by polybasic peptides are mediated by acidic residues in the 55–64 region of the beta-subunit. A study with calmodulin as phosphorylatable substrate. Biochemistry 33: 4336–4342
48. Boldyreff B, Meggio F, Pinna LA, Issinger O-G (1993) Reconstruction of normal and hyperactive forms of casein kinase-2 by variably mutated beta-subunits. Biochemistry 32: 12672–12677
49. Boldyreff B, Meggio F, Pinna LA, Issinger O-G (1994) Protein kinase CK2 structure-function relationship: effects of the beta subunit on reconstruction and activity. Cell Mol Biol Res 40:391–399
50. Hinrichs MV, Gatica M, Allende CC, Allende JE (1995) Site-direct mutants of the beta subunit of protein kinase CK2 demonstrate the important role of Pro-58. FEBS Lett 368:211–214
51. Yamaki M, Umehara T, Chimura T, Horikoshi M (2001) Cell death with predominant apoptotic features in Saccharomyces cerevisiae mediated by deletion of the histone chaperone ASF1/CIA1. Genes Cells 6:1043–1054
52. Hu F, Alcasabas AA, Elledge SJ (2001) Asf1 links Rad53 to control of chromatin assembly. Genes Dev 15:1061–1066
53. Munakata T, Adachi N, Yokoyama N, Kuzuhara T, Horikoshi M (2000) A human homologue of yeast anti-silencing factor has histone chaperone activity. Genes Cells 5:221–233
54. Sillje HHW, Nigg EA (2001) Identification of human Asf1 chromatin assembly factors as substrates of Tousled-like kinases. Curr Biol 11:1068–1073
55. Umehara T, Chimura T, Ichikawa N, Horikoshi M (2002) Polyanionic stretch-deleted histone chaperone cia1/Asf1 is functional both in vivo and in vitro. Genes Cells 7:59–73
56. Riera M, Peracchia G, de Nadal E, Ariño J, Pages M (2001) Maize protein kinase CK2: regulation and functionality of three beta regulatory subunits. Plant J 25:365–374

57. Olsen BB, Rasmussen T, Niefind K, Issinger O-G (2008) Biochemical characterization of CK2α and α' paralogues and their derived holoenzymes: evidence for the existence of a heteroterameric CK2α'-holoenzyme forming trimeric complexes. Mol Cell Biochem 316:37–47
58. Janeczko M, Orzeszko A, Kazimierczuk Z, Szyszka R, Baier A (2012) CK2α and CK2α' subunits differ in their sensitivity to 4,5,6,7-tetrabromo- and 4,5,6,7-tetraiodo-1H-benzimidazole derivatives. Eur J Med Chem 47:345–350
59. Jach M, Kubiński K, Zieliński R, Sajnaga E, Szyszka R (2005) Asf1—yeast inductor of apoptosis inhibits CK2 activity *in vitro*. Cell Mol Biol Lett 10(Suppl 2):121–122

Impressions from the Conformational and Configurational Space Captured by Protein Kinase CK2

Jennifer Hochscherf, Alexander Schnitzler, Olaf-Georg Issinger, and Karsten Niefind

Abstract As key components of cellular regulation and signal transduction, eukaryotic protein kinases (EPKs) are strictly regulated. Sophisticated control mechanisms of EPKs are necessary which typically include conformational changes of the enzymes in critical regions. Local structural plasticity is therefore a prerequisite of normal EPK function. Protein kinase CK2, a member of the CMGC family of EPKs, was regarded as an exception from this rule for a long time due to its constitutive activity (lack of an inactive state) and due to its structural rigidity in typical EPK control regions of its catalytic subunit CK2α like the activation segment and the helix αC. Gradually, however, several cases of inherent local plasticity within CK2α were detected, and questions about their crosstalk and their functional significance became an issue. It is very likely now that structural plasticity and dynamics is more important for CK2 function than believed previously. Novel interpretation methods of crystallographic data even confirm an allosteric communication between the ATP site and CK2β-binding site of CK2α which had been only hypothetically postulated before. Similarly, local mobilities of CK2α are subject to modern computational approaches suggesting conformational equilibria in solution as assumed previously. In summary, CK2 structural biology has reached now a mature phase in which sophisticated modern techniques overcome the limitations of classical crystallography so that structural dynamics rather than single "snapshots" is investigated.

Keywords Protein kinase CK2 • Casein kinase 2 • Eukaryotic protein kinases • CMGC kinases • Signal transduction pathways • Structural plasticity • Structural dynamics • Constitutive activity • Hinge region • Allosteric communication

---

J. Hochscherf • O.-G. Issinger
KinaseDetect ApS, Skovvej 22, DK-6340 Kruså, Denmark

A. Schnitzler • K. Niefind (✉)
Department of Chemistry, Institute of Biochemistry, University of Cologne, Otto-Fischer-Str. 12-14, D-50674 Cologne, Germany
e-mail: Karsten.Niefind@uni-koeln.de

K. Ahmed et al. (eds.), *Protein Kinase CK2 Cellular Function in Normal and Disease States*, Advances in Biochemistry in Health and Disease 12, DOI 10.1007/978-3-319-14544-0_2

1 CK2: A Protein Kinase of Limited Internal Mobility

"Conformational plasticity" as a basis of regulation and functional control was described as a hallmark of eukaryotic protein kinases (EPKs) more than a decade ago [1]. In recent reviews, this feature received particular attention and emphasis: EPKs were referred to as "dynamic molecular switches" [2] or as "allosteric macromolecular switches" [3] that received plasticity-correlated regulatability during evolution in contrast to similarly folded but more simple small-molecule kinases [4]. In particular, EPKs acquired an activation segment as part of their C-terminal domains (C-lobes) [4]. This activation segment is often the subject of (auto)phosphorylation and adapt able in its conformational state [1, 4], and it is assisted in this regulatorily relevant plasticity by the helix αC, the main control element of the N-terminal domain (N-lobe). Apart from regulation, large conformational changes within EPKs were also discussed to be necessary steps within the catalytic cycle itself [5].

Against this background, protein kinase CK2—a heterotetrameric complex of two catalytic chains (CK2α) attached to an obligatory homodimer of regulatory subunits (CK2β)[6]—was (and essentially still is) regarded as an apparent exception: it is no subject of regulatory phosphorylation at the activation segment of CK2α and does not dispose of a clearly defined inactive state. On a structural level, this "constitutive activity" was underpinned by a remarkable conformational rigidity of the activation segment and the helix αC. Both of them are fixed by the likewise structurally invariant N-terminal segment of CK2α via intramolecular interactions that resemble functionally the intermolecular contacts of the cyclin proteins with their corresponding cyclin-dependent kinases [7].

In the first review on CK2 structural biology published in 2009 [8], the structural conservation of the ensemble of activation segment, helix αC and N-terminal segment was emphasised by systematic structural comparisons including all 46 CK2α structures published at that time. The same analysis, however, illustrated that there are zones of structural variability as well (Fig. 1), some of them are typical loop regions in which flexibility is not unusual; others are with a clear conformational consensus among EPKs so that the plasticity discovered in CK2α (however not in all CK2α orthologues [9]) came as a surprise.

In subsequent years—with increasing numbers of structures—the picture of CK2 as an enzyme being conformationally constrained, where other EPKs are adaptable, but simultaneously disposing of regions of unusual adaptability, intensified and was addressed in a number of original and review publications [10–15], among them a recent metadynamics study to investigate conformational states separated by fairly high energy barriers [16]. Therefore, the local conformational and configurational variability of CK2 was the main subject of our contributions to the 7th International Conference on Protein Kinase CK2 in September 2013 in Lublin, Poland.

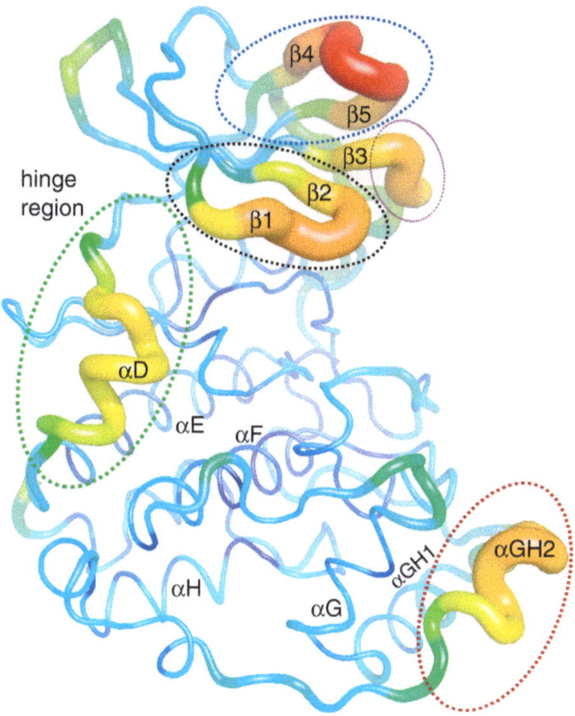

Fig. 1 CK2α structure with regions of increased mobility. The crystallographic B-factors of the atoms are encoded by the colour and the thickness of the tube. Rigid regions are drawn in *blue colour* with thin tubes while high-mobility regions are illustrated as thick and *yellow* to *red tubes*. The picture was created from PDB file 2ZJW using PYMOL [52]

2 Indicators and Study Methods of Structural Plasticity

Basically, X-ray crystallography is of limited use for the investigation of conformational dynamics in proteins because the crystalline state of a protein requires and causes strong restraints on atomic mobility. However, some phenomena and approaches are able to relativise these restrictions:

- B-factors: even under crystalline conditions, the individual atoms retain a certain degree of mobility. In crystallographic refinement, it is taken into account by the so-called B-factor, also known as the atomic temperature factor. The B-factor of an atom is proportional to the square of its mean displacement, i.e. a (hypothetical) ideally frozen atom has a B-factor of 0 while strong mobility correlates with high B-factors. Noteworthy, these B-factors reflect the inherent mobilities of the proteins and are no crystal-packing artefacts [17]. In Fig. 1, regions of human CK2α with high B-factors are illustrated by thick tubes and red colour.

- Polymorphism: many proteins crystallise in different crystal forms so that several snapshots of the conformational space are captured. This phenomenon is even enhanced by a sort of "functional polymorphism", i.e. the fact that the protein can exist and can be crystallised in different functional states, e.g. in complex with various ligands like substrates, inhibitors or other interaction partner. Recent CK2-related examples for a systematic investigation of structural polymorphism are the works of Papinutto et al. [14] and of Klopffleisch et al. [15].
- Starting from one or several experimentally determined conformations, computational methods (molecular dynamics) can be used to explore the conformational space quite efficiently and to expand in this way the structure-based understanding. A valuable example of this approach in the case of CK2 was published by Gouron et al. [16].
- In recent years, sophisticated methods were developed to regard a protein crystal structure even more than just as a "snapshot", but to derive conformational dynamic information directly from the (properly calculated and scaled) electron densities [18, 19]. In this context, room-temperature crystallography might experience a revival [20].

3 In Front of and Behind the β4β5 Loop

The β4β5 loop is a mobile structure element of CK2α with particularly high temperature factors (blue ellipse in Fig. 1). Like other EPKs [21], it harbours a hydrophobic surface cavity at either side: the "N-lobe cap" [21] and the "PIF pocket" [22] (Fig. 2). The plasticity of the β4β5 loop correlates with the occupation states of these two cavities.

3.1 The PIF-Pocket Region of CK2α Is Intramolecularly Plugged by an Absolutely Conserved Trp Side Chain

The intimate hook-up of the N-terminal segment with the helix αC and activation segment was the most conspicuous feature of the first CK2α structure [7]. It inspired Sarno et al. [23] to investigate the importance of the N-terminal segment by a set of point and N-terminal deletion mutants of human CK2α. Remarkably, the mutants $CK2\alpha^{\Delta 2\text{-}24}$ and $CK2\alpha^{\Delta 2\text{-}30}$ were inactive, but catalytic activity was partially rescued by CK2β [23].

Sarno et al.'s [23] results show that an active conformation of the helix αC and the activation segment does not absolutely depend on a network of contacts to the N-terminal segment but can be stabilised by CK2β as well. How CK2β manages this is not understood, in particular, since the CK2 holoenzyme structure (Fig. 3a) [6] revealed that CK2β does not directly touch the helix αC and the activation segment.

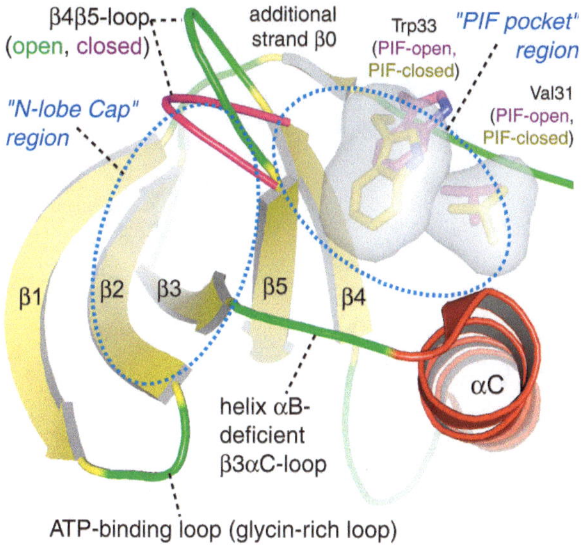

Fig. 2 "PIF pocket" and "N-lobe cap": the two N-lobal cavities flanking the adaptable β4β5 loop The majority of the picture was drawn with the PDB file 3NSZ which shows the PIF pocket filled with the side chains of Trp33 and Val31. The magenta-coloured parts representing on open PIF pocket, but a closed β4β5 loop are extracted from PDB file 2PVR. The figure was generated with PYMOL [52]

Recent observations with CK2α' [24, 25], however, and unpublished mutational data from our group suggest that the rescue effect of CK2β on CK2$\alpha^{\Delta 2\text{-}24}$ and CK2$\alpha^{\Delta 2\text{-}30}$ might have to do with the PIF pocket [22] of CK2α. "PIF" means"PDK1 interacting fragment". PIF is a region within enzymes of a certain AGC kinase subfamily which are phosphorylated and activated by 3-phosphoinositide-dependent protein kinase 1 (PDK1). The PIF fragment of these enzymes binds to PDK1's PIF pocket providing thus the prime example for an external occupation of the PIF pocket. The global analysis of Thompson et al. [21] revealed a more general relevance of the PIF pocket for EPKs: it is a hydrophobic surface cavity that can be used for docking of intra- or intermolecular peptide regions for the purpose of substrate recruitment or activation, the latter being caused by the stabilisation of the adjacent helix αC in its active conformation.

In the case of CK2α, the PIF pocket is mainly occupied by the side chain of a tryptophan residue which is absolutely conserved among all CK2α sequences reported so far (Trp33 of human CK2α, Fig. 2). This Trp side chain can be plugged into the PIF pocket very deeply (PIF closed), as it is the case in the two human CK2α' structures published to date [24, 25], or it can be located a bit more outside as found in most other CK2α structures (PIF open). So far, human CK2α is the only CK2α orthologue to be found with both extreme positions of Trp33 (Fig. 2). It may be possible to replace Trp33 completely from the PIF pocket by suitable small molecules and with consequences for the enzyme that are difficult to predict. Changing

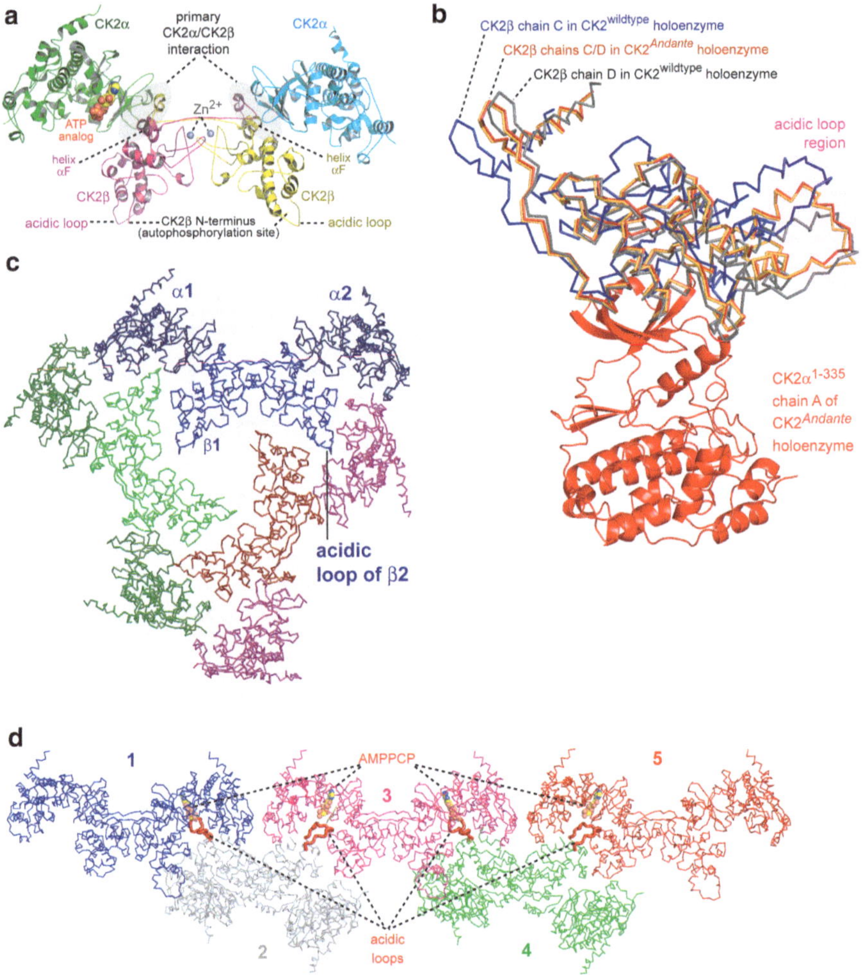

Fig. 3 The primary CK2α/CK2β interaction constituting the CK2 holoenzyme is adaptable which allows the formation of ring-like and linear higher-order aggregates. (**a**) Heterotetrameric CK2$\alpha_2\beta_2$ holoenzyme. (**b**) Superimposition of four CK2α subunits from two different CK2$\alpha_2\beta_2$ holoenzyme structures [6, 33] with attached CK2β subunits to illustrate the adaptability of the primary CK2α/CK2β contact. (**c**) Trimers of CK2$\alpha_2\beta_2$ heterotetramers identified in a hexagonal crystal form [6, 31]. The acidic loop of one CK2β subunit per heterotetramer approaches the positively charged substrate-binding region of CK2α from a neighbouring CK2$\alpha_2\beta_2$ holoenzyme complex. Packing of CK2$\alpha_2\beta_2$ tetramers in this way requires a certain deviation from the twofold symmetry and the adaptability illustrated in part **b** of the figure. (**d**) Linear aggregation of highly symmetrical CK2$\alpha_2\beta_2$ heterotetramers in a monoclinic crystal form [33]. In this arrangement, the acidic loops of all CK2β subunits are embedded in the substrate-binding regions of neighbouring CK2α chains. Part **a**, **b** and **d** of the figure were reprinted from the Journal of Molecular Biology, Vol. 424, Schnitzler, A. et al., "The protein kinaseCK2^{Andante} holoenzyme structure supports proposed models of autoregulation and *trans*-autophosphorylation", pp 1871–1882, 2014, with kind permission from Elsevier. Part **c** of the figure was reprinted from Molecular and Cellular Biochemistry, Vol. 274, Niefind, K. and Issinger, O.-G., "Primary and secondary interactions between CK2α and CK2β lead to ring-like structures in the crystals of the CK2 holoenzyme" pp 3–14, 2005, with kind permission from Springer Science+Business Media

Trp33 to alanine in human CK2α has nearly no effect on the K_M value for ATP and the k_{cat} value for the kinase reaction with an artificial peptide substrate; however, the Trp33Ala mutant was significantly more susceptible to activation by CK2β (J. Hochscherf, A. Köhler, unpublished data) which resembles the behaviour of the mutants CK2αΔ2-24 and CK2α$^{Δ2-30}$ observed by Sarno et al. [23].

3.2 *The "N-Lobe Cap" of CK2α Coordinates CK2β and Serves as a Mobile Hinge of the CK2α/CK2β Interaction*

The "N-lobe cap", i.e. the front cavity of the β4β5 loop in EPKs [21], is a prominent functional site of CK2α since it serves for docking of CK2β via the primary CK2α/CK2β interaction [6] (Fig. 3a). CK2β binding to CK2α is an enthalpically driven process [26] that requires the β4β5 loop in an extended ("open") conformation (Fig. 2), whereas in CK2β-unbound form the β4β5 loop can switch to a closed state [27, 28]. The closed β4β5 loop partly occupies the "N-lobe cap" cavity (Fig. 2) and contributes to the coordination of small molecules with a certain CK2β-antagonistic effect [29]. For coordination of peptidic CK2β competitors, however, the β4β5 loop has to be open like in the CK2 holoenzyme as recently shown in the complex structure of human CK2α$^{1-335}$ with a cyclic peptide (Fig. 4a) [30].

But not only the β4β5 loop itself has conformational plasticity depending on the occupation state of its neighbouring cavities and the crystalline environment. Rather, already the first CK2$α_2β_2$ holoenzyme structure (Fig. 3a) [6] revealed that the whole primary CK2α/CK2β interaction is not rigid but adaptable. The structure originated from a hexagonal crystal packing in which the CK2$α_2β_2$ tetramers associate to trimeric rings (Fig. 3c) [31]. To fit into this arrangement, the two CK2α chains are attached in significantly different orientations to the CK2β dimer (Fig. 3b). This asymmetry is no inherent property of the CK2$α_2β_2$ holoenzyme; rather, in two recent studies with CK2$α_2β_2$ tetramers in monoclinic crystalline environments, the twofold symmetry of the CK2$α_2β_2$ holoenzyme is well preserved [32, 33]. Thus, three different orientations of the CK2β dimer relative to CK2α are known to date (Fig. 3b) providing the impression that the N-lobe cap cavity acts like a hinge to permit adaptability and mobility to a certain degree.

Whether and how these observations can be functionally interpreted is a subject of debate. While Lolli et al. [32] regarded the higher symmetrical form of the CK2$α_2β_2$ tetramer *per se* as the CK2 state of maximum activity, we emphasised the adaptability of the primary CK2α/CK2β contact as a prerequisite of the CK2$α_2β_2$ holoenzyme's propensity to integrate into ring-like [31] or linear [33] higher-molecular aggregates (Fig. 3c/d) of reduced activity. The ensembles illustrated in Fig. 3c/d were found in two different crystal forms, but linear and ring-like states of the CK2 holoenzyme exist in solution as well where they are strongly effected by the concentration of salts and of polycationic substances [34]. This downregulatory aggregation propensity is such a prominent feature of the CK2 holoenzyme that it inspired Poole et al. [35] to classify CK2 as a "constitutively inactive" kinase.

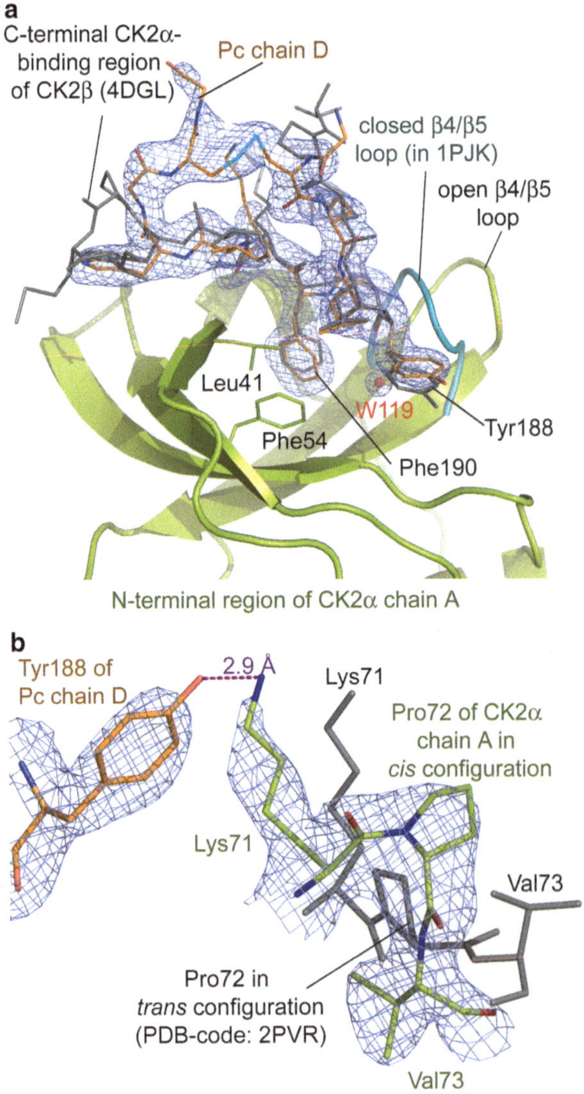

Fig. 4 The structure of CK2α in complex with a CK2β-competitive cyclic peptide revealed that Pro72 within the β3αC loop can switch to the *cis*-configuration (**a**) The CK2β interface of CK2α binds a cyclic peptide [53] which is embedded in its final electron density [30]. For comparison, CK2β in the $CK2\alpha_2\beta_2$ holoenzyme structure 4DGL was overlaid. (**b**) Configurational plasticity of Pro72 that enables a contact of Lys71 to a critical side chain of the cyclic peptide (Tyr188). Both parts of the figure were reprinted from ACS Chemical Biology, Vol. 8, Raaf, J. et al., "First structure of protein kinase CK2 catalytic subunit with an effectiveCK2β-competitive ligand", pp 901–907, 2013, with kind permission from the American Chemical Society

4 Proline *Cis-/Trans*-Isomerisation at the β3αC Loop

The preferred configuration of a peptide bond with its partial double bond character is *trans*, but *cis-/trans*-isomerisation of peptide bonds is a well-known phenomenon, in particular if proline provides the amide nitrogen atom [36]. Since the process requires breaking of a (partial) double bond, it is by definition a change of the configuration rather than the conformation as it is incorrectly designated in many publications. *Cis-/trans*-isomerisation in polypeptide chains often occurs as part of the protein folding process; CK2α, e.g., always contains a stable *cis*-peptide, namely, in a loop region of its C-terminal helical domain (Pro231 of human CK2α). With one *cis*-peptide among 17 proline residues, the *cis*-proline frequency in human CK2α is close to the average determined from a data set of 571 nonredundant protein structures [37].

Yet, *cis-/trans*-isomerisation happens in mature proteins as well where it is assumed to be an "intrinsic molecular switch" [38] without the need for covalent modifications. Often special enzymes—the peptidyl prolyl isomerases—are required to enable and control the process which is sometimes dependent on the phosphorylation of a neighbouring side chain and which is regarded as an important "molecular timer" [39]. In addition, enzyme-independent "native state proline isomerisation" [38] was observed and is likely to be functionally relevant.

In summary, there is an amazingly large literature about both enzyme-catalysed and spontaneous *cis-/trans*-plasticity at prolyl peptide bonds. Against this background, it was less surprising than we believed when we recently identified a so far unknown *cis*-proline in the N-terminal domain of CK2α (PDB 4IB5, Fig. 4b) [30]. More precisely, it was found at Pro72 within the β3αC loop (magenta-coloured ellipse of Fig. 1) of human CK2α$^{1\text{-}335}$. In all previous CK2α structures, a *trans*-proline had been observed at this position. Noteworthy, the preparation procedure involving bacterial gene expression and chromatographic purification did not significantly differ.

The functional meaning of a *cis-/trans*-plasticity at Pro72 is unclear so far. Pro72 is an absolutely conserved residue of CK2α, and the β3αC loop of CK2α is unconventional among EPKs because of its shortness missing the helix αB of the canonical protein kinase fold [40]. The aforesaid structure [30] suggested a correlation with the occupation of the N-lobe cap by a cyclic peptide mimicking the binding region of CK2β, since only with the *cis*-configuration at Pro72 the preceeding Lys71 side chain could form a hydrogen bond with a critical tyrosine side chain of the cyclic peptide (Fig. 4b). To probe the hypothesis that Pro72 and its ability to switch to the *cis*-configuration play a role in binding of the cyclic peptide or even CK2β, we created a point mutant Pro72Ala of CK2α$^{1\text{-}335}$. Preliminary and unpublished data of this mutant show no significant changes in the binding profile. Unexpectedly, however, it was thermostabilised and folded much more efficiently in the bacterial expression system than the wild type.

5 The Hinge/Helix αD Region: A Case of "Cracking"?

The interdomain hinge/helix αD region of CK2α (green ellipse in Fig. 1, Fig. 5) has two remarkable features otherwise unknown from EPKs:

– The predominant conformation of this zone is different from other EPKs, leaving more space at the ATP-binding site and contributing thus to the "dual-cosubstrate specificity" of the enzyme [41, 42]. This fact was already discussed in the first

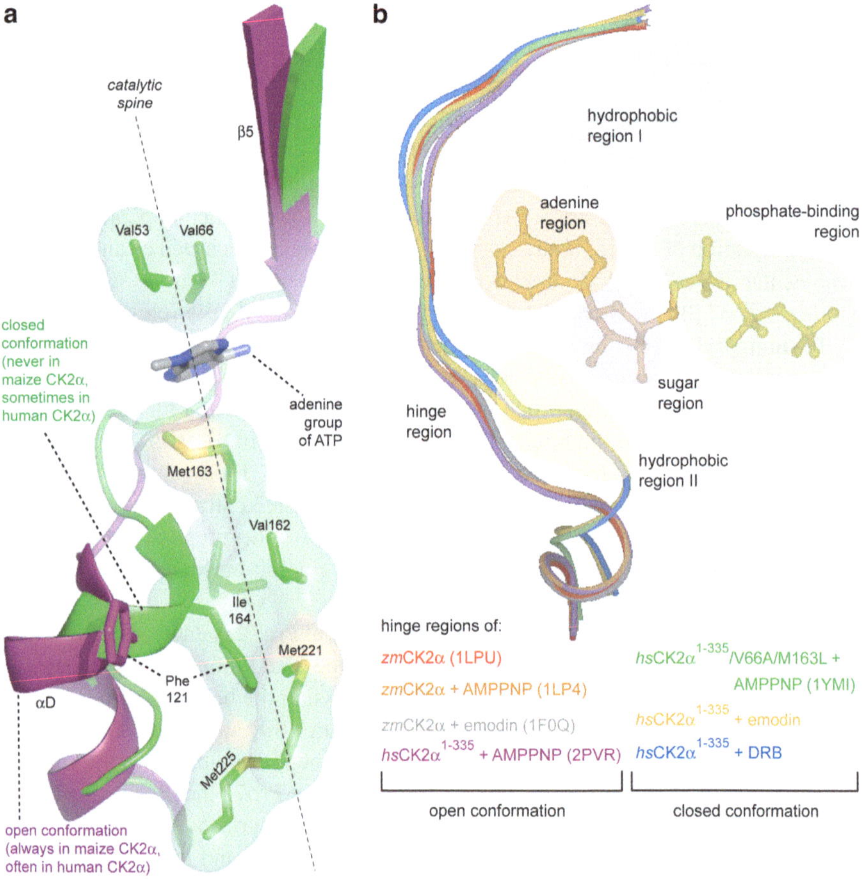

Fig. 5 Conformational ambiguity of the hinge/helix αD region next to the ATP site. (**a**) The hinge/helix αD region of CK2α was found with an open (*purple*) or with a closed conformation (*green/yellow*). The catalytic spine (indicated by a molecular surface) is fully assembled only in the closed conformation which so far occurred exclusively with human CK2α; in this conformation, the Phe121 side chain completes the catalytic spine. (**b**) The closed conformation of the hinge/helix αD region partly occupies the hydrophobic region II, i.e. one of the five sections defined by Traxler and Furet [44] in their protein kinase pharmacophore model. This picture was a part of a talk presented by one of the authors at the 5th International Conference on Protein Kinase CK2 2007 in Padua (Italy)

structure publication of CK2α [7] when only a handful of comparison structures were available, but it received a more profound basis by the spine concept of Taylor and Kornev [4] according to which any fully active protein kinase requires two "spines", i.e. two stacks of hydrophobic side chains crossing the interdomain interface. Consistent to this concept, CK2α has a "regulatory spine" which was never observed in a disassembled state, but Taylor and Kornev [4] left CK2α out when they illustrated the second, the "catalytic spine", which includes a side chain from the hinge/helix αD region. In other words, with respect to the regulatory spine, CK2α is a perfect representative of a typical EPK, but concerning the catalytic spine, it is an outlier [12, 13] (magenta-coloured trace in Fig. 5a) and once again a "challenge to canons" [43].

– The hinge/helix αD region of CK2α is not always stable in its non-canonical conformation; rather, it can switch to the EPK-canonical catalytic spine state (green trace in Fig. 5a). This conformational plasticity was noticed first in 2005 [42] and presented to the community at the 5th International Conference on Protein Kinase CK2 in 2007 in Padua, Italy, where it was discussed in the light of the protein kinase pharmacophore model of Traxler and Furet [44] (Fig. 5b). In 2008, the first report published [9] that CK2α orthologues differ in this respect: while maize CK2α sticks to the non-canonical and space-providing "open" hinge/helix αD conformation (Fig. 5a/b), human CK2α is able to exist with the EPK-typical "closed" conformation as well.

A number of follow-up questions were addressed in the subsequent years: (1) Which internal or external restraints and in particular which ligands favour either the open or the closed hinge/helix αD conformation? (2) Why do some CK2α orthologues exist always and stably with an open hinge/helix αD conformation while others can switch to the closed state? (3) Which of the two conformations is the (more) active one? (4) Can the particular features of the hinge/helix αD region be exploited for the design of CK2α-targeting drug, i.e. for improving the selectivity by addressing the unique open conformation? (5) Is the plasticity of the hinge/helix αD region coupled to the flexibility of other regions like the β4β5 loop or the ATP-binding loop?

One of the problems in this context became recently apparent by an analysis in which we described a strong artificial influence of the crystallisation conditions on the conformation of the hinge/helix CK2α region [15]: CK2α structures going back to crystallisation media dominated by kosmotropic salts like ammonium sulphate or sodium citrate as precipitants always have the "closed" hinge/helix αD conformation; this makes sense insofar as kosmotropic salts typically support hydrophobic interactions which means in this context that they induce the completion of the hydrophobic catalytic spine by burying the otherwise solvent-exposed aromatic side chain of Phe121 in a hydrophobic cavity formed by the other catalytic spine side chains (Fig. 5a). In other words, whenever CK2α was crystallised by a kosmotropic salt as a salting-out agent, the hinge/helix αD region is dominated by this medium and adopts the EPK-canonical conformation irrespective of any ligand. Noteworthy, maize CK2α that seems to have particular sequence adaptations in this region in favour of the open hinge/helix αD conformation [12] was never crystallised by a kosmotropic salt.

Hence, it is not trivial to distinguish between such artificial restraints on the hinge/helix αD conformation and genuine effects of ATP-competitive ligands. Nevertheless, Battistutta and Lolli [12] derived a rule that can be regarded as well established now: ATP-site ligands without polar interactions to the backbone of the hinge region (e.g. the inhibitor emodin [9, 14, 27]) do not prefer either of the two conformations, whereas ligands that exploit such hinge interactions for binding (like ATP as the prototype) have a clear preference for the open hinge/helix αD conformation.

CK2β was a further ligand of CK2α suspected to be capable of stabilising the open hinge/helix αD conformation [11]. This hypothesis was based on the first $CK2\alpha_2\beta_2$ holoenzyme structure [6] in which both CK2α chains have an open hinge/helix αD conformation, but due to the non-proximity of CK2β and the hinge/helix αD region, it lacked a convincing mechanistic rationale. In fact, recent $CK2\alpha_2\beta_2$ tetramer [32] and CK2α/cyclic peptide structures [30] with closed hinge/helix αD conformation ruled out the notion that the occupation of CK2α's N-lobe cap by CK2β or by a CK2β analogue induces the open hinge/helix αD conformation.

Recent molecular dynamic simulations have revealed that the closed hinge/helix αD conformation is energetically even more stable than the open one [16]. The highest activation energy barrier between the two states is about 33.5 kJ/mol and depends largely on the rotation of the aforementioned Phe121 side chain (Fig. 5a) [16]. This is consistent with stable populations in solution coexisting in a dynamic equilibrium and with the possibility to stabilise either of them by specific ligands.

In many human CK2α structures, however, the whole hinge/helix αD region is characterised by a high degree of disorder, i.e. by high crystallographic B-factors and disrupted electron density maps. Noteworthy, this zone is not a typical surface loop for which this kind of disorder is not unusual. Rather, it looks like an example of local protein unfolding or "cracking" [45], a phenomenon that is assumed to play a key role for the change of a protein between two functionally relevant states by lowering the activation barrier via entropy reduction. In the case of epithelial growth factor receptor kinase, molecular dynamic simulations accompanied by hydrogen/deuterium exchange studies suggested that cracking of the hinge/helix αD region is important for the transition to catalytically inactive conformations [46]. How general cracking of the hinge/helix αD region is for EPKs and in particular whether it plays a role in the function of CK2α is an open and a very interesting question.

6 The Glycine-Rich ATP-Binding Loop: The Classical Case of Structural Plasticity

Apart from the hinge/helix αD region, the ATP-binding loop, which connects the β-strands 1 and 2 and stabilises via highly conserved glycine residues the negatively charged triphospho moiety of the cosubstrate, is the second functionally important high-mobility zone (black ellipse in Fig. 1) investigated in the aforementioned molecular dynamics study [16]. The rationale for this selection was a model [11] according to which CK2β-unbound CK2α might exist in a dynamic equilibrium between three conformationally and functionally different states (Fig. 6a): (i) fully active with open hinge/helix αD region and stretched ATP-binding loop; (ii) fully inactive [10] with a closed hinge/helix αD region, with a collapsed ATP-binding loop blocking the ATP site and with a shortcut of both elements via an ionic interaction between Arg47 and Asp120; and (iii) partially active with a closed hinge/helix αD region but with a stretched ATP-binding loop so that binding of ATP is possible [42] albeit in a not fully productive way.

The computational study [16] confirmed that the two principal states of the ATP-binding loop (stretched and collapsed as illustrated in Fig. 6b) have approximately the same free energy. Although the highest activation barrier along the path between them is only about 13 kJ/mol, a spontaneous collapse of the stretched ATP-binding loop during a 40 ns calculation could not be observed [16]. Nevertheless, among CK2α crystal structures, a collapse of the ATP-binding loop [27] and in particular the bending down of Arg47 into the ATP site (Fig. 6b) were observed more than once. Most recently, it was found (at least in one of three CK2α chains unambiguously) in the complex structure of human CK2α with a cyclic CK2β-mimetic peptide [30].

Insofar the existence of CK2α conformations with a collapsed ATP-binding loop is assured. It is well possible that CK2β by touching the back of the β1/β2 region (Fig. 6c) [6, 32, 33]—noteworthy, Phe54 in strand β2 is one of the "hot spots" of the CK2α/CK2β interaction [47] as visible in Fig. 4a—activates CK2α by stabilising the stretched ATP-binding loop and by preventing Arg47 from blocking the access to the ATP site [10]. In accordance with this notion, a recent study, in which conformational dynamics was derived from a sophisticated redefinition of electron density noise level followed by a systematic analysis of concerted side chain movements in some structures of CK2α and other EPKs, has revealed such an allosteric communication pathway between the ATP site and the CK2β-binding site of CK2α [19] (Fig. 6d).

7 The CMGC-Typical αGH Insert Remains Puzzling in its Function

As a member of the CMGC family of EPKs [48], CK2α contains a CMGC-characteristic helical insertion between the helices αG and αH of the EPK consensus fold [40]. In all CK2α structures, this insert has high B-factors (red ellipse in Fig. 1). Five years ago, we developed some ideas about possible functions of this region in the context of substrate binding and recognition based on homology to other CMGC family EPKs [8]; none of those notions was confirmed since that time, but falsification did not happen either. So the function of this high-mobility region remains enigmatic. For more insight, structures of enzyme/substrate complexes have to be awaited.

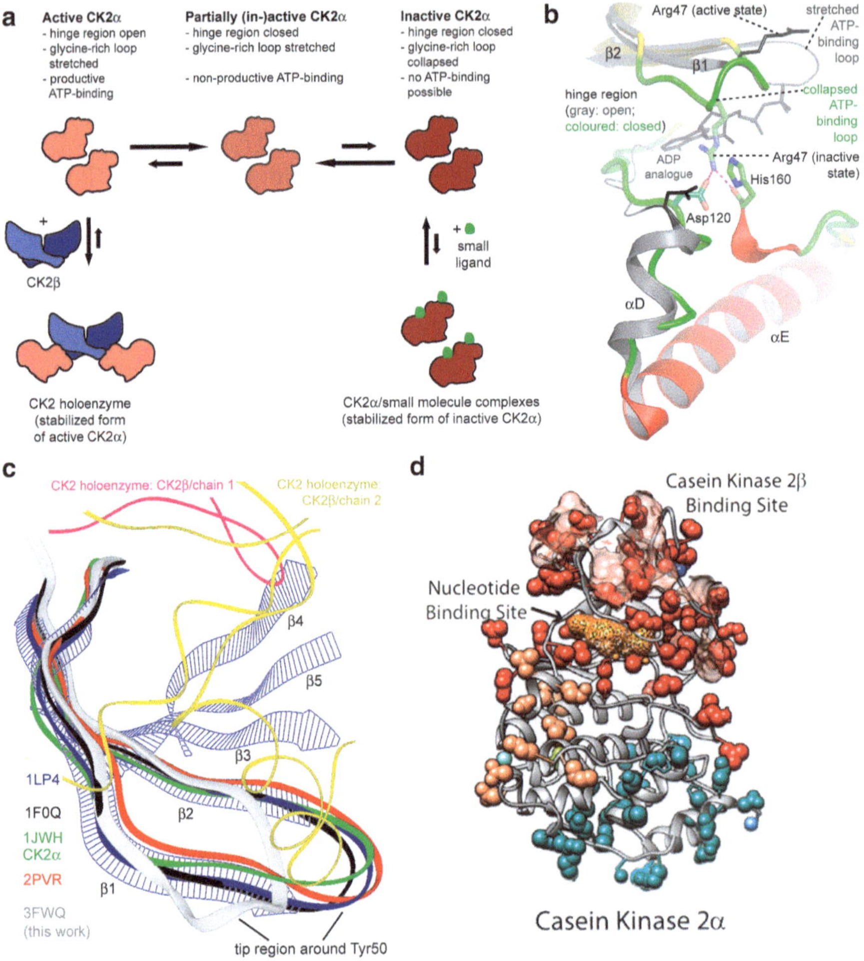

Fig. 6 The plasticity of ATP-binding loop as potential basis of the allosteric activation of CK2α by CK2β. (**a**) Model of CK2α regulation by CK2β and by small metabolites, published in a basic form after a structure of human CK2α with collapsed ATP-binding loop had been discovered [10] and in an extended version (including a partially active state of CK2α) shortly afterwards [11]. Reprinted from Biochimica and Biophysica Acta, Vol. 1804, Niefind, K. and Issinger, O.-G., "Conformational plasticity of the catalytic subunit of protein kinase CK2 and its consequences for regulation and drug design", pp484–492, 2010, with kind permission from Elsevier. (**b**) Shortcut between the ATP-binding loop (after collapse) and the hinge/helix αD region (after switching to the closed conformation) preventing access to the ATP site [10]; the largest conformational change is seen in Arg47 which directly interacts with Asp120 and His160 while it points to the opposite direction in fully active CK2α with extended ATP-binding loop. (**c**) Illustration of the first structure-based hypothesis [10] on how CK2β might activate CK2α by stabilising the ATP-binding loop from behind and thus preventing its collapse into the ATP site. Reprinted from the Journal of Molecular Biology, Vol. 386, Raaf et al., "First inactive conformation of CK2α, the catalytic subunit of protein kinase CK2", pp 1212–1221, 2009, with kind permission from Elsevier. (**d**) Allosteric communication path between the ATP site and the CK2β-binding site suggested by the concerted side chain mobility in that region [19]; the authors of this study describe a sophisticated procedure to detect the side chain polymorphism in protein crystal structures and apply it to CK2α and to some other eukaryotic protein kinases in an apo form and in an ATP-bound form, respectively. Reprinted from the Proceedings of the National Academy of Science USA, Vol. 111, Lang, P. T. et al., "Protein structural ensembles are revealed by redefining X-ray electron density noise", pp 237–242, 2014, with kind permission from the Proceedings of the National Academy of Science USA

8 CK2 Structural Biology: From Basic Insight to Application and Back!

After in an initial phase some fundamental CK2 structures had been published [6, 7, 49], CK2 crystallography entered an "application phase" already several years ago when CK2α/inhibitor structures became more and more important as valuable tools for drug development [50].While efforts in structure-based design of CK2 inhibitors is ongoing, the interest in the structural bases of fundamental properties of CK2 is revitalised. Aggregation and autoregulation phenomena of the CK2 holoenzyme are structurally investigated [31–33]; more species than so far are taken into account in CK2 structural biology [51]. We believe that more information about the conformational and configurational dynamics of the enzyme will enrich these efforts in a most fruitful way.

Acknowledgement The contributions of Anja Asendorf, Anna Köhler and Nicole Splett are gratefully acknowledged. The work was supported by a grant from the Deutsche Forschungsgemeinschaft (DFG) to KNI (NI 643/4–1).

References

1. Huse M, Kuriyan J (2002) The conformational plasticity of protein kinases. Cell 109: 275–282
2. Taylor SS, Keshwani MM, Steichen JM, Kornev AP (2012) Evolution of the eukaryotic protein kinases as dynamic molecular switches. Philos Trans R Soc Lond B Biol Sci 367: 2517–2528
3. Taylor SS, Ilouz R, Zhang P, Kornev AP (2012) Assembly of allosteric macromolecular switches: lessons from PKA. Nat Rev Mol Cell Biol 13:646–658
4. Taylor SS, Kornev AP (2011) Protein kinases: evolution of dynamic regulatory proteins. Trends Biochem Sci 36:65–77
5. Jura N, Zhang X, Endres NF, Seeliger MA, Schindler T, Kuriyan J (2011) Catalytic control in the EGF receptor and its connection to general kinase regulatory mechanisms. Mol Cell 42:9–22
6. Niefind K, Guerra B, Ermakowa I, Issinger O-G (2001) Crystal structure of human protein kinase CK2: insights into basic properties of the CK2 holoenzyme. EMBO J 20:5320–5331
7. Niefind K, Guerra B, Pinna LA, Issinger O-G, Schomburg D (1998) Crystal structure of the catalytic subunit of protein kinase CK2 from *Zea mays* at 2.1 Å resolution. EMBO J 17: 2451–2462
8. Niefind K, Raaf J, Issinger O-G (2009) Protein kinase CK2: from structures to insights. Cell Mol Life Sci 66:1800–1816
9. Raaf J, Klopffleisch K, Issinger O-G, Niefind K (2008) The catalytic subunit of human protein kinase CK2 structurally deviates from its maize homologue in complex with the nucleotide competitive inhibitor emodin. J Mol Biol 377:1–8
10. Raaf J, Issinger O-G, Niefind K (2009) First inactive conformation of CK2α, the catalytic subunit of protein kinase CK2. J Mol Biol 386:1212–1221
11. Niefind K, Issinger O-G (2010) Conformational plasticity of the catalytic subunit of protein kinase CK2 and its consequences for regulation and drug design. Biochim Biophys Acta 1804:484–492

12. Battistutta R, Lolli G (2011) Structural and functional determinants of protein kinase CK2α: facts and open questions. Mol Cell Biochem 356:67–73
13. Bischoff N, Raaf J, Olsen B, Bretner M, Issinger O-G, Niefind K (2011) Enzymatic activity with an incomplete catalytic spine: insights from a comparative structural analysis of human CK2α and its paralogous isoform CK2α'. Mol Cell Biochem 356:57–65
14. Papinutto E, Ranchio A, Lolli G, Pinna LA, Battistutta R (2012) Structural and functional analysis of the flexible regions of the catalytic alpha-subunit of protein kinase CK2. J Struct Biol 177:382–391
15. Klopffleisch K, Issinger OG, Niefind K (2012) Low-density crystal packing of human protein kinase CK2 catalytic subunit in complex with resorufin or other ligands: a tool to study the unique hinge-region plasticity of the enzyme without packing bias. Acta Crystallogr D68: 883–892
16. Gouron A, Milet A, Jamet H (2014) Conformational flexibility of human casein kinase catalytic subunit explored by metadynamics. Biophys J 106:1134–1141
17. Artymiuk PJ, Blake CC, Grace DE, Oatley SJ, Phillips DC, Sternberg MJ (1979) Crystallographic studies of the dynamic properties of lysozyme. Nature 280:563–568
18. Lang PT, Ng HL, Fraser JS, Corn JE, Echols N, Sales M, Holton JM, Alber T (2010) Automated electron-density sampling reveals widespread conformational polymorphism in proteins. Protein Sci 19:1420–1431
19. Lang PT, Holton JM, Fraser JS, Alber T (2014) Protein structural ensembles are revealed by redefining X-ray electron density noise. Proc Natl Acad Sci U S A 111:237–242
20. Fraser JS, van den Bedem H, Samelson AJ, Lang PT, Holton JM, Echols N, Alber T (2011) Accessing protein conformational ensembles using room-temperature X-ray crystallography. Proc Natl Acad Sci U S A 108:16247–16252
21. Thompson EE, Kornev AP, Kannan N, Kim C, Ten Eyck LF, Taylor SS (2009) Comparative surface geometry of the protein kinase family. Protein Sci 18:2016–2026
22. Biondi RM, Cheung PC, Casamayor A, Deak M, Currie RA, Alessi DR (2000) Identification of a pocket in the PDK1 kinase domain that interacts with PIF and the C-terminal residues of PKA. EMBO J 19:979–988
23. Sarno S, Ghisellini P, Pinna LA (2002) Unique activation mechanism of protein kinase CK2: the N-terminal segment is essential for constitutive activity of the catalytic subunit but not of the holoenzyme. J Biol Chem 277:22509–22514
24. Nakaniwa T, Kinoshita T, Sekiguchi Y, Tada T, Nakanishi I, Kitaura K, Suzuki Y, Ohno H, Hirasawa A, Tsujimoto G (2009) Structure of human protein kinase CK2α2 with a potent indazole-derivative inhibitor. Acta Crystallogr F65:75–79
25. Bischoff N, Olsen B, Raaf J, Bretner M, Issinger O-G, Niefind K (2011) Structure of the human protein kinase CK2 catalytic subunit CK2α' and interaction thermodynamics with the regulatory subunit CK2β. J Mol Biol 407:1–12
26. Raaf J, Brunstein E, Issinger O-G, Niefind K (2008) The interaction of CK2α and CK2β, the subunits of protein kinase CK2, requires CK2β in a preformed conformation and is enthalpically driven. Protein Sci 17:2180–2186
27. Battistutta R, Sarno S, De Moliner E, Papinutto E, Zanotti G, Pinna LA (2000) The replacement of ATP by the competitive inhibitor emodin induces conformational modifications in the catalytic site of protein kinase CK2. J Biol Chem 275:29618–29622
28. Ermakova I, Boldyreff B, Issinger O-G, Niefind K (2003) Crystal structure of a C-terminal deletion mutant of human protein kinase CK2 catalytic subunit. J Mol Biol 330:925–934
29. Raaf J, Brunstein E, Issinger OG, Niefind K (2008) The CK2α/CK2β interface of human protein kinase CK2 harbors a binding pocket for small molecules. Chem Biol 15:111–117
30. Raaf J, Guerra B, Neundorf I, Bopp B, Issinger O-G, Jose J, Pietsch M, Niefind K (2013) First structure of protein kinase CK2 catalytic subunit with an effective CK2β-competitive ligand. ACS Chem Biol 8:901–907
31. Niefind K, Issinger O-G (2005) Primary and secondary interactions between CK2α and CK2β lead to ring-like structures in the crystals of the CK2 holoenzyme. Mol Cell Biochem 274:3–14

32. Lolli G, Ranchio A, Battistutta R (2014) Active form of the protein kinase CK2 $\alpha_2\beta_2$ holoenzyme is a strong complex with symmetric architecture. ACS Chem Biol 9:366–371
33. Schnitzler A, Olsen BB, Issinger O-G, Niefind K (2014) The protein kinase CK2*Andante* holoenzyme structure supports proposed models of autoregulation and *trans-autophosphorylation*. J Mol Biol 426:1871–1882
34. Valero E, De Bonis S, Filhol O, Wade RH, Langowski J, Chambaz EM, Cochet C (1995) Quaternary structure of casein kinase 2. Characterization of multiple oligomeric states and relation with its catalytic activity. J Biol Chem 270:8345–8352
35. Poole A, Poore T, Bandhakavi S, McCann RO, Hanna DE, Glover CV (2005) A global view of CK2 function and regulation. Mol Cell Biochem 274:163–170
36. Craveur P, Joseph AP, Poulain P, de Brevern AG, Rebehmed J (2013) *Cis-trans* isomerization of omega dihedrals in proteins. Amino Acids 45:279–289
37. Jabs A, Weiss MS, Hilgenfeld R (1999) Non-proline *cis* peptide bonds in proteins. J Mol Biol 286:291–304
38. Andreotti AH (2003) Native state proline isomerization: an intrinsic molecular switch. Biochemistry 42:9515–9524
39. Lu KP, Finn G, Lee TH, Nicholson LK (2007) Prolyl *cis-trans* isomerization as a molecular timer. Nat Chem Biol 3:619–629
40. Knighton DR, Zheng JH, Ten Eyck LF, Ashford VA, Xuong NH, Taylor SS, Sowadski JM (1991) Crystal structure of the catalytic subunit of cyclic adenosine monophosphate-dependent protein kinase. Science 253:407–414
41. Niefind K, Pütter M, Guerra B, Issinger O-G, Schomburg D (1999) GTP plus water mimic ATP in the active site of protein kinase CK2. Nat Struct Biol 6:1100–1103
42. Yde CW, Ermakova I, Issinger OG, Niefind K (2005) Inclining the purine base binding plane in protein kinase CK2 by exchanging the flanking side-chains generates a preference for ATP as a cosubstrate. J Mol Biol 347:399–414
43. Pinna LA (2002) Protein kinase CK2: a challenge to canons. J Cell Sci 115:3873–3878
44. Traxler P, Furet P (1999) Strategies toward the design of novel and selective protein tyrosine kinase inhibitors. Pharmacol Ther 82:195–206
45. Miyashita O, Onuchic JN, Wolynes PG (2003) Nonlinear elasticity, proteinquakes, and the energy landscapes of functional transitions in proteins. Proc Natl Acad Sci U S A 100: 12570–12575
46. Shan Y, Arkhipov A, Kim ET, Pan AC, Shaw DE (2013) Transitions to catalytically inactive conformations in EGFR kinase. Proc Natl Acad Sci U S A 110:7270–7275
47. Raaf J, Bischoff N, Klopffleisch K, Brunstein E, Olsen BB, Vilk G, Litchfield DW, Issinger O-G, Niefind K (2011) Interaction between CK2α and CK2β, the subunits of protein kinase CK2: thermodynamic contributions of key residues on the CK2α surface. Biochemistry 50:512–522
48. Niefind K, Yde CW, Ermakova I, Issinger O-G (2007) Evolved to be active: sulfate ions define substrate recognition sites of CK2alpha and emphasise its exceptional role within the CMGC family of eukaryotic protein kinases. J Mol Biol 370:427–438
49. Chantalat L, Leroy D, Filhol O, Nueda A, Benitez MJ, Chambaz EM, Cochet C, Dideberg O (1999) Crystal structure of the human protein kinase CK2 regulatory subunit reveals its zinc finger-mediated dimerization. EMBO J 18:2930–2940
50. Battistutta R (2009) Protein kinase CK2 in health and disease: structural bases of protein kinase CK2 inhibition. Cell Mol Life Sci 66:1868–1889
51. Liu H, Wang H, Teng M, Li X (2014) The multiple nucleotide-divalent cation binding modes of *Saccharomyces cerevisiae* CK2α indicate a possible co-substrate hydrolysis product (ADP/GDP) release pathway. Acta Crystallogr D70:501–513
52. The PyMOL Molecular Graphics System, Version 1.7, Schrödinger, LLC
53. Laudet B, Barette C, Dulery V, Renaudet O, Dumy P, Metz A, Prudent R, Deshiere A, Dideberg O, Filhol O, Cochet C (2007) Structure-based design of small peptide inhibitors of protein kinase CK2 subunit interaction. Biochem J 408:363–373

Structural Basis of CK2 Regulation by Autoinhibitory Oligomerization

Graziano Lolli and Roberto Battistutta

Abstract The mechanism of regulation of CK2 differs from those common to other eukaryotic protein kinases and is not entirely established yet. Nowadays, several crystal structures of the tetrameric $\alpha_2\beta_2$ holoenzyme are available, supporting a structural model of an autoinhibitory regulation by oligomerization proposed several years before on the basis of biochemical, biophysical and functional data. Monoclinic crystal forms of the holoenzyme reveal the symmetric architecture of the "free" isolated active tetramers. The dimension and the nature of the α/β interfaces configure the symmetric holoenzyme as a strong complex that does not spontaneously dissociate in solution, in accordance with the low dissociation constant ($\approx$4 nM). Hexagonal crystal forms of the CK2 holoenzyme show an asymmetric arrangement of the two catalytic α-subunits around the obligate β_2 regulatory subunits. These asymmetric $\alpha_2\beta_2$ tetramers are organised in trimeric rings and filaments that correspond to the inactive forms of the enzyme, whereby the β-subunit plays an essential role in the formation of inactive polymeric assemblies. The derived structural model of (down)regulation by aggregation contributes to the interpretation of many available biochemical and functional data, although it awaits for a more comprehensive validation at the cellular level. These findings pave the way for the design of novel strategies aimed at the modulation of the CK2 activity.

Keywords Crystal structure • Symmetric CK2 holoenzyme • Asymmetric CK2 holoenzyme • Trimeric CK2 holoenzyme • Oligomerization • Filaments • Activity regulation • Autoinhibition • Crystal contacts • α-Subunit • β-Subunit • $\alpha_2\beta_2$ tetramer • α/β interfaces • CK2β acidic loop • CK2α p+1 loop • CK2β C-terminal tail • Electrostatic interactions • Polyamines • Autophosphorylation

G. Lolli (✉) • R. Battistutta (✉)
Department of Chemical Sciences, University of Padua, via Marzolo 1, 35131 Padua, Italy

Venetian Institute for Molecular Medicine (VIMM), via Orus 2, 35129 Padua, Italy
e-mail: graziano.lolli@unipd.it; roberto.battistutta@unipd.it

K. Ahmed et al. (eds.), *Protein Kinase CK2 Cellular Function in Normal and Disease States*, Advances in Biochemistry in Health and Disease 12, DOI 10.1007/978-3-319-14544-0_3

1 The First Asymmetric $\alpha_2\beta_2$ Holoenzyme Crystal Structure

The first crystallographic structure of the CK2 holoenzyme was determined in 2001 (PDB 1JWH) [1], joining those already known of the isolated α- and β-subunits [2, 3], a great achievement finally solving the puzzling CK2 quaternary arrangement. CK2 tetramer has a butterfly shape with a central β-dimer and the two α-subunits, one on each side and not in contact between them (Fig. 1a). This was also the first structure including the full-length β-subunit that can be divided into a body and a C-terminal tail, the last one not included in the previous structures. CK2β C-terminal tail has no contact with the body of its own chain but binds both the other CK2β monomer and one of the α-subunits (Fig. 1a). This tail contributes significantly to the stability of the β-dimer. The β/β interface increases from 543 $Å^2$ in the structure of the isolated and truncated β-dimer to 1,766 $Å^2$ in the structure of the holoenzyme, in agreement with the obligate nature of the β-dimer. The authors also highlighted that this increase in the interface requires the CK2β tail in a specific conformation, which is stabilised by the involvement of the same tail in large contacts with the α-subunits and hence depends on the existence of the complete tetramer. The CK2β tail has then a synergistic character, stabilising both the β/β and the α/β contacts. At the same time, the authors pointed out that this region was not well defined in the electron density and then the very C-terminal amino acids (from Asn206 to Arg215) were left out from the final model, while in the preceding region (from Phe190 to Ser205) the conformations of peptide groups and side chains remain questionable.

Each α-subunit binds both subunits of the β-dimer, the body of one chain and the tail of the other (Fig. 1a); this last interaction is exploring a larger interface (491 $Å^2$ vs. 336 $Å^2$) and being considered the major determinant of the α/β stability. The structure of the catalytic α-subunit is not significantly affected by the binding of the β-dimer, maintaining its active conformation and confirming previous observations that CK2β is an environment- and substrate-dependent modulator of CK2α activity rather than an on/off switch. It was also noted that the tetramer is asymmetric with a relative rotation of 16.4° at the two α/β interfaces for the two halves of the tetramer (Fig. 1a). As a consequence, the two α/β interfaces are different (Fig. 1b) and this was interpreted as an intersubunit flexibility. In the basis of such observation, together with an average size for the α/β interface of 832 $Å^2$, authors proposed that the CK2 tetramer is a transient heterocomplex.

Finally, in the same paper it was pointed out that the inhibitory effect of the CK2β acidic loop (Asp55–Asp64) observed in previous works [4] cannot be exerted intramolecularly since the distance between this and the closest CK2α basic cluster (the potential substrate binding site) was >30 Å.

A few years later, the same authors analysed the crystal packing in the previous structure, showing that the contact between the β-acidic loop and the α-basic cluster was indeed present but between different tetramers [5]. They identified a new CK2 organisation, composed by self-consistent ring-like trimers of tetramers, where the core of the interaction between different tetramers in $(\alpha_2\beta_2)_3$ is formed by the interface between the β-acidic loop and the α p+1 loop (Arg191–Lys198) at the entrance

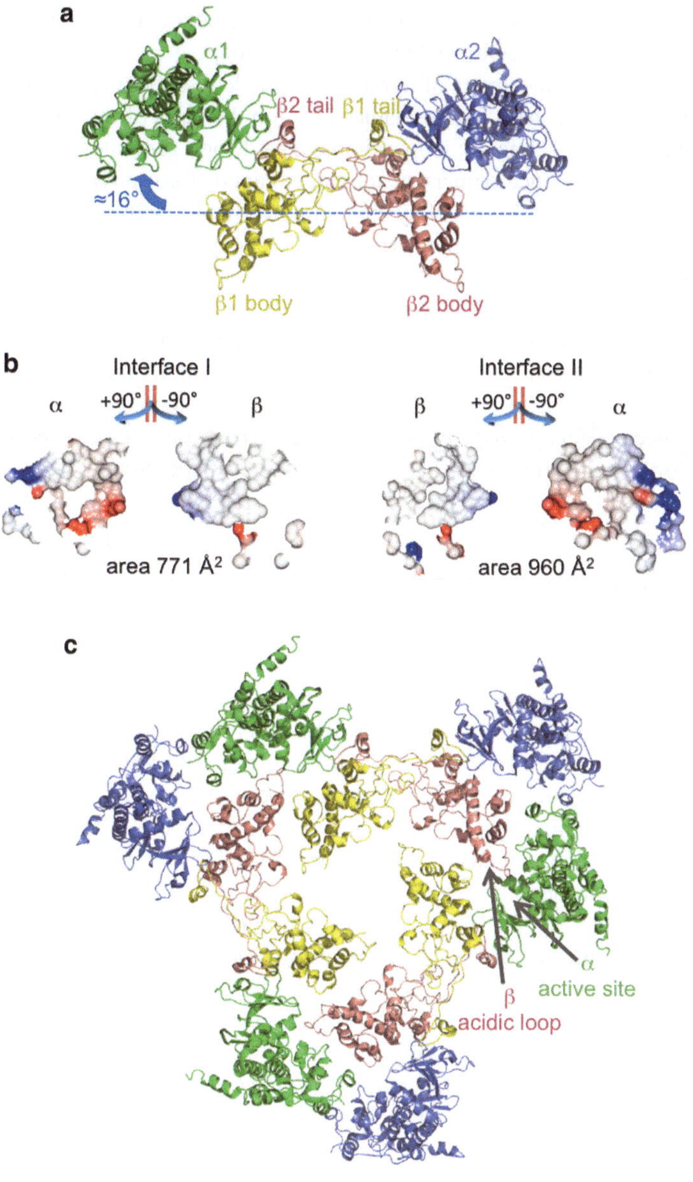

Fig. 1 Structure of the CK2 holoenzyme (PDB, 1JWH) [1]. (**a**) The CK2 holoenzyme has a butterfly shape with the two catalytic α-subunits not in contact between them. The CK2β tails have a fundamental role in the holoenzyme both by strengthening the β-dimer and by largely contributing to the α/β interface. The holoenzyme is asymmetric with a relative rotation of 16.4° for the two halves of the tetramer. (**b**) The asymmetric assembly generates two different α/β interfaces. Adapted with permission from [28]. Copyright (2014) American Chemical Society. (**c**) The holoenzyme assembly in circular trimers of tetramers has been identified in the crystallographic packing [5]. The core of the interaction is between the β-acidic loop and the α p+1 loop at the entrance of the catalytic site

of the α active site (Fig. 1c). As a consequence, three out of six catalytic subunits of the trimeric ensemble are blocked by the β-acidic loop.

This organisation in trimers of tetramers was able to explain a number of previous observations not easily interpretable solely on the basis of the tetrameric structure. Mutations in the β-acidic loop or in the α p+1 loop generate hyperactive CK2 holoenzyme [4, 6], since trimers of tetramers cannot form anymore. Polyamines like spermine activate CK2 [7–9] by interacting with the CK2β acidic loop and then disrupting the trimers of tetramers. The CK2 holoenzyme exhibits maximum activity at about 300 mM NaCl or KCl, a property of the holoenzyme and not of the isolated CK2α [10]. The holoenzyme also needs a minimal NaCl concentration of 250 mM NaCl to bind to phosphocellulose, while it aggregates under low salt conditions [11]. Since the interaction between the β-acidic loop and the α p+1 loop is electrostatic in nature, it is strongly affected by ionic strength. Finally, CK2 holoenzyme has been observed in different ring-like and linear polymeric forms connected via reversible equilibria affected by ionic strength and spermine [12, 13].

2 The Second Asymmetric $\alpha_2\beta_2$ Holoenzyme Crystal Structure

In 2012, we presented a new structure of the holoenzyme (4DGL) crystallised in the same space group and unit cell of the previous 1JWH structure but with higher overall quality [14]. While confirming many features of the 1JWH structure, the 4DGL structure shows significant improvements in important regions such as the α/β interface and the C-terminal tail of the β-subunit. In particular, β-Phe190 leans on the outer surface of the α-subunit N-terminal lobe in 1JWH, while in 4DGL it inserts deeply into a hydrophobic cavity in the α-subunit, known as the secondary or allosteric binding site (Fig. 2a, b). The α/β interaction is then significantly strengthened in accordance with the almost complete abrogation of the binding for the mutant β-Phe190Ala [15]. Also the β C-terminus was built differently with an upstream shift in sequence of two amino acids and extended the αG helix from residues 196–200 in 1JWH to residues 194–207 in 4DGL (Fig. 2b). This again contributes to refine the α/β interface to a better energetic profile (interface increased from 832 Å^2 in 1JWH to 1,010 Å^2 in 4DGL) (Fig. 2c). Finally, the β C-terminus was built to the last Arg215 amino acid.

As already noted by Niefind and Issinger [5], the acidic loop of the β-subunit from one tetramer is in contact with the p+1 loop of the α-subunit from another tetramer (Fig. 3a). The interface area is 651 Å^2, larger than in 1JWH (538 Å^2), and greater than normal crystallographic contacts, which have a mean interface of 285 Å^2 and rarely exceed 500 Å^2 [16]. This interaction is essential for the organisation of circular trimers of tetramers as previously observed (Fig. 1c), with a positive cooperative effect for their formation and a large interface of 1,953 Å^2 (651 Å^2×3).

The newly traced C-terminus of CK2β is instead responsible for a different intertetrameric interaction. This region protrudes from one tetramer and contacts the

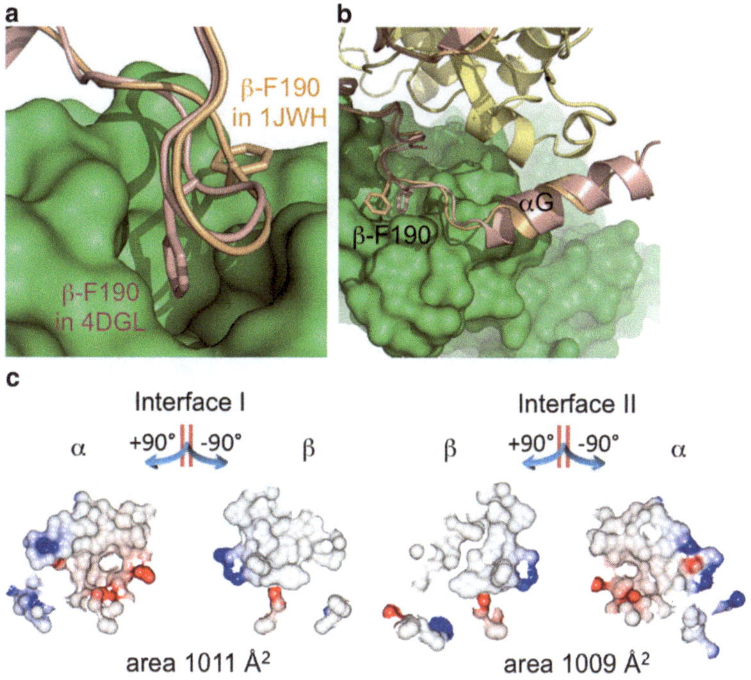

Fig. 2 Details of the α/β interaction in the improved structure of the CK2 holoenzyme (4DGL) [14]. (**a**) In 4DGL, β-Phe190 (*salmon*) inserts into a deep hydrophobic pocket on the CK2α surface (*green*), differently from the 1JWH structure (*orange*). Adapted with permission from [14]. Copyright (2012) American Chemical Society. (**b**) In 4DGL, β C-terminus was built to the last Arg215 and region 194–205 was better modelled with respect to the 1JWH structure; colour code as in (**a**). Adapted with permission from [14]. Copyright (2012) American Chemical Society. (**c**) In the improved 4DGL structure, the two α/β interfaces are larger than in 1JWH (Fig. 1b). Adapted with permission from [28]. Copyright (2014) American Chemical Society

C-terminal lobe of an α-subunit of a neighbouring tetramer ending in its ATP pocket (Fig. 3a, b). The interface is large (753 Å^2) and is complemented by the interaction between the N-terminal lobe and the C-terminal lobe of two α chains from the same two tetramers. The overall interface of 1,254 Å^2 is then quite extended and again much larger than a normal crystallographic contact. It shows a number of polar contacts, which again makes this interaction susceptible to ionic strength changes. This new interaction is fundamental for a piling organisation of trimers of tetramers (Fig. 3c), and the newly identified polymeric form can represent the filamentous CK2 aggregates described in literature [12, 13]. Moreover, the β C-terminus inserts into the ATP-binding pocket of a catalytic subunit, competing with ATP and stabilising a nonproductive conformation of the α catalytic residues (Fig. 3b). This polymeric form is then expected to be completely inactivated by the combination of the interactions of the α-subunits with the acidic loops and C-termini of β-subunits (Fig. 3a) and by steric hindrance.

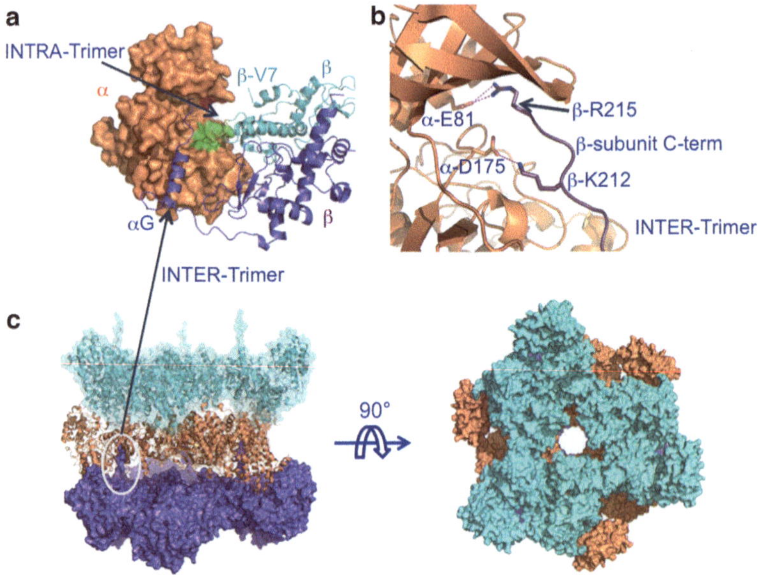

Fig. 3 Oligomeric organisation in the 4DGL structure. Adapted with permission from [14]. Copyright (2012) American Chemical Society. (**a**) Different inter-tetrameric contacts can be identified. The interaction between the β-acidic loop (β-subunit in *cyan* with its acidic loop in *yellow*) and the α p+1 loop (α-subunit in *orange* with its p+1 loop in *green*) is responsible for the assembly in circular trimers of tetramers. The interaction between the αG and the C-terminal tail of a different β-subunit (*purple*) and the α C-terminal lobe and active site (*orange*) is responsible for a piling organisation of trimers of tetramers. The position of the first amino acid visible at the β N-terminus (Val7) is compatible with the autophosphorylation of the upstream residues Ser2 and Ser3. Phosphorylated residues could then interact with the α-basic cluster (*red*). (**b**) Insertion of the β C-terminus into the α active site disrupts the correct orientation of α residues involved in nucleotide binding. (**c**) In the filamentous organisation, the C-termini of three β-subunits are plugged into the following trimer

This new structure and the hypothesised organisation in trimers of tetramers and subsequently in filaments allow expanding the experimental data previously reported in support of the trimeric organisation by Niefind and Issinger [5]. FRET experiments revealed the existence in vivo of two subpopulations of “fast” and “slow-moving” CK2 and showed that different CK2β dimers can be in close molecular proximity [17]. The downregulating effect proposed for the CK2β C-terminal tail is in accordance with other in vivo observations. CK2 activity is increased in prostate cancer cells by CDK1 phosphorylation of β-Ser209 [18] that engages α-Glu230 in the filamentous organisation, which will then be weakened by the phosphorylation event. Phosphorylation of β-Thr213 by Chk1 is observed in the free β-subunits but not in the tetrameric holoenzyme [19], where the residue is masked in the filamentous oligomers. The downregulatory polymerisation mechanism mediated by the β-subunits is also supported by the observation that transfection of HeLa cells with a degradation-resistant CK2β strongly inhibits proliferation and counteracts the promotion of proliferation exerted by the isolated CK2α [20].

It is well known that CK2 autophosphorylates on residues Ser2 and Ser3 of the β-chain [21]. As already observed by Niefind and Issinger [5] and further discussed in [14], the β N-terminus is close to an α catalytic site in the trimeric organisation, so that one tetramer could phosphorylate the neighbouring one (Fig. 3a). Observations in favour of this trans-autophosphorylation mechanism inside the trimeric organisation are the following: autophosphorylation (a) is abolished at high ionic strength [22]and (b) is strongly inhibited by neutralisation of residues in the β-acidic loop [23]; (c) although being in trans, it shows an intramolecular kinetics [24]; and (d) no autophosphorylation occurs in the holoenzyme with the α′ paralog, a holoenzyme that does not form oligomers [25]. We also proposed that the β N-terminus, once phosphorylated, could engage the nearby αC basic cluster of the α-subunit from the neighbouring tetramer, further stabilising the ring-like organisation. Autophosphorylation is (e) extensively observed in cells [21], (f) stabilising the enzyme in vivo [26] and (g) having an inhibitory effect, since treatment of CK2 with phosphatases increases its activity [27].

3 The Symmetric $\alpha_2\beta_2$ Holoenzyme Crystal Structures

Very recently, we added to the above picture three additional crystal structures of the $\alpha_2\beta_2$ holoenzyme [28]. In both the 1JWH and 4DGL structures, the $\alpha_2\beta_2$ tetramer is asymmetric. Notably, in the new structures (4MD7, 4MD8 and 4MD9), obtained in monoclinic (rather than hexagonal) space groups, the tetramer is symmetric (Fig. 4a). This reveals that the asymmetry observed in the previous structures is not an intrinsic structural property of the monomeric holoenzyme but rather is generated by the incorporation of tetramers into trimeric rings, whose assembly is not compatible with the symmetric architecture (Fig. 4b). The analysis of the crystal packing of the monoclinic form of the holoenzyme does not seem to support the possibility of stable oligomeric or filamentous forms. Unlike in the hexagonal lattice, where discrete trimeric rings of tetramers are clearly recognisable, in the monoclinic form all contacts between different tetramers have a pure crystallographic nature. Indeed, there is one interaction between the acidic loop of the β-subunit and the basic clusters of the α-subunit of a symmetric tetramer with a significant interface area (around 750 Å^2). However, this contact is not responsible for the formation of any closed, self-consistent, oligomeric form of tetramers, nor of a filament of tetramers. Rather, it is responsible for the propagation of two parallel layers of tetramers in two dimensions for the entire crystal lattice (i.e. infinitively), giving raise to a sort of infinite two-dimensional bilayer of tetramers. Hence, it has the characteristic of a crystal lattice contact, making highly improbable, in our opinion, the existence of stable discrete oligomeric or filamentous forms based on this interaction. We then concluded that the new symmetric structures represent the organisation of the "free", active, monomeric CK2 holoenzyme, the previous 1JWH and 4DGL showing instead the structure of inactive trimers of tetramers. However, it has to be noted that some of the inter-tetrameric interactions typical of the monoclinic crystal packing, observed also for the CK2^{andante} holoenzyme mutant [29],

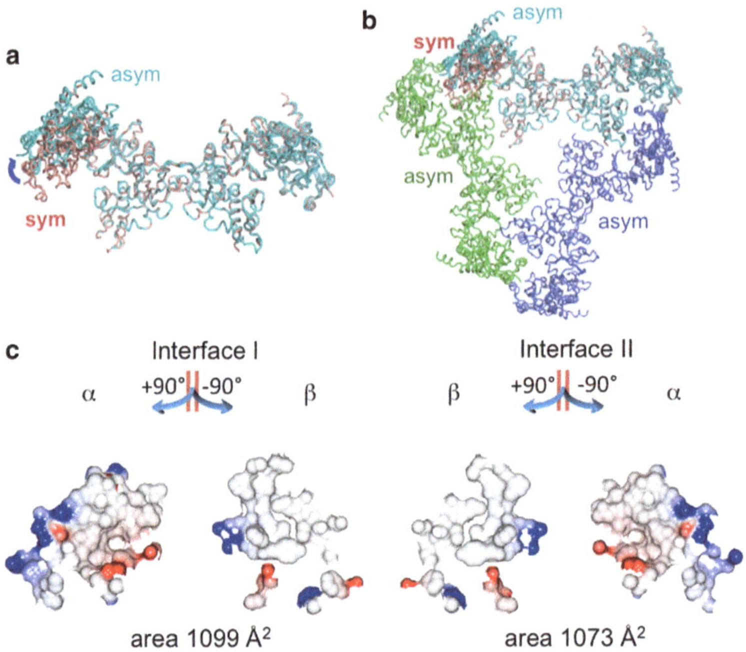

Fig. 4 Structural features of the symmetric tetrameric holoenzyme. Adapted with permission from [28]. Copyright (2014) American Chemical Society. (**a**) Structure of symmetric tetramer in salmon (4MD7) [28] superposed to the asymmetric 4DGL tetramer (*cyan*). (**b**) In 4DGL, the assembly of three tetramers (*green*, *cyan* and *purple*) in trimers of tetramers induces the distortion not present in the symmetric tetramer (*salmon*). (**c**) In the symmetric tetramer, the two α/β interfaces are very similar in size and nature

have been postulated to be at the basis of other possible (linear) assemblies with physiological significance [30].

The comparison of symmetric and asymmetric tetramers in terms of their α/β interfaces and then of their stability is of particular relevance since the possibility that the CK2 holoenzyme coexists in equilibrium with its free constituent subunits can represent a regulatory mechanism for its biological function. In 1JWH, the two α/β interfaces are significantly different in size, one being larger (960 $Å^2$) and the other significantly smaller (771 $Å^2$) [1] (Fig. 1b). This suggested that the CK2 holoenzyme might be a transient heterocomplex, spontaneously dissociating in vivo. This seems in contrast with the well-known elevated stability in vitro of the complex ($K_D \approx 4$ nM) [31, 32]. In 4DGL, the better-traced β C-terminal tail generated larger interfaces with similar sizes (1,009 and 1,011 $Å^2$); however, due to the asymmetric assembly, the two interfaces are different in nature with different residues involved [14] (Fig. 2b).

The new symmetric structures have very similar α/β interfaces both in size (1,099 and 1,073 $Å^2$) and chemical nature (Fig. 4c). This was also confirmed by PISA ('protein interfaces, surfaces and assemblies' service at the European Bioinformatics

Table 1 α/β interface parameters in symmetric $\alpha_2\beta_2$ holoenzyme

Parameter	Value	Comment
Mean interface area	1,086 Å²	740–810 Å² for weak transient complexes
f_{np} (nonpolar fraction of interface area)	70.5 %	Largely hydrophobic
P-value (PISA solvation energy gain)	0.109	Largely hydrophobic
R_p (residue propensity score)	9.8	Typical of 'strong' complexes
f_{bu} (fraction of fully buried atoms)	30 %	High shape complementarity
S_c (shape complementarity score)	0.68	High shape complementarity
I_{gap} (gap volume index)	1.6	High shape complementarity
L_D (atomic density index)	38.7	High shape complementarity

Adapted with permission from [28]. Copyright (2014) American Chemical Society

Institute [33]) that predicted a dissociation pattern in $2\alpha+\beta_2$ from monoclinic structures but in $\alpha\beta_2+\alpha$ for both 1JWH and 4DGL as a consequence of their asymmetry. The interface size is the most commonly used parameter to infer the stability of a complex; indeed, also its chemical and sterical nature (hydrophobicity and complementarity) needs to be considered [34]. Evaluating a number of parameters (reported in Table 1), we proposed that the CK2 holoenzyme configures as a non-obligate permanent heterocomplex. In obligate complexes, the constitutive protomers are not stable in the isolated form and they simultaneously fold and bind to their partner [35, 36], as in the case of the CK2β dimer. Obligate complexes have interface areas generally much larger than 1,000 Å² [37]. The CK2 holoenzyme is obviously a non-obligate complex. Permanent protein-protein interactions are strong, with nanomolar or subnanomolar dissociation constants, and irreversible, normally only perturbed by proteolysis [35, 36]. Transient complexes can instead dissociate in their constituent subunits and can be distinguished as weak or strong [35, 36]. In strong transient complexes, a molecular trigger (i.e. phosphorylation or effector molecule) drives the equilibrium through a large conformational change between a high-affinity state (low- or subnanomolar K_D) and a low-affinity state (micromolar K_D). Weak transient complexes have instead micromolar dissociation constants and are characterised by a dynamic equilibrium where interaction is formed and broken continuously; their interfaces are small (mean interface for weak transient complexes was determined in 740 Å² by Nooren and Thornton [37] and in 810 Å² by Dey and co-workers [34]). Given those data, with a $K_D \approx 4$ nM and an interface area of 1,086 Å², CK2 cannot be considered a weak transient complex. It is however impossible to distinguish between permanent and strong transient complexes solely on the basis of the dissociation constant and the interface parameters [34]. Despite that, since no molecular triggers able to alter the affinity of the CK2 protomers have been identified so far and no significant structural rearrangements of the individual subunits are observed during the formation of the CK2 holoenzyme complex, the $\alpha_2\beta_2$ tetramer should be considered a permanent complex, until proven otherwise.

Indeed, isolated α- and β-subunits have been observed in cells [38]. However, as underlined above, crystallographic data, the K_D and the stability of the enzyme in vitro strongly argue against the possibility that the CK2 $\alpha_2\beta_2$ holoenzyme can spontaneously dissociate in solution. Instead, the existence of "free", isolated α- and β_2-subunits can be associated to different factors such as unbalanced expression, differential localization or transport mechanism and to competitive binding with other partners, like the A-Raf, c-Mos and Chk1 kinases, all reported to interact with CK2β possibly using the same interface explored by CK2α [39]. As already stated, molecular triggers capable to dissociate the complex have not been identified so far. We also excluded by crystallography that this molecular switch could reside in the phosphorylation of the CK2α C-terminal tail (phosphorylated by Cdk1 at mitosis [40]).

4 Working Model of CK2 Regulation by Autoinhibitory Oligomerization

The overall regulatory model emerging from structural studies and in accordance with many biochemical, biophysical and activity data is depicted in Fig. 5. When co-localised in time and space, CK2α and CK2β dimers spontaneously assemble into an active, symmetric and permanent complex with a $K_D \approx 4$ nM. Enzymatic

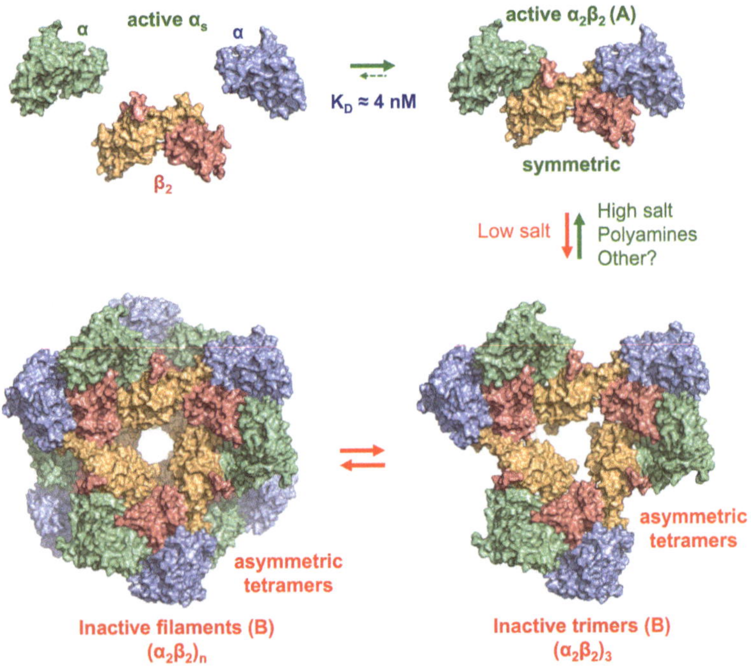

Fig. 5 Proposed model of CK2 regulation. Reprinted with permission from [28]. Copyright (2014) American Chemical Society. CK2α and CK2β subunits readily assemble into active and symmetric tetramers. Activity is turned off through organisation in inactive trimers and filaments. Activity is restored upon a depolymerization process induced by variations in the physical-chemical environment, action of substrates or effectors or other events

activity is turned on/off by the equilibrium between the active tetramers and the inactive asymmetric trimers of tetramers that can further polymerise in higher supramolecular assemblies. These represent the latent form of the kinase in accordance with the observation that most of CK2 in crude liver extract is inactive [41]. As for the majority of protein kinases, activity is restored only upon necessity, in this case by a depolymerization process generating active free tetramers. This process can be triggered by different events like β dephosphorylation, physical-chemical variations of the environment (pH, ionic strength, etc.) and/or action of substrates themselves or other effectors (highly charged macromolecules or low-molecular-weight compounds). When the stimuli end, CK2 tetramers are assembled back in trimers and then in fibrous polymers, with the turning off of the catalytic activity. The β-subunit is fundamental in this process, driving both the assembly in trimers of tetramers (via the acidic loop) and in filaments (via its C-terminus), fully deserving the name of "regulatory" subunit.

Acknowledgements This work was supported by Italian Miur (PRIN 2008, R.B.), by FEBS (Distinguished Young Investigator Award, GL) and by the University of Padua (Progetto Giovani GRIC101044, GL).

References

1. Niefind K, Guerra B, Ermakowa I, Issinger OG (2001) Crystal structure of human protein kinase CK2: insights into basic properties of the CK2 holoenzyme. EMBO J 20:5320–5331
2. Niefind K, Guerra B, Pinna LA et al (1998) Crystal structure of the catalytic subunit of protein kinase CK2 from Zea mays at 2.1 A resolution. EMBO J 17:2451–2462
3. Chantalat L, Leroy D, Filhol O et al (1999) Crystal structure of the human protein kinase CK2 regulatory subunit reveals its zinc finger-mediated dimerization. EMBO J 18:2930–2940
4. Boldyreff B, Meggio F, Pinna LA, Issinger OG (1993) Reconstitution of normal and hyperactivated forms of casein kinase-2 by variably mutated beta-subunits. Biochemistry 32:12672–12677
5. Niefind K, Issinger OG (2005) Primary and secondary interactions between CK2alpha and CK2beta lead to ring-like structures in the crystals of the CK2 holoenzyme. Mol Cell Biochem 274:3–14
6. Sarno S, Vaglio P, Marin O et al (1997) Basic residues in the 74-83 and 191-198 segments of protein kinase CK2 catalytic subunit are implicated in negative but not in positive regulation by the beta-subunit. Eur J Biochem 248:290–295
7. Hathaway GM, Traugh JA (1984) Kinetics of activation of casein kinase II by polyamines and reversal of 2,3-bisphosphoglycerate inhibition. J Biol Chem 259:7011–7015
8. Mamrack MD (1989) Stimulation of enzymatic activity in filament preparations of casein kinase II by polylysine, melittin, and spermine. Mol Cell Biochem 85:147–157
9. Leroy D, Heriché JK, Filhol O et al (1997) Binding of polyamines to an autonomous domain of the regulatory subunit of protein kinase CK2 induces a conformational change in the holoenzyme. A proposed role for the kinase stimulation. J Biol Chem 272:20820–20827
10. Grankowski N, Boldyreff B, Issinger OG (1991) Isolation and characterization of recombinant human casein kinase II subunits alpha and beta from bacteria. Eur J Biochem 198:25–30
11. Hathaway GM, Traugh JA (1979) Cyclic nucleotide-independent protein kinases from rabbit reticulocytes. Purification of casein kinases. J Biol Chem 254:762–768
12. Glover CV (1986) A filamentous form of Drosophila casein kinase II. J Biol Chem 261: 14349–14354

13. Valero E, De Bonis S, Filhol O et al (1995) Quaternary structure of casein kinase 2. Characterization of multiple oligomeric states and relation with its catalytic activity. J Biol Chem 270:8345–8352
14. Lolli G, Pinna LA, Battistutta R (2012) Structural determinants of protein kinase CK2 regulation by autoinhibitory polymerization. ACS Chem Biol 7:1158–1163
15. Laudet B, Barette C, Dulery V et al (2007) Structure-based design of small peptide inhibitors of protein kinase CK2 subunit interaction. Biochem J 408:363–373
16. Carugo O, Argos P (1997) Protein-protein crystal-packing contacts. Protein Sci 6:2261–2263
17. Theis-Febvre N, Martel V, Laudet B et al (2005) Highlighting protein kinase CK2 movement in living cells. Mol Cell Biochem 274:15–22
18. Gotz C, Kartarius S, Schetting S, Montenarh M (2005) Immunologically defined subclasses of the protein kinase CK2 beta-subunit in prostate carcinoma cell lines. Mol Cell Biochem 274:181–187
19. Kristensen LP, Larsen MR, Hojrup P et al (2004) Phosphorylation of the regulatory beta-subunit of protein kinase CK2 by checkpoint kinase Chk1: identification of the in vitro CK2beta phosphorylation site. FEBS Lett 569:217–223
20. French AC, Luscher B, Litchfield DW (2007) Development of a stabilized form of the regulatory CK2beta subunit that inhibits cell proliferation. J Biol Chem 282:29667–29677
21. Litchfield DW, Lozeman FJ, Cicirelli MF et al (1991) Phosphorylation of the beta subunit of casein kinase II in human A431 cells. Identification of the autophosphorylation site and a site phosphorylated by p34cdc2. J Biol Chem 266:20380–20389
22. Pagano MA, Sarno S, Poletto G et al (2005) Autophosphorylation at the regulatory beta subunit reflects the supramolecular organization of protein kinase CK2. Mol Cell Biochem 274:23–29
23. Boldyreff B, Meggio F, Pinna LA, Issinger OG (1994) Efficient autophosphorylation and phosphorylation of the beta-subunit by casein kinase-2 require the integrity of an acidic cluster 50 residues downstream from the phosphoacceptor site. J Biol Chem 269:4827–4831
24. Meggio F, Pinna LA (1984) Subunit structure and autophosphorylation mechanism of casein kinase-TS (type-2) from rat liver cytosol. Eur J Biochem 145:593–599
25. Olsen BB, Boldyreff B, Niefind K, Issinger OG (2006) Purification and characterization of the CK2alpha′-based holoenzyme, an isozyme of CK2alpha: a comparative analysis. Protein Expr Purif 47:651–661
26. Zhang C, Vilk G, Canton DA, Litchfield DW (2002) Phosphorylation regulates the stability of the regulatory CK2beta subunit. Oncogene 21:3754–3764
27. Agostinis P, Goris J, Pinna LA, Merlevede W (1987) Regulation of casein kinase 2 by phosphorylation/dephosphorylation. Biochem J 248:785–789
28. Lolli G, Ranchio A, Battistutta R (2014) Active form of the protein kinase CK2 $\alpha_2\beta_2$ holoenzyme is a strong complex with symmetric architecture. ACS Chem Biol 9:366–371
29. Schnitzler A, Olsen BB, Issinger OG, Niefind K (2014) The protein kinase CK2(Andante) holoenzyme structure supports proposed models of autoregulation and trans-autophosphorylation. J Mol Biol 426:1871–1882
30. Poole A, Poore T, Bandhakavi S et al (2005) A global view of CK2 function and regulation. Mol Cell Biochem 274:163–170
31. Raaf J, Guerra B, Neundorf I et al (2013) First structure of protein kinase CK2 catalytic subunit with an effective CK2β-competitive ligand. ACS Chem Biol 8:901–907
32. Raaf J, Bischoff N, Klopffleisch K et al (2010) Interaction between CK2α and CK2β, the subunits of protein kinase CK2: thermodynamic contributions of key residues on the CK2α surface. Biochemistry 50:512–522
33. Krissinel E, Henrick K (2007) Inference of macromolecular assemblies from crystalline state. J Mol Biol 372:774–797
34. Dey S, Pal A, Chakrabarti P, Janin J (2010) The subunit interfaces of weakly associated homodimeric proteins. J Mol Biol 398:146–160
35. Nooren IM, Thornton JM (2003) Diversity of protein-protein interactions. EMBO J 22:3486–3492

36. Perkins JR, Diboun I, Dessailly BH et al (2010) Transient protein-protein interactions: structural, functional, and network properties. Structure 18:1233–1243
37. Nooren IM, Thornton JM (2003) Structural characterisation and functional significance of transient protein-protein interactions. J Mol Biol 325:991–1018
38. Filhol O, Nueda A, Martel V et al (2003) Live-cell fluorescence imaging reveals the dynamics of protein kinase CK2 individual subunits. Mol Cell Biol 23:975–987
39. Filhol O, Martiel JL, Cochet C (2004) Protein kinase CK2: a new view of an old molecular complex. EMBO Rep 5:351–355
40. St-Denis NA, Derksen DR, Litchfield DW (2009) Evidence for regulation of mitotic progression through temporal phosphorylation and dephosphorylation of CK2alpha. Mol Cell Biol 29:2068–2081
41. Baggio B, Pinna LA, Moret V, Siliprandi N (1970) A simple procedure for the purification of rat liver phosvitin kinase. Biochim Biophys Acta 212:515–517

Maize RNA-Binding Protein ZmTGH: A New Partner for CK2β1 Regulatory Subunit

Isabel Cristina Vélez-Bermúdez and Marta Riera

Abstract In plants, as in humans, splicing regulation controls plant growth and development. Plant CK2 is involved in multiple signaling pathways. The maize CK2β1 regulatory subunit is preferentially accumulated in nuclear speckles, and little is known about the role of CK2 in these nuclear structures that contain splicing regulators among other proteins. Here we perform a yeast two-hybrid screening using CK2β1 as a bait, and between other putative partners, we found an RNA-binding protein that presents high homology with *Arabidopsis* TGH, an evolutionary conserved protein required for proper plant development. In addition, transient co-transformation of both proteins in maize embryo cells confirms that maize CK2β1 regulatory subunit and ZmTGH colocalize in nuclear speckles. The interaction and colocalization of both proteins suggest a possible role of CK2 modulating proteins involved in splicing processes in plants.

Keywords Maize • RNA-binding protein • Nuclear speckles • Splicing regulators

1 Introduction

In humans, more than 95 % of intron-containing genes process their precursor mRNA through alternative splicing (AS), and approximately 5 % of genetic diseases are caused by mutations affecting splicing or AS [1]. Some splicing regulators, as human RNPS1 splicing activator, are controlled by CK2 phosphorylation [2]. In plants splicing and AS are involved in multiple processes affecting plant growth, development, and responses to external cues, but little is known about the role of CK2 in the regulation of these processes [3]. As its animal counterpart, plant CK2 is a key enzyme crucial for plant growth and development and is known to be involved in the regulation of multiple pathways as light signaling, circadian rhythm, cell-cycle regulation, DNA damage, salicylic acid-mediated defense, and abiotic stress responses [4, 5].

I.C. Vélez-Bermúdez • M. Riera (✉)
Center for Research in Agricultural Genomics (CRAG), CSIC-IRTA-UAB-UB Consortium, Campus UAB - Edifici CRAG, Bellaterra (Cerdanyola del Vallès), 08193 Barcelona, Spain
e-mail: marta.riera@cragenomica.es

K. Ahmed et al. (eds.), *Protein Kinase CK2 Cellular Function in Normal and Disease States*, Advances in Biochemistry in Health and Disease 12, DOI 10.1007/978-3-319-14544-0_4

A higher number of genes encoding for CK2α/β subunits have been found in plants compared to animal genomes, since they are organized in multigenic families composed by up to four genes. In plants, as in animals, CK2 holoenzyme constitutes also a heterotetramer where the two CK2β subunits associate as a stable dimer in the core of the holoenzyme, whereas the two CK2α are located in the external part without interacting among themselves [6]. However, the maize holoenzyme presents less stability and specific activity than its maize counterpart [7], reinforcing the hypothesis of functionality of the individual subunits alone or through interaction with other partners. In line with this, CK2 subunits have been identified in different cell compartments: CK2α subunits have been described in the nucleus (nucleoplasm), nucleolus, and chloroplasts [8, 9]. Recently, it has been found in *Arabidopsis* that the plastid CK2 isoform (CKA4) presents specific roles in response to ABA and heat stress [10]. In *Arabidopsis* and maize, the CK2β subunits can be found in both nuclei (nuclear speckles) and some isoforms also are found in cytoplasm [8, 9, 11]. In a previous work [12], using bimolecular fluorescence complementation (BiFC) assays in agroinfiltrated tobacco leaves, we have demonstrated that interaction between maize CK2α1 and CK2β1 split YFP-tagged proteins is located not only in the nucleus (nucleoplasm and nuclear speckles) but also in cytoplasmic aggregates, suggesting that CK2 subunits shuttle between nucleus/nucleolus and cytoplasm as described in animals [13]. Thus, differential subcellular localization of plant individual CK2α/β subunits versus CK2α/β interaction suggests functional specificity for each CK2 isoform. In the same work, we also hypothesized that CK2β1 aggregation in nuclear speckles prevents fast degradation by proteasome because within this structure the protein is assumed to be tightly complexed and less accessible to degradation machinery. However, little is known about the composition of these nuclear speckles in plants. Nuclear speckles are located in the interchromatin space and contain mainly splicing regulators but also transcription factors and are often located near transcription sites, and the localization and dynamics of these structures can be regulated by the phosphorylation status of its components [14].

To further elucidate the role of CK2β in plants, we performed a yeast two-hybrid screening using CK2β1 as bait, and we isolated several putative partners, between them; here we focus in an RNA-binding protein that present high homology with *Arabidopsis* TGH [15]. We performed transient transformation of embryo cells and we confirmed that in maize CK2β1 and ZmTGH colocalize in nuclear speckles.

2 Materials and Methods

2.1 Yeast Two-Hybrid Library Screening

For yeast two-hybrid screening, the *S. cerevisiae* strain HF7c (MATa, ura3–52, his3–200, ade2–101, lys2–801, trp1–901, leu2–3112gal4–542, gal80–538, LYS2:: GAL1UAS–GAL1TATA–HIS3, URA3::GAL417MERS(3X)–CYC1TATA–LacZ) was transformed with the GAL4 binding domain/CK2β1 fusion (pGBT9-CK2β1)

[16] according to the manufacturer (Matchmaker, Clontech). Positive clones were selected in plates lacking tryptophan and transformed with the HybriZap two-hybrid library. The HybriZap two-hybrid vector system library was constructed from poly(A)+ RNA maize stressed leaf of the inbred line W64A, according to the manufacturer (Matchmaker, Clontech Laboratories Inc., Palo Alto, CA, USA). Transformants were selected in –Leu, –Trp, and –His plates containing 1 mM 3-amino-1,2,4,-triazole. Purified colonies were tested for β-galactosidase activity using filter assays, according to the manufacturer (Matchmaker, Clontech). Plasmids from His+ LacZ+ colonies were isolated and electroporated into *E. coli*, and the DNA sequence of the inserts was determined.

2.2 *ZmTGH Isolation and Cloning*

The full-length cDNA of ZmTGH was amplified by PCR using specific primers (TGH_F GCCATGGGGGTCGACAGCGACGACG and TGH_R CCCATGG CATCTACATGATCTTTTCTTTTGTGC) and cloned into pGJ1425 vector under the control of a CaMV 35S promoter and fused in the 3′ region with the dsRED using NcoI site.

2.3 *Transient Expression of GFP Fusions in Maize Embryo Cells*

For maize transformation immature maize embryos about 1 mm long were aseptically dissected from the ears of field-grown maize plants (AxBxB73) after 10 days of pollination (10 DAP) and were transiently transfected with CK2β1-GFP [6] and ZmTGH-dsRED (this paper) fusion constructs by particle bombardment using the Biolistic PDS-1000/He particle delivery system (Bio-Rad) as has been described in [12]. After 24 h embryos were stained with DAPI and examined by FV1000 confocal microscopy (Olympus).

3 Results and Discussion

Previous results overexpressing maize CK2β1-GFP in onion cells and tobacco leaves show that this protein is highly prone to aggregation in nuclear speckles [12]. The identification of CK2β partners will be helpful to better understand its role and its accumulation in these nuclear structures. With this aim, we performed a yeast two-hybrid screening of a cDNA library from maize stressed leaves using CK2β1 as a bait, and we isolated several proteins (Fig. 1). Mainly we found the other CK2 subunits (both CK2α and CK2β) but also transcription factors and proteins involved in RNA processing.

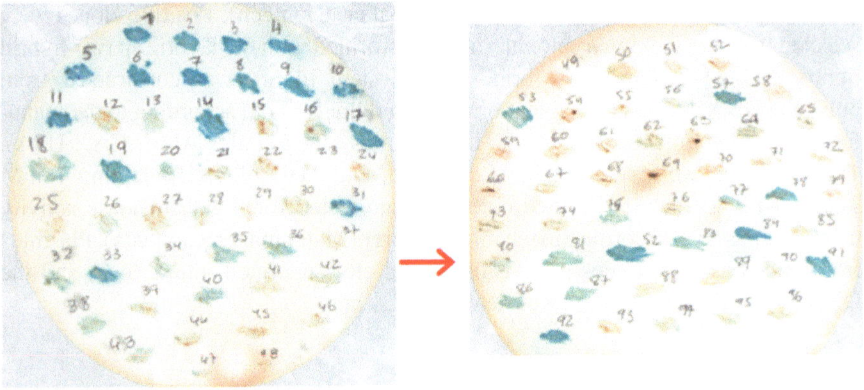

Fig. 1 β-Galactosidase assay performed on a filter of 96 clones isolated in the yeast two-hybrid screening of maize stressed leaves using maize CK2β1 as a bait. Clone 80 corresponds to ZmTGH

One of the selected CK2 putative interactors corresponds to an orthologue of *Arabidopsis* RNA-processing protein TOUGH (TGH). TGH is an evolutionary conserved protein found in many eukaryotic organisms. AtTGH interacts with the transcription initiation factor TATA-box-binding protein 2 and colocalizes with the splicing regulator SRp34 to subnuclear particles. Recently, it has been shown that it belongs to a complex that processes primary transcripts of miRNAs into miRNAs in *Arabidopsis* [15, 17]. The ZmTGH (accession number GRMZM2G131482) transcript contains 16 exons and encodes for a protein of 989 amino acids. The alignment with AtTGH protein is shown in Fig. 2, where conserved domains (TGH, G-patch, and SWAP) are highlighted. The TGH is specific of TGH proteins and the G-patch and the suppressor-of-white-apricot (SWAP) domains are found in proteins involved in RNA binding or processing.

It has been also shown that AtTGH is localized in nuclear speckles, thus, to investigate if CK2β1 and ZmTGH are located in the same nuclear structures; we perform colocalization assays in immature embryo cells using ZmCK2β1-GFP and ZmTGH-dsRED fusions. As shown in Fig. 3, both proteins colocalize in nuclear speckles, located in the interchromatin spaces.

CK2 is known to phosphorylate a large number of substrates with roles in transcription and splicing in animals and in yeast, where it has been shown that CK2β may be involved in the modulation of pre-mRNA processing dynamics of ribosomal protein genes under stress conditions [18]. In animals, CK2 relocalizes to speckle-like structures in the nucleus of cells subjected to elevated temperatures [19]. However, less is known about the role of CK2 in regulation of splicing process in plants. In *Arabidopsis* another kinase, CDKC2, dynamically associates with spliceosomal components in a transcription- and kinase-dependent manner [20]. The number of CK2 substrates described in plants is ten times lower than the animal substrates, but the partners/substrates described suggest a role of this kinase in regulation of transcriptional processes or stress response [4, 5]. The interaction and colocalization in nuclear speckles of maize CK2β1 and ZmTGH suggest a possible

Fig. 2 Protein sequence alignment between *Zea mays* ZmTGH (GRMZM2G131482) and *Arabidopsis* AtTGH (At5g23080). The conserved domains TGH, G-path, and SWAP are highlighted in boxes *red*, *blue*, and *green*, respectively

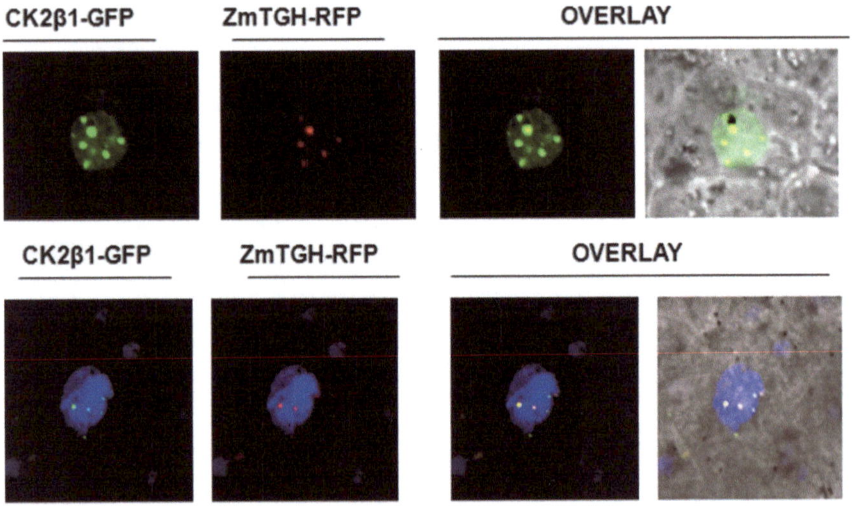

Fig. 3 (**a** and **b**) Colocalization of CK2β1-GFP and ZmTGH-dsRED. Ten DAP embryo cells were transformed by particle bombardment with the indicated constructs (60×). GFP and RFP (dsRED) are shown in epifluorescence images (*left*). *Yellow color* in the *right panels* epifluorescence and bright-field images (merged with epifluorescence) indicates overlap between RFP and GFP signals. The nucleus of the embryo cells in *panel b* is stained with DAPI

role of CK2 modulating proteins involved in splicing processes. In maize, whereas CK2β1 alone is located in nuclear speckles, the presence of CK2α in these structures is observed only when both subunits (CK2α/β) are interacting [12]. Thus, further experiments are needed to elucidate the role of CK2 in these nuclear structures and if this role involves only CK2β subunits or also the catalytic CK2α.

References

1. Nissim-Rafinia M, Kerem B (2002) Splicing regulation as a potential genetic modifier. Trends Genet 18(3):123–127
2. Trembley J, Tatsumi S, Sakashita E, Loyer P, Slaughter C, Suzuki H, Endo H, Kidd V, Mayeda A (2005) Activation of pre-mRNA splicing by human RNPS1 is regulated by CK2 phosphorylation. Mol Cell Biol 25(4):1446–1457
3. Staiger D, Brown J (2013) Alternative splicing at the intersection of biological timing, development, and stress responses. Plant Cell 25(10):3640–3656
4. Riera M, Vélez-Bermúdez IC, Legnaioli T, Pagès M (2013) Specific features of plant CK2, in protein kinase CK2 (ed L. A. Pinna), John Wiley & Sons, Inc., Oxford, UK. doi: 10.1002/9781118482490.ch9
5. Mulekar JJ, Huq E (2013) Expanding roles of protein kinase CK2 in regulating plant growth and development. J Exp Bot. doi:10.1093/jxb/ert401

6. Graham KC, Litchfield DW (2000) The regulatory beta subunit of protein kinase CK2 mediates formation of tetrameric CK2 complexes. J Biol Chem 275(7):5003–5010
7. Riera M, Pages M, Issinger OG, Guerra B (2003) Purification and characterization of recombinant protein kinase CK2 from Zea mays expressed in Escherichia coli. Protein Expr Purif 29(1):24–32
8. Riera M, Figueras M, Lopez C, Goday A, Pages M (2004) Protein kinase CK2 modulates developmental functions of the abscisic acid responsive protein Rab17 from maize. Proc Natl Acad Sci U S A 101(26):9879–9884
9. Salinas P, Fuentes D, Vidal E, Jordana X, Echeverria M, Holuigue L (2006) An extensive survey of CK2 alpha and beta subunits in Arabidopsis: multiple isoforms exhibit differential subcellular localization. Plant Cell Physiol 47(9):1295–1308
10. Wang Y, Chang H, Hu S, Lu X, Yuan C, Zhang C, Wang P, Xiao W, Xiao L, Xue G-P, Guo X (2014) Plastid casein kinase 2 knockout reduces abscisic acid (ABA) sensitivity, thermotolerance, and expression of ABA- and heat-stress-responsive nuclear genes. J Exp Bot. doi:10.1093/jxb/eru190
11. Perales M, Portolés S, Más P (2006) The proteasome-dependent degradation of CKB4 is regulated by the Arabidopsis biological clock. Plant J 46(5):849–860
12. Riera M, Irar S, Vélez-Bermúdez IC, Carretero-Paulet L, Lumbreras V, Pagès M (2011) Role of plant-specific N-terminal domain of maize CK2b1 subunit in CK2B functions and holoenzyme regulation. PLoS One 6(7):e21909
13. Filhol O, Nueda A, Martel V, Gerber-Scokaert D, Benitez MJ, Souchier C, Saoudi Y, Cochet C (2003) Live-cell fluorescence imaging reveals the dynamics of protein kinase CK2 individual subunits. Mol Cell Biol 23(3):975–987
14. Reddy A, Day I, Göhring J, Barta A (2012) Localization and dynamics of nuclear speckles in plants. Plant Physiol 158(1):67–77
15. Calderon-Villalobos L, Kuhnle C, Dohmann E, Li H, Bevanb M, Schwechheimer C (2005) The evolutionarily conserved TOUGH protein is required for proper development of Arabidopsis thaliana. Plant Cell 17(9):2473–2485
16. Riera M, Peracchia G, de Nadal E, Arino J, Pages M (2001) Maize protein kinase CK2: regulation and functionality of three beta regulatory subunits. Plant J 25(4):365–374
17. Ren G, Xie M, Dou Y, Zhang S, Zhang C, Yu B (2012) Regulation of miRNA abundance by RNA binding protein TOUGH in Arabidopsis. Proc Natl Acad Sci U S A 109(31):12817–12821
18. Bergkessel M, Whitworth GB, Guthrie C (2011) Diverse environmental stresses elicit distinct responses at the level of pre-mRNA processing in yeast. RNA 17(8):1461–1478
19. Gerber DA, Souquere-Besse S, Puvion F, Dubois M-F, Bensaude O, Cochet C (2000) Heat-induced relocalization of protein kinase CK2: implication of CK2 in the context of cellular stress. J Biol Chem 275(31):23919–23926
20. Kitsios G, Alexiou KG, Bush M, Shaw P, Doonan JH (2008) A cyclin-dependent protein kinase, CDKC2, colocalizes with and modulates the distribution of spliceosomal components in Arabidopsis. Plant J 54(2):220–235

Part II
CK2 Control of Organismal and Cellular Functions

CK2 in Organ Development, Physiology, and Homeostasis

Charina E. Ortega, Lawrence Prince-Wright, and Isabel Dominguez

Abstract The serine/threonine kinase CK2 is associated with a wide variety of cellular processes, including cell growth, cell proliferation, and cell apoptosis. These cellular processes are vital for proper cell function, organogenesis, embryogenesis, and organ homeostasis. Indeed, CK2 is essential for embryonic development in different model organisms; however, the cellular, biochemical, molecular, and signaling mechanisms that CK2 utilizes to control embryo development are poorly defined. In this chapter, we review the effect of *CK2α* knockout in the developing organs and review the literature to understand the role of CK2α in organ formation. Published studies show roles for CK2 in the control of embryonic organ development, as well as a role for CK2α in organ homeostasis and physiology. In addition, CK2 also plays a role in a number of organ diseases.

Keywords CK2 • Organ development • Physiology • Disease • Expression • Homeostasis • Histology • Knockout • Morphology • Serine/threonine • Wnt • Embryo • Organ function • Germ layers

1 Introduction

CK2 is a serine/threonine protein kinase whose function is associated with a wide variety of cellular processes [1]. Several roles have been defined for CK2 within a cell, but CK2 is most often associated with cell proliferation [2] and cell growth [2]. Indeed, upregulation of CK2 expression has been linked with uncontrolled cellular proliferation and transformation [3–6], and its consequent implication in human cancers [7]. CK2 also plays an important role in cell survival [8], morphology [5], and angiogenesis [9].

C.E. Ortega (✉) • L. Prince-Wright • I. Dominguez (✉)
Department of Medicine, Boston University School of Medicine, Boston, MA 02118, USA
e-mail: ceortega@bu.edu; lhprinc@bu.edu; isdoming@bu.edu

K. Ahmed et al. (eds.), *Protein Kinase CK2 Cellular Function in Normal and Disease States*, Advances in Biochemistry in Health and Disease 12, DOI 10.1007/978-3-319-14544-0_5

In mammals, the CK2 kinase protein is encoded by two different genes *CK2α* and *CK2α′*. CK2 can function as either a monomeric kinase or a tetrameric holoenzyme, composed of two CK2α/α′ subunits and two CK2β regulatory subunits. Within the holoenzyme, CK2β's role is twofold; CK2β regulates the ability of the complex to bind to the cell membrane and changes CK2α/α′ substrate specificity [10]. It is relevant to note that various types of tissues and tumor samples have an unequal ratio of catalytic proteins to regulatory ones, indicating that these proteins may also play roles independent of the tetrameric complex [4, 11].

CK2 genes are a common component in the genome of vertebrates [12], invertebrates [13], and plants [14]. As such, CK2 is a key factor in embryonic development [1, 15]. In mice, knockout embryos for *CK2α* are lethal by embryonic day 11 (E11) and have apparent gross morphological cardiac and neural tube defects [16–18]. Mouse embryos knockout for *CK2β* die at E6.5; however, the precise defects resulting in death are not known [19]. Knockout embryos for *CK2α′* result in infertile male mice, with defective spermatozoa morphogenesis [20]. CK2 may mediate some of its effects on embryo development through its role as a positive regulator of the Wnt/β-catenin pathway, which regulates cellular differentiation and growth [1]. For example, CK2 is required for Wnt/β-catenin-dependent activation of target genes that are responsible for dorsal axis formation during embryogenesis [21].

CK2 transcripts and proteins are differentially expressed in animal models [15]. For example, in the mouse embryo, both CK2α and CK2β are expressed at higher levels in both neuroepithelial and epithelial cells compared to other cell types [22]. CK2α and CK2β are also expressed in connective tissue, skeletal muscle, and cartilage [22]. We have no data on CK2α′ expression during embryo development, but in adult mice CK2α′ expression is limited to the brain and testis [20]. Interestingly, in mouse development, CK2 expression is maximal during organogenesis and diminishes by birth, suggesting a key role for CK2 in organ formation [23].

Our previous studies have shown the overall effect of mouse *CK2α* deletion in whole embryos (Fig. 1) [16–18]. In this book chapter, we review the histological features of several developing organs in *CK2α* knockouts at E9.5 and E10.5. We also describe the importance of CK2 in organ physiology and homeostasis. Due to the large amount of published data, we have summarized the best known examples for each selected organ. We apologize to colleagues for not having mentioned some data.

2 Histological Analysis of *CK2α* Knockouts

During embryo development, three germ layers will form that give rise to different tissues and organs: ectoderm, endoderm, and mesoderm. The ectoderm will give rise to organs like the central nervous system, the endoderm to the gut and associated organs, and the mesoderm to muscle and skeletal tissues. Here we describe the histological features of *CK2α* knockouts at days E9.5 and E10.5 in organs derived from these germ layers (Figs. 2 and 3).

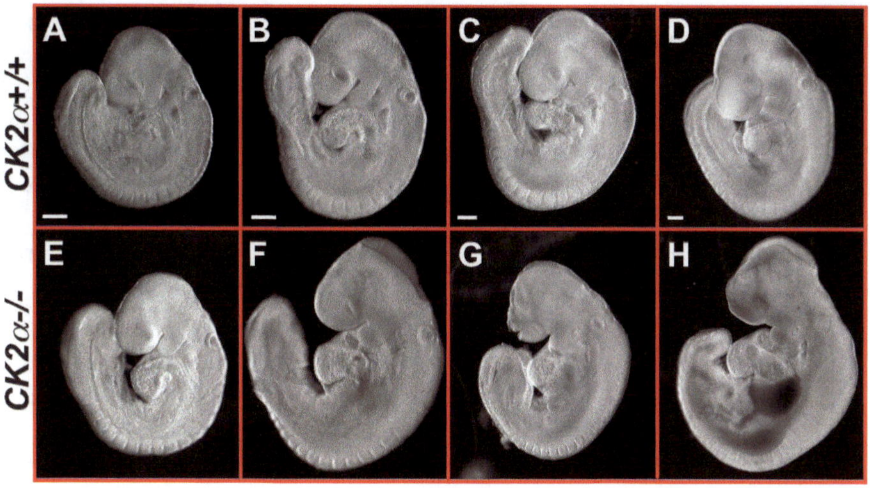

Fig. 1 Macroscopic analysis of E9 to E10 embryos. *Left* views of wild type and *CK2α–/–* embryos at somite pairs 16 (E9, **a**, **e**), 19 (E9, **b**, **f**), 21 (E9.5, **c**, **g**), and 25–26 (E10, **d**, **h**). Representative photographs. Scale bar: 250 μm

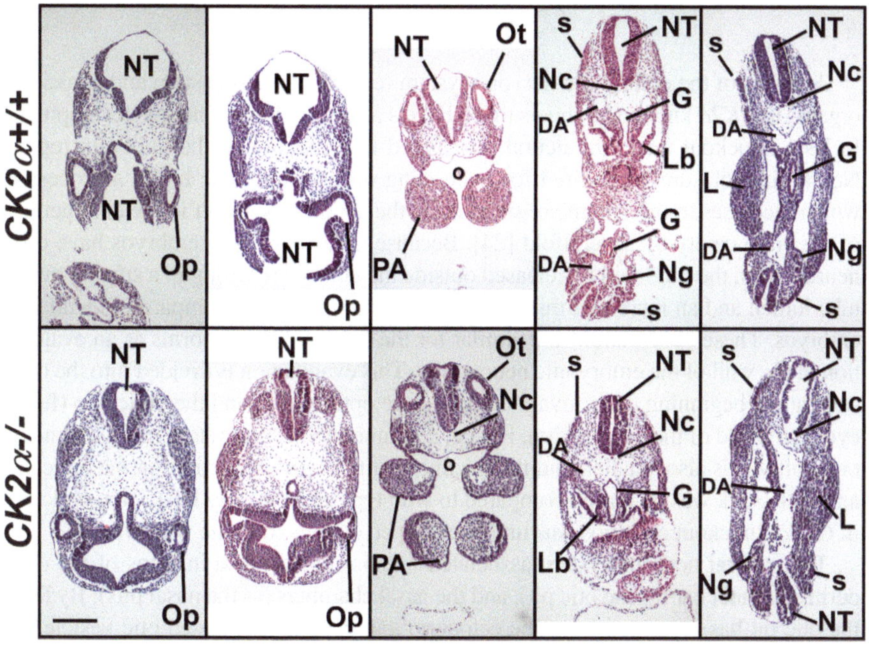

Fig. 2 Microscopic analysis of E9.5 embryo sections (23 somite pairs). Bright-field photographs of 6 μm sections of wild type and *CK2α–/–* embryos were taken at ×4 magnification. Sections displayed from anterior (*left*) to posterior (*right*). Dorsal is on the *top*. Representative photographs. Scale bar: 500 μm. Abbreviations: *DA* dorsal aorta, *G* gut, *L* limb bud, *Lb* lung bud, *Nc* notochord, *NT* neural tube, *Ng* nephrogenic cord tissue, *o* oropharynx, *Ot* otic vesicle, *Op* optic vesicle, *PA* pharyngeal arch, *s* somite

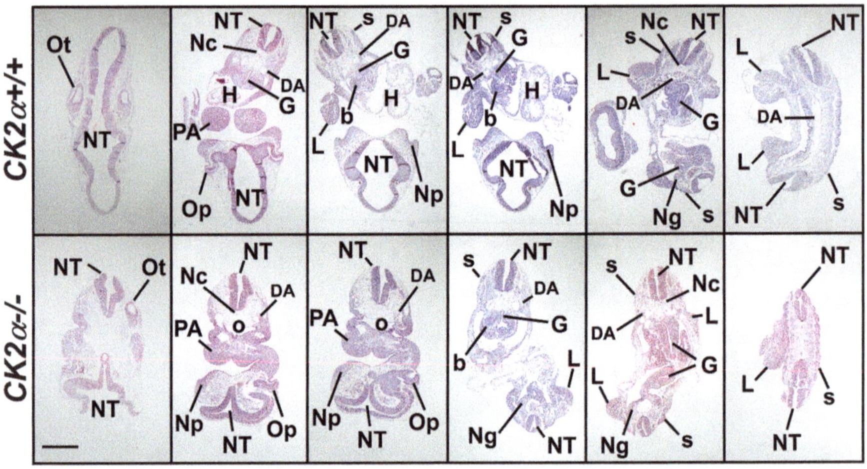

Fig. 3 Microscopic analysis of E10.5 embryo sections (34 somite pairs). Bright-field photographs of 6 μm sections of wild type and *CK2α−/−* embryos were taken at ×4 magnification. Sections displayed from anterior (*left*) to posterior (*right*). Dorsal is on the *top*. Representative photographs. Scale bar: 500 μm. Abbreviations: *b* bronchus, *DA* dorsal aorta, *G* gut, *H* heart tube, *L* limb bud, *Nc* notochord, *Np* nasal pit, *NT* neural tube, *Ng* nephrogenic cord tissue, *o* oropharynx, *Ot* otic vesicle, *Op* optic vesicle, *PA* pharyngeal arch, *s* somite

Analysis of the embryonic nervous system (central nervous system and associated organs) in *CK2α* knockouts shows that CK2α is a key regulator of neural development. In E9.5 knockout mice, the neural tube failed to close around the midbrain region. Neural tube closure is required for ballooning of the embryonic brain, a process in which increases in the volume of the fluid in the neural tube result in the enlargement of the brain cavity by severalfold [24]. Because *CK2α* knockout embryos have open neural tubes, the fluid will be released outside the embryo resulting in a smaller neural tube lumen and an increased thickness of the neuroepithelium compared to wild type embryos. These observations are similar for the future eye, as it forms as an evagination of the wall of the embryonic neural tube. This evagination is divided into the optic stalk at the beginning of the evagination (future optic nerve) and the optic cup (future eye) at the end of the evagination. Hence, the lumen of the optic stalk in *CK2α* knockout embryos is also smaller, and the thickness of the neuroepithelium in the optic cup and optic stalk was increased compared to wild type embryos. By E10.5, the optic cup in *CK2α* knockout embryos was underdeveloped compared to the wild type.

Two central nervous system-associated organs start as invaginations of the ectoderm, the inner ear (as the otic pit), and the nasal chambers (as the nasal pits). By E9.5, the otic pit has separated from the ectoderm and become a rounded otic vesicle that will later acquire a more irregular morphology (oval-like with protrusions). At E9.5, the otic vesicle in *CK2α* knockouts is rounder, and the lumen of the otic vesicle is smaller than those in wild type embryos. In addition, the cells that form the otic vesicle seem disorganized compared to wild type embryos. At E10.5, the otic vesicle is still rounder and smaller in *CK2α* knockouts compared to wild type embryos. Similar to the otic vesicle, cells in the nasal pit seem disorganized compared to wild type embryos.

In endodermal derivatives, the oropharynx has a different morphology in *CK2α* knockouts compared to wild type. This may be due to the malformation of adjacent tissues such as the pharyngeal arches and the neural tube. At E9.5, cells forming the lung bud had a more cuboidal morphology in *CK2α* knockouts compared to wild type embryos. Similarly, at E10.5, cells lining the embryonic bronchus also had a more cuboidal morphology.

As for mesodermal derivatives, *CK2α* knockouts display abnormal hearts, somites, and nephric tissue. In *CK2α* knockouts, the heart tube appears distended and has a thinner myocardium, reduced trabeculation, and thinner endocardium lining compared to wild type [1]. In addition, the division between the heart chambers in the *CK2α* knockout is not as apparent as in the wild type. The somites (precursors of body muscles and vertebra) in *CK2α* knockout mice at E9.5 are hypoplastic, particularly the posterior somites [18]. In addition, our histological analysis showed that the cellular organization of the somites is abnormal.

Similar to other developing organ cavities, the lumen of the mesonephric tubules appears smaller in *CK2α* knockouts, and the morphology of the cells in the tubule is altered.

The forelimb and hindlimb buds are hypoplastic in *CK2α* knockouts [18]; however, we did not observe an apparent difference in the epithelium or mesenchyme of the limb buds compared to wild type.

The pharyngeal arches are structures made of ectoderm, endoderm, and mesoderm as well as neural crest cells, and are the precursors of the cranial nerves, the skeletal and muscle elements in the face, and the pharynx. At E9.5, most of the *CK2α* knockout embryos only had one hypoplastic first pharyngeal arch, and some failed to develop any pharyngeal arches (wild type embryos have two pharyngeal arches at this stage) [18]. By E10.5, the total number and size of the pharyngeal arches was smaller in *CK2α* knockouts compared to wild type embryos [18]. At both E9.5 and E10.5, the endodermal lining of the pharyngeal arches was disorganized in *CK2α* knockouts compared to wild type. In addition, *CK2α* knockout embryos had an increased pharyngeal arch artery lumen. Similarly, at E10.5, the lumen of the dorsal aorta appears increased in the *CK2α* knockouts. The trigeminal ganglia, a neural crest cell derivative that will form head-specific sensory neurons, are smaller in E10.5 knockout mice.

All together, these data show that the morphogenesis of a number of organs is abnormal and suggest that cell morphology is altered in *CK2α* knockouts.

3 CK2 in Ectoderm-Derived Organs

3.1 Central Nervous System and Neuronal Development

A number of studies demonstrate the important effect of CK2 genes on the development and function of the brain and neurons. For example, CK2β is essential for central nervous system development in mouse, particularly the telencephalon (the embryonic cerebral hemispheres). CK2β is required for neural progenitor cell proliferation and for their differentiation to oligodendrocytes (a type of glial cells that function to support neurons). CK2β may be acting through phosphorylation of Olig2, a transcription factor important for embryonic oligodendrocyte precursor cell specification, as Olig2 is a CK2β-dependent CK2 substrate [25]. In addition, CK2β is partially necessary for neurite growth [26]. During neuronal development, *CK2α* has been shown to be expressed earlier than *CK2α′* [27]. In neurons, CK2α′ is expressed during dendrite maturation and synaptogenesis [27]. CK2α′ is associated with microtubules and found concentrated in neuronal cell bodies and dendrites, in a similar distribution to that of phospho-MAP1B, a microtubule-associated protein linked to axonal elongation, neuronal migration, and axonal guidance [27]. The CK2 phosphorylation site in MAP1B remains unknown [28]. CK2 also regulates proper formation of hippocampal dendritic spines, doorknob extensions in a neuron's dendrite that receive postsynaptic input and whose density may be related to mental agility. During the formation of the dendritic spines, it is possible that CK2 is phosphorylating PACSIN 1 (protein kinase C and casein kinase 2 substrate in neurons 1) promoting its dissociation from Rac1, a protein that promotes proper formation of spines [29]. These data show that CK2 regulate neuronal morphogenesis. In addition, CK2 is also important in regulating neuronal survival probably inducing the expression of the anti-apoptotic gene *Mcl-1* in the hippocampal CA1 neurons that are involved in memory [30].

In the adult brain, CK2α and CK2α′ proteins are expressed throughout; however, CK2α levels are significantly higher than CK2α′ (8:1 ratio). CK2α levels are more uniformly expressed than those of CK2α′ in the different mouse brain regions studied. Importantly, CK2α′ levels are highest in the cortex and hippocampus, while CK2α levels are highest in the striatum [31]. In addition, CK2α′ is increased during the late postnatal neocortical maturation period, the same time when synaptogenesis and dendrite maturation occur in mature neurons [32]. In the brain, CK2 controls neuronal function. CK2 alters the Ca^{2+} sensitivity of small conductance Ca^{2+}-activated K^+ (SK) channels in the neural tube and also in the auditory nerves. Neurotransmitters such as noradrenaline decrease SK channel activity in a CK2-dependent manner through phosphorylation of SK-bound calmodulin (CaM) [33, 34]. In addition, CK2 also regulates the interaction between sodium channels and ankyrin. The interaction of sodium channels with ankyrin is important to form and maintain the axonal initial segment (AIS) of neurons where action potentials are generated [35]. CK2α and CK2α′ proteins concentrate in the AIS in an ankyrin-dependent manner. Ultimately, CK2 is also required for proper microtubule cytoskeleton function, which in turn is necessary for CK2 localization to the AIS [36]. Not only does CK2 alter channel sensitivity but CK2 is also rapidly activated in the synapses and alters *N*-methyl-D-aspartate (NMDA) receptor function. The phosphorylation by CK2 of NMDA receptor and other proteins contributes to synaptic plasticity [37]. Because CK2 is involved in NMDA receptor regulation, and this receptor is required for pain hypersensitivity caused by immunosuppressants such as cyclosporine A, CK2 inhibition may help in reducing immunosuppressant-induced pain hypersensitivity [38].

In addition to neuron and neural tube development and neuronal function, CK2 is also important in adult diseases. For example, upregulation of CK2 in the paraventricular nucleus in the hypothalamus causes diminished SK channel activity that in turn may regulate hypertension, at least in animal models [39]. In Alzheimer's disease (AD), amyloid-β increases CK2 activity that in turn phosphorylates components of the fast axonal transport (FAT) altering their activity. It is believed that deficits in FAT can eventually cause the degeneration of neurons in AD [40]. Interestingly, the localization of CK2 is altered in AD neurons, particularly in tangle-bearing neurons. In these neurons, CK2 is localized to the somatodendritic body and on the neurofibrillary tangle (NFT), aggregates of highly phosphorylated cytoskeletal components that appear in AD patients [41, 42]. In addition, CK2 can phosphorylate huntingtin, the gene underlying Huntington disease (HD) [43]. CK2 phosphorylation may localize huntingtin to the nucleus where it may exert its deleterious effects [44]. CK2 may be also involved in Parkinson disease (PD), a disease characterized by loss of motor function caused by the death of neurons producing dopamine, which controls motor function. Knocking out *CK2α* in spiny neurons negatively affects its dopaminergic function suggesting a role for CK2 in motor control [45]. In addition, CK2 phosphorylates α-synuclein, a protein that will form aggregates in preclinical PD [46]. CK2 expression correlates with increased α-synuclein phosphorylation even in preclinical PD. Therefore, CK2 could be a therapeutic target in AD, HD, and PD and other diseases with dopamine imbalance.

3.2 Eye

CK2 plays a role in eye development, particularly regulating various proteins affecting the formation of the eye. Previous studies focused on eye development in *Drosophila melanogaster* and mammalian models.

CK2 regulates eye morphogenesis. For example, in *Drosophila melanogaster* retinal patterning, the Notch signaling pathway leads to the expression E(spl) proteins in non-sensory organ precursor cells, thereby extinguishing their neuronal potential. One of the proteins in the E(spl) locus is the basic-helix-loop-helix repressor M8, a protein phosphorylated by CK2 on the C-terminus (Ser^{159}). This phosphorylation is proposed to open the conformation of M8 and allow repression of genes involved in neuronal potential [47–49]. CK2 is also important in mammalian retinas. In human and mouse adult retinas, CK2 is expressed in astrocytes, which are cells important for normal development of the retinal vasculature. In rat retina, CK2 is also expressed in proliferating retinal progenitor cells within the neuroblastic layer of the developing retina, the proliferative zone of the inner optic cup. In the neuroblastic layer, the neural progenitor cells are attached to the retinal inner and outer limiting membranes on both sides of the proliferative zone of neural cells. In this neuroblastic layer, the nucleus moves during the cell cycle, with the help of microtubules, from the outer margin of the cell toward the inner basal margin of the cell, where DNA is replicated. CK2 is a powerful microtubule stabilizer in the developing retina. CK2 inhibition by the highly selective inhibitor of CK2, TBB (4, 5,6,7-tetrabromobenzotriazole), causes the tubulin cytoskeleton along the basal-apical axis of the retinal tissue to become disorganized and DNA synthesis in retinal progenitor cells to be reduced [50].

In addition to regulating eye development, CK2 is proposed to play a key inhibitory role in angiogenesis and apoptosis in the retina. Chemical inhibition of CK2 suppresses angiogenesis and hematopoietic stem cell recruitment to retinal neovascularization sites, in a mouse model of proliferative retinopathy. In line with this, in human diabetic retinas, transcripts for all CK2 proteins decreased coinciding with increased apoptosis. These studies labeled CK2 as a "master regulator" of many critical signaling pathways in angiogenesis [9, 51]. One of the potential targets of CK2's effect in angiogenesis is vascular endothelial growth factor (VEGF). For example, CK2 is involved in VEGF upregulation in endothelial cells induced by oxidized phospholipids. Oxidized phospholipids are known to induce proangiogenic and proinflammatory cellular effects. Importantly, overexpression of VEGF is a cause of pathological neovascularization that is characteristic of eye diseases such as age-related macular degeneration. Since CK2 inhibition reduces VEGF production significantly, it suggests that CK2 inhibitors may be used for therapy of pathological angiogenesis in the eye [52].

In addition, CK2 is involved in light adaptation in the eye. Important for dark adaptation of photoreceptor cells is the reduction of centrin/retinal-specific G-protein complexes that control the transport of signaling proteins. The transport of proteins in photoreceptor cells occurs through the cilium and plays a key role in the organization and function of photoreceptor cells [53]. CK2 and centrins co-localize in the photoreceptor cell cilium. In dark-adapted photoreceptor cells, CK2 phosphorylates centrin isoforms in mouse models, which strongly reduces the binding affinities between centrins and retinal-specific G-protein transducin, a signal-amplifying protein, allowing it to flow through the cilium [54, 55].

CK2 is also involved with degradation and function of important eye proteins. Within the eye, gap junctions are important channels that allow the passage of small molecules between two adjacent cells in order to maintain metabolic activities and homeostasis. This communication is particularly important between the cells inside the lens and cells on the lens' surfaces. An important structural component of gap junctions is connexins, a family of membrane proteins. Connexin phosphorylation regulates its stabilization, intracellular trafficking, gap channel assembly, and gap junctional communication. CK2 is proposed to phosphorylate Ser^{363} in connexin 45.6 leading to connexin degradation and destabilization in chick lens fibers [56].

3.3 Auditory Organs

CK2 plays a role in auditory function, particularly in regulating the activity of various channels involved in auditory sensory transduction. For example, CK2 is involved in the regulation of small conductance Ca^{2+}-activated K^+ (SK) channels. These channels provide a hyperpolarizing K^+ conductance that is fundamental for a range of physiological processes including the control of the hearing organ when they are activated due to elevated levels of intracellular Ca^{2+}. CK2 and protein phosphatase 2A form a complex with the cytoplastic domains of these SK channels, and they phosphorylate/dephosphorylate the Ca^{2+} sensor calmodulin (CaM) changing the Ca^{2+} sensitivity range. The phosphorylation of CaM causes an acceleration of SK channel deactivation, while dephosphorylation causes the inverse effect. As a result, CK2 phosphorylation of CaM may change the control of auditory sensory transduction [57]. In addition to SK channels, CK2 phosphorylation also influences the activity of the voltage-dependent Kv3.1 potassium channel. This channel is expressed in neurons of the medial nucleus of the trapezoid body in the auditory brainstem that is involved in the perception of sound duration and localization. Basal phosphorylation of the Kv3.1 channel influences the basal properties of the current (e.g., its voltage dependence of activation and inactivation), enabling the channel to work as a high-threshold current. This basal phosphorylation may be regulated by CK2 [58].

3.4 *Olfactory Organs*

Few studies report the effects of CK2 in the olfactory bulb and olfactory sensory and receptor neurons. For example, CK2 regulates odorant adaptation by phosphorylating the enzyme heme oxygenase 2 (HO2) that subsequently cleaves heme to generate carbon monoxide (CO), a gaseous neurotransmitter. Endogenously produced CO increases cyclic guanosine monophosphate levels (cGMP) to regulate neurotransmission [59].

CK2 is also involved in the function of olfactory sensory neurons. In these neurons, ciliary protein transport promotes the compartmentalization of signaling molecules in cilia that allows for rapid and efficient activation of the odorant cascade. CK2 phosphorylates a major intracellular trafficking protein (PACS-1) that localizes to the base of human respiratory cilia and controls the localization of ciliary proteins. CK2 also phosphorylates PACS-1 cargo, including CNGB1b, the subunit of the olfactory cyclic-nucleotide-gated (CNG) channel. CNGB1b and PACS-1 are necessary components of the olfactory signal transduction cascade that is enriched in cilia. Therefore, CK2 is required for maintenance of CNG channel in cilia and for subsequent olfactory function in olfactory sensory neurons [60].

4 CK2 in Endoderm-Derived Organs

4.1 *Oropharynx*

Little is known about the role of CK2 in the digestive tract, but it may regulate pathogen-induced damage. CK2α′ positively regulates the capacity of *Candida albicans* to damage endothelial and oral epithelial cells in vitro in both disseminated and oropharyngeal candidiasis. In both types, *C. albicans* adheres to and penetrates the host cells (endocytosis), causing the destruction and loss of these cells. Mutant strains of *C. albicans* that lack *CK2α′* have decreased host cell-damaging properties, but only in oral epithelial cells were they endocytosed poorly. This indicates that CK2 may regulate invasin-like surface proteins in *C. albicans* that are required for oral epithelial cell endocytosis [61].

4.2 Lung

In the lung and associated blood vessels, CK2 plays a role in homeostasis and diseases. CK2α′ protein and activity are upregulated in the artery's smooth muscle cells during pulmonary artery remodeling in hypoxic conditions. Hypoxia-induced pulmonary artery remodeling is associated with hypertension in diseases such as emphysema. CK2α′ may be required to diminish the levels of the transcription factor CREB, which stimulates smooth muscle cell proliferation and migration in vitro [62]. It is interesting to note that CK2 may be also involved in blood pressure regulation by phosphorylating the angiotensin-converting enzyme (ACE) in endothelial cells [63]. ACE converts angiotensin I into angiotensin II, a key factor in the increase of blood pressure via vasoconstriction, sympathetic nervous stimulation, increased aldosterone biosynthesis, and renal actions. CK2 phosphorylation of ACE increases its localization in the plasma membrane and perhaps its secretion [64].

CK2 may be important in platelet-activating factor (PAF)-induced systemic anaphylaxis. PAF is involved in the induction of plasma extravasation by promoting vasodilation and increasing vascular permeability, the most important contributor to fatal reactions of anaphylaxis in mouse models. CK2 activity is increased by PAF, and CK2 is required for PAF-induced anaphylaxis upstream of a signaling cascade involving PTEN/PI3K/Akt/eNOS [65]. Therefore, CK2 inhibition could be a potential target to treat anaphylaxis.

CK2 regulates respiratory tract pathogen infection and clearance. For example, CK2β presence inhibits influenza A virus infection in epithelial lung cells [66]. Likewise, CK2 enhances phagocytosis of *Streptococcus pneumoniae* through surfactant protein A (SP-A) and macrophage scavenger receptor A (SR-A). SP-A, a major constituent of the surfactant that lines the pulmonary alveoli, plays an important role in the innate immune system in the lungs. SR-A interacts with a number of ligands including Gram-negative and -positive bacteria. SP-A increases cell surface localization of SR-A in a CK2-dependent manner and enhances the phagocytosis of particular bacteria such as *S. pneumoniae* by alveolar macrophages [67].

Lung inflammation in patients with chronic obstructive pulmonary diseases and asthma is cigarette smoke-induced and shows decreased histone deacetylase 2 (HDAC2) level/expression. Cigarette smoke extract exposure in bronchial and airway epithelial cells leads to HDAC2 phosphorylation by a CK2-mediated mechanism, resulting in decreased HDAC2 activity and increased HDAC2 degradation. Therefore, CK2 inhibition may be a potential target to inhibit abnormal inflammation in bronchial and small airway cells in response to cigarette smoke [68].

5 CK2 in Mesoderm-Derived Organs

5.1 Heart

CK2 is important for cardiac muscle formation and function. For example, CK2 phosphorylates muscle FHOD3, a formin homology domain-containing protein involved in actin filament nucleation and elongation and downregulated in heart disease. CK2-dependent phosphorylation of muscle FHOD3 inhibits its degradation and becomes more localized to the myofibrils where it will regulate filament assembly and maintenance [69–71].

Not only does CK2 regulate heart development, but CK2 may also play a positive role in anti-apoptotic mechanisms in the heart. In addition to its anti-hypertrophic function, ARC is a key anti-apoptotic regulator in the heart. CK2 can phosphorylate ARC at Thr^{149}, enabling ARC to translocate from the cytoplasm to the mitochondria. In the mitochondria, ARC binds to caspase-8 or caspase-2, thereby enabling ARC to inhibit in the myocardium. Elevated apoptosis is found in human hearts that suffered myocardial infarction and heart failure and also in animal models of hypertrophy and heart failure [72].

CK2 also regulates cardiac hypertrophy. For example, CK2α′ induces cardiac hypertrophy through phosphorylation and subsequent degradation of p27, an important anti-hypertrophic cell cycle regulator in adult cardiomyocytes. CK2α′ activity is stimulated and is necessary for cardiomyocyte hypertrophy, in a p27 degradation manner [73, 74]. Another mechanism utilized by CK2 to promote hypertrophy is activation of histone deacetylases. HDACs were first thought to be anti-hypertrophic mediators; however, more recently, it has been shown that HDACs mediate hypertrophy. HDAC2 phosphorylation and activation are required for hypertrophy, and this phosphorylation may be mediated by CK2α during cardiac hypertrophy [75]. In contrast, angiotensin and other hypertrophic stimuli reduced CK2 activity in cardiomyocytes correlating with decreased ARC (apoptosis repressor with caspase recruitment domain) phosphorylation and catalase expression levels. Hypertrophic stimuli led to carbonylation and inhibition of CK2. Decreased CK2 activity causes ARC to lose its anti-hypertrophic function due to the reduction of Thr^{149} phosphorylation by CK2 [76]. These contradictory findings indicate that the use of CK2 as a potential therapeutic target in heart disease needs to be further researched.

5.2 Muscle

During myogenesis, CK2 phosphorylation is necessary for myf-5 activity. Myf-5 is a muscle-specific transcription factor important for early myogenic determination and differentiation of developing epaxial muscle [77, 78].

CK2 is also a key kinase in myogenic proliferation and differentiation, by phosphorylating Pax3, a myogenic transcription factor essential for early skeletal muscle development. In proliferating myoblasts, CK2 phosphorylates Pax3 at Ser^{205} subsequently enhancing GSK3β-dependent phosphorylation of Pax3 at Ser^{201}. Later, during myogenic differentiation, CK2 phosphorylates Pax3 at Ser^{205} subsequently enhancing the CK2-dependent phosphorylation of Pax3 at Ser^{209} [79].

CK2 is important for muscle regeneration and progression of myogenic progenitor cells. During normal myoblast cell differentiation, the expression of CK2α and CK2α′ remained unchanged while CK2β transcript was downregulated. CK2β expression is essential for cell cycle progression during muscle regeneration. Cell cycle progression regulation by CK2 may depend on Sds3 (suppressor of defective silencing 3) phosphorylation. A CK2-Sds3-Foxk1 cascade combined with CK2 phosphorylation and inhibition of p21 and p27, cell cycle inhibitor genes, may control skeletal cellular proliferation and gene expression during muscle regeneration. In support of this observation, *CK2α′* knockouts have impaired skeletal muscle regeneration [80].

CK2 transcripts and activity levels were studied in muscle diseases such as limb girdle muscular dystrophy, mitochondrial myopathy, and neurogenic atrophy. In general, CK2β levels were lower in diseased compared to muscle tissue control. Only a few samples of each disease had abnormally elevated CK2α transcript levels, and CK2 activity was only elevated in a few samples of limb girdle muscular dystrophy. In this study, since samples originated from different anatomical muscles, this could have affected the results [81].

5.3 *Kidney*

CK2 inhibition may have a protective role in immunosuppressant-induced nephrotoxicity after organ transplantation. Two mechanisms have been proposed. CK2 phosphorylates KAP (kidney androgen-regulated protein), a gene expressed in kidney proximal tubule cells that protects against cyclosporin A (CsA) toxicity. CK2 phosphorylated KAP is degraded by calpain, a protease, therefore enhancing the susceptibility of proximal tubule cells to CsA [82]. In addition, CK2 expression is upregulated in rodent model of chronic CsA-induced nephropathy, which may increase apoptosis [83].

Interestingly, CK2 is also a key molecule in the progression of glomerulonephritis (GN), a renal inflammation that leads to loss of renal function. CK2 was overexpressed in the kidneys of GN rodent models. CK2 inhibition reverts renal dysfunction and histological progression of GN. Therefore, inhibiting CK2 may be therapeutic in GN processes [84].

Similar to the cilia in the olfactory sensory neurons, mutations in ciliary proteins in the kidneys can eventually cause renal complications. For example, nephrocystin is mutated in nephronophthisis (NPHP), a group of renal cystic diseases.

CK2 phosphorylates nephrocystin, which helps its binding to PACS-1, an intracellular trafficking protein that localizes the base of the cilia, thus targeting nephrocystin to the base of the monocilia of polarized kidney epithelial cells. Defects in trafficking are proposed to eventually cause NPHP [85].

5.4 *Extremities*

As in muscle, CK2 may regulate limb bud formation through regulation of *myf-5*, a muscle-specific transcription factor important for myogenic determination and differentiation [77, 78]. Later in development, CK2 may play a role in mineralization through phosphorylation of matrix proteins. It was shown that inhibition of CK2 activity decreased the rate of mineral deposition in a time-dependent manner, where CK2 effects were seen only early during mineralized matrix formation. Therefore, CK2 is involved in the calcification process in mesenchymal cells [86].

6 CK2 in Pharyngeal Arch Derivatives

The effect of CK2 on the pharyngeal arches is poorly studied. A kinase activity in avian branchial arches, thought to be CK2, phosphorylates the protein treacle, the product of the *TCOF1* gene that is mutated in Treacher Collins syndrome (TCS). TCS is a craniofacial disorder that affects the embryonic branchial arches leading to face abnormalities [87].

7 Conclusion

CK2 plays a key role in embryonic organ development and organ physiology, homeostasis, and disease (Table 1). Importantly, we found some commonalities in the mechanisms that CK2 utilized to regulate these functions and processes. During embryonic development, CK2 may utilize similar genes to develop organs such as the muscle and the limb bud (i.e., *myf-5*). For proper cell morphology in both the central nervous system (CNS) and eye, CK2 regulated cytoskeletal components (i.e., microtubules). We also found that CK2 controls cell function through microtubule and myofibril protein phosphorylation in the CNS and in muscle, respectively (i.e., neurotransmission and contraction, respectively). In the CNS and its associated organs such as the auditory organ, CK2 also regulates neurotransmission by regulating K^+ and Na^+ channels. CK2 can also impact organ function through increased intracellular transport to the cilia in the optic and olfactory organs and in the kidney (through centrins and PACS-1).

Table 1 Proposed roles of CK2 in organ and cell development, physiology and disease

	Ectoderm-derived organs				Endoderm-derived organs		Mesoderm-derived organs				
	CNS	Eye	Auditory	Olfactory	Oropharynx	Lung	Heart	Muscle	Kidney	Extremities	Pharyngeal arches
Organ and cell development	Required for neural progenitor cell proliferation, survival, and differentiation Important for neurite growth and formation of hippocampal dendritic spines	Allows repression of genes with neuronal potential Essential for the organization of proliferating retinal progenitor cells Inhibits angiogenesis and apotosis in retina				CK2α′ may diminish CREB levels	Phosphorylates muscle FHOD3 Regulates filament assembly and maintenance	Phosphorylates myf-5 and Pax3 Essential for cell-cycle progression during muscle regeneration (Sds3 phosphorylation) CK2α′ required for skeletal muscle regeneration		Phosphorylates myf-5 Decreases the rate of mineral deposition	
Organ function	Alters Ca2+ sensitivity of SK channels Regulates interaction between Na+ channels and ankyrin.	Involved in light adaptation in the eye through regulating centrin isoforms that control the transport of signaling proteins through cilium	Regulates SK channels Phosphorylates CaM Influences Kv3.1 potassium channel	Regulates odorant adaptation through CO Controls the localization of ciliary proteins Phosphorylates the traffickingi protein PACS-1 and its cargo					Phosphorylates nephrocystin, a PACS-1 binding protein		

Diseases	May regulate hypertension Involved in AD: β-Amloid increases CK2 activity that phosphosphorylates components of FAT Involved in HD: Phosphorylates huntingtin Involved in Parkinson's disease Controls dopaminergic function Involved in NMDA receptor function	In human diabetic retinas, CK2 protein transcripts are decreased			CK2α′ regulates the capacity of *C. Albicans* to damage endothelial and oral epithelial cells	CK2 regulates blood pressure Regulates respiratory tract, pathogen infection and clearance: Inhibits Influenza A infection in epithelial lung cells and enhances phagocytosis of *S. pneumonaia* t	Induces cardiac hypertrophy t Anti-apoptotic regulator in the heart through ARC	Altered levels in Limb Girdle Dystrophy, mitochondrial myopathy, and neurogenic atrophy	Inhibition reverts renal dysfunction of GN Protective role in immunosuppressant-induced nephrotoxicity after organ transplantation		Phosphoryates the protein treacle, the product of the TCOF1 gene mutated in TCS

Just as CK2 regulates normal cell and organ function, CK2 may also be involved in disease in organs derived from all germ cell layers. For example, CK2 plays a pro-survival role in the CNS and in the eye and heart in diseased conditions (diabetes, infarction). CK2 can also regulate immunosuppressant-induced damage in the kidneys and CNS. In addition, it minimizes pathogen damage due to fungal, bacterial, and viral infections (*C. albicans*, *S. pneumoniae*, and influenza A). In addition, CK2 also plays an important role in diseases not reviewed here such as cancer [4, 88]. In some instances, an embryonic substrate of CK2 drives cancer progression. For example, CK2 phosphorylation of Pax3 also helps the development of alveolar rhabdomyosarcoma, a childhood solid muscle tumor. CK2 is already being investigated as a cancer therapeutic target, and it could be also a target for other diseases mentioned above.

In summary, CK2 is involved in organogenesis and organ function, but further research is needed to understand the specific molecular mechanisms controlled by CK2 during organogenesis and organ function and their similarities.

Acknowledgments We would like to thank Julie Cha for her help with the histological analysis of *CK2α* knockout embryos. The support by the National Institutes of Health (NIH GMS) and the American Heart Association (AHA) is acknowledged (grants to I. D.). This project was supported by Boston University's Undergraduate Research Opportunities Program (UROP) Student Research Award (to C.O.).

Note added in proof CK2 can also control the adaptive immune responses [88].

References

1. Dominguez I, Sonenshein GE, Seldin DC (2009) Protein kinase CK2 in health and disease: CK2 and its role in Wnt and NF-kappaB signaling: linking development and cancer. Cell Mol Life Sci 66(11–12):1850–1857. doi:10.1007/s00018-009-9153-z
2. Litchfield DW (2003) Protein kinase CK2: structure, regulation and role in cellular decisions of life and death. Biochem J 369(Pt 1):1–15. doi:10.1042/BJ20021469
3. Landesman-Bollag E, Romieu-Mourez R, Song DH, Sonenshein GE, Cardiff RD, Seldin DC (2001) Protein kinase CK2 in mammary gland tumorigenesis. Oncogene 20(25):3247–3257. doi:10.1038/sj.onc.1204411
4. Tawfic S, Yu S, Wang H, Faust R, Davis A, Ahmed K (2001) Protein kinase CK2 signal in neoplasia. Histol Histopathol 16(2):573–582
5. Canton DA, Litchfield DW (2006) The shape of things to come: an emerging role for protein kinase CK2 in the regulation of cell morphology and the cytoskeleton. Cell Signal 18(3):267–275. doi:10.1016/j.cellsig.2005.07.008
6. Seldin DC, Landesman-Bollag E, Farago M, Currier N, Lou D, Dominguez I (2005) CK2 as a positive regulator of Wnt signalling and tumourigenesis. Mol Cell Biochem 274(1–2):63–67
7. Unger GM, Davis AT, Slaton JW, Ahmed K (2004) Protein kinase CK2 as regulator of cell survival: implications for cancer therapy. Curr Cancer Drug Targets 4(1):77–84
8. Ahmad KA, Wang G, Unger G, Slaton J, Ahmed K (2008) Protein kinase CK2—a key suppressor of apoptosis. Adv Enzyme Regul 48:179–187. doi:10.1016/j.advenzreg.2008.04.002

9. Kramerov AA, Saghizadeh M, Caballero S, Shaw LC, Li Calzi S, Bretner M, Montenarh M, Pinna LA, Grant MB, Ljubimov AV (2008) Inhibition of protein kinase CK2 suppresses angiogenesis and hematopoietic stem cell recruitment to retinal neovascularization sites. Mol Cell Biochem 316(1–2):177–186. doi:10.1007/s11010-008-9831-4
10. Bibby AC, Litchfield DW (2005) The multiple personalities of the regulatory subunit of protein kinase CK2: CK2 dependent and CK2 independent roles reveal a secret identity for CK2beta. Int J Biol Sci 1(2):67–79
11. Ortega CE, Seidner Y, Dominguez I (2014) Mining CK2 in Cancer. PLoS One 26;9(12):e115609. doi:10.1371/journal.pone.0115609. eCollection 2014
12. Ahmed K, Issinger OG, Niefind K (2011) Protein kinase CK2: a catalyst for biology, medicine and structural biochemistry. Mol Cell Biochem 356(1–2):1–3. doi:10.1007/s11010-011-0940-0
13. Russo GL, Tosto M, Mupo A, Castellano I, Cuomo A, Tosti E (2004) Biochemical and functional characterization of protein kinase CK2 in ascidian Ciona intestinalis oocytes at fertilization. Cloning and sequence analysis of cDNA for alpha and beta subunits. J Biol Chem 279(31):33012–33023. doi:10.1074/jbc.M401085200
14. Riera M, Figueras M, Lopez C, Goday A, Pages M (2004) Protein kinase CK2 modulates developmental functions of the abscisic acid responsive protein Rab17 from maize. Proc Natl Acad Sci U S A 101(26):9879–9884. doi:10.1073/pnas.0306154101
15. Alvarez LM, Revuelta-Cervantes J, Dominguez I (2013) CK2 in embryonic development. Protein kinase CK2. Wiley-Blackwell, Ames, IA
16. Lou DY, Dominguez I, Toselli P, Landesman-Bollag E, O'Brien C, Seldin DC (2008) The alpha catalytic subunit of protein kinase CK2 is required for mouse embryonic development. Mol Cell Biol 28(1):131–139. doi:10.1128/MCB.01119-07
17. Seldin DC, Lou DY, Toselli P, Landesman-Bollag E, Dominguez I (2008) Gene targeting of CK2 catalytic subunits. Mol Cell Biochem 316(1–2):141–147. doi:10.1007/s11010-008-9811-8
18. Dominguez I, Degano IR, Chea K, Cha J, Toselli P, Seldin DC (2011) CK2alpha is essential for embryonic morphogenesis. Mol Cell Biochem 356(1–2):209–216. doi:10.1007/s11010-011-0961-8
19. Buchou T, Vernet M, Blond O, Jensen HH, Pointu H, Olsen BB, Cochet C, Issinger OG, Boldyreff B (2003) Disruption of the regulatory beta subunit of protein kinase CK2 in mice leads to a cell-autonomous defect and early embryonic lethality. Mol Cell Biol 23(3):908–915
20. Xu X, Toselli PA, Russell LD, Seldin DC (1999) Globozoospermia in mice lacking the casein kinase II alpha′ catalytic subunit. Nat Genet 23(1):118–121. doi:10.1038/12729
21. Dominguez I, Mizuno J, Wu H, Song DH, Symes K, Seldin DC (2004) Protein kinase CK2 is required for dorsal axis formation in Xenopus embryos. Dev Biol 274(1):110–124. doi:10.1016/j.ydbio.2004.06.021
22. Mestres P, Boldyreff B, Ebensperger C, Hameister H, Issinger OG (1994) Expression of casein kinase 2 during mouse embryogenesis. Acta Anat (Basel) 149(1):13–20
23. Schneider HR, Reichert GH, Issinger OG (1986) Enhanced casein kinase II activity during mouse embryogenesis. Identification of a 110-kDa phosphoprotein as the major phosphorylation product in mouse embryos and Krebs II mouse ascites tumor cells. Eur J Biochem 161(3):733–738
24. Desmond ME, Schoenwolf GC (1986) Evaluation of the roles of intrinsic and extrinsic factors in occlusion of the spinal neurocoel during rapid brain enlargement in the chick embryo. J Embryol Exp Morphol 97:25–46
25. Huillard E, Ziercher L, Blond O, Wong M, Deloulme JC, Souchelnytskyi S, Baudier J, Cochet C, Buchou T (2010) Disruption of CK2beta in embryonic neural stem cells compromises proliferation and oligodendrogenesis in the mouse telencephalon. Mol Cell Biol 30(11):2737–2749. doi:10.1128/MCB.01566-09
26. Ulloa L, Diaz-Nido J, Avila J (1994) Depletion of catalytic and regulatory subunits of protein kinase CK2 by antisense oligonucleotide treatment of neuroblastoma cells. Cell Mol Neurobiol 14(5):407–414

27. Moreno FJ, Diaz-Nido J, Jimenez JS, Avila J (1999) Distribution of CK2, its substrate MAP1B and phosphatases in neuronal cells. Mol Cell Biochem 191(1–2):201–205
28. Villarroel-Campos D, Gonzalez-Billault C (2014) The MAP1B case: an old MAP that is new again. Dev Neurobiol. doi:10.1002/dneu.22178
29. Schael S, Nuchel J, Muller S, Petermann P, Kormann J, Perez-Otano I, Martinez SM, Paulsson M, Plomann M (2013) Casein kinase 2 phosphorylation of protein kinase C and casein kinase 2 substrate in neurons (PACSIN) 1 protein regulates neuronal spine formation. J Biol Chem 288(13):9303–9312. doi:10.1074/jbc.M113.461293
30. Chang CM, Chao CC (2013) Protein kinase CK2 enhances Mcl-1 gene expression through the serum response factor-mediated pathway in the rat hippocampus. J Neurosci Res 91(6):808–817. doi:10.1002/jnr.23212
31. Ceglia I, Flajolet M, Rebholz H (2011) Predominance of CK2alpha over CK2alpha′ in the mammalian brain. Mol Cell Biochem 356(1–2):169–175. doi:10.1007/s11010-011-0963-6
32. Diaz-Nido J, Mizuno K, Nawa H, Marshak DR (1994) Regulation of protein kinase CK2 isoform expression during rat brain development. Cell Mol Biol Res 40(5–6):581–585
33. Adelman JP, Maylie J, Sah P (2012) Small-conductance Ca2+-activated K+ channels: form and function. Annu Rev Physiol 74:245–269. doi:10.1146/annurev-physiol-020911-153336
34. Faber ES (2009) Functions and modulation of neuronal SK channels. Cell Biochem Biophys 55(3):127–139. doi:10.1007/s12013-009-9062-7
35. Brechet A, Fache MP, Brachet A, Ferracci G, Baude A, Irondelle M, Pereira S, Leterrier C, Dargent B (2008) Protein kinase CK2 contributes to the organization of sodium channels in axonal membranes by regulating their interactions with ankyrin G. J Cell Biol 183(6):1101–1114. doi:10.1083/jcb.200805169
36. Sanchez-Ponce D, Munoz A, Garrido JJ (2011) Casein kinase 2 and microtubules control axon initial segment formation. Mol Cell Neurosci 46(1):222–234. doi:10.1016/j.mcn.2010.09.005
37. Kimura R, Matsuki N (2008) Protein kinase CK2 modulates synaptic plasticity by modification of synaptic NMDA receptors in the hippocampus. J Physiol 586(13):3195–3206. doi:10.1113/jphysiol.2008.151894
38. Hu YM, Chen SR, Chen H, Pan HL (2014) Casein kinase II inhibition reverses pain hypersensitivity and potentiated spinal N-methyl-d-aspartate receptor activity caused by calcineurin inhibitor. J Pharmacol Exp Ther 349(2):239–247. doi:10.1124/jpet.113.212563
39. Pachuau J, Li DP, Chen SR, Lee HA, Pan HL (2014) Protein kinase CK2 contributes to diminished small conductance Ca -activated K channel activity of hypothalamic pre-sympathetic neurons in hypertension. J Neurochem. doi:10.1111/jnc.12758
40. Kanaan NM, Pigino GF, Brady ST, Lazarov O, Binder LI, Morfini GA (2013) Axonal degeneration in Alzheimer's disease: when signaling abnormalities meet the axonal transport system. Exp Neurol 246:44–53. doi:10.1016/j.expneurol.2012.06.003
41. Iimoto DS, Masliah E, DeTeresa R, Terry RD, Saitoh T (1990) Aberrant casein kinase II in Alzheimer's disease. Brain Res 507(2):273–280
42. Saitoh T, Iimoto D (1989) Aberrant protein phosphorylation and cytoarchitecture in Alzheimer's disease. Prog Clin Biol Res 317:769–780
43. Caron NS, Desmond CR, Xia J, Truant R (2013) Polyglutamine domain flexibility mediates the proximity between flanking sequences in huntingtin. Proc Natl Acad Sci U S A 110(36):14610–14615. doi:10.1073/pnas.1301342110
44. Atwal RS, Desmond CR, Caron N, Maiuri T, Xia J, Sipione S, Truant R (2011) Kinase inhibitors modulate huntingtin cell localization and toxicity. Nat Chem Biol 7(7):453–460. doi:10.1038/nchembio.582
45. Rebholz H, Zhou M, Nairn AC, Greengard P, Flajolet M (2013) Selective knockout of the casein kinase 2 in d1 medium spiny neurons controls dopaminergic function. Biol Psychiatry 74(2):113–121. doi:10.1016/j.biopsych.2012.11.013
46. Ferrer I, Martinez A, Blanco R, Dalfo E, Carmona M (2011) Neuropathology of sporadic Parkinson disease before the appearance of parkinsonism: preclinical Parkinson disease. J Neural Transm 118(5):821–839. doi:10.1007/s00702-010-0482-8

47. Eastwood K, Yin C, Bandyopadhyay M, Bidwai A (2011) New insights into the Orange domain of E(spl)-M8, and the roles of the C-terminal domain in autoinhibition and Groucho recruitment. Mol Cell Biochem 356(1–2):217–225. doi:10.1007/s11010-011-0996-x
48. Karandikar UC, Shaffer J, Bishop CP, Bidwai AP (2005) Drosophila CK2 phosphorylates Deadpan, a member of the HES family of basic-helix-loop-helix (bHLH) repressors. Mol Cell Biochem 274(1–2):133–139
49. Karandikar UC, Trott RL, Yin J, Bishop CP, Bidwai AP (2004) Drosophila CK2 regulates eye morphogenesis via phosphorylation of E(spl)M8. Mech Dev 121(3):273–286. doi:10.1016/j.mod.2004.01.008
50. Carneiro AC, Fragel-Madeira L, Silva-Neto MA, Linden R (2008) A role for CK2 upon interkinetic nuclear migration in the cell cycle of retinal progenitor cells. Dev Neurobiol 68(5):620–631. doi:10.1002/dneu.20613
51. Kramerov AA, Saghizadeh M, Pan H, Kabosova A, Montenarh M, Ahmed K, Penn JS, Chan CK, Hinton DR, Grant MB, Ljubimov AV (2006) Expression of protein kinase CK2 in astroglial cells of normal and neovascularized retina. Am J Pathol 168(5):1722–1736. doi:10.2353/ajpath.2006.050533
52. Pollreisz A, Afonyushkin T, Oskolkova OV, Gruber F, Bochkov VN, Schmidt-Erfurth U (2013) Retinal pigment epithelium cells produce VEGF in response to oxidized phospholipids through mechanisms involving ATF4 and protein kinase CK2. Exp Eye Res 116:177–184. doi:10.1016/j.exer.2013.08.021
53. Schmitt A, Wolfrum U (2001) Identification of novel molecular components of the photoreceptor connecting cilium by immunoscreens. Exp Eye Res 73(6):837–849. doi:10.1006/exer.2001.1086
54. Trojan P, Krauss N, Choe HW, Giessl A, Pulvermuller A, Wolfrum U (2008) Centrins in retinal photoreceptor cells: regulators in the connecting cilium. Prog Retin Eye Res 27(3):237–259. doi:10.1016/j.preteyeres.2008.01.003
55. Trojan P, Rausch S, Giessl A, Klemm C, Krause E, Pulvermuller A, Wolfrum U (2008) Light-dependent CK2-mediated phosphorylation of centrins regulates complex formation with visual G-protein. Biochim Biophys Acta 1783(6):1248–1260. doi:10.1016/j.bbamcr.2008.01.006
56. Yin X, Jedrzejewski PT, Jiang JX (2000) Casein kinase II phosphorylates lens connexin 45.6 and is involved in its degradation. J Biol Chem 275(10):6850–6856
57. Bildl W, Strassmaier T, Thurm H, Andersen J, Eble S, Oliver D, Knipper M, Mann M, Schulte U, Adelman JP, Fakler B (2004) Protein kinase CK2 is coassembled with small conductance Ca(2+)-activated K+ channels and regulates channel gating. Neuron 43(6):847–858. doi:10.1016/j.neuron.2004.08.033
58. Macica CM, Kaczmarek LK (2001) Casein kinase 2 determines the voltage dependence of the Kv3.1 channel in auditory neurons and transfected cells. J Neurosci 21(4):1160–1168
59. Boehning D, Moon C, Sharma S, Hurt KJ, Hester LD, Ronnett GV, Shugar D, Snyder SH (2003) Carbon monoxide neurotransmission activated by CK2 phosphorylation of heme oxygenase-2. Neuron 40(1):129–137
60. Jenkins PM, Zhang L, Thomas G, Martens JR (2009) PACS-1 mediates phosphorylation-dependent ciliary trafficking of the cyclic-nucleotide-gated channel in olfactory sensory neurons. J Neurosci 29(34):10541–10551. doi:10.1523/JNeurosci.1590-09.2009
61. Chiang LY, Sheppard DC, Bruno VM, Mitchell AP, Edwards JE Jr, Filler SG (2007) Candida albicans protein kinase CK2 governs virulence during oropharyngeal candidiasis. Cell Microbiol 9(1):233–245. doi:10.1111/j.1462-5822.2006.00784.x
62. Garat CV, Crossno JT Jr, Sullivan TM, Reusch JE, Klemm DJ (2010) Thiazolidinediones prevent PDGF-BB-induced CREB depletion in pulmonary artery smooth muscle cells by preventing upregulation of casein kinase 2 alpha′ catalytic subunit. J Cardiovasc Pharmacol 55(5):469–480. doi:10.1097/FJC.0b013e3181d64dbe
63. Kohlstedt K, Shoghi F, Muller-Esterl W, Busse R, Fleming I (2002) CK2 phosphorylates the angiotensin-converting enzyme and regulates its retention in the endothelial cell plasma membrane. Circ Res 91(8):749–756

64. Fyhrquist F, Metsarinne K, Tikkanen I (1995) Role of angiotensin II in blood pressure regulation and in the pathophysiology of cardiovascular disorders. J Hum Hypertens 9 Suppl 5:S19–S24
65. Kang NI, Yoon HY, Kim HA, Kim KJ, Han MK, Lee YR, Hwang PH, Soh BY, Shin SJ, Im SY, Lee HK (2011) Protein kinase CK2/PTEN pathway plays a key role in platelet-activating factor-mediated murine anaphylactic shock. J Immunol 186(11):6625–6632. doi:10.4049/jimmunol.1100007
66. Marjuki H, Scholtissek C, Yen HL, Webster RG (2008) CK2beta gene silencing increases cell susceptibility to influenza A virus infection resulting in accelerated virus entry and higher viral protein content. J Mol Signal 3:13. doi:10.1186/1750-2187-3-13
67. Kuronuma K, Sano H, Kato K, Kudo K, Hyakushima N, Yokota S, Takahashi H, Fujii N, Suzuki H, Kodama T, Abe S, Kuroki Y (2004) Pulmonary surfactant protein A augments the phagocytosis of Streptococcus pneumoniae by alveolar macrophages through a casein kinase 2-dependent increase of cell surface localization of scavenger receptor A. J Biol Chem 279(20):21421–21430. doi:10.1074/jbc.M312490200
68. Adenuga D, Yao H, March TH, Seagrave J, Rahman I (2009) Histone deacetylase 2 is phosphorylated, ubiquitinated, and degraded by cigarette smoke. Am J Respir Cell Mol Biol 40(4):464–473. doi:10.1165/rcmb.2008-0255OC
69. Iskratsch T, Lange S, Dwyer J, Kho AL, dos Remedios C, Ehler E (2010) Formin follows function: a muscle-specific isoform of FHOD3 is regulated by CK2 phosphorylation and promotes myofibril maintenance. J Cell Biol 191(6):1159–1172. doi:10.1083/jcb.201005060
70. Iskratsch T, Ehler E (2011) Formin-g muscle cytoarchitecture. Bioarchitecture 1(2):66–68. doi:10.4161/bioa.1.2.15467
71. Iskratsch T, Reijntjes S, Dwyer J, Toselli P, Degano IR, Dominguez I, Ehler E (2013) Two distinct phosphorylation events govern the function of muscle FHOD3. Cell Mol Life Sci 70(5):893–908. doi:10.1007/s00018-012-1154-7
72. Tan WQ, Wang JX, Lin ZQ, Li YR, Lin Y, Li PF (2008) Novel cardiac apoptotic pathway: the dephosphorylation of apoptosis repressor with caspase recruitment domain by calcineurin. Circulation 118(22):2268–2276. doi:10.1161/Circulation.107.750869
73. Force T (2008) The weakness of a big heart. Nat Med 14(3):244–245. doi:10.1038/nm0308-244
74. Hauck L, Harms C, An J, Rohne J, Gertz K, Dietz R, Endres M, von Harsdorf R (2008) Protein kinase CK2 links extracellular growth factor signaling with the control of p27(Kip1) stability in the heart. Nat Med 14(3):315–324. doi:10.1038/nm1729
75. Eom GH, Cho YK, Ko JH, Shin S, Choe N, Kim Y, Joung H, Kim HS, Nam KI, Kee HJ, Kook H (2011) Casein kinase-2alpha1 induces hypertrophic response by phosphorylation of histone deacetylase 2 S394 and its activation in the heart. Circulation 123(21):2392–2403. doi:10.1161/Circulation.110.003665
76. Murtaza I, Wang HX, Feng X, Alenina N, Bader M, Prabhakar BS, Li PF (2008) Down-regulation of catalase and oxidative modification of protein kinase CK2 lead to the failure of apoptosis repressor with caspase recruitment domain to inhibit cardiomyocyte hypertrophy. J Biol Chem 283(10):5996–6004. doi:10.1074/jbc.M706466200
77. Winter B, Kautzner I, Issinger OG, Arnold HH (1997) Two putative protein kinase CK2 phosphorylation sites are important for Myf-5 activity. Biol Chem 378(12):1445–1456
78. Kablar B, Rudnicki MA (2000) Skeletal muscle development in the mouse embryo. Histol Histopathol 15(2):649–656
79. Iyengar AS, Loupe JM, Miller PJ, Hollenbach AD (2012) Identification of CK2 as the kinase that phosphorylates Pax3 at Ser209 in early myogenic differentiation. Biochem Biophys Res Commun 428(1):24–30. doi:10.1016/j.bbrc.2012.09.141
80. Shi X, Seldin DC, Garry DJ (2012) Foxk1 recruits the Sds3 complex and represses gene expression in myogenic progenitors. Biochem J 446(3):349–357. doi:10.1042/BJ20120563
81. Heuss D, Klascinski J, Schubert SW, Moriabadi T, Lochmuller H, Hashemolhosseini S (2008) Examination of transcript amounts and activity of protein kinase CK2 in muscle lysates of different types of human muscle pathologies. Mol Cell Biochem 316(1–2):135–140. doi:10.1007/s11010-008-9819-0

82. Tornavaca O, Sarro E, Pascual G, Bardaji B, Montero MA, Salcedo MT, Plana M, Lopez-Hellin J, Itarte E, Meseguer A (2011) KAP degradation by calpain is associated with CK2 phosphorylation and provides a novel mechanism for cyclosporine A-induced proximal tubule injury. PLoS One 6(9):e25746. doi:10.1371/journal.pone.0025746
83. Son YK, Lee SM, An WS, Kim KH, Rha SH, Rho JH, Kim SE (2013) The role of protein kinase CK2 in cyclosporine-induced nephropathy in rats. Transplant Proc 45(2):756–762. doi:10.1016/j.transproceed.2012.02.050
84. Yamada M, Katsuma S, Adachi T, Hirasawa A, Shiojima S, Kadowaki T, Okuno Y, Koshimizu TA, Fujii S, Sekiya Y, Miyamoto Y, Tamura M, Yumura W, Nihei H, Kobayashi M, Tsujimoto G (2005) Inhibition of protein kinase CK2 prevents the progression of glomerulonephritis. Proc Natl Acad Sci U S A 102(21):7736–7741. doi:10.1073/pnas.0409818102
85. Schermer B, Hopker K, Omran H, Ghenoiu C, Fliegauf M, Fekete A, Horvath J, Kottgen M, Hackl M, Zschiedrich S, Huber TB, Kramer-Zucker A, Zentgraf H, Blaukat A, Walz G, Benzing T (2005) Phosphorylation by casein kinase 2 induces PACS-1 binding of nephrocystin and targeting to cilia. EMBO J 24(24):4415–4424. doi:10.1038/sj.emboj.7600885
86. Boskey AL, Doty SB, Kudryashov V, Mayer-Kuckuk P, Roy R, Binderman I (2008) Modulation of extracellular matrix protein phosphorylation alters mineralization in differentiating chick limb-bud mesenchymal cell micromass cultures. Bone 42(6):1061–1071. doi:10.1016/j.bone.2008.01.016
87. Jones NC, Farlie PG, Minichiello J, Newgreen DF (1999) Detection of an appropriate kinase activity in branchial arches I and II that coincides with peak expression of the Treacher Collins syndrome gene product, treacle. Hum Mol Genet 8(12):2239–2245
88. Ulges A, Klein M, Gerlitzki B, Hoffmann M, Grebe N, Staudt V, Stergiou N, Bohn T, Brühl TJ, Muth S, Yurugi H, Rajalingam K, Schild H, Schmitt E, Bopp T. Reuter S, Bellinghausen I, Probst HC, Tuettenberg A, Hahn S, Jonuleit H, Reißig S, Waisman A, Haben I, Breloer M, Zipp F, Beilhack A, Buchou T, Filhol-Cochet O, Boldyreff B (2015) Protein kinase CK2 enables regulatory T cells to suppress excessive TH2 responses in vivo. Nat Immunol Jan 19. doi:10.1038/ni.3083. [Epub ahead of print]

Protein Kinase CK2: A Window into the Posttranslational Regulation of the E(spl)/HES Repressors from Invertebrates and Vertebrates

Adam T. Majot, Tyler R. Sizemore, Mohna Bandyopadhyay, Lucas M. Jozwick, and Ashok P. Bidwai

Abstract The basic helix-loop-helix (bHLH) repressors encoded by the *Drosophila Enhancer of split Complex* (*E(spl)C*) are the terminal effectors of Notch signaling, a pathway that is highly conserved through Metazoa. Although the E(spl) proteins are structurally conserved, one region that exhibits length and sequence heterogeneity is the C-terminal domain (CtD), which links the b/HLH domains to the terminal WRPW tetrapeptide, facilitating recruitment of the corepressor Groucho. Consequently, the CtD has been largely thought to act as a nonfunctional linker. However, studies are revealing that this region is not only key to controlling E(spl) repressor activity (*cis*-inhibition) but is subject to sophisticated regulation through posttranslational modifications (PTM). These modifications are mediated by protein kinases, phosphatase(s), and accessory factors, which together regulate phospho-occupancy, conferring spatial and temporal control over E(spl) protein activities and levels. We suggest that E(spl)M8 is a paradigm for understanding the regulation of mammalian E(spl) homologues by PTM. In the case of E(spl)M8, repressor activity first requires multisite phosphorylation led by CK2, with steady state control provided by the phosphatase PP2A. The later participation of additional kinases would activate a phosphodegron, enabling timely clearance of the protein. Controlled activation and deactivation may both be essential for repeated rounds of Notch signaling, employing different E(spl) repressors. This mode of regulation likely impacts a preponderance of E(spl) members, underscoring its importance to Notch signaling. PTM therefore imposes greater functional diversity among the E(spl) proteins, which are themselves differentially expressed during development. Aberrant posttranslational modification of the human E(spl) homologues, the HES proteins, may underlie diverse developmental disorders and cancer, both of which have been linked to defects in Notch signaling.

A.T. Majot • T.R. Sizemore • M. Bandyopadhyay • L.M. Jozwick • A.P. Bidwai (✉)
Department of Biology, West Virginia University, Life Sciences Building,
53 Campus Drive, Morgantown, WV 26506, USA
e-mail: abidwai@wvu.edu

K. Ahmed et al. (eds.), *Protein Kinase CK2 Cellular Function in Normal and Disease States*, Advances in Biochemistry in Health and Disease 12, DOI 10.1007/978-3-319-14544-0_6

Keywords CK2 • MAPK • Phosphorylation • Notch signaling • Neurogenesis • bHLH • E(spl) • HES • Repressors

1 Protein Kinase CK2

CK2 is a Ser/Thr protein kinase that has been highly conserved throughout eukaryotic evolution [1–3]. This enzyme is composed of catalytic (α) and non-catalytic (β) subunits, which associate both in vivo and in vitro to form the tetrameric α2β2 holoenzyme. On its own, the CK2α subunit is competent to recognize and phosphorylate most target proteins [4], whereas CK2β, in large part, modulates catalytic activity. The substrate specificity of CK2, an acidophilic protein kinase, is well defined and best described as S/T-E/D-x-D/E [5, 6]; the presence of additional Asp/Glu residues either N- or C-terminal to the phosphoacceptor(s) further enhances phosphorylation of target proteins. This recognition site is invariably devoid of the basic amino acids Lys/Arg, and the presence of these residues within the consensus abrogates targeting by CK2. Moreover, the presence of phospho-Ser (pSer) or phospho-Thr (pThr) C-terminal to the primary phosphoacceptor also stimulates CK2 activity toward a particular protein [7], revealing that this enzyme can act in cooperation with other protein kinases (hierarchical phosphorylation) or can progressively phosphorylate a target protein on its own. An excellent and perhaps extreme example is Nopp140, a nuclear/cytoplasmic shuttling protein that is phosphorylated at >70 sites by CK2 [8]. It should also be noted that in some instances, CK2 has been found to target Tyr residues [9, 10], raising the prospect that this enzyme may, in fact, regulate a broader spectrum of proteins.

Since its discovery in 1953 [11, 12], many targets of CK2 have been identified by biochemistry, in vivo studies, and through computational analyses of genome sequences, the latter greatly facilitated by its unique recognition site in combination with the observation that acidic micro-domains are typically solvent accessible. These studies reveal that the CK2 phosphoproteome is extraordinarily large [13–16] and encompasses virtually all aspects of cell biology such as cell proliferation [17], cell-cycle control [18–20], regulation of transcription [21, 22], heterochromatin formation [23], cell signaling and animal development (see below), the circadian clock [24–26], and possibly in learning and memory [27]. Importantly, CK2 is the etiological agent for theileriosis, a B- and T-cell lymphoproliferative disorder in cattle infected with the parasite *Theileria parva* [28–31], and accordingly targeted overexpression of CK2 in the mouse model elicits the development of lymphomas [32, 33]. In addition, CK2 has been closely linked both to the development of tumors and to metastasis [34–37], reasons for which substantial efforts are underway to develop specific inhibitors of this enzyme. Given the large number of interactions in vivo and its roles in cell proliferation and cancer, loss of CK2 elicits cell lethality in all unicellular/metazoan organisms that have been experimentally tested, such as yeast [17, 38, 39], fruit flies [26, 40], and mammals [41].

2 CK2 Regulation of Signaling and Development

Given its ability to target diverse cell biological processes and act as an oncogene, significant efforts have sought to understand how CK2 regulates cell signaling and animal development. Earlier efforts revealed that signaling by insulin and epidermal growth factor (EGF) were associated with the activation of CK2 [42, 43] and that this enzyme directly regulates wingless/Wnt signaling [44–47] and the Hedgehog pathway [48]. Expectedly, CK2 regulates animal development, which hinges on the coordinated activities of a handful of signaling pathways. In addition, this enzyme plays a vital role in embryonic development via the targeting of proteins such as Antennapedia [49], Engrailed [50], Cut [51], Odd Skipped [52], β-Catenin [53], and Orb [54]. Despite its importance to eukaryotes, the "Gordian knot" in our understanding of CK2 is its regulation in vivo. No second messengers have so far been identified that modulate CK2 activity in vivo, and despite being thought to be constitutively active, this enzyme does not phosphorylate its targets in a constitutive manner. One such example is yeast cdc37, which is phosphorylated in precise coordination with the cell cycle [19]. The possibility remains open that CK2 is regulated by the dynamic formation of molecular complexes with its targets in coordination with scaffolds, other kinases, and phosphatases.

3 *Drosophila* CK2 and the Notch Signaling Pathway

Studies in our lab using the fruit fly *Drosophila* as a model organism are revealing the importance of CK2 to the Notch signaling pathway. Here we first briefly review the structure and composition of *Drosophila* CK2 and later discuss previous and ongoing studies that are beginning to reveal a level of complexity in Notch signaling that was previously unrecognized. We focus on the Notch pathway in *Drosophila*, a preeminent genetic model that has been instrumental in the identification of the core components and regulators of this pathway and has laid the foundations for our understanding of the mechanisms of Notch signaling, its importance to the development of other animals, and its association with disease states (see below).

3.1 Drosophila CK2

Drosophila CK2 was first purified from embryos and, like its mammalian counterpart, was shown to be a tetrameric α2β2 holoenzyme with the ability to form filamentous structures [55, 56]. It has, in fact, been suggested that these filaments may represent an inactive state of CK2, perhaps hinting at a mode of regulation of this enzyme in vivo. Unlike mammals, the *Drosophila* genome encodes for a single CK2α subunit, whereas it contains multiple CK2β subunits that are encoded by independent genes. These include the dominant CK2β subunit and the β-like protein Stellate (Ste) [57, 58], both X-linked, and the autosomal β′ [59] and SSL [60] subunits. Of these, expression of the CK2β gene appears the most complex; it gives rise to multiple splice variants, which encode protein isoforms with distinct C-terminal tails that differ in their ability to rescue the effects of loss of the CK2β gene [61, 62]. To date, genetic analysis has been conducted on the CK2α, CK2β, and Stellate. Loss of CK2α elicits lethality at the first larval stage [26], whereas that of CK2β is embryo lethal [25, 62]. The *Stellate* locus appears to be a unique case. Normally, this gene is maintained in a silent state due to silencing by the Y-linked *Suppressor of Stellate* (*Su(Ste)*) locus (also known as *crystal*), and accordingly, in XO males (lacking the Y chromosome), *Stellate* is massively derepressed in the testis, and the protein accumulates as star-shaped crystals that disrupt spermatocytes and elicit sterility. In all cases, the β′, SSL, and Stellate proteins appear competent to associate with CK2α in vitro and form an active tetrameric holoenzyme. However, it remains unknown if these alternative isoforms alter CK2 substrate specificity in a tissue-specific manner.

3.2 Notch Signaling

Identified almost a hundred years ago, Notch is a ubiquitous metazoan signaling pathway whose components and mechanisms of signaling are highly conserved [63–69]. Earlier studies revealed that Notch was vital for neurogenesis, but it has since been realized that this pathway regulates diverse aspects of animal development, including that of the early embryo, as well as the heart [70], vertebrae [71–73], blood vessels [74], and immune system [75], and also plays roles in the adult animal via the regulation of circadian rhythms [76, 77].

The core components of the *Drosophila* Notch signaling pathway consist of the Notch receptor, its ligands Delta or Serrate, the transcription factor Suppressor of Hairless (Su(H)), and its transcriptional targets, the genes of the *Enhancer of split Complex* (*E(spl)C*, Fig. 1f). The *E(spl)C* encodes seven basic helix-loop-helix (bHLH) repressors (Mδ, Mγ, Mβ, M3, M5, M7, and M8) and their corepressor Groucho [78–81], all with clearly identified structural and functional homologues in vertebrates, which have been named the Hairy-E(spl)-like (HES) or the Hairy-E(spl)-related (HER) proteins [82–84]. While deceptively simple with regard to its core components, studies over the years have uncovered that Notch signaling is dynamically regulated by glycosylation, protein traffic and endocytosis [85, 86], ubiquitylation [87], and proteolytic cleavage, and these have revealed an astonishing level of biological complexity [85]. One area that has been considered the simpler part of this pathway is the E(spl)/HES/HER proteins that are the terminal readout(s) of Notch signaling and which have been considered to function in a largely dosage-dependent manner.

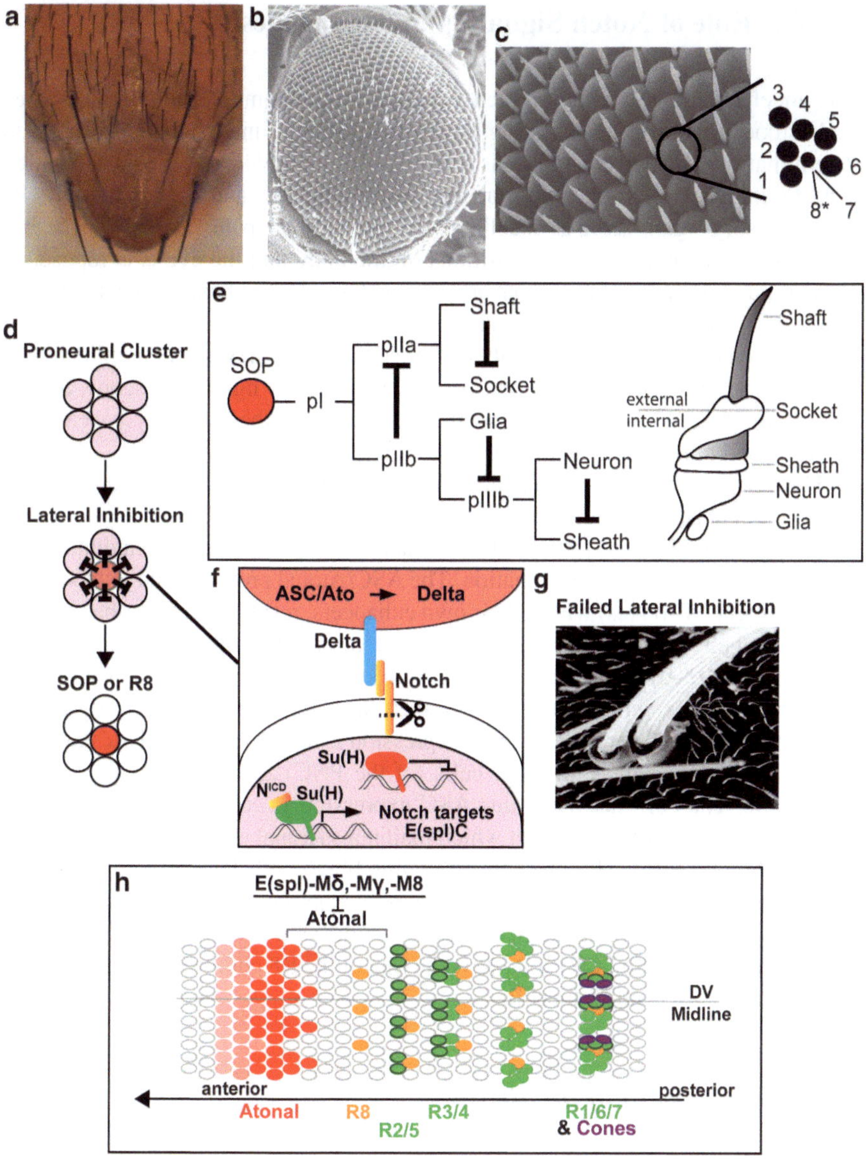

Fig. 1 Notch signaling during fly neurogenesis. (**a**) The adult notum is marked with characteristic positioning of macro- and microchaetae. (**b**, **c**) The adult retina is comprised of ~700 hexagonally arrayed facets, each comprised of eight photoreceptors. The R7 rhabdomere sits directly above the R8. (**d**) R8 and SOP specification requires Notch signaling to restrict proneural clusters to single cells that maintain neural competency. (**e**) Bristle formation requires a series of asymmetric cell divisions. (**f**) During lateral inhibition, presumptive R8s and SOPs stimulate Notch signaling in neighboring cells. Resultant E(spl)C expression diminishes proneural gene expression. (**g**) Failed lateral inhibition results in multiple SOPs and twinned bristles. (**h**) Retinal differentiation requires Atonal, whose expression is progressively narrowed to individualized cells, R8s, by Notch signaling. The R8 recruits R1–R7 photoreceptors through stereotyped, inductive signaling. Notch is also required for rotation around the DV midline and specification of the R7 and cones

4 The Role of Notch Signaling in Neurogenesis

The mechanisms by which Notch regulates development are, perhaps, best understood in neurogenesis during which this pathway mediates binary cell fate determination. The two external sensory organs in *Drosophila* that have been widely used to analyze Notch signaling are the sensory bristles and the compound eye. These sensory organs serve as excellent readouts of Notch pathway activity because the bristles occupy stereotyped positions on the body and the eye is composed of ommatidia (unit eyes or facets) that are arranged in a precise hexagonal pseudocrystalline array (see Fig. 1a, b). Importantly, changes in Notch pathway activity perturb the arrangement/position and number of bristles and the hexagonal architecture of the eye. Both of these perturbations are easy to detect by light microscopy, and the underlying defects in cell type specification can then be precisely revealed by sophisticated molecular/genetic and immunological assays.

The process of neural cell fate determination begins with the expression of bHLH transcriptional activators encoded by the genes of the *achaete-scute complex* (*ASC*, [88, 89]) or *atonal* [90, 91]. While the ASC activators drive bristle formation, Atonal is the primary factor for eye formation. The ASC/Atonal proteins bind to hexameric DNA elements (E-boxes) within their own enhancers, resulting in a feedback stimulation of their expression, a process called proneural autoactivation or "self-stimulation" [92–97]. This initial expression of ASC/Atonal occurs in groups of equipotential cells, which have been termed the "proneural clusters" (PNCs, see Fig. 1d). Later, however, ASC/Atonal expression is maintained in only a single cell from each PNC, whereas all other cells (of the PNC) are forced to adopt a non-neural fate. The cell that maintains ASC expression is called the bristle sensory organ precursor (SOP), while that which maintains Atonal expressivity in the eye is the R8 photoreceptor. The mechanism of this binary cell fate determination also drives oogenesis, myogenesis, hematopoiesis, etc., processes that are Notch dependent. Because the nature of Notch signaling is somewhat distinct between the bristle (selection of the bristle SOP) and eye (selection of the R8 photoreceptor), we briefly discuss differences and similarities.

4.1 Development of the Bristle

Each bristle is composed of five cell types (neuron, glial, sheath, shaft, and socket) that arise from successive divisions of a single SOP (Fig. 1e, adapted from [98]). Following its selection from an ASC-positive PNC, the SOP goes on to form the pI cell, which then undergoes three asymmetric cell divisions, all of which are dependent on Notch signaling. The cumulative result of these asymmetric divisions is the specification of four distinct cell types (see above) that together form a functional bristle sensory organ. Consistent with its reiterative role in the bristle, loss of Notch elicits distinct phenotypes. First, loss of Notch in the PNC (prior to SOP selection) results in the formation of supernumerary (juxtaposed) SOPs (Fig. 1g), which manifest in the adult as ectopic bristles. Second, loss of Notch during division of the pI cell results in the specification of two pIIb cells, which go on to form duplicated neurons and sheath cells. Because of the loss of the pIIa cell fate, the shaft and socket cells fail to form, which phenotypically results in the loss of the external structures (Fig. 1e). Third, loss of Notch after formation of the pIIa cell results in the replacement of the socket cell with a shaft cell; this manifests as split bristles. Fourth, loss of Notch after formation of the pIIIb cell results in replacement of the sheath cell with a neuron; this manifests as a "dual neuron" phenotype. Numerous studies over the years have revealed that ectopic expression of any E(spl) member leads to the loss of the bristle SOP and its progeny cell types. However, technical challenges have stymied a determination of which specific E(spl) member (or combination) mediates the effects of endogenous Notch signaling at each distinct stage of bristle ontogeny. Of relevance to the multiple roles for Notch in this sensory organ, targeted knockdown of CK2 has been shown to elicit the specification of ectopic bristles as well as split and missing bristles [99], phenotypes consistent with roles for this kinase in SOP selection from a PNC, and the asymmetric divisions of the pIIa and pIIIb cells, respectively (summarized in Fig. 1e). The relevance of these effects of CK2 to the E(spl) repressors is discussed below.

4.2 Development of the Eye

The *Drosophila* eye is composed of ~750 facets, each of which contains 8 photoreceptors (called retinula cells R1–R8, numbered according to their clockwise orientation in the adult eye; see Fig. 1b, c). Facets are patterned from a wave of differentiation, termed the morphogenetic furrow (MF), which sweeps from posterior to anterior across the larval eye disc. The passage of the MF is accompanied by the induction of the proneural transcriptional activator Atonal, whose expression is required for both continued MF propagation and differentiation of the R8 cell type, which is the first photoreceptor specified in each facet. The R8, once specified, sequentially recruits to itself all other cell types (R1–R7, cones, pigment cells) of the facet. Thus, the initial positioning of R8s in the MF is critical to proper patterning of the adult eye superstructure. MF movement across the eye disc occurs over a period of ~48 h, thereby providing a spatial and temporal perspective on specification of R8 cells and all secondary cell types.

R8 specification requires two modes of Notch signaling that are separated by time, space, and the manner of signaling [100]. Hedgehog signaling first elicits Atonal expression in a broad swath of cells (akin to PNCs) at the anterior margin of the MF (Fig. 1h), whereupon Notch is required for induction of high-level Atonal [101, 102]. This phase of Notch signaling does not involve expression of the E(spl) proteins, as it is Su(H) independent [103]. In contrast, at the posterior margin of the MF, Notch elicits the expression of E(spl) proteins, which then restrict Atonal expressivity to a single cell—the future R8 photoreceptor. This process contrasts to that which produces the bristle in that the R8 is mitotically arrested and relies upon recruitment of surrounding cells to construct facets. Consequently, loss of the *atonal* gene leads to the absence of all R8 cells and abolishes eye development. In addition to R8s, lateral inhibition by E(spl) proteins is used reiteratively to specify R1–R7 cell types, cones, and the interommatidial bristle (IOB) whose developmental pathway shares many, but not all, of the genetic determinants that also specify the bristles [104, 105].

Consistent with this model for biphasic Notch signaling, loss of E(spl) proteins impairs refinement of the Atonal-positive PNCs and results in the specification of extra juxtaposed R8 cells (Fig. 1d). This over-specification of the R8 cells results in a "rough eye" phenotype, an outcome also seen with the overexpression of Atonal [91]. These findings reveal that a precise balance between the activities of Atonal versus E(spl) proteins is vital for proper R8 birth and for eye development. As in the case of bristle development (see above), knockdown of CK2 elicits ectopic R8s and rough eyes [99], revealing that this protein kinase plays a similar role in R8 selection.

4.3 Selection of the Bristle SOP and R8 Photoreceptor

The mechanisms by which Notch signaling resolves a PNC to generate a single SOP or R8 are strikingly similar. During this process, Notch drives the determination of binary cell fates (SOP vs. non-SOP and R8 vs. non-R8). This process called lateral inhibition [106, 107] occurs in an E(spl)-dependent manner [66, 108, 109]. Specifically, the future SOP/R8 expresses the Notch ligand Delta at a level sufficient to activate Notch in adjacent cells of a PNC (Fig. 1f). In these responding cells, expression of *E(spl)C* ensues, and these bHLH proteins then inhibit ASC/Atonal activities. Therefore, the broad expression of ASC/Atonal becomes progressively refined such that a single ASC/Atonal-positive cell remains from each PNC. Accordingly, loss of Notch or its downstream components leads to the emergence of extra (juxtaposed) SOPs and neuronal hypertrophy. Moreover, forced overexpression of E(spl) repressors elicits dominant inhibition of the SOP fate, revealing that these bHLH proteins are the terminal effectors of this pathway. The conservation of the E(spl) proteins, and their importance to diversifying Notch signaling, is illustrated by the conservation of the HES/HER bHLH proteins in vertebrates. It is, however, important to note that E(spl) proteins are unlikely to act in a redundant manner, because individual members exhibit tissue-specific expression patterns [110] that reflect different architecture of the binding sites for Su(H) [111].

Although the mode of signaling within a PNC occurs in a juxtacrine manner, it should be noted that Notch signaling can occur over several cell diameters, mediated through filopodia [112, 113].

5 CK2 Regulation of Notch Signaling

Extensive studies have explored the mechanisms by which E(spl)/HES antagonize SOP and R8 fate through the repression of ASC and Atonal. Stereotypical E(spl)/HES bHLH repressors contain a highly conserved architecture consisting of a bHLH, Orange (a second HLH), and an ultimate tetrapeptide WRPW. The N-terminus of the bHLH is comprised of a clustering of basic residues that mediate E(spl)-DNA interaction. The bHLH also facilitates homo- and/or heterodimerization between E(spl) proteins. From the bHLH, a linker of variable length leads to a second HLH, referred to as the Orange domain. Orange promotes interaction with non-E(spl)/HES proteins [114], and WRPW mediates recruitment of the corepressor Groucho. Between the Orange and WRPW is a region of length and compositional heterogeneity, which we call the C-terminal domain (CtD). As discussed below, the CtD appears key to the repressive activities of certain E(spl) members, and within it is a resident phosphorylation domain (P-domain) that conserves sites for several protein kinases (see below).

Despite almost 30 years of investigations on repression by E(spl) proteins, no universal mechanism has emerged. It was initially proposed that E(spl) proteins bind DNA and elicit repression of target genes upon recruitment of Groucho [115], target gene recognition being conferred by DNA elements called the N-boxes [116, 117]. However, this model fell out of favor because, unlike the genes of the *ASC*, no N-boxes have been identified in the enhancers of *atonal* [118], a gene vital for eye development [90, 91] and whose expression pattern is refined by E(spl) proteins [100]. An alternative and prevailing model proposed that protein–protein interactions regulate repression, dispensing the role for DNA-binding [119, 120]. In this model, E(spl) proteins directly interact with enhancer-bound ASC/Atonal proteins, and this (physical) interaction is mediated through the Orange domain that forms the second HLH motif. This model is strongly supported by the findings that E(spl) proteins directly interact with ASC/Atonal in biochemical and yeast two-hybrid assays [121–125]. The observation that CK2 phosphorylates E(spl)-M8, E(spl)-M7, and E(spl)-M5 [126] has revealed the importance of posttranslational modification (PTM) in the regulation of these Notch effectors, raised the prospect of nonredundancy, and begun to unravel the mechanism by which a dominant mutation in the *m8* gene (*E(spl)D*) abolishes eye development.

6 Relevance of CK2 Regulation of M8 and the Eye Defects of the *E(spl)D* Mutation

The *E(spl)D* mutation, identified in 1956 [127, 128] based on its ability to abolish eye formation, truncates the M8 protein (referred to as M8*) and deletes the CtD (Fig. 2), including the WRPW motif [124, 129]. Despite not binding Groucho, M8*

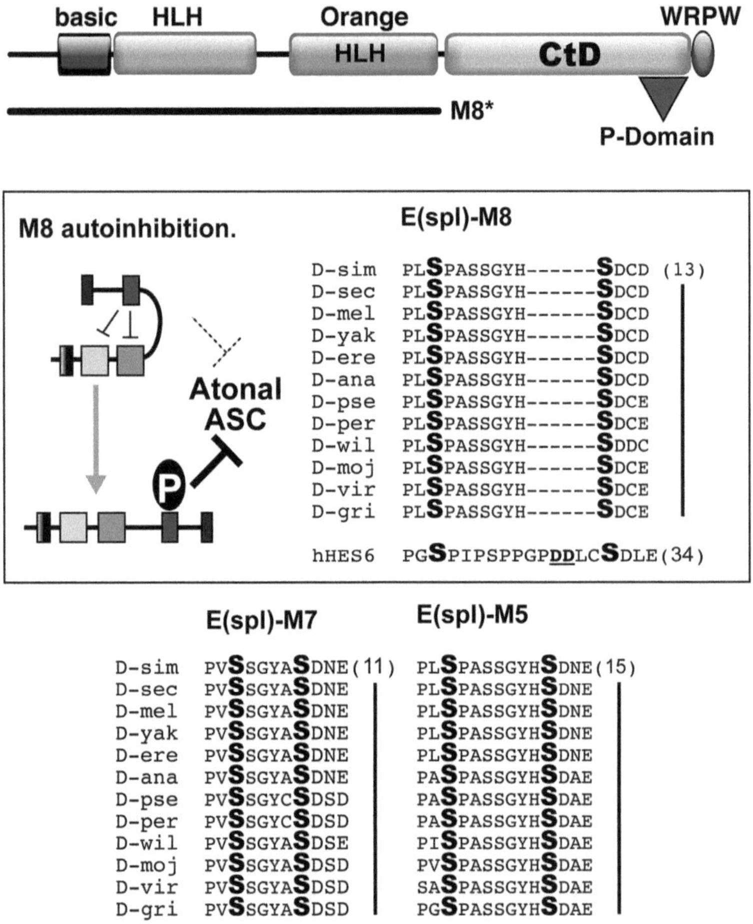

Fig. 2 Conserved sites for CK2 in E(spl)-M8, E(spl)-M7, and E(spl)-M5. (*Top*) Schematic of the E(spl) proteins and location of functional domains. Note that the Orange domain is also predicted to form an HLH motif. The M8* protein is encoded by the dominant mutation *E(spl)D*. Alignment of M8, M7, and M5 from 12 *Drosophila* species illustrating the conserved CK2 and MAPK sites (*highlighted in bold*). Note that human HES6 (hHES6) also contains a similarly located P-domain that conserves the CK2 and MAPK sites. Inset next to M8 sequences shows a model for regulation of repressor activity by M8. The number in parenthesis after the sequences denotes n residues between the P-domain and the C-terminal WRPW motif. *Drosophila* species are abbreviated as *D. simulans* (D-sim), *D. sechellia* (D-sec), *D. melanogaster* (D-mel), *D. yakuba* (D-yak), *D. erecta* (D-ere), *D. ananassae* (D-ana), *D. pseudoobscura* (D-pse), *D. persimilis* (D-per), *D. willistoni* (D-wil), *D. mojavensis* (D-moj), *D. virilis* (D-vir), and *D. grimshawi* (D-gri)

displays exacerbated repression of Atonal, thereby blocking R8 birth and eye development. Such a developmental defect could not be conferred simply by forced overexpression of full-length M8 [125, 130] or by a variant lacking just the WRPW motif [131]. It was presciently proposed by the late Dr. José Campos-Ortega that the region between the Orange domain and WRPW may have a regulatory influence on E(spl)–M8 activity [130], but the reasons for the hyperactivity of M8* remained unresolved. We had identified that CK2 phosphorylates M8/M5/M7 (see Fig. 2) at a highly conserved site in the region deleted in M8* [126]. Remarkably, in the case of M8, replacement of the CK2 phosphoacceptor (Ser^{159}) with Asp resulted in a form of the protein that mimicked the eye defects of M8* (*E(spl)D*), and both M8-S^{159}D and M8* displayed equivalent interactions with Atonal that were not seen with wild-type M8 or its CK2 refractory variant M8-S^{159}A [125]. We thus proposed that the CtD of M8 auto-inhibits the Orange domain (see inset next to M8 in Fig. 2); phosphorylation by CK2 alleviates this interaction to then permit repressor activity. The *E(spl)D* mutation thus removes this critical auto-inhibitory (control) region, resulting in a protein that is CK2 independent and hyperactive. Indeed, forced overexpression of just the 56 amino acid CtD peptide rescues the eye defects of *E(spl)D*, whereas a CtD peptide that replaces the CK2 site (Ser^{159}) with Asp was nonfunctional [132]. These findings provided strong evidence that intramolecular interactions regulate M8 activity in a CK2-responsive manner in vivo.

The importance of the CK2 site is highlighted by its invariant positioning in E(spl)-M8, E(spl)-M5, and E(spl)-M7 proteins from 12 *Drosophila* species that diverged ~50×10^6 years (myr) ago (Fig. 2). This remarkably conserved P-domain resides in the CtD, which is otherwise variable in length and sequence between the individual E(spl) proteins (data not shown). The subsequent finding that a similar P-domain is present in human HES6 (hHES6, see Fig. 2), which is regulated in an identical manner by CK2 [133], renders it likely that a highly conserved regulatory mechanism controls E(spl)/HES activity. While these studies established the importance of CK2 to Notch pathway output, it was not applicable to other members such as Mδ, Mβ, and M3, because these proteins lacked an identifiable CK2 consensus site.

7 Four *Drosophila* E(spl) Members Are Targeted by CK2

To better understand the importance of PTM in the regulation of E(spl) proteins, we conducted detailed bioinformatics analyses, which revealed that a fourth member E(spl)-Mγ contains CK2 sites that appear significantly different from those in M5, M7, and M8. This is best highlighted by sequence alignments of E(spl)-Mγ from 12 species representing ~50 myr of *Drosophila* evolution (Fig. 3a, b). Remarkably, E(spl)-Mγ from all 12 species harbor three highly conserved consensus sites for CK2 (Fig. 3a). Two of these predicted sites, Thr^{45} and Ser^{48}, are located within the loop of the HLH domain adjacent to the N-terminal basic domain. Of these, Ser^{48} appears to be the stronger site as it meets the consensus for CK2, while Thr^{45} may potentially be targeted by CK2 in a hierarchical manner, but only upon prior phosphorylation at Ser^{48}. It is of interest to note that of the seven E(spl) bHLH proteins in *Drosophila*, E(spl)-Mγ is the sole member that contains potential site(s) for CK2 within the HLH domain (data not shown). The third and strongest predicted CK2

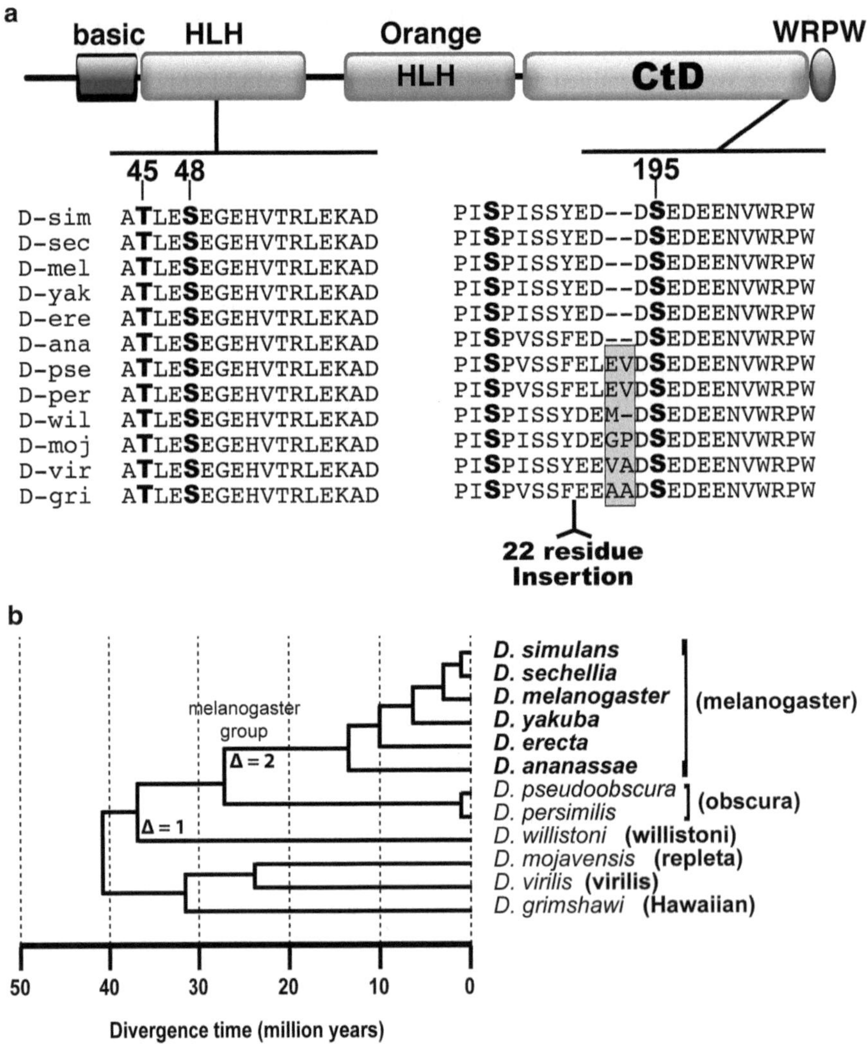

Fig. 3 Conserved sites for CK2 in E(spl)-Mγ. (**a**) Schematic of the Mγ protein and location of functional domains. Alignment of Mγ from 12 *Drosophila* species illustrating the conserved CK2 and MAPK sites (*highlighted in bold*); the C-terminal CK2 site meets the strict consensus. The *gray box* denotes the insertion of one/two amino acids in the Mγ protein from select *Drosophila* species. Note that all Mγ isoforms harbor an insertion of 22 residues that separates the CK2 and MAPK sites. (**b**) Phylogenetic relationship between the *Drosophila* species analyzed in panel A. Note loss of the 2 residues in the melanogaster group

site (Ser^{195}) is located in close proximity to the C-terminal WRPW motif (Fig. 3a), and once again it is a unique feature of E(spl)-Mγ isoforms. In addition, this site is notable because Ser^{195} is flanked by a number of acidic amino acids (Asp/Glu), a hallmark of "high-likelihood" targets of CK2. Although the E(spl)-Mγ protein from several *Drosophila* species harbor the insertion of one/two amino acids N-terminal to Ser^{195}, in no case is the $DS^{195}EDEE$ motif perturbed (see Fig. 3a, the numbering is based on the *D. melanogaster* sequence). Biochemical analysis is revealing that CK2 primarily targets Ser^{195} (Jozwick and Bidwai, unpublished), and the evaluation of the consequences of this modification on E(spl)-Mγ activity in vivo is currently underway. Thus, the CK2 site is a conserved feature of the P-domain in four E(spl) proteins. In addition, the P-domains of M8, M7, M5 (Fig. 2), and Mγ (Fig. 3) conserve additional Ser residues, one of which meets the consensus for MAPK, an enzyme implicated in R8 selection (Sect. 10, see below).

The known evolutionary relationships between the 12 species representing seven groups of fruit flies [134] reveal that this insertion/deletion of the two amino acids closely correlates to specific branches of the *Drosophila* family (Fig. 3b). This comparison reveals that the two-residue insertion was a feature of the ancestral E(spl)-Mγ protein in the Hawaiian group (*D. grimshawi*), virilis group (*D. virilis*), repleta group (*D. mojavensis*), and obscura group (*D. pseudoobscura* and *D. persimilis*), with a one residue deletion unique to the willistoni group (*D. willistoni*). Both residues were lost ~13 myr ago with the emergence of the melanogaster group. That this region of E(spl)-Mγ has been subject to insertion/deletion suggests that the one/two-residue insertion (in the aforementioned ancestral species) is unlikely to alter the overall structure of the protein or is vital for function.

8 Molecular Synteny of the *E(spl)C* Across *Drosophila* Species

The *E(spl)C* in *Drosophila* encompasses ~60 kb. The existence of multiple E(spl) proteins within a single locus presents an opportunity to determine the strictness of conservation of this locus across the genus, with respect to both the number of homologues and their positions relative to each other. Our analysis (Fig. 4a) reveals remarkable molecular synteny of the *E(spl)C* across 12 *Drosophila* species, indicating that this locus has remained essentially unchanged over 50 myr. In all species examined, the order and direction of transcription of individual E(spl) transcripts are the same. In the case of *D. melanogaster*, the *E(spl)C* is located on chromosome III such that the *mδ* gene is centromere proximal, whereas *groucho* is telomere proximal (Fig. 4a). This arrangement is shared in the ancestral Hawaiian group and those most closely related to it, the virilis, repleta, and willistoni groups. In contrast, the obscura group and three members of the melanogaster group, *D. yakuba*, *D. erecta*, and *D. ananassae*, display an inversion of the entire *E(spl)C* such that *groucho* is now centromere proximal and *mδ* is telomere proximal. It therefore seems to be the case that two inversion events have occurred, the first ~37 myr ago and the second ~3 myr ago. This extraordinary synteny of the *E(spl)C* and that four members of this locus exhibit essentially invariant CK2 consensus sites, which make for a strong evolutionary argument that these bHLH proteins likely perform distinct functions during development.

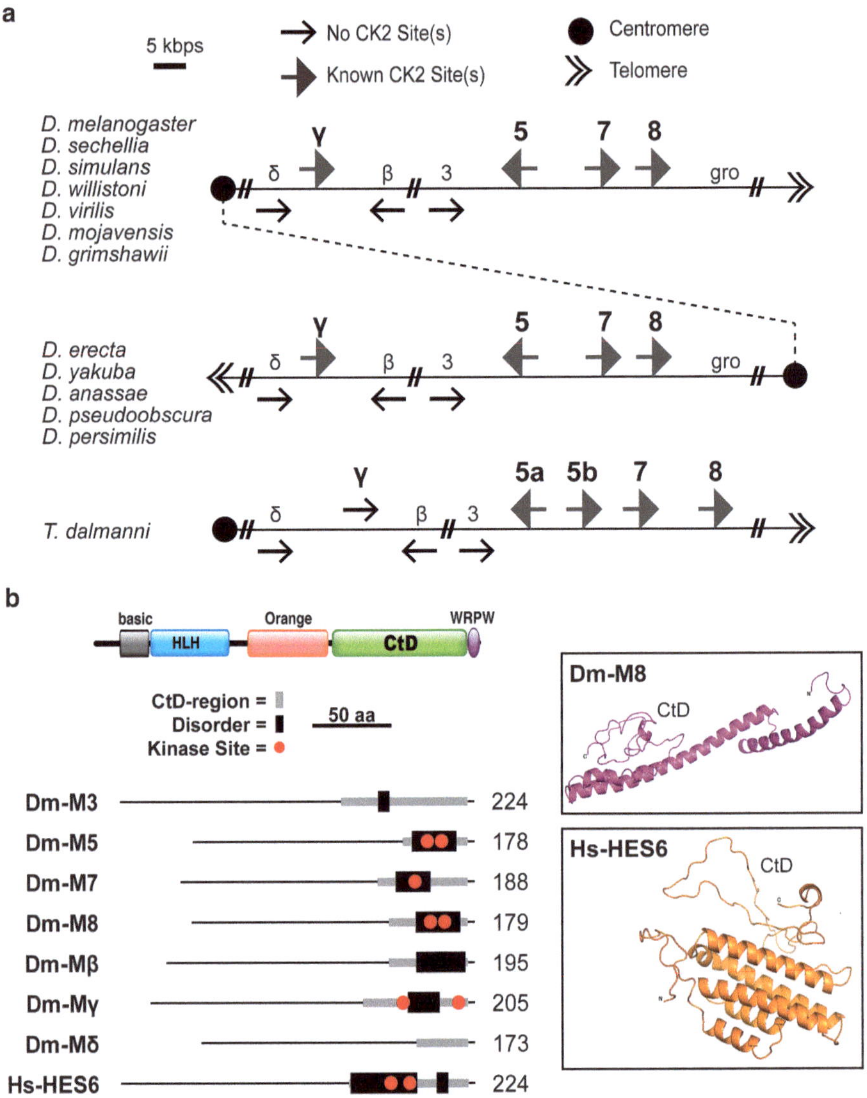

Fig. 4 Intrinsic disorder in the CtD of *Drosophila* E(spl) and human HES6. (**a**) Molecular synteny of the *E(spl)C*. Those members that conserve the CK2 sites are shown as *block arrows*, and the *dotted line* denotes a complete inversion of the locus, although the order of the transcription units remains. (**b**) The indicated proteins were analyzed for disordered regions. The CtD is *shaded gray*, the region of disorder is in *black*, and sites for CK2 and MAPK are denoted as *red dots*. The length of the proteins and scale bar are indicated. Regions of intrinsic disorder reported are those with a >70 % disorder probability output. In all cases, regions with low probability of disorder or low scores are omitted. Abbreviations are Dm, *D. melanogaster*, and Hs, *Homo sapiens*. Inset on the right: I-TASSER predictions of the M8 and HES6 proteins. Note that the CtD regions in both proteins are predicted to be unstructured. The accession numbers for the sequences are E(spl)-M3, FBpp0084332; E(spl)-M5, FBpp0084352; E(spl)-M7, FBpp0084334; E(spl)-M8, FBpp0084335; E(spl)-Mβ, FBpp0084355; E(spl)-Mγ, FBpp0084329; E(spl)-Mδ, FBpp0084328; and human HES6, BAA96082

To further illustrate the importance of conservation at this locus, we look to flies more distantly related to *Drosophila*. Stalk-eyed flies and tsetse flies, which are isolated from the *Drosophila* lineage by 60–100 myr, share strikingly similar homologous *E(spl)* loci [135]. The genome of *T. dalmanni*, a stalk-eyed fly, encodes eight E(spl) bHLHs, all of which directly correspond to *Drosophila* counterparts in addition to a duplication of M5 (Fig. 4a). Aside from this duplication, the order and arrangement of each gene is conserved. Similarly, the tsetse fly, *G. morsitans*, encodes five E(spl) bHLHs, lacking M3 and M5. Of note, CK2 sites are also conserved in select members from stalk-eyed and tsetse flies. The remarkable stability of this locus and the CtD/P-domain strongly suggests that each E(spl) member serves a distinct developmental role.

9 Intrinsic Disorder in the E(spl)/HES Proteins

It is becoming increasingly appreciated that certain regions within diverse proteins have a high propensity for "intrinsic disorder" (ID). Originally considered an oddity, this feature is being identified in numerous proteins that regulate transcription, cell signaling, and development [136–139]. In addition, these regions of ID often correlate with sites for regulation by PTM such as phosphorylation. We thus sought to correlate the presence of ID regions with sites for phosphorylation among the seven *Drosophila* E(spl) repressors and human HES6. Sequences were analyzed by a number of web-based programs, and the results are shown schematically in Fig. 4b. Remarkably, the predominant region of ID in all *Drosophila* E(spl) proteins and human HES6 localizes to the CtD. Another smaller region of ID localizes to the loop within the HLH motif, or in the middle of the Orange domain, which is also predicted to form an HLH domain (not shown). However, these regions of ID are not found in all isoforms, and their presence in a loop is to be expected in an HLH motif. The major region of disorder in the CtDs is variable in both length and position (Fig. 4b). Notable among these are M8/M5/M7 and HES6 where the sites for phosphorylation (known and predicted) are located within the disordered region. In contrast, although M3 and Mβ contain regions with ID, neither harbors sites for CK2, MAPK, or other protein kinases (data not shown). Mδ lacks an ID region or sites for phosphorylation. In contrast, Mγ appears unique in that the CK2 and MAPK sites flank the ID region. In the absence of defined 3D structures of any E(spl)/HES/HER proteins, predictions of ID are, at best, speculative. To bolster the case, we used I-TASSER, a computational approach to predict 3D structures [140]. This approach has proven adept at accurately predicting the structures of numerous proteins. Indeed, we find that the CtDs of M8 and Mγ are unstructured (Fig. 4b), as are those for M5 and M7 (data not shown). The possibility is high that phosphorylation effects on Mγ are unlikely to be similar to those with M5, M7, M8, and HES6.

It is unsurprising that ID predominates within the CtDs of E(spl) members. As is to be expected, E(spl) architecture offers little variation in either the bHLH or Orange domains. It is worth speculating that the CtD may have originally served as a placeholder, allowing distance between the Orange domain and WRPW, to prevent steric inhibition of Groucho interaction. Once this locus had become amplified, the CtD, under no selective pressures but to be of a sufficient number of residues, may have been subject to accelerated changes through evolution. In such a case, it is plausible that the CtD could have assumed regulatory roles and, furthermore, that those regulatory roles could become modulated by the acquisition of kinase consensus sites. Such kinase sites would then become fixed in a select group of repressors, each under differing evolutionary constraints, as seen across *Drosophila*. In support of these possibilities, phospho-occupancy and the length of the CtD do influence the activities of the Orange domain and that of the WRPW motif [141].

10 The CK2 Site, the P-Domain, and Multisite Phosphorylation of E(spl) Members

The high conservation of the CK2 site among fly M8, M5, M7, Mγ and human HES6 is noteworthy, because it reveals the presence of additional conserved Ser residues (see Figs. 2 and 3). In the case of HES6, it has been reported that the PxSP motif is targeted by MAPK [142], an effector of receptor tyrosine kinase (RTK) signaling. Although the developmental consequences of phosphorylation of HES6 by MAPK are presently unknown, a number of observations in *Drosophila* point to a role in the regulation of E(spl) proteins. For example, epidermal growth factor receptor (EGFR) signaling is known to be active during R8 selection [143], as is active (dual-phosphorylated) MAPK [144], and loss of *Drosophila* EGFR leads to the emergence of extra, supernumerary R8s, a phenotype akin to that upon loss of E(spl) or CK2 (see above). Additionally, removal of the canonical EGFR/MAPK target protein pointed has no effect of R8 development [145], further suggesting that RTK signaling impinges upon noncanonical substrate such as E(spl). These findings thus raise the prospect that E(spl)-M8 activity is regulated through multisite phosphorylation and that PTM affords a means to fine-tune M8 activity in vivo. Indeed, our ongoing studies are revealing that the hyperactive CK2-mimetic M8-S^{159}D variant is rendered inactive by replacement of Ser^{151} (of the PxSP motif) with Ala (Bandyopadhyay and Bidwai, unpublished). The possibility is thus high that two regulatory inputs (CK2 and MAPK) modulate Notch pathway activity through phosphorylation of M8. The potential roles for the additionally conserved Ser residues in the P-domain of M8 (Fig. 2) are discussed below.

11 Control of E(spl) Activity Through Multisite Phosphorylation

The close clustering of the CK2 and MAPK sites in M8 (Fig. 5a) raises immediate questions on their order of modification, their effects on repressor activity, and the potential roles for the two additionally conserved Ser residues that lie between the two kinase sites. In this section, we discuss the potential interplay between the four Ser residues of the P-domain in control over M8 activity.

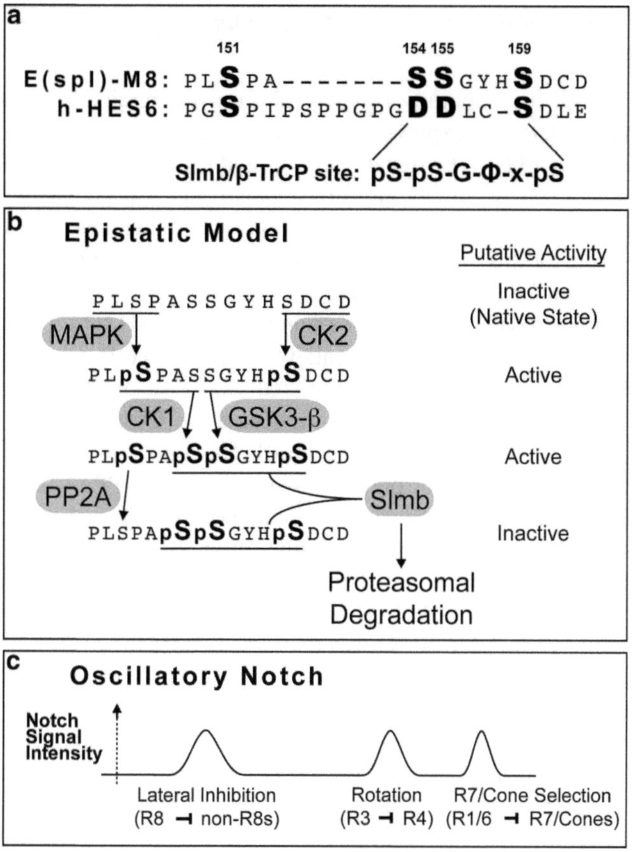

Fig. 5 Model for regulation of M8 through multisite phosphorylation. (**a**) E(spl)-M8 CtD matches the canonical Slmb-consensus site. Human HES6 (h-HES6) may encode a noncanonical motif for recognition by β-TrCP. Phosphoserine (denoted as pS) can be substituted by Asp or Glu. (**b**) CK2 and MAPK phosphorylations activate M8 and prime for modification by CK1 and GSK3-β. M8 downregulation is facilitated by PP2A (dephosphorylation) and Slmb (ubiquitylation). Although CK2 and MAPK potentiate M8 repressor activity, the immediate effect of phosphorylation by CK1 and GSK3-β is unclear. (**c**) Notch signaling oscillates throughout *Drosophila* retinal differentiation, similar to cycles of Notch activity observed in somitogenesis

The accumulation of evidence for multisite phosphorylation now hints at the possibility of a sophisticated regulation of E(spl) activity. Thus far, we have implicated CK2 as a potentiator of M8-mediated repression of proneural gene activity and, by proxy, the repressive capabilities of Mγ, M5, and M7. Given the elaborate measures required to activate M8, it stands to reason that subsequent modifications may have opposite effects. In such a scenario, two likely downregulatory mechanisms emerge: dephosphorylation via phosphatase activity and phosphodegron-mediated protein degradation.

11.1 PP2A Opposes M8 Activity via Dephosphorylation

M8 serves a distinct, temporally isolated role in its repression of Atonal, because neither early nor late M8 expression perturbs patterning of the developing retina (Fig. 5c, see [131, 132]). The dynamism required of cells of the eye disc strongly suggests that M8 is subject to a shutoff mechanism that is as potent and rapid as the processes governing its activation. Interestingly, CK2 interacts with PP2A [146], supporting a role for this phosphatase as an antagonist of either CK2 or, potentially, other local sites on CK2 target proteins.

The PP2A holoenzyme features three subunits, A, B, and C [147–149]. A-subunit acts as a scaffold, C-subunit provides catalytic activity, and B-subunit confers substrate specificity to the C-subunit, whose shallow active site precludes binding to target proteins. Previous studies have implicated the C-subunit microtubule star (Mts) in opposition to Notch signaling [150]. However, the *Drosophila* genome encodes a number of known or putative B-subunits to the PP2A holoenzyme. We have recently reported the identification of Widerborst (Wdb) as the PP2A B-subunit that recognizes M8 as substrate [151]. Unique among B-subunits, Wdb loss of function interacts with multiple Notch alleles, in all cases resulting in enhanced Notch signal output. In support of Wdb as a direct modifier of HES activity, the CK2-phosphomimetic variant M8-S^{159}D was expressed under the control of a heterologous, Notch-independent transactivator. In congruence with the modification of Notch alleles, Wdb modulates the eye defects elicited by M8-S^{159}D; activity was enhanced by Wdb loss of function but suppressed by Wdb gain of function. Such evidence supports a direct role for Wdb and PP2A in mitigating M8 repressor activity and similar roles in the regulation of HES members with similar CtD architecture such as Mγ, M5, and M7 (see Figs. 2 and 3).

11.2 Phosphodegron-Mediated Clearance of M8

Interestingly, a large-scale *Drosophila* screen of protein–protein interactions [152] revealed M8 as an interaction partner with supernumerary limbs (Slmb). Slmb, the ortholog of mammalian β-TrCP, is the F-box component of the SCF E3-ubiquitin ligase that is responsible for the identification of substrate fated for degradation via the 26S proteasome. Slmb has been shown to interact with the acidic D/E-D/E-G-ϕ-x-D/E sequence motif, where D/E can be equally represented by pSer/pThr. The regulation of several conserved signaling pathways relies upon the modification or degradation of transcription factors that bear such phosphodegrons, such as cubitus interruptus (Ci) of the Hedgehog pathway and β-catenin of the Wg/Wnt pathway [153] and, more broadly, the cell-cycle regulator Cdc25a [154]. Notably, Slmb lacks a documented interaction with any core components of the Notch signaling pathway.

The Slmb-consensus motif occupies residues 154–159 of the M8-CtD (SSGYHS, Fig. 5a), and a similar, acid-rich motif is present in human HES6. The most conservative definition of the Slmb consensus requires phosphorylation of Ser^{154}, Ser^{155}, and Ser^{159} of M8-CtD. As previously discussed, Ser^{159} is modified by CK2, and Ser^{151}, located just three residues upstream of the putative phosphodegron, may be modified through EGFR/MAPK, leaving the modification of residues 154 and 155 to question. Using the dually phosphorylated M8-CtD ($pSer^{151+159}$) as a template, the conserved kinases CK1 and GSK3-β could modify both Ser^{154} and Ser^{155}, respectively, eliciting formation of the Slmb phosphodegron.

11.3 Cooperativity of Wdb and Slmb Interactions

The suggestion of proteasomal degradation of several E(spl) members elicits several questions. Under most circumstances, phosphatase activity would appear to contraindicate the formation of a phosphodegron. Clearance of M8 as cells are emerging from the MF would downregulate Notch signaling following completion of lateral inhibition; thus, phosphatase activity against any phosphorylated residue of the degron would predictably enhance E(spl) activity. However, Wdb opposes $M8\text{-}S^{159}D$ function, arguing against the possibility that PP2A opposes modification at the CK2 site. Thus, Ser^{151} is the only suitable target for phosphatase activity that reconciles the tandem employment of phosphodegron formation and local dephosphorylation, allowing both pathways to contribute to reducing E(spl) activity.

We therefore propose a mechanism (Fig. 5b) where CK2 and MAPK first phosphorylate residues 151 and 159, which prime for subsequent phosphorylation of residues 154 and 155. With the completion of the phosphodegron, Ser^{151} can then be dephosphorylated with no apparent impact on Slmb interaction. Thus, Wdb and Slmb might cooperate to extinguish M8 through two steps: first, Wdb dephosphorylates Ser^{151} to reduce M8 activity and second, Slmb facilitates ubiquitylation and proteasomal degradation. Furthermore, we find it no coincidence that this putative phosphodegron exists in the CtDs of those E(spl) members that may be subject to repeated, sequence-specific phosphorylation events (see below). In the least, those members that require phospho-activation (a temporally regulated event) might also require a robust method to ensure a timely shutdown.

12 CK2, Eye Development, and the Nonredundancy of E(spl) Proteins

As stated above, the selection of the R8 photoreceptor from a cluster of Atonal-positive cells closely correlates with the expression of three E(spl) proteins, i.e., M8, Mδ, and Mγ, and in only those cells destined for the non-R8 fate [100].

To confer the R8 fate, Atonal must directly interact with the product of the *daughterless* gene [155], and dimers of Atonal+Daughterless [156] then activate transcription of the downstream gene *senseless*, which is essential for R8 cell differentiation [157]. It seems counterintuitive that M8, Mδ, and Mγ would be redundant for repression of Atonal, because they markedly differ in the CtD length and sequence, its region of ID, and the organization/spacing of the kinase sites for regulation by PTM. In addition, genetic evidence supports the aforementioned argument. For example, forced expression of only Mδ can repress Atonal activity, albeit weakly [100], but this protein is unlikely to be the primary repressor of Atonal because animals homozygous for mutation of the *m*δ transcription unit do not display sustained Atonal activity or significantly perturbed eyes [158]. This is in contrast to M8, whose mutation *(E(spl)D)* severely impairs eye development by dominantly repressing Atonal. Overexpression of Mγ during R8 photoreceptor patterning elicits no eye defects (Jozwick and Bidwai, unpublished), in a manner akin to wild-type M8. Moreover, E(spl) proteins form homo-/heterodimers and directly interact with Atonal, Daughterless, and Senseless in yeast two-hybrid assays. The possibility thus remains that homo-/heterodimers of E(spl) proteins interact in a target-specific (differential) manner with Atonal, Daughterless, or Senseless in vivo, an aspect that cannot be discerned in explicit two-hybrid assays. Systematic biochemical and genetic analyses will be required to identify which E(spl) proteins form heterodimers with Mγ and if phosphorylation influences their binding in a preferential manner with Atonal, Daughterless, or Senseless.

In summary, preponderances of E(spl) proteins are targets of CK2 and conserve a site for MAPK, implicating these protein kinases in regulating Notch signaling and serving as a node between the Notch and EGFR pathways. These sites of phosphorylation co-localize with regions that are predicted to be intrinsically disordered but are positioned differently in each isoform, raising the prospect that these bHLH repressors do not serve redundant (anti-neural) functions. Although the roles of multisite phosphorylation (of M7, M5, Mγ, and HES6) remain to be resolved during development, our studies suggest that PTM is likely to be more central to these proteins than has been recognized [159]. Future efforts are needed to clarify the roles of phosphorylation in dimerization of E(spl) proteins, their ability to bind and repress proneural factors such as Atonal, Daughterless, and Senseless, and determine if these modifications occur in a tissue-specific manner in the developing nervous system and elsewhere.

We believe that our findings have implications on the oscillatory nature of Notch signaling. For example, somitogenesis requires, in addition to RTK signaling, employment of a "*Notch clock*," in which periodic activation of the Notch pathway allows for proper segmentation of somites from unspecified mesoderm [71]. It stands to reason that both periodic Notch signaling and, therefore, cyclical E(spl)/HES expression are anciently conserved characteristics of bilaterian development. Routine, wavelike modulation of E(spl)/HES expression would likely be accompanied by mechanisms to ensure timely downregulation of E(spl)/HES function, such as those that we have discussed. PTM regulation of E(spl)/HES proteins would then constitute an anciently conserved mode of regulating this aspect of periodic

Notch activity. Interestingly, differentiation of the *Drosophila* eye disc features periodic Notch signaling, first in R8 resolution and twice again in determination of R1–R7 photoreceptors (Fig. 5c). Even though the eye disc may appear to be distinct, the underlying mechanisms are unlikely to be fundamentally different. Such similarities strengthen the likelihood that our findings derived from *Drosophila* neurogenesis have implications on E(spl)/HES in somitogenesis and other contexts requiring repeated (pulsatile) Notch signaling.

Given the close associations between CK2, MAPK, and cancer, it will be of interest to determine if altered activities of these kinases result in defective Notch signaling [160], which has been linked to broad-spectrum diseases, such as Alagille syndrome [161] and Williams–Beuren syndrome, and disorders affecting the immune system [75], cardiac development [162], and cancer [37, 163, 164], and is increasingly being recognized as an important target for therapeutic intervention [165].

Acknowledgments Supported by a grant from the National Institutes of Health (EY015718) to A. P. B.

References

1. Pinna LA (2002) Protein kinase CK2: a challenge to canons. J Cell Sci 115(Pt 20):3873–3878
2. Litchfield DW (2003) Protein kinase CK2: structure, regulation and role in cellular decisions of life and death. Biochem J 369:1–15
3. Issinger OG (1993) Casein kinases: pleiotropic mediators of cellular regulation. Pharmacol Ther 59:1–30
4. Bidwai AP, Hanna DE, Glover CVC (1992) The free catalytic subunit of casein kinase II is not toxic in vivo. J Biol Chem 267:18790–18796
5. Kuenzel EA, Krebs EG (1985) A synthetic substrate specific for casein kinase II. Proc Natl Acad Sci U S A 82:737–741
6. Kuenzel EA et al (1987) Substrate specificity determinants for casein kinase II as deduced from studies with synthetic peptides. J Biol Chem 262:9136–9140
7. Hrubey TW, Roach PJ (1990) Phosphoserine in peptide substrates can specify casein kinase II action. Biochem Biophys Res Commun 172:190–196
8. Meier UT, Blobel G (1992) Nopp 140 shuttles on tracks between nucleolus and cytoplasm. Cell 70:127–138
9. Wilson LK et al (1997) Casein kinase II catalyzes tyrosine phosphorylation of yeast nucleolar immunophilin Fpr3. J Biol Chem 272:12961–12967
10. Marin O et al (1999) Tyrosine versus serine/threonine phosphorylation by protein kinase casein kinase-2. J Biol Chem 274:29260–29265
11. Kennedy EP (1992) Sailing to Byzantium. Annu Rev Biochem 61:1–28
12. Burnett G, Kennedy EP (1954) The enzymatic phosphorylation of proteins. J Biol Chem 211:969–980
13. Salvi M et al (2009) Extraordinary pleiotropy of protein kinase CK2 revealed by weblogo phosphoproteome analysis. Biochim Biophys Acta 1793(5):847–859
14. Meggio F, Pinna LA (2003) One-thousand-and-one substrates of protein kinase CK2. FASEB J 17:349–368
15. Bidwai AP (2000) Structure and function of casein kinase II. Recent Res Devel Mol Cell Biol 1:51–82

16. Glover CVC (1998) On the physiological role of casein kinase II in Saccharomyces cerevisiae. Prog Nucleic Acid Res Mol Biol 59:95–133
17. Padmanabha R et al (1990) Isolation, sequencing, and disruption of the yeast CKA2 gene: casein kinase II is essential for viability in Saccharomyces cerevisiae. Mol Cell Biol 10:4089–4099
18. Hanna DE, Rethinaswamy A, Glover CVC (1995) Casein kinase II is required for cell cycle progression during G1 and G2/M in Saccharomyces cerevisiae. J Biol Chem 270:25905–25914
19. Bandhakavi S et al (2003) Genetic interactions among ZDS1,2, CDC37, and protein kinase CK2 in Saccharomyces cerevisiae. FEBS Lett 554(3):295–300
20. Pepperkok R et al (1994) Casein kinase II is required for transition of G0/G1, early G1, and G1/S phases of the cell cycle. J Biol Chem 269:6986–6991
21. Ghavidel A, Schultz MC (2001) TATA binding protein-associated CK2 transduces DNA damage signals to the RNA polymerase II transcriptional machinery. Cell 106:575–584
22. Luscher B et al (1990) Myb DNA binding inhibited by phosphorylation at a site deleted during oncogenic activation. Nature 344:517–522
23. Zhao T, Eissenberg JC (1999) Phosphorylation of heterochromatin protein 1 by casein kinase II is required for efficient heterochromatin binding in Drosophila. J Biol Chem 274:15095
24. Tamaru T et al (2009) CK2alpha phosphorylates BMAL1 to regulate the mammalian clock. Nat Struct Mol Biol 16(4):446–448
25. Akten B et al (2003) A role for CK2 in the Drosophila circadian oscillator. Nat Neurosci 6:251–257
26. Lin JM et al (2002) A role for casein kinase 2alpha in the Drosophila circadian clock. Nature 420(6917):816–820
27. Sanz-Clemente A et al (2010) Casein kinase 2 regulates the NR2 subunit composition of synaptic NMDA receptors. Neuron 67(6):984–996
28. ole-MoiYoi OK (1989) Theileria parva: an intracellular protozoan parasite that induces reversible lymphocyte transformation. Exp Parasitol 69:204–210
29. ole-MoiYoi OK et al (1993) Evidence for the induction of casein kinase II in bovine lymphocytes transformed by the intracellular protozoan parasite Theileria parva. EMBO J 12:1621–1631
30. ole-MoiYoi OK et al (1992) Cloning and characterization of the casein kinase II alpha subunit gene from the lymphocyte-transforming intracellular protozoan parasite Theileria parva. Biochemistry 31:6193–6202
31. ole-MoiYoi OK (1995) Casein kinase II in theileriosis. Science 267:834–836
32. Kelliher MA, Seldin DC, Leder P (1996) Tal-1 induces T cell acute lymphoblastic leukemia accelerated by casein kinase II alpha. EMBO J 15:5160–5166
33. Seldin DC, Leder P (1995) Casein kinase II alpha transgene-induced murine lymphoma: relation to theileriosis in cattle. Science 267:894–897
34. Ahmed K (1999) Nuclear matrix and protein kinase CK2 signaling. Crit Rev Eukaryot Gene Expr 9:329–336
35. Munstermann U et al (1990) Casein kinase II is elevated in solid human tumours and rapidly proliferating non-neoplastic tissue. Eur J Biochem 189:251–257
36. Prowald K, Fischer H, Issinger OG (1984) Enhanced casein kinase II activity in human tumor cell cultures. FEBS Lett 176:479–483
37. Deshiere A et al (2013) Unbalanced expression of CK2 kinase subunits is sufficient to drive epithelial-to-mesenchymal transition by Snail1 induction. Oncogene 32(11):1373–1383
38. Snell V, Nurse P (1994) Genetic analysis of cell morphogenesis in fission yeast-a role for casein kinase II in the establishment of polarized growth. EMBO J 13:2066–2074
39. Roussou I, Dretta G (1994) The Schizosaccharomyces pombe casein kinase II alpha and beta subunits: evolutionary conservation and positive role for the beta subunits. Mol Cell Biol 14:576–586
40. Jauch E et al (2002) In vivo functional analysis of Drosophila protein kinase casein kinase 2 (CK2) beta-subunit. Gene 298:29–39

41. Buchou T et al (2003) Disruption of the regulatory beta subunit of protein kinase CK2 in mice leads to a cell-autonomous defect and early embryonic lethality. Mol Cell Biol 23:908–915
42. Litchfield DW, Dobrowolska G, Krebs EG (1994) Regulation of casein kinase II by growth factors: a reevaluation. Cell Mol Biol Res 40(5/6):373–381
43. Sommercorn J et al (1987) Activation of casein kinase II in response to insulin and to epidermal growth factor. Proc Natl Acad Sci U S A 84:8834–8838
44. Song DH, Sussman DJ, Seldin DC (2000) Endogenous protein kinase CK2 participates in Wnt signaling in mammary epithelial cells. J Biol Chem 275(31):23790–23797
45. Willert K et al (1997) Casein kinase 2 associates with and phosphorylates Dishevelled. EMBO J 16:3089–3096
46. Gao Y, Wang HY (2006) Casein kinase 2 is activated and essential for Wnt/beta-catenin signaling. J Biol Chem 281(27):18394–18400
47. Seldin DC et al (2005) CK2 as a positive regulator of Wnt signalling and tumourigenesis. Mol Cell Biochem 274(1–2):63–67
48. Jia H et al (2010) Casein kinase 2 promotes Hedgehog signaling by regulating both smoothened and Cubitus interruptus. J Biol Chem 285(48):37218–37226
49. Jaffe L, Ryoo H-D, Mann RS (1997) A role for phosphorylation by casein kinase II in modulating Antennapedia activity in Drosophila. Genes Dev 11:1327–1340
50. Bourbon HM et al (1995) Phosphorylation of the Drosophila engrailed protein at a site outside its homeodomain enhances DNA binding. J Biol Chem 270:11130–11139
51. Coqueret O et al (1998) DNA binding by Cut homeodomain proteins Is down-modulated by Casein Kinase II. J Biol Chem 273(5):2561–2566
52. Goldstein RE et al (2005) An eh1-like motif in odd-skipped mediates recruitment of groucho and repression in vivo. Mol Cell Biol 25(24):10711–10720
53. Dominguez I et al (2004) Protein kinase CK2 is required for dorsal axis formation in Xenopus embryos. Dev Biol 274(1):110–124
54. Wong LC et al (2011) The functioning of the Drosophila CPEB protein Orb is regulated by phosphorylation and requires casein kinase 2 activity. PLoS One 6(9):e24355
55. Glover CVC (1986) A filamentous form of Drosophila casein kinase II. J Biol Chem 261:14349–14354
56. Glover CVC, Shelton ER, Brutlag DL (1983) Purification and characterization of a type II casein kinase from Drosophila melanogaster. J Biol Chem 258(5):3258–3265
57. Saxena A, Padmanabha R, Glover CVC (1987) Isolation and sequencing of cDNA clones encoding a and b subunits of Drosophila melanogaster casein kinase II. Mol Cell Biol 7(10):3409–3417
58. Bozzetti MP et al (1995) The Ste locus, a component of the parasitic cry-ste system of Drosophila melanogaster, encodes a protein that forms crystals in primary spermatocytes and mimics properties of the beta subunit of casein kinase II. Proc Natl Acad Sci U S A 92:6067–6071
59. Bidwai AP, Zhao WF, Glover CVC (1999) A gene located at 56 F1-2 in Drosophila melanogaster encodes a novel metazoan beta-like subunit of casein kinase II. Mol Cell Biol Res Commun 1:21–28
60. Karandikar U et al (2003) The Drosophila SSL gene is expressed in larvae, pupae, and adults, exhibits sexual dimorphism, and mimics properties of the b subunit of casein kinase II. Biochem Biophys Res Commun 301:941–947
61. Bidwai AP et al (2000) Multiple, closely spaced alternative 5′ exons in the DmCKIIbeta gene of Drosophila melanogaster. Mol Cell Biol Res Commun 3:283–291
62. Jauch E et al (2006) The Drosophila melanogaster DmCK2beta transcription unit encodes for functionally non-redundant protein isoforms. Gene 374:142–152
63. Greenwald I (2012) Notch and the awesome power of genetics. Genetics 191(3):655–669
64. D'Souza B, Meloty-Kapella L, Weinmaster G (2010) Canonical and non-canonical Notch ligands. Curr Top Dev Biol 92:73–129
65. Artavanis-Tsakonas S, Muskavitch MA (2010) Notch: the past, the present, and the future. Curr Top Dev Biol 92:1–29

66. Tien AC, Rajan A, Bellen HJ (2009) A Notch updated. J Cell Biol 184(5):621–629
67. Kopan R (2002) Notch: a membrane-bound transcription factor. J Cell Sci 115(Pt 6): 1095–1097
68. Bray SJ (2006) Notch signalling: a simple pathway becomes complex. Nat Rev Mol Cell Biol 7(9):678–689
69. Ilagan MX, Kopan R (2007) Notch signaling pathway. Cell 128(6):1246
70. de la Pompa JL, Epstein JA (2012) Coordinating tissue interactions: notch signaling in cardiac development and disease. Dev Cell 22(2):244–254
71. Bajard L, Oates AC (2012) Breathe in and straighten your back: Hypoxia, Notch, and Scoliosis. Cell 149:255–256
72. Louvi A, Artavanis-Tsakonas S (2006) Notch signalling in vertebrate neural development. Nat Rev Neurosci 7(2):93–102
73. Pierfelice T, Alberi L, Gaiano N (2011) Notch in the vertebrate nervous system: an old dog with new tricks. Neuron 69(5):840–855
74. Beets K et al (2013) Robustness in angiogenesis: notch and BMP shaping waves. Trends Genet 29(3):140–149
75. Maillard I, Pear WS (2007) Immunology. Keeping a tight leash on Notch. Science 316(5826):840–842
76. Wu MN, Raizen DM (2011) Notch signaling: a role in sleep and stress. Curr Biol 21(10):397–398
77. Seugnet L et al (2011) Notch signaling modulates sleep homeostasis and learning after sleep deprivation in drosophila. Curr Biol 21(10):835–840
78. Oellers N, Dehio M, Knust E (1994) BHLH proteins encoded by the enhancer of split complex of drosophila negatively interfere with transcriptional activation mediated by proneural genes. Mol Gen Genet 244:465–473
79. Klambt C et al (1989) Closely related transcripts encoded by the neurogenic gene complex Enhancer of split of Drosophila melanogaster. EMBO J 8:203–210
80. Knust E et al (1992) Seven genes of the enhancer of split complex of Drosophila melanogaster encode helix-loop-helix proteins. Genetics 132:505–518
81. Delidakis C, Artavanis-Tsakonas S (1991) The enhancer of split [E(spl)] locus of Drosophila encodes seven independent helix-loop-helix proteins. Proc Natl Acad Sci U S A 89:8731–8735
82. Sun H, Ghaffari S, Taneja R (2007) bHLH-Orange transcription factors in development and cancer. Transl Oncogenomics 2:105–118
83. Dambly-Chaudiere C, Vervoort M (1998) The bHLH genes in neural development. Int J Dev Biol 42(3):269–273
84. Massari ME, Murre C (2000) Helix-loop-helix proteins: regulators of transcription in eukaryotic organisms. Mol Cell Biol 20:429–440
85. Pratt EB et al (2011) The cell giveth and the cell taketh away: an overview of Notch pathway activation by endocytic trafficking of ligands and receptors. Acta Histochem 113(3): 248–255
86. Yamamoto S, Charng WL, Bellen HJ (2010) Endocytosis and intracellular trafficking of Notch and its ligands. Curr Top Dev Biol 92:165–200
87. Weinmaster G, Fischer JA (2011) Notch ligand ubiquitylation: what is it good for? Dev Cell 21(1):134–144
88. Modolell J, Campuzano S (1998) The achaete-scute complex as an integrating device. Int J Dev Biol 42(3):275–282
89. Campos-Ortega JA (1998) The genetics of the Drosophila achaete-scute gene complex: a historical perspective. Int J Dev Biol 42:291–297
90. Jarman AP et al (1994) Atonal is the proneural gene for Drosophila photoreceptors. Nature 369:398–400
91. White N, Jarman A (2000) Drosophila atonal controls photoreceptor R8-specific properties and modulates both receptor tyrosine kinase and Hedgehog signalling. Development 127(8): 1681–1689

92. Reeves N, Posakony JW (2005) Genetic programs activated by proneural proteins in the developing Drosophila PNS. Dev Cell 8(3):413–425
93. Kiefer JC (2005) Proneural factors and neurogenesis. Dev Dyn 234:808–813
94. Jarman AP, Ahmed I (1998) The specificity of proneural genes in determining Drosophila sense organ identity. Mech Dev 76(1–2):117–125
95. Bertrand N, Castro DS, Guillemot F (2002) Proneural genes and the specification of neural cell types. Nat Rev Neurosci 3(7):517–530
96. Van Doren M et al (1992) Spatial regulation of proneural gene activity: auto- and cross-activation of achaete is antagonized by extramacrochaetae. Genes Dev 6(12B):2592–2605
97. Culi J, Modolell J (1998) Proneural gene self-stimulation in neural precursors: an essential mechanism for sense organ development that is regulated by Notch signaling. Genes Dev 12:2036–2047
98. Okabe M et al (2001) Translational repression determines a neuronal potential in Drosophila asymmetric cell division. Nature 411(6833):94–98
99. Bose A et al (2006) Drosophila CK2 regulates lateral-inhibition during eye and bristle development. Mech Dev 123:649–664
100. Ligoxygakis P et al (1998) A subset of notch functions during drosophila eye development require Su(H) and E(spl) gene complex. Development 125:2893–2900
101. Dominguez M (1999) Dual role for Hedgehog in the regulation of the proneural gene atonal during ommatidia development. Development 126(11):2345–2353
102. Suzuki T, Saigo K (2000) Transcriptional regulation of atonal required for Drosophila larval eye development by concerted action of Eyes absent, Sine oculis and Hedgehog signaling independent of Fused kinase and Cubitus interruptus. Development 127:1531–1540
103. Baker NE, Yu S, Han D (1996) Evolution of proneural atonal expression during distinct regulatory phases in the developing Drosophila eye. Curr Biol 6(10):1290–1301
104. Cooper MTD, Bray SJ (1999) Frizzled regulation of Notch signaling polarizes cell fate in the Drosophila eye. Nature 397:526–530
105. Tomlinson A, Mavromatakis YE, Struhl G (2011) Three distinct roles for notch in Drosophila R7 photoreceptor specification. PLoS Biol 9(8):e1001132
106. Axelrod JD (2010) Delivering the lateral inhibition punchline: it's all about the timing. Sci Signal 3(145):pe38
107. Simpson P (1990) Lateral inhibition and the development of the sensory bristles of the adult peripheral nervous system of Drosophila. Development 109(3):509–519
108. Ehebauer M, Hayward P, Martinez-Arias A (2006) Notch signaling pathway. Sci STKE 2006(364):cm7
109. Ehebauer M, Hayward P, Arias AM (2006) Notch, a universal arbiter of cell fate decisions. Science 314(5804):1414–1415
110. Cooper MTD et al (2000) Spatially restricted factors cooperate with notch in the regulation of enhancer of split genes. Dev Biol 221:390–403
111. Cave JW, Xia L, Caudy M (2011) Differential regulation of transcription through distinct Suppressor of Hairless DNA binding site architectures during Notch signaling in proneural clusters. Mol Cell Biol 31(1):22–29
112. Cohen M et al (2010) Dynamic filopodia transmit intermittent Delta-Notch signaling to drive pattern refinement during lateral inhibition. Dev Cell 19(1):78–89
113. Milan M, Cohen SM (2010) Notch signaling: filopodia dynamics confer robustness. Curr Biol 20(18):R802–R804
114. Dawson SR et al (1995) Specificity for the hairy/enhancer of split basic helix-loop-helix (bHLH) proteins maps outside the bHLH domain and suggests two separable modes of transcriptional repression. Mol Cell Biol 15:6923–6931
115. Paroush Z et al (1994) Groucho is required for Drosophila neurogenesis, segmentation, and sex determination and interacts directly with hairy related bHLH proteins. Cell 79:805–815
116. Jennings BH, Tyler DM, Bray SJ (1999) Target specificities of drosophila enhancer of split basic helix-loop-helix proteins. Mol Cell Biol 19(7):4600–4610

117. Jimenez G, Ish-Horowicz D (1997) A chimeric Enhancer of split transcriptional activator drives neural development and achaete-scute expression. Mol Cell Biol 17(8):4355–4362
118. Sun Y, Jan L, Jan Y (1998) Transcriptional regulation of atonal during development of the Drosophila peripheral nervous system. Development 125(18):3731–3740
119. Baker NE (2004) Atonal points the way- protein-protein interactions and developmental biology. Dev Cell 7(5):632–634
120. Giagtzoglou N et al (2003) Two modes of recruitment of E(spl) repressors onto target genes. Development 130:259–270
121. Alifragis P et al (1997) A network of interacting transcriptional regulators involved in Drosophila neural fate specification revealed by the yeast two-hybrid system. Proc Natl Acad Sci U S A 94:13099–13104
122. Gigliani F et al (1996) Interactions among the bHLH domains of the proteins encoded by the enhancer of split and achaete-scute gene complexes of Drosophila. Mol Gen Genet 251:628–634
123. Nagel AC, Preiss A (1999) Notch spl is deficient for inductive processes in the eye, and E(spl) D enhances split by interfering with proneural activity. Dev Biol 208:406–415
124. Nagel A, Yu Y, Preiss A (1999) Enhancer of Split [E(spl)D] is a Gro-independent, hypermorphic mutation in Drosophila. Dev Genet 25:168–179
125. Karandikar U et al (2004) Drosophila CK2 regulates eye morphogenesis via phosphorylation of E(spl)M8. Mech Dev 121:273–286
126. Trott RL et al (2001) Drosophila melanogaster casein kinase II interacts with and phosphorylates the basic-helix-loop-helix (bHLH) proteins m5, m7, and m8 derived from the Enhancer of split complex. J Biol Chem 276:2159–2167
127. Welshons WJ (1956) Dosage experiments with split mutations in the presence of an enhancer of split. Drosophila Inform Serv 30:157–158
128. Welshons WJ (1965) Analysis of a gene in Drosophila. Science 150:1122–1129
129. Tietze K, Oellers N, Knust E (1992) Enhancer of SplitD, a dominant mutation of Drosophila, and its use in the study of functional domains of a helix-loop-helix protein. Proc Natl Acad Sci U S A 89(13):6152–6156
130. Giebel B, Campos-Ortega JA (1997) Functional dissection of the Drosophila enhancer of split protein, a suppressor of neurogenesis. Proc Natl Acad Sci U S A 94:6250–6254
131. Kahali B et al (2009) On the mechanism underlying the divergent retinal and bristle defects of M8* (E(spl)D) in Drosophila. Genesis 47:456–468
132. Kahali B et al (2010) Evidence that the C-terminal domain (CtD) autoinhibits neural repression by Drosophila E(spl)M8. Genesis 48:44–55
133. Gratton M-O et al (2003) Hes6 promotes cortical neurogenesis and inhibits Hes1 transcription repression activity by multiple mechanisms. Mol Cell Biol 23(19):6922–6935
134. Beverly SM, Wilson AC (1984) Molecular evolution in Drosophila and the higher Diptera II. A time scale for fly evolution. J Mol Evol 21:1–13
135. Baker RH, Kuehl JV, Wilkinson GS (2011) The Enhancer of split complex arose prior to the diversification of schizophoran flies and is strongly conserved between Drosophila and stalk-eyed flies (Diopsidae). BMC Evol Biol 11:354
136. Tompa P (2012) Intrinsically disordered proteins: a 10-year recap. Trends Biochem Sci 37(12):509–516
137. Goh GK, Dunker AK, Uversky VN (2008) Protein intrinsic disorder toolbox for comparative analysis of viral proteins. BMC Genomics 9(Suppl 2):S4
138. Garza AS, Ahmad N, Kumar R (2009) Role of intrinsically disordered protein regions/ domains in transcriptional regulation. Life Sci 84(7–8):189–193
139. Dunker AK et al (2008) Function and structure of inherently disordered proteins. Curr Opin Struct Biol 18(6):756–764
140. Roy A, Kucukural A, Zhang Y (2010) I-TASSER: a unified platform for automated protein structure and function prediction. Nat Protoc 5(4):725–738
141. Eastwood K et al (2011) New insights into the Orange domain of E(spl)-M8, and the roles of the C-terminal domain in autoinhibition and Groucho recruitment. Mol Cell Biochem 356:217–225

142. Belanger-Jasmin S et al (2007) Inhibition of cortical astrocyte differentiation by Hes6 requires amino- and carboxy-terminal motifs important for dimerization and phosphorylation. J Neurochem 103(5):2022–2034
143. Lesokhin AM et al (1999) Several levels of EGF receptor signaling during photoreceptor specification in wild-type, ellipse, and null mutant Drosophila. Dev Biol 205:129–144
144. Kumar JP et al (2003) Nuclear translocation of activated MAP kinase is developmentally regulated in the developing Drosophila eye. Development 130(16):3703–3714
145. Yang L, Baker NE (2003) Cell cycle withdrawal, progression, and cell survival regulation by EGFR and its effectors in the differentiating Drosophila eye. Dev Cell 4(3):359–369
146. Heriche JK et al (1997) Regulation of protein phosphatase 2A by direct interaction with casein kinase 2alpha. Science 276(5314):952–955
147. Eichhorn PJ, Creyghton MP, Bernards R (2009) Protein phosphatase 2A regulatory subunits and cancer. Biochim Biophys Acta 1795(1):1–15
148. Millward T, Zolnierowicz S, Hemmings B (1999) Regulation of protein kinase cascades by protein phosphatase 2A. Trends Biochem Sci 24:186–191
149. Xu Y et al (2006) Structure of the protein phosphatase 2A holoenzyme. Cell 127(6): 1239–1251
150. Kunttas-Tatli E et al (2009) Functional dissection of Timekeeper (Tik) implicates opposite roles for CK2 and PP2A during Drosophila neurogenesis. Genesis 47:647–658
151. Bose A, Majot AT, Bidwai AP (2014) The Ser/Thr phosphatase PP2A regulatory subunit widerborst inhibits notch signaling. PLoS One 9(7):e101884
152. Guruharsha KG et al (2012) Drosophila protein interaction Map (DPiM): a paradigm for metazoan protein complex interactions. Fly (Austin) 6(4):246–253
153. Jiang J, Struhl G (1998) Regulation of the Hedgehog and Wingless signalling pathways by the F-box/WD40-repeat protein Slimb. Nature 391(6666):493–496
154. Isoda M et al (2009) The extracellular signal-regulated kinase-mitogen-activated protein kinase pathway phosphorylates and targets Cdc25A for SCF beta-TrCP-dependent degradation for cell cycle arrest. Mol Biol Cell 20(8):2186–2195
155. Brown NL et al (1996) Daughterless is required for drosophila photoreceptor cell determination, eye morphogenesis, and cell cycle progression. Dev Biol 179:65–78
156. Powell LM et al (2004) The proneural proteins Atonal and Scute regulate neural target genes through different E-box binding sites. Mol Cell Biol 24(21):9517–9526
157. Nolo R, Abbott LA, Bellen HJ (2000) Senseless, a Zn finger transcription factor, is necessary and sufficient for sensory organ development in Drosophila. Cell 102(3):349–362
158. The I et al (1997) Rescue of a Drosophila NF1 mutant phenotype by protein kinase A. Science 276:791–794
159. Fortini ME (2009) Notch Signaling: The core pathway and Its posttranslational regulation. Dev Cell 16:633–647
160. Gridley T (2003) Notch signaling and inherited disease syndromes. Hum Mol Genet 12(1):R9–R13
161. McDaniell R et al (2006) NOTCH2 mutations cause Alagille syndrome, a heterogeneous disorder of the notch signaling pathway. Am J Hum Genet 79(1):169–173
162. MacGrogan D, Nus M, de la Pompa JL (2010) Notch signaling in cardiac development and disease. Curr Top Dev Biol 92:333–365
163. Qiao L, Wong BC (2009) Role of Notch signaling in colorectal cancer. Carcinogenesis 30(12):1979–1986
164. Bolos V, Grego-Bessa J, de la Pompa JL (2007) Notch signaling in development and cancer. Endocr Rev 28(3):339–363
165. Andersson ER, Lendahl U (2014) Therapeutic modulation of Notch signalling–are we there yet? Nat Rev Drug Discov 13(5):357–378

Tissue-Specific Functions and Regulation of Protein Kinase CK2

Sabrina Welker, Christina Servas, Meng Rui, Claudia Götz, and Mathias Montenarh

Abstract CK2 is a ubiquitously expressed protein kinase which seems to be required for the viability of cells. An increasing number of CK2 substrates raise the questions how CK2 can phosphorylate so many different substrates and how these phosphorylation reactions are regulated. One key answer to these questions seems to be the subcellular localization of CK2 and its individual subunits. Here, we have addressed another strategy for the regulation of CK2, i.e. a tissue-specific expression, subcellular localization and tissue-specific substrates. In the case of pancreatic β-cells, we show that there is a hormonal impact by insulin which is similar to the hormonal regulation by androgens in prostate cancer cells. Furthermore, the subcellular localization of CK2 is specifically regulated by the metabolic status, i.e. glucose concentration in the pancreatic β-cells. Also the enzymatic activity as well as the binding to a pancreas-specific transcription factor Pdx-1 is modulated by the glucose concentration in pancreatic β-cells. Furthermore, the CK2 kinase activity seems to be an important survival factor for pancreatic β-cells. Thus, one key for the understanding of the pleiotropic effects of CK2 seems to be the determination of the regulation of CK2 in specific tissues.

Keywords Protein kinase • CK2 • Phosphorylation • Pancreas • Islet • Beta cells • Transcription factor

1 Introduction

In 1954 Burnett and Kennedy first described protein kinase CK2 (formerly known as casein kinase 2) in rat liver extracts [1]. Meanwhile protein kinase CK2 has been investigated in many organisms from yeast to man. It turned out that CK2 is a pleiotropic and ubiquitously expressed serine and threonine kinase, which is highly conserved during evolution. The predominant form of CK2 in mammalian cells is a

S. Welker • C. Servas • M. Rui • C. Götz • M. Montenarh (✉)
Medizinische Biochemie und Molekularbiologie, Universität des Saarlandes,
Gebäude 44, 66424 Homburg, Germany
e-mail: Mathias.Montenarh@uks.eu

K. Ahmed et al. (eds.), *Protein Kinase CK2 Cellular Function in Normal and Disease States*, Advances in Biochemistry in Health and Disease 12, DOI 10.1007/978-3-319-14544-0_7

holoenzyme consisting of two catalytic CK2α or CK2α′ subunits and two non-catalytic β subunits [2, 3]. There is increasing evidence that the individual subunits have functions different from the holoenzyme. Although CK2α and CK2α′ genes are located on different chromosomes, both proteins share 90 % sequence identity in the N-terminal 330 amino acids [4]. In the C terminus CK2α has a 20-amino-acid-long tail not present in CK2α′. CK2α is phosphorylated by cdk1 in a cell cycle-dependent manner [5, 6]. These phosphorylation sites are all located within the unique region of the CK2α subunit. This may be the reason for the functional differences between CK2α and CK2α′. CK2β is autophosphorylated by CK2α and also by cdk1 in a cell cycle-dependent manner [7]. A number of studies have shown that CK2 can also be phosphorylated by other protein kinases [6–10]. It remains, however, to be elucidated whether phosphorylation of the subunits indeed plays a role in the regulation of CK2. Crystallographic data shed more light on the composition of the holoenzyme where a dimer of CK2β binds two CK2α or CK2α′ [11–14]. Experiments in yeast and mice have demonstrated that CK2α and CK2β are essential for viability [15–18]. In contrast CK2α′ knockout mice are viable, but male mice are infertile due to a defect in spermatogenesis [19, 20]. These results indicate a specific function of CK2α and CK2α′ which cannot be compensated by either of the two. This specialization cannot be explained by major differences in the enzymatic activity of CK2α and CK2α′ because both proteins have identical catalytic sites. CK2 seems to have some features which are not shared with other protein kinases. CK2 belongs to the rather few kinases that can use either GTP or ATP as phosphate donors. Furthermore, CK2 mainly phosphorylates serine or threonine residues within a cluster of acidic residues [21–23]. A minimal consensus sequence for CK2 has been defined as S/T-X-X-D/E, where X can be any amino acid. Interestingly, phosphoserine or phosphotyrosine can be used as an acidic residue within the consensus sequence indicating that CK2 might play a role in hierarchical phosphorylation events. So far, CK2 is regarded as a constitutively active, second-messenger-independent protein kinase [24, 25]. There are now many reports on cellular regulators of CK2 [26–29]. Although we are far away from understanding how CK2 is regulated, an interaction of CK2 with cellular proteins or other factors in the cells seems to be one regulatory mechanism for CK2.

One other regulation for CK2 seems to be its subcellular localization. CK2 was found from the nucleus to the plasma membrane, and it seems that CK2 phosphorylates substrates at specific subcellular localizations [30]. Furthermore, binding partners of CK2 might help to target CK2 to specific localization within the cell where the enzyme meets certain substrates.

2 Tissue-Specific Expression and Regulation of CK2

CK2 is not only distributed in various locations in the cell as mentioned above but also at different stages of development and in all tissues of higher organisms. In a detailed study Maridor et al. investigated the relative distribution of mRNAs for

CK2α, α′ and β subunits in chicken tissues [31]. Most strikingly the expression of CK2β was not balanced with the expression of CK2α and CK2α′. Furthermore, most tissues were found to express one of the two catalytic subunits predominantly. In 1994 Mestres et al. detected CK2 in nearly all organs of the mouse embryo [32]. It was further shown that the level of CK2 transcripts usually correlates well with the protein expression profile. In the skin, however, high levels of transcripts were found, whereas the protein expression level was low. It was further shown that the rat brain in general expressed higher levels of CK2 than other tissues. Thus, this observation was an early indication for a tissue-specific expression of CK2 which might indicate a different and specific activity in certain tissues and organs [33–35]. Furthermore, the CK2α′ subunit is more abundant in neurons than in glia cells which supports the idea about individual functions of CK2α and CK2α′ and which might indicate tissue-specific activities for each of these subunits. As reported by several authors, beside the brain, high activities and expression levels were found for rat and mouse testes [36–38]. For other organs and tissues, there are controversial reports about the CK2 expression level or CK2 activity which might be due to differences in the methods applied for the analysis [19, 36–41].

CK2 is not only highly abundant in the brain; there are also a number of substrates for CK2 in synaptic and nuclear compartments, which have implications for information storage, synaptic plasticity, synaptic transmission and neuritogenesis and development [42]. CK2 kinase activity was found in all brain regions such as the cortex, septum, hippocampus, caudate-putamen, thalamus, olfactory bulb, cerebellum and spinal cord [40, 43–45]. The expression of the CK2 subunits was mainly analyzed by the immune reactivity in different areas of the brain [40, 46]. Blanquet already described a long list of potential and confirmed substrates of CK2 such as signalling molecules, cytoskeleton and structural proteins, transcription factors and DNA replication factors which play a role in the nervous system [42]. Furthermore, there is accumulating evidence that CK2 plays a role in neurodegenerating diseases, although the mode of action is not yet well understood [46–49]. Interestingly, immunohistochemical studies on Lewy bodies, one of the hallmarks of Parkinson's disease, revealed positive staining for CK2β but not for CK2α [50], indicating individual functional roles of CK2 subunits in the brain and brain disorders. This observation is further supported by recently published experiments where CK2β was shown not to be essential for central nervous system development [51].

Another tissue that has been thoroughly investigated for CK2 and its specific role in proliferation regulation is the prostate. Already in 1971 Ahmed and his coworkers started studies on the role of CK2 in the prostate [52]. It was rapidly shown that alterations in the CK2 activity depending on the androgenic status were due to a rapid shuttling of CK2 from chromatin structures into the nucleus and to the nuclear matrix [53, 54]. Furthermore, androgen withdrawal in rats was followed by a rapid loss of CK2 from nuclear compartments [55] and induction of apoptosis. Administration of androgens resulted in a rapid shuttling of CK2 to the nuclear matrix which goes along with elevated cell growth and suppression of apoptosis. These data are clear indications for a tissue-specific signalling on CK2.

3 Expression, Subcellular Localization and Regulation of CK2 in Pancreatic β-Cells

An implication of CK2 in another tissue, namely, the pancreatic β-cells, was first described in 1997 [56]. Zhang and Kim showed that a glucose-induced expression of the acetyl-CoA carboxylase is partly blocked in the presence of CK2α. This inhibitory effect was not observed in the absence of glucose or at low glucose concentration. It turned out that the transcription factor Sp1 regulates the promoter of acetyl-CoA carboxylase. Sp1 is a substrate for CK2, and Sp1 phosphorylation by CK2 downregulates its DNA binding activity. On the other hand, elevated glucose concentration in β-cells of the pancreas is the signal for the synthesis of insulin and its release from β-cells. Whether there is a link between the regulation of the transcription factor Sp1 by CK2 phosphorylation and the insulin production and secretion was not addressed. It was, however, shown by Sommercorn et al. that insulin stimulates CK2 kinase activity in pre-adipocytes and not in pancreatic β-cells [57, 58]. Years later a very similar stimulation of the CK2 kinase activity was demonstrated also in pancreatic β-cells without affecting the CK2 protein level [59].

Further support for a role of CK2 as a mediator of the insulin metabolism came from experiments which demonstrated that CK2 phosphorylated the transcription factor Pdx-1 at threonine 231 and serine 232. The sequence around these two phosphorylation sites is highly conserved in the animal kingdom indicating a functional unit of general importance. Pdx-1 is a crucial factor for the glucose-stimulated insulin synthesis. It turned out that inhibition of the CK2 kinase activity stimulates the transactivation of the insulin gene [60]. The same results were obtained with a Pdx-1 mutant where the CK2 phosphorylation sites were mutated to alanine residues indicating that Pdx-1 phosphorylation by CK2 might play a role in the regulation of the transcription of insulin. It was further shown that inhibition of CK2 activity led to an increase in the amount of insulin which was released from the pancreatic β-cells.

Pancreatic β-cells are highly sensitive to alterations in the glucose level. Therefore, it was an interesting question whether CK2 might also be involved in the glucose-mediated signalling in this type of cells. It was recently shown that CK2α is co-localized with Pdx-1 in primary mouse islets [61]. Moreover, further experiments showed that Pdx-1 and CK2α not only co-localized, but they can also be co-precipitated from extracts of primary mouse islet. A Duolink® in situ proximity ligation assay (PLA) confirmed binding of CK2α to Pdx-1 in pancreatic islets. Immunofluorescence studies revealed that the subcellular localization of CK2 and of Pdx-1 was glucose sensitive, i.e. in the absence of glucose, Pdx-1 and CK2 reside in the cytoplasm, whereas after addition of glucose, both proteins were translocated into the nucleus. Kinetic experiments revealed that Pdx-1 and CK2 were translocated independently and with a different kinetic into the nucleus. Only under high glucose concentration CK2α and Pdx-1 were bound to each other which was detected by co-immunoprecipitation as well as by PLA. The glucose sensitivity seems to be restricted to glucose-sensitive pancreatic β-cells, because in

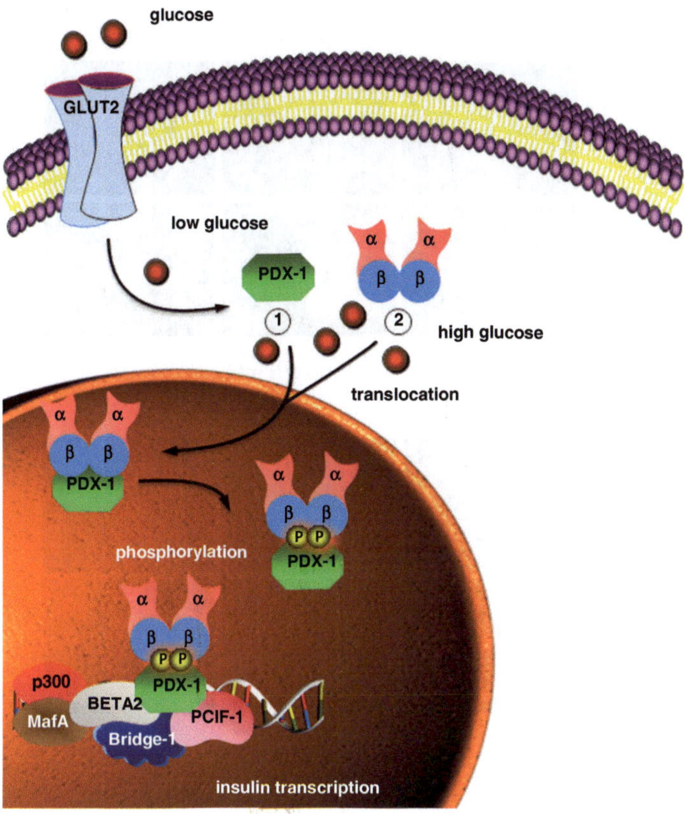

Fig. 1 Regulation of the subcellular localization and binding of Pdx-1 to CK2 depending on the glucose concentration in pancreatic β-cells. *Arrows* show a stimulatory effect

glucose-insensitive pancreatic β-cells, CK2α and Pdx-1 were both located in the nucleus at low and high glucose concentration [61] (Fig. 1). In order to analyze the subcellular localization of CK2α in totally different tissues, we used the hormone-sensitive prostate cancer cell line LNCaP [62] and embryonal endothelial cells (eEND) [63]. As shown in Fig. 2a, b in both cell lines under low and high glucose concentrations, CK2α was located in the cell nucleus. Thus, these results further support the idea that glucose is a regulator for the subcellular localization of Pdx-1 and CK2 in glucose-sensitive cells.

Pdx-1 has been shown to be phosphorylated by CK2 at threonine 231 and serine 232 [60]. Therefore, in the next step we asked whether the CK2 phosphorylation of Pdx-1 might have an influence on the subcellular localization of Pdx-1. MIN6 cells were incubated without glucose for 8 h. As expected Pdx-1 and CK2α were found in the cytoplasm (Fig. 3a). Then, cells were incubated with 25 mM glucose in the presence of 50 μM TBB to inhibit CK2 enzyme activity. Cells were analyzed for the localization of Pdx-1 and CK2α after 2, 4 and 24 h. As shown in Fig. 3b, both Pdx-1 and CK2α were found exclusively in the nucleus after 4 h. Co-localization of both

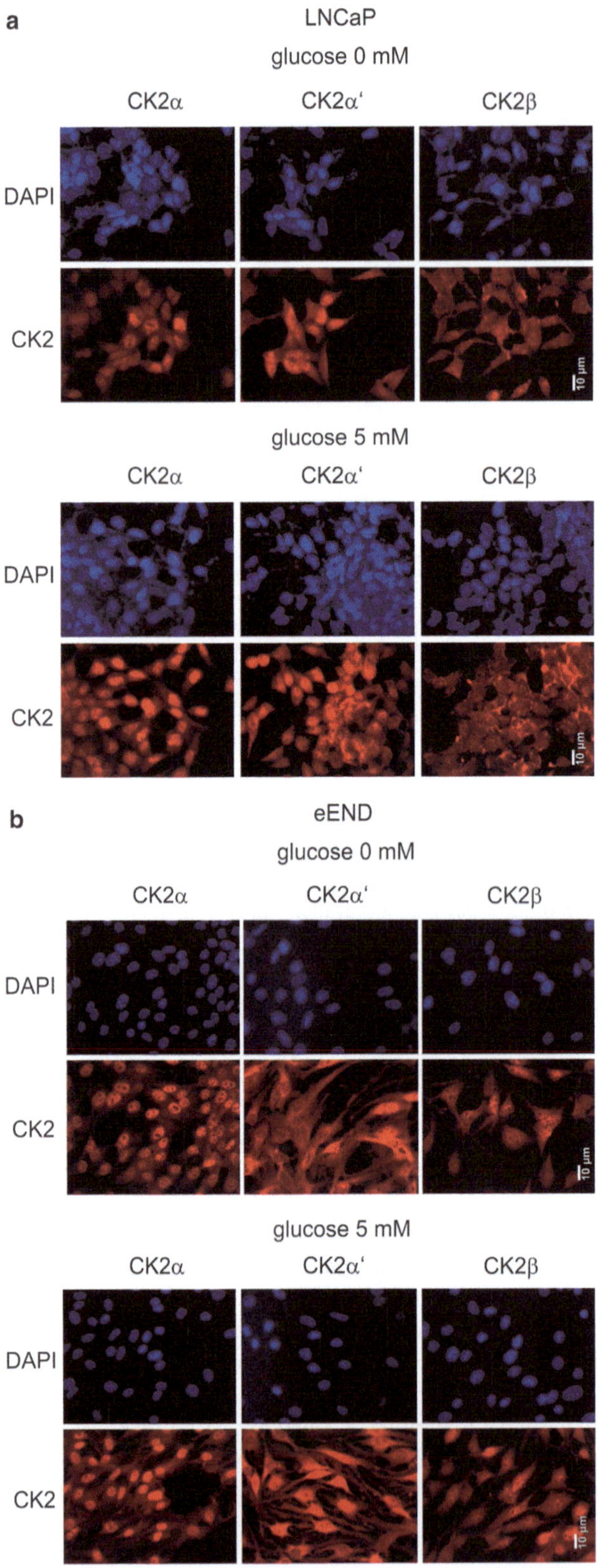
a
LNCaP
glucose 0 mM
CK2α
CK2α'
CK2β
DAPI
CK2
10 µm
glucose 5 mM
CK2α
CK2α'
CK2β
DAPI
CK2
10 µm
b
eEND
glucose 0 mM
CK2α
CK2α'
CK2β
DAPI
CK2
10 µm
glucose 5 mM
CK2α
CK2α'
CK2β
DAPI
CK2
10 µm

proteins was demonstrated by merging the green fluorescence for Pdx-1 with the red fluorescence for CK2α which resulted in a yellow colour of the co-localized proteins.

In another experiment we transfected Pdx-1 in its wild type (myc-Pdx-1_{WT}) or in its mutant conformation (myc-Pdx-$1_{T231A/S232A}$) into the pancreatic β-cell line βTC-3. We analyzed the subcellular localization of the proteins by immunofluorescence studies using an antibody directed to the myc-tag (Fig. 4). We never found a difference in the localization of wild type or mutant-Pdx1 which supports the data obtained with a CK2 inhibitor.

Thus, we have found that the subcellular localization of Pdx-1 and CK2 is dependent on a glucose stimulus in glucose-sensitive pancreatic β-cells and that the phosphorylation of Pdx-1 by CK2 presumably does not have an influence on its localization.

Diabetes type 2 is characterized by elevated levels of blood glucose owing to insulin resistance in the adipose tissue, muscle, liver and/or impaired insulin secretion from pancreatic β-cells. To adapt to the increased metabolic load caused by insulin resistance, pancreatic islets usually respond with increased β-cell proliferation as well as by enhancing insulin production. Upon elongated exposure to high glucose, β-cells adaption fails and there is a progressive decline in β-cell function and survival. The underlying mechanisms for β-cell death remain to be clarified. Here, we analyzed the fate of β-cells in the presence of high glucose. In order to study an influence of CK2 on cell viability, we counted pancreatic β-cells INS-1 after treatment with different glucose concentrations and in the presence or absence of the CK2 inhibitor CX-4945. As shown in Fig. 5a, we found a slight reduction in the percentage of living cells after 24 h. There was no further drop in the percentage of living cells after 48 and 64 h. In the presence of CX-4945, we found, however, a considerable drop in the percentage of living cells after 48 h and even more pronounced after 64 h. In the next step we addressed the question whether this drop in survival of the cells might be due to apoptosis. As a marker of apoptosis, we decided

Fig. 2 (**a**) Subcellular localization of CK2 under 0 mM (**a**) and 5 mM (**b**) glucose in LNCaP cells. LNCaP cells [71] were starved in glucose-free RPMI 1640 medium (Gibco) supplemented with 1 % glutamine and 10 % (v/v) fetal calf serum for 5 hours before they were treated with 0 mM or 5 mM glucose. After 4 h the cells on the coverslips were stained with rabbit antibody #26 against CK2α [72], rabbit antibody #30 against CK2α′ [72], rabbit antibody #269 against CK2β [72] and DAPI for nuclei staining. As secondary antibody, the FITC-labelled Alexa Fluor™ 594 was used. Immunofluorescence was analyzed with a ZEISS Axioskop: Ocular (10×), objective (Plan APOCHROMAT 40×Oil). (**b**) Subcellular localization of CK2 under 0 mM and 5 mM glucose in eEND cells. eEND cells [73] were starved in glucose-free Dulbecco's Modified Eagle's medium (Gibco) containing 1 % glutamine supplemented with 10 % (v/v) fetal calf serum for 5 h before they were treated with 0 mM or 5 mM glucose. After 4 h the cells on the coverslips were stained with antibodies against CK2α, CK2α′, CK2β and DAPI for nuclei staining. As secondary antibody, the FITC-labelled Alexa Fluor™ 594 was used. Immunofluorescence was analyzed with a ZEISS Axioskop. Ocular (10×), objective (Plan APOCHROMAT 40×Oil)

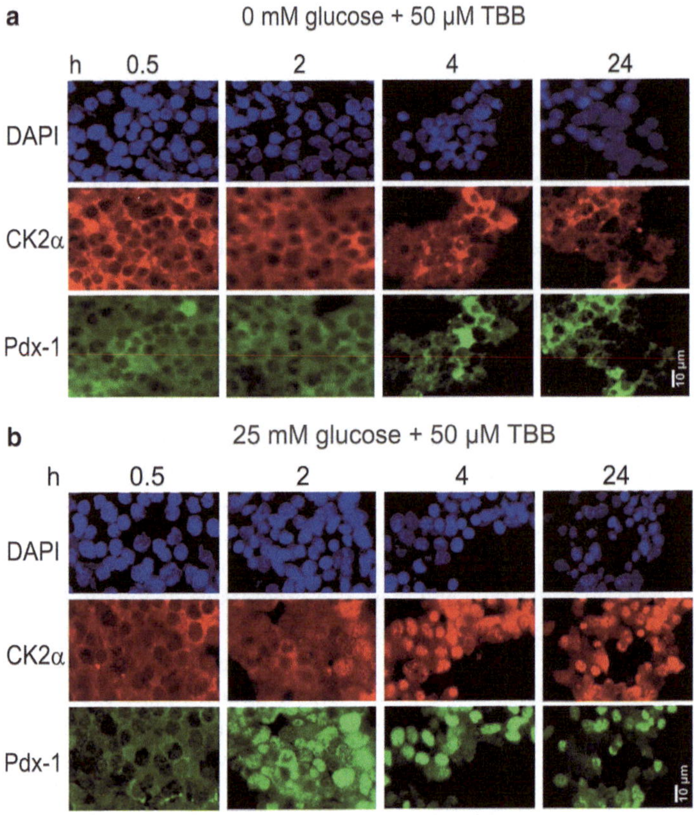

Fig. 3 Subcellular localization of Pdx-1 and CK2α under 0 mM (**a**) and 25 mM (**b**) glucose in MIN6 cells after treatment with 50 μM TBB. MIN6 cells [74] were cultured on coverslips and incubated in glucose-free DMEM containing 1 % glutamine supplemented with 15 % fetal calf serum and 100 μM β-mercaptoethanol for 8 h before incubation with 0 mM and 25 mM glucose and parallel treatment of 50 μM TBB. After 0.5, 2, 4 and 24 h, the cells on the coverslips were stained with the mouse monoclonal antibody against Clone 267712 (R & D Systems, Wiesbaden) Pdx-1 and CK2α (#26) as well as DAPI for nuclei staining. As secondary antibodies, the FITC-labelled Alexa Fluor™ 488 and Alexa Fluor™ 594 were used. Immunofluorescence was analyzed with a ZEISS Axioskop. Ocular (10×), objective (Plan APOCHROMAT 40×Oil)

to analyze the cleavage of poly ADP-ribose polymerase (PARP cleavage) as a late event in apoptosis. As shown in Fig. 5b, in the presence of normal glucose concentration, there was no PARP cleavage for up to 48 h. In the presence of high glucose, PARP cleavage was already observed after 48 h under 25 mM glucose. Most interestingly, in the presence of normal glucose concentration and in the presence of 25 mM glucose, there was clearly a signal for the cleavage product of PARP when cells were in addition treated with the CK2 inhibitor CX-4945. These data show that treatment of pancreatic β-cells with high glucose concentrations leads to apoptosis

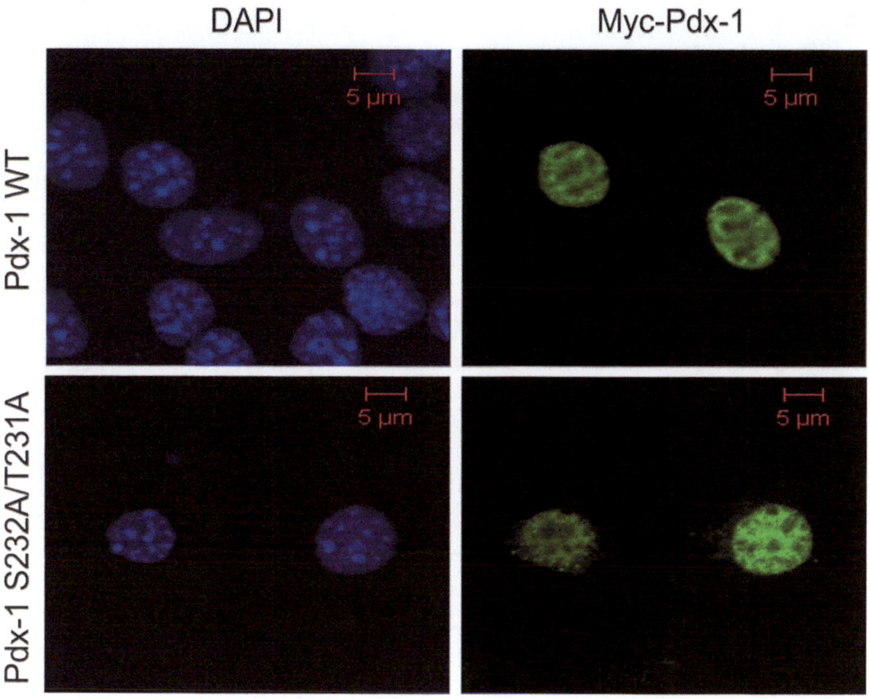

Fig. 4 Subcellular localization of myc-Pdx-1$_{WT}$ and myc-Pdx-1$_{T231A/S232A}$ in pancreatic β-cells. βTC-3 cells [75] were maintained in Dulbecco's Modified Eagle's Medium (Gibco) containing 5.5 mM D-glucose and 1 % glutamine supplemented with 15 % (v/v) fetal calf serum and 100 μM β-mercaptoethanol (β-MSH). Cells were transfected with myc-tagged Pdx-1$_{WT}$ and Pdx-1$_{T231A/S232A}$ constructs. Forty-eight hours post-transfection, cells were incubated with a mouse anti-myc antibody followed by FITC-labelled secondary antibody Alexa Fluor™ 488. DAPI was used as an indication of the nucleus. Immunofluorescence was analyzed with a ZEISS LSM 510 confocal microscope

of pancreatic β-cells and that CK2 inhibition accelerates apoptosis. It was previously shown that 2.5 mM and 5 mM glucose stimulated CK2 kinase activity, whereas 25 mM glucose already reduced the enzyme activity [61]. These results further support our hypothesis that CK2 inhibition contributes a synergistic effect to high glucose conditions and potentiates apoptosis.

4 Discussion

CK2 is a pleiotropically expressed protein kinase. This enzyme is absolutely essential for survival of cells which was demonstrated by the embryonic lethality of CK2α and CK2β knockout mice [17, 18]. One reason for the embryonic lethality might be the great number of cellular substrates which is already known to be

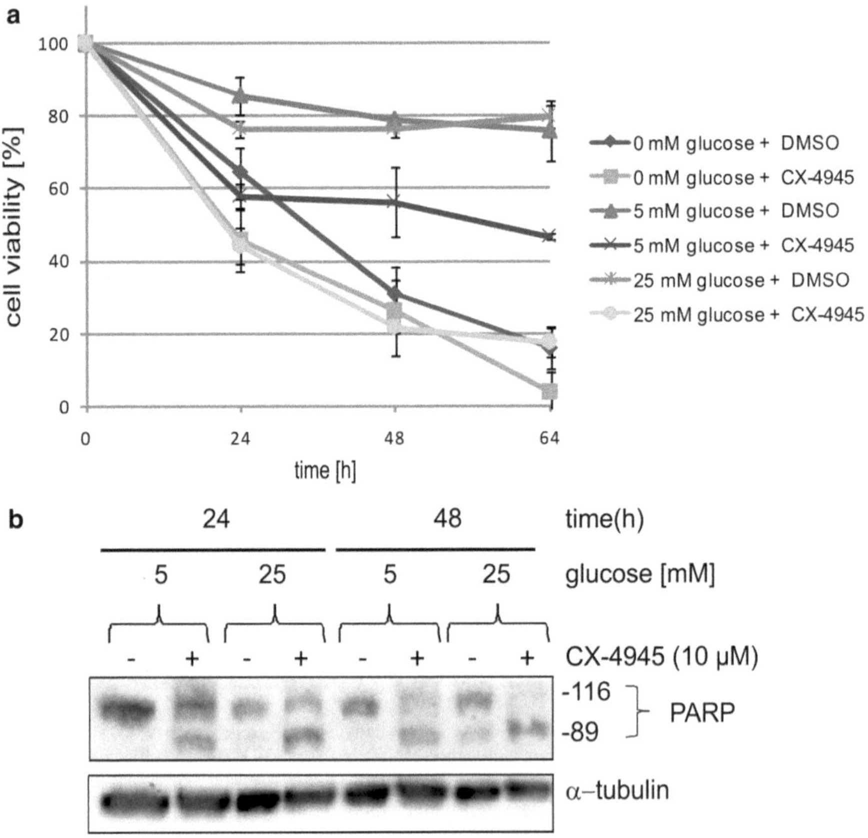

Fig. 5 (**a**) Viability of INS-1 cells in media with different glucose concentrations and inhibition of CK2 activity. INS-1 cells [73] were starved in glucose-free RPMI 1640 medium (Gibco) containing 10 % (v/v) fetal calf serum and 50 μM β-mercaptoethanol (β-MSH) as well as 1 mM sodium pyruvate. Cells were incubated for 24, 48 and 64 h in RPMI 1640 media with 0 mM, 5 mM or 25 mM glucose. CK2 activity was inhibited by 10 μM CX-4945 (Selleckchem, Munich, Germany) or as control, cells were treated with DMSO, respectively. Cells were harvested and washed with PBS, and living cells were detected with the Luna™ automated cell counter (Logos Biosystems, Anyang City, Korea). For counting dead cells the cell suspension was mixed with 20 μl of the diazo dye trypan blue and counted. The graph shows the mean values of three counts. The standard deviation is indicated. (**b**) PARP cleavage in INS-1 cells in media with high glucose concentrations and CK2 inhibition. INS-1 cells were incubated in media with 5 mM or 25 mM glucose with 10 μM CX-4945 or DMSO as a control for 24 and 48 h. The cell extracts were analyzed on a 10 % SDS-polyacrylamide gel and transferred on a PVDF membrane by Western blotting. The blot was stained with antibodies against PARP (#95425, Cell Signalling, Frankfurt, Germany) and α-tubulin (Clone DM1A, Sigma-Aldrich, Munich, Germany)

phosphorylated by CK2 [23]. Two intriguing questions were: How can CK2 phosphorylate so many different substrates in the cell and how are these phosphorylation reactions regulated? One possible answer to these questions might be the subcellular localization of CK2 in various cellular compartments, which seems to be highly dynamic [64, 65]. By this dynamic process, CK2 can meet different substrates depending on their subcellular localization. Furthermore, there is increasing evidence for a tissue-specific expression of CK2 and their substrates. It may also be that CK2 phosphorylates specific substrates only in a particular tissue. Here, we have summarized recent findings about the specific role of CK2 in β-cells of the pancreas which might have implications for diabetes.

We have shown that the subcellular localization of CK2 is regulated by the glucose concentration, i.e. under low glucose conditions, CK2 is located in the cytoplasm, whereas under normal and high glucose, CK2 is found in the nucleus. Under the same conditions also the transcription factor Pdx-1 translocates from the cytoplasm into the nucleus. Moreover, this translocation of Pdx-1 and of CK2 seems to be independent of each other and independent from the phosphorylation of Pdx-1 by CK2. Binding of Pdx-1 to CK2 was only detected under high glucose conditions. A re-localization of CK2 depending on the androgen level was reported for prostate cells [66, 67]. In this case CK2 was translocated from the nucleoplasm and chromatin to the nuclear matrix. These two examples might indicate that tissue-specific signals are implicated in directing CK2 to distinct places in the cell where CK2 might meet specific substrates. One of these specific substrates in pancreatic β-cells is the transcription factor Pdx-1.

Pharmacological inhibition of CK2 leads to apoptosis of cancer cells [68, 69], whereas normal healthy cells are unaffected [70]. Here, we demonstrated that high glucose led to apoptosis of pancreatic β-cells. Furthermore, inhibition of CK2 seems to accelerate this process, and moreover, also at low glucose concentration, CK2 inhibition triggers apoptosis of pancreatic β-cells. These results indicate that CK2 inhibition synergizes with high glucose concentration with respect to apoptosis induction. Thus, from the data known so far for pancreatic β-cells, CK2 seems to be implicated in the regulation of the synthesis of insulin. Elevated insulin production may induce an ER stress response. In the case of prolonged insulin synthesis, the ER stress response resists. In the case of a severe ER stress, cells go into apoptosis. One has to bear in mind these consequences for the pancreatic β-cells and insulin production when CK2 inhibitors are used for the treatment of cancer patients.

References

1. Burnett G, Kennedy EP (1954) The enzymatic phosphorylation of proteins. J Biol Chem 211:969–980
2. Gietz RD, Graham KC, Litchfield DW (1995) Interactions between the subunits of casein kinase II. J Biol Chem 270:13017–13021
3. Litchfield DW, Slominski E, Lewenza S, Narvey M, Bosc DG, Gietz RD (1996) Analysis of interactions between the subunits of protein kinase CK2. Biochem Cell Biol 74:541–547
4. Litchfield DW (2003) Protein kinase CK2: structure, regulation and role in cellular decisions of life and death. Biochem J 369:1–15
5. Bosc DG, Slominski E, Sichler C, Litchfield DW (1995) Phosphorylation of casein kinase II by p34^{cdc2}—Identification of phosphorylation sites using phosphorylation site mutants *in vitro*. J Biol Chem 270:25872–25878
6. Litchfield DW, Lüscher B, Lozeman FJ, Eisenman RN, Krebs E (1992) Phosphorylation of CK II by p34cdc2 in vitro and at mitosis. J Biol Chem 267:13943–13951

7. Litchfield DW, Lozeman FJ, Cicirelli MF, Harrylock M, Ericsson LH, Piening CJ, Krebs EG (1991) Phosphorylation of the beta subunit of casein kinase II in human A431 cells. Identification of the autophosphorylation site and a site phosphorylated by p34cdc2. J Biol Chem 266:20380–20389
8. Ackerman P, Glover CV, Osheroff N (1990) Stimulation of casein kinase II by epidermal growth factor: relationship between the physiological activity of the kinase and the phosphorylation state of its beta subunit. Proc Natl Acad Sci U S A 87:821–825
9. Mulner-Lorillon O, Cormier P, Labbe J-C, Doree M, Pouhle R, Osborne H, Belle R (1990) M-phase-specific cdc2 protein kinase phosphorylates the β subunit of casein kinase II and increases casein kinase II activity. Eur J Biochem 193:529–534
10. Carroll D, Santoro N, Marshak DR (1988) Regulating cell growth: casein-kinase-II-dependent phosphorylation of nuclear oncoproteins. Cold Spring Harb Symp Quant Biol 53(Pt 1):91–95
11. Lolli G, Pinna LA, Battistutta R (2012) Structural determinants of protein kinase CK2 regulation by autoinhibitory polymerization. ACS Chem Biol 7:1158–1163
12. Niefind K, Battistutta R (2013) Structural bases protein kinase CK2 function and inhibition. In: Pinna LA (ed) Protein kinase CK2. John Wiley & Sons Inc., Hoboken, NJ, pp 3–75
13. Niefind K, Guerra B, Ermakowa I, Issinger OG (2001) Crystal structure of human protein kinase CK2: insights into basic properties of the CK2 holoenzyme. EMBO J 20:5320–5331
14. Niefind K, Raaf J, Issinger OG (2009) Protein kinase CK2: from structures to insights. Cell Mol Life Sci 66:1800–1816
15. Glover CVC (1998) On the physiological role of casein kinase II in Saccharomyces cerevisiae. Prog Nucleic Acid Res Mol Biol 59:95–133
16. Rethinaswamy A, Birnbaum MJ, Glover CVC (1998) Temperature-sensitive mutations of the CAK1 gene reveal a role for casein kinase II in maintenance of cell polarity in Saccharomyces cerevisiae. J Biol Chem 273:5869–5877
17. Buchou T, Vernet M, Blond O, Jensen HH, Pointu H, Olsen BB, Cochet C, Issinger OG, Boldyreff B (2003) Disruption of the regulatory β subunit of protein kinase CK2 in mice leads to a cell-autonomous defect and early embryonic lethality. Mol Cell Biol 23:908–915
18. Lou DY, Dominguez I, Toselli P, Landesman-Bollag E, O'Brien C, Seldin DC (2008) The alpha catalytic subunit of protein kinase CK2 is required for mouse embryonic development. Mol Cell Biol 28:131–139
19. Xu X, Toselli PA, Russell LD, Seldin DC (1999) Globozoospermia in mice lacking the casein kinase II α′ catalytic subunit. Nat Genet 23:118–121
20. Escalier D, Silvius D, Xu X (2003) Spermatogenesis of mice lacking CK2α′: Failure of germ cell survival and characteristic modifications of the spermatid nucleus. Mol Reprod Dev 66:190–201
21. Kuenzel EA, Mulligan JA, Sommercorn J, Krebs EG (1987) Substrate specificity determinants for casein kinase II as deduced from studies with synthetic peptides. J Biol Chem 262: 9136–9140
22. Salvi M, Sarno S, Cesaro L, Nakamura H, Pinna LA (2009) Extraordinary pleiotropy of protein kinase CK2 revealed by weblogo phosphoproteome analysis. Biochim Biophys Acta 1793:847–859
23. Meggio F, Pinna LA (2003) One-thousand-and-one substrates of protein kinase CK2? FASEB J 17:349–368
24. Litchfield DW, Dobrowolska G, Krebs EG (1994) Regulation of casein kinase II by growth factors: a reevaluation. Cell Mol Biol Res 40:373–381
25. Hathaway GM, Traugh JA (1982) Casein kinases—multipotential protein kinases. Curr Top Cell Regul 21:101–127
26. Montenarh M (2010) Cellular regulators of protein kinase CK2. Cell Tissue Res 342: 139–146
27. Montenarh M (2003) Protein kinase CK2: in search for its regulation. Adv Clin Exp Med 12:15–22

28. Montenarh M, Götz C (2013) The interactome of protein kinase CK2. In: Pinna LA (ed) Protein kinase CK2. John Wiley & Sons Inc., Hoboken, NJ, pp 76–116
29. Guerra B, Boldyreff B, Sarno S, Cesaro L, Issinger OG, Pinna LA (1999) CK2: A protein kinase in need of control. Pharmacol Ther 82:303–313
30. Faust M, Montenarh M (2000) Subcellular localization of protein kinase CK2: a key to its function? Cell Tissue Res 301:329–340
31. Maridor G, Park W, Krek W, Nigg EA (1991) Casein kinase II. cDNA sequences, developmental expression, and tissue distribution of mRNAs for alpha, alpha′, and beta subunits of the chicken enzyme. J Biol Chem 266:2362–2368
32. Mestres P, Boldyreff B, Ebensperger C, Hameister H, Issinger O-G (1994) Expression of casein kinase 2 during mouse embryogenesis. Acta Anat 149:13–20
33. Diaz-Nido J, Mizuno K, Nawa H, Marshak DR (1994) Regulation of protein kinase Ck2 isoform expression during rat brain development. Cell Mol Biol Res 40:581–585
34. Diaz-Nido J, Armes-Portela R, Avila J (1992) Increase in cytoplasmic casein kinase II-type activity accompanies neurite growth after DNA synthesis. J Neurochem 587:1820–1828
35. Ulloa L, Diaz-Nido J, Avila J (1993) Depletion of casein kinase II by antisense oligonucleotide prevents neuritogenesis in neuroblastoma cells. EMBO J 12:1633–1640
36. Yutani Y, Tei Y, Yukioka M, Inoue A (1982) Occurrence of NI and NII type protein kinases in the nuclei from various tissues of the rat. Arch Biochem Biophys 218:409–420
37. Singh TJ, Huang K (1985) Glycogen synthase (casein) kinase I: tissue distribution and subcellular localization. FEBS Lett 190:84–88
38. Guerra B, Siemer S, Boldyreff B, Issinger OG (1999) Protein kinase CK2: evidence for a protein kinase CK2β subunit fraction, devoid of the catalytic CK2α subunit, in mouse brain and testicles. FEBS Lett 462:353–357
39. Nakajo S, Hagiwara T, Nakajo K, Nakamura Y (1986) Tissue distribution of casein kinases. Biochem Int 13:701–707
40. Girault A-J, Hemmings HC, Zorn SH, Gustafson EL, Greengard P (1990) Characterization in mammalian brain of a DARPP-32 serine kinase identical to casein kinase II. J Neurochem 55:1772–1783
41. Stalter G, Siemer S, Becht E, Ziegler M, Remberger K, Issinger O-G (1994) Asymmetric expression of protein kinase CK2 in human kidney tumors. Biochem Biophys Res Commun 202:141–147
42. Blanquet PR (2000) Casein kinase 2 as a potentially important enzyme in the nervous system. Prog Neurobiol 60:211–246
43. Moreno FJ, Díaz-Nido J, Jiménez JS, Avila J (1999) Distribution of CK2, its substrate MAP1B and phosphatases in neuronal cells. Mol Cell Biochem 191:201–205
44. De Camilli P, Greengard P (1986) Synapsin I: a synaptic vesicle-associated neuronal phosphoprotein. Biochem Pharmacol 35:4349–4357
45. Martin ME, Alcazar A, Salinas M (1990) Subcellular and regional distribution of casein kinase II and initiation factor 2 activities during rat brain development. Int J Dev Neurosci 8:47–54
46. Iimoto DS, Masliah E, Deteresa R, Terry RD, Saitoh T (1990) Aberrant casein kinase II in Alzheimer's disease. Brain Res 507:273–280
47. Masliah E, Iimoto DS, Mallory M, Albright T, Hansen L, Saitoh T (1992) Casein kinase II alteration precedes tau accumulation in tangle formation. Am J Pathol 140:263–268
48. Baum L, Masliah E, Iimoto DS, Hansen LA, Halliday WC, Saitoh T (1992) Casein kinase II is associated with neurofibrillary tangles but is not an intrinsic component of paired helical filaments. Brain Res 573:126–132
49. Jin LW, Saitoh T (1995) Changes in protein kinases in brain aging and Alzheimer's disease. Implications for drug therapy. Drugs Aging 6:136–149
50. Ryu MY, Kim DW, Arima K, Mouradian MM, Kim SU, Lee G (2007) Localization of CKII beta subunits in Lewy bodies of Parkinson's disease. J Neurol Sci 266:9–12

51. Huillard E, Ziercher L, Blond O, Wong M, Deloulme JC, Souchelnytskyi S, Baudier J, Cochet C, Buchou T (2010) Disruption of CK2beta in embryonic neural stem cells compromises proliferation and oligodendrogenesis in the mouse telencephalon. Mol Cell Biol 30:2737–2749
52. Ahmed K, Ishida H (1971) Effect of testosterone on nuclear phosphoproteins of rat ventral prostate. Mol Pharmacol 7:323–327
53. Ahmed K, Tawfic S (1994) Mechanism of intracellular regulation of protein kinase CK2: role of stimulus-mediated subnuclear association. Cell Mol Biol Res 40:539–545
54. Ahmed K, Yenice S, Davis A, Goueli SA (1993) Association of casein kinase 2 with nuclear chromatin in relation to androgenic regulation of rat prostate. Proc Natl Acad Sci U S A 90:4426–4430
55. Yu SH, Wang HM, Davis A, Ahmed K (2001) Consequences of CK2 signaling to the nuclear matrix. Mol Cell Biochem 227:67–71
56. Zhang S, Kim KH (1997) Protein kinase CK2 down-regulates glucose-activated expression of the acetyl-CoA carboxylase gene. Arch Biochem Biophys 338:227–232
57. Sommercorn J, Mulligan JA, Lozeman FJ, Krebs E (1987) Activation of casein kinase II in response to insulin and to epidermal growth factor. Proc Natl Acad Sci U S A 84:8834–8838
58. Sommercorn J, Krebs EG (1987) Induction of casein kinase II during differentiation of 3T3-L1 cells. J Biol Chem 262:3839–3843
59. Meng R, Götz C, Montenarh M (2010) The role of protein kinase CK2 in the regulation of the insulin production of pancreatic islets. Biochem Biophys Res Commun 401:203–206
60. Meng R, Al-Quobaili F, Müller I, Götz C, Thiel G, Montenarh M (2010) CK2 phosphorylation of Pdx-1 regulates its transcription factor activity. Cell Mol Life Sci 67:2481–2489
61. Welker S, Götz C, Servas C, Laschke MW, Menger MD, Montenarh M (2013) Glucose regulates protein kinase CK2 in pancreatic ß-cells and its interaction with PDX-1. Int J Biochem Cell Biol 45:2786–2795
62. Horoszewicz JS, Leong SS, Kawinski E, Karr JP, Rosenthal H, Chu TM, Mirand EA, Murphy GP (1983) LNCaP model of human prostatic carcinoma. Cancer Res 43:1809–1818
63. Bussolino F, De RM, Sica A, Colotta F, Wang JM, Bocchietto E, Padura IM, Bosia A, DeJana E, Mantovani A (1991) Murine endothelioma cell lines transformed by polyoma middle T oncogene as target for and producers of cytokines. J Immunol 147:2122–2129
64. Martel V, Filhol O, Nueda A, Gerber D, Benitez MJ, Cochet C (2001) Visualization and molecular analysis of nuclear import of protein kinase CK2 subunits in living cells. Mol Cell Biochem 227:81–90
65. Theis-Febvre N, Martel V, Laudet B, Souchier C, Grunwald D, Cochet C, Filhol O (2005) Highlighting protein kinase CK2 movement in living cells. Mol Cell Biochem 274:15–22
66. Tawfic S, Ahmed K (1994) Growth stimulus-mediated differential translocation of casein kinase 2 to the nuclear matrix. Evidence based on androgen action in the prostate. J Biol Chem 269:24615–24620
67. Yenice S, Davis AT, Goueli SA, Akdas A, Limas C, Ahmed K (1994) Nuclear casein kinase 2 (CK-2) acitivity in human normal, benign hyperplastic, and cancerous prostate. Prostate 24:11–16
68. Ahmad KA, Wang G, Unger G, Slaton J, Ahmed K (2008) Protein kinase CK2- A key suppressor of apoptosis. Adv Enzyme Regul 48:179–187
69. Trembley JH, Wu J, Unger GM, Kren BT, Ahmed K (2013) CK2 suppression of apoptosis and its implication in cancer biology and therapy. In: Pinna LA (ed) Protein kinase CK2. John Wiley & Sons Inc., Hoboken, NJ, pp 319–343
70. Intemann J, Saidu NEB, Schwind L, Montenarh M (2014) ER stress signaling in ARPE-19 cells after inhibition of protein kinase CK2 by CX-4945. Cell Signal 26:1567–1575
71. Horoszewicz JS, Leong SS, Chu TM, Wajsman ZL, Friedman M, Papsidero L, Kim U, Chai LS, Kakati S, Arya SK, Sandberg AA (1980) The LNCaP cell line–a new model for studies on human prostatic carcinoma. Prog Clin Biol Res 37:115–132
72. Faust M, Schuster N, Montenarh M (1999) Specific binding of protein kinase CK2 catalytic subunits to tubulin. FEBS Lett 462:51–56

73. Asfari M, Janjic D, Meda P, Li G, Halban PA, Wollheim CB (1992) Establishment of 2-mercaptoethanol-dependent differentiated insulin-secreting cell lines. Endocrinology 130:167–178
74. Ishihara H, Asano T, Tsukuda K, Katagiri H, Inukai K, Anai M, Kikuchi M, Yazaki Y, Miyazaki JI, Oka Y (1993) Pancreatic beta cell line MIN6 exhibits characteristics of glucose metabolism and glucose-stimulated insulin secretion similar to those of normal islets. Diabetologia 36:1139–1145
75. Efrat S (1999) Genetically engineered pancreatic beta-cell lines for cell therapy of diabetes. Ann N Y Acad Sci 875:286–293

CK2 Function in the Regulation of Akt Pathway

Cristina Girardi and Maria Ruzzene

Abstract CK2 and Akt (also known as PKB) are two antiapoptotic protein kinases, which phosphorylate hundreds of substrates. While CK2 is constitutively active, Akt, under physiological conditions, is activated by external growth/proliferation signals. However, in cancer, on the one hand, the mechanism of Akt activation is frequently dysregulated and, on the other, CK2 is usually overexpressed. Both kinases are therefore involved in tumorigenesis and considered promising drug targets.

CK2 profoundly intersects the Akt signaling: it phosphorylates several components of the Akt pathway, producing positive effects on stimulatory elements while blocking the inhibitory ones. The final outcome is that CK2 potentiates the Akt survival message.

Here we summarize the major contact points between the two kinases, focusing on the direct connection due to the CK2-dependent phosphorylation of Akt1 Ser129, but also on different levels of integration, spanning from the Akt activation mechanism at plasma membrane to the downstream effectors. It will be also highlighted how this orchestrated circuitry involving Akt and CK2 is aberrantly amplified under certain pathological circumstances, offering sites for therapeutic intervention. However, we will also review recent findings concerning specific regulation of different Akt isoforms by CK2, which can explain some of their divergent functions and warn against a too cursory pharmacological targeting of CK2/Akt, which instead can be planned only once a comprehensive picture of their connections in a specific tumor cell type has been depicted.

Keywords CK2 • Akt • Casein kinase 2 • CKII • PKB • Akt isoforms • PKB isoforms • Ser129 Akt1 • Akt2 • PTEN • PI3K • PML

C. Girardi • M. Ruzzene (✉)
Department of Biomedical Sciences, University of Padova,
Via U. Bassi 58/b, Padova 35131, Italy
e-mail: maria.ruzzene@unipd.it

K. Ahmed et al. (eds.), *Protein Kinase CK2 Cellular Function in Normal and Disease States*, Advances in Biochemistry in Health and Disease 12, DOI 10.1007/978-3-319-14544-0_8

Abbreviations

AML	Acute myeloid leukemia
B-ALL	B-cell acute lymphoblastic leukemia
CLL	Chronic lymphocytic leukemia
CML	Chronic myeloid leukemia
Fbxo9	F-box only protein 9
GF	Growth factor
NEP	Neprilysin
PAF	Platelet-activating factor
PD-1	Receptor programmed death 1
PH	Pleckstrin homology
PHLPP	Pleckstrin homology domain leucine-rich repeat protein phosphatase
PI3K	Phosphoinositide 3-kinase
PIP_3	Phosphatidylinositol-3,4,5-trisphosphate
Plk3	Polo-like kinase 3
PML	Promyelocytic leukemia gene
PP2A	Protein phosphatase 2A
T-ALL	T-cell acute lymphoblastic leukemia
Tel2	Telomere maintenance 2
TIF-1A	Transcription initiation factor I
Tti1	Tel2-interacting protein 1

1 Introduction

Protein kinase CK2 is a ubiquitous and constitutively active Ser/Thr protein kinase, usually present in cells as a tetrameric holoenzyme composed of two catalytic and two regulatory subunits. α and the less represented isoform α′ are the catalytic subunits, while the β subunits form a dimer which has a major role in determining substrate selection [1]. CK2 is highly pleiotropic and phosphorylates hundreds of substrates, impinging on practically all cellular processes [2]; however, it has a recognized major function as an antiapoptotic agent. Several targets have been identified by which CK2 promotes cell survival and resistance to apoptosis (as reviewed in [3–6]); among them, there are different components of the PI3K/Akt pathway.

Akt, also known as PKB, is another prosurvival kinase, whose signaling is frequently deregulated in cancer [7, 8]. Three different isoforms of Akt exist, named Akt1 (or PKBα), Akt2 (or PKBβ), and Akt3 (or PKBγ); they share the same activation mechanism [9], which relies on the phosphorylation of two key residues, Thr308 in the catalytic domain and Ser473 at the C-terminal hydrophobic tale (numbering refers to Akt1). Under physiological conditions, the Akt pathway is triggered by external growth/proliferation stimuli which, activating PI3K, promote an increased concentrations of PIP_3; this binds to the PH domain of Akt and recruits it to the plasma membrane, where it can be activated by the action of the upstream kinase PDK1 (which phosphorylates Thr308) and of the complex mTORC2 (which phosphorylates Ser473). An alternative Ser473 kinase has been proposed, DNA-PK, whose action would be particularly important following DNA damage [10]. Once Akt is phosphorylated on Thr308 and Ser473 and the structural inhibition exerted by the PH domain is removed [11], the enzyme can move to its targets (largely represented by nuclear proteins), phosphorylate them, and produce its effects. These are mainly related to cell growth, proliferation, and survival, although important metabolic processes are also under the control of Akt, with different roles according to the specific Akt isoform (see below). As crucial elements of the PI3K/Akt signaling under healthy conditions, several phosphatases have been identified as able to turn off the pathway. In particular, the lipid phosphatase PTEN dephosphorylates PIP_3, stopping the signal at an upstream level, while the Ser/Thr phosphatases PP2A and PHLPP1/2 have been shown to dephosphorylate the Akt residues Thr308 and Ser473, respectively [12]. Not surprisingly, the complexity of this signaling mechanism, while allowing fine tuning, also offers numerous possibilities of malfunctioning; indeed, many cancers are characterized by upregulated PI3K/Akt signaling, and oncogenic mutations of individual components are also known [13].

As mentioned above, Akt signaling is among the targets that CK2 exploits to execute its survival message. There are several cross points between the two kinases, with the global effect of potentiating the Akt signaling. This implies that even in the absence of gain-of-function mutations of Akt pathway components, its signaling can become hyperactive under cellular conditions where CK2 is aberrantly high, as typically occurs in cancer cells.

Here we review the different levels of interconnection between CK2 and Akt, which span along the whole pathway, starting from the plasma membrane and getting the nucleus.

2 The Paradox of PTEN Regulation by CK2

As mentioned above, PTEN is a lipid phosphatase which reverses the action of PI3K, thus preventing an excessive and untimely activation of PI3K/Akt signaling. Indeed, PTEN is considered a tumor suppressor, and several loss-of-function tumorigenic mutations have been identified [14, 15].

The phosphorylation of PTEN by CK2 has been known for several years [16]; the phosphorylated sites have been identified at the C-terminus [17], the main targets being Ser370, Ser385, and Thr366 (this latter identified as phosphorylated by other kinases as well, see below). Their phosphorylation was initially related to an enhanced stability of the protein, which becomes more resistant to proteasome degradation. However, the C-terminal tail of PTEN has the double role of maintaining protein stability and inhibiting PTEN activity [18], and it was later evident that the phosphorylation of the C-terminal sites by CK2 was accompanied by a reduced catalytic activity. Consistently, overexpression of CK2 has been evoked to explain why samples from T-ALL patients showed constitutive hyperactivation of the PI3K/Akt pathway without PTEN gene alterations and with PTEN protein levels paradoxically higher than in normal T cells [19]. Similar data were reported for adult B-ALL specimens [20]. In summary, CK2, by phosphorylating PTEN, maintains a huge amount of an inactive form of the enzyme, thus providing potentiation of the PI3K signaling. CK2 overexpression can therefore be crucial for all those tumors where abnormal Akt activity is accompanied by higher, instead of lower, levels of PTEN protein compared to normal cells [21]. The scenario is further complicated by the observation that other kinases may contribute to the phosphorylation of the PTEN C-terminal sites: it has been found that GSK3β phosphorylates Ser362 and Thr366, with a hierarchical mechanism requiring a priming action of CK2 [22], and a negative-feedback loop has been proposed for the action of GSK3β, which is inhibited by the Akt-mediated phosphorylation. Moreover, it has been found that also Plk3 phosphorylates Thr366 and Ser370 [23]. These other kinases are therefore expected to cross the action of CK2 in the regulation of PTEN function. However, the understanding of how their signals are integrated, and to which extent their effects synergize, is still largely unknown.

The direct phosphorylation of PTEN is not the only mechanism by which CK2 counteracts the function of PTEN: it has been recently reported [24] that an interacting partner of PTEN, neprilysin (NEP), is a substrate of CK2. NEP is a membrane metalloproteinase whose N-terminal cytoplasmic domain recruits PTEN to its site of action. When NEP is phosphorylated at Ser6 by CK2, it becomes unable to interact with PTEN, which fails to localize at the plasma membrane and is no more able to dephosphorylate PIP_3. We can therefore hypothesize that high levels of CK2 can cause Akt hyperactivation also by determining an excessive level of NEP phosphorylation.

The action of CK2 at the PTEN level has several physiological and pathological consequences. For example, it has been reported that the receptor programmed death 1 (PD-1), an inhibitor of T-cell receptor responses, acts by inhibiting CK2 expression and activity, thus preventing the phosphorylation of the PTEN C-terminal sites [25]. Kang and coauthors [26] demonstrated that the CK2/PTEN connection is crucial for the Akt activation during PAF-mediated anaphylactic shock, and, for this reason, they propose CK2 as a possible target for the control of this pathological condition.

It has been also proposed that the signaling mediated by stimulation of leptin and adiponectin receptors requires the inhibition of PTEN due to its CK2-dependent phosphorylation; the adiponectin receptor (AdipoR1) has been also shown to physically interact with the regulatory β subunit of CK2 [27, 28].

As schematically depicted in Fig. 1, a multifaceted role is therefore recognized to CK2 in controlling PTEN functions; in cancer cells, where CK2 activity is abnormally high, this level of CK2 action can be crucial to sustain Akt activity, as recently reviewed in [29].

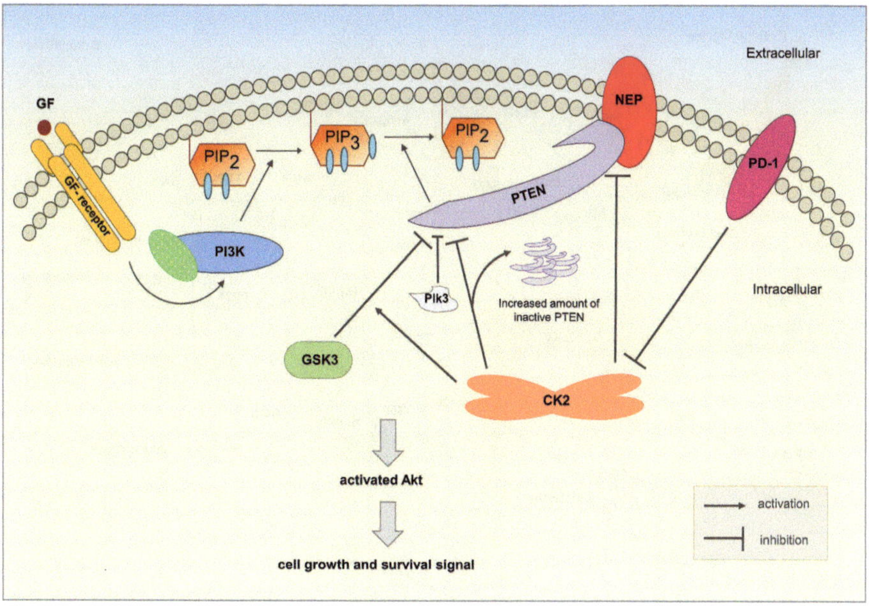

Fig. 1 Regulation of PTEN functions by CK2. The PIP_3 concentration balance due to the action of PI3K and PTEN is represented. The indicated sites of CK2 intervention have the global effect of maintaining a high level of PIP_3, with consequent potentiation of Akt signaling

3 The Direct CK2/Akt Cross Talk (Fig. 2)

Our previous work [30] has demonstrated that Akt1 is a direct target of CK2, both in vitro and in vivo; CK2 phosphorylates Akt1 at Ser129, a site located in the linker region, which connects the PH and the catalytic domain. The phosphorylation of this site, in addition to the canonical ones at Thr308 and Ser473, confers higher catalytic activity to Akt1, as judged from the phosphorylation of some of its substrates, such as GSK3, FHKR, and AFX, implying that downregulation of CK2 negatively affects Akt pathway even in PTEN-null cells [30]. These findings suggest that CK2 is a general and valuable target in order to counteract a hyperactive PI3K/Akt signaling, and in fact several papers have been published showing that treatment of different cancer cells with CK2 inhibitors produces the downregulation of Akt signaling. Without the pretension to be exhaustive, we can mention studies in blood cancer cells such as AML [31, 32], CLL [33], CML [34], and T-ALL [35], in prostate cancer cells [36–38], in breast cancer cells [39], in lung cancer cells [40–42], in glioblastoma [43, 44], in skin squamous cell carcinoma [41], or in thyroid cancer cells [45]. At present, the level of Akt1 phosphorylation at Ser129, checked by means of appropriate phospho-specific antibodies, is widely used as a reliable report of the cellular activity of CK2; being this site rapidly dephosphorylated upon blockage of CK2 activity [46], its analysis is particularly useful in studies on CK2 inhibitors.

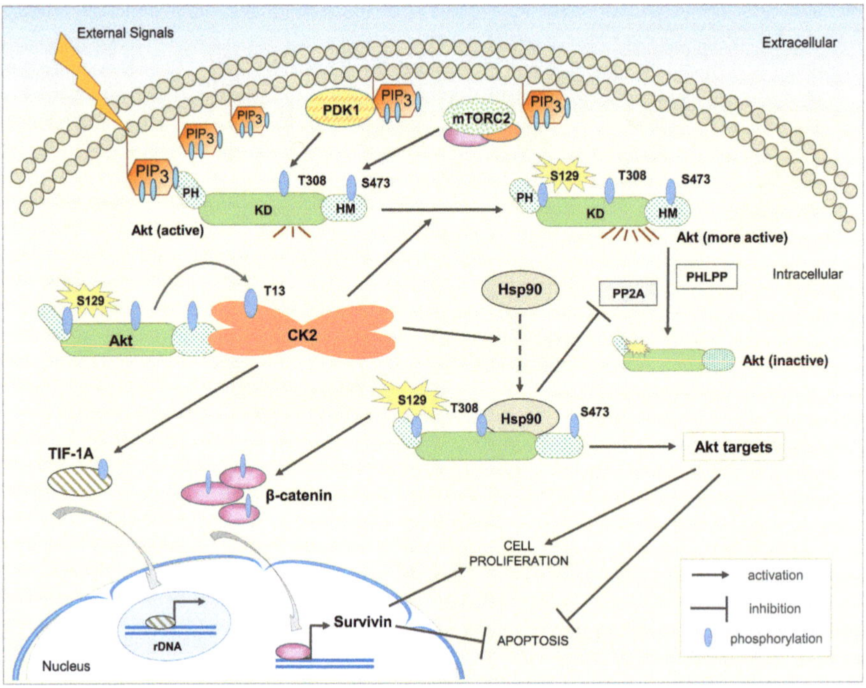

Fig. 2 Direct connections between CK2 and Akt. The mechanism of Akt activation is depicted, with full activity reached upon Thr308, Ser473, and Ser129 phosphorylation. Also shown are the effects of Ser129 in mediating Hsp90 recruitment and prevention of dephosphorylation by PP2A, Ser129 importance in the activation of β-catenin functions, and the Akt-mediated phosphorylation of α CK2 at Thr13 that regulates its activity toward TIF-IA. The Akt molecule is in green, showing its domain organization (*PH* pleckstrin homology domain, *KD* kinase domain, *HM* hydrophobic motif)

Although the exact mechanism by which phosphorylation at Ser129 activates Akt is still poorly understood, a number of observations and consequences have been reported. For example, we found that phospho-Ser129 stabilizes the association of Akt with Hsp90 [47]; since the complex of Akt with this chaperone protein is known to protect Thr308 from dephosphorylation by PP2A [48], we concluded that Akt phosphorylation by CK2 promotes its maintenance in an active form. Ser129 phosphorylation has been later demonstrated to be required for the Akt-mediated upregulation of the β-catenin pathway: in particular, the increase in the β-catenin transcriptional activity, its nuclear localization, and the consequent overexpression of survivin are prevented when a mutant of Akt1, not phosphorylatable at Ser129, is expressed [49].

Phosphorylation of Akt1 on Ser129 was also revealed as a key element of the ephrin B3-mediated signaling, a pathway frequently deregulated in cancer [40].

Apart from these indirect effects of the CK2-dependent phosphorylation of Akt, the site location suggests a possible prominent role in Akt structural and functional features: Ser129 is in the linker region, a flexible 40-amino-acid segment, which is expected to play a function in the regulation mechanism by which the PH domain can be "in" or "out" with respect to the catalytic domain, conferring close (inactive) or open (active) conformation states, respectively [50, 51]. In future, the determination of the PH/linker structure and of tridimensional connections to the other parts of the Akt molecule will be necessary to better understand the regulation mechanism driven by Ser129 phosphorylation. It will also help in clarifying to which extent this phosphorylation is somehow dependent and/or correlates to those of the canonical sites, Thr308 and Ser473. Our preliminary results would suggest that, despite the constitutive activity of CK2, its phosphorylation of Akt is slightly increased under conditions of Akt stimulation, when also Thr308 and Ser473 are phosphorylated (unpublished results).

Another mechanism by which CK2 would control Akt is a physical interaction between the two kinases, which have been demonstrated to form a stable complex where both individual α and β subunits of CK2 bind to Akt; it has been proposed that this interaction enhances Akt kinase activity independently of CK2-mediated phosphorylation [52]. Therefore, CK2 would play a double direct role in the regulation of Akt activity, exploiting two distinct mechanisms: protein phosphorylation and protein-protein interaction.

Further complexity in the interconnections between CK2 and Akt has been recently added by the report that CK2 itself is phosphorylated by Akt at Thr13. Interestingly, only once its Thr13 is phosphorylated, CK2 can phosphorylate transcription initiation factor I (TIF-IA), which is an essential player in the transcription of ribosomal DNA. The authors therefore propose that TIF-IA represents a key element in the concerted mechanism by which Akt and CK2 regulate rRNA synthesis [53].

As it will be discussed below, the abundance of cooperative actions and interconnections of CK2 and Akt with the common aim of promoting cell survival suggests that the exploitation of their combined inhibition may represent a promising strategy for cancer therapy.

4 Other Levels of CK2/Akt Cross Talk

Both CK2 and Akt are highly pleiotropic enzymes, counting more than 300 and 100 substrates, respectively [2, 8]. In this multitude of targets, several additional points of connection between the two signaling pathways are expected, besides the well-known intersections described above. Among the many findings already reported in this regard, we focus on the following ones.

4.1 *The PML Connection*

PML is a tumor suppressor whose phosphorylation by CK2 is known to promote its ubiquitin-mediated proteasomal degradation [54], and this is a proven mechanism by which CK2 favors tumorigenesis. However, PML is also crucial in the regulation of Akt activity, since it interacts with the two Akt phosphatases PP2A and PHLPP2. It has been reported that PML specifically recruits PP2A and Akt into PML nuclear bodies, allowing the dephosphorylation of the Akt activation site Thr308 [55]; Trotman and coauthors also confirmed their results by showing that PML-null cells accumulate active Akt into the nucleus due to impaired PP2A phosphatase activity toward Akt. More recently, a similar mechanism has been described for PHLPP2, the phosphatase active toward phospho-Ser473 of Akt1: PML interacts with PHLPP2 in the nucleus and provides a scaffold platform for the functional interaction of the phosphatase with Akt [37]. Therefore, both in the case of PP2A and of PHLPP2, the binding to PML has the effect of promoting Akt dephosphorylation and inactivation and represents a crucial element of the anti-survival and tumor suppressor program driven by PML (Fig. 3a). It is evident that, whenever CK2 activity is too high and PML is committed to proteolysis, as typically occurs in tumor cells, Akt is maintained more phosphorylated and more active (Fig. 3b).

4.2 *The Hsp90/Cdc37 Connection*

The chaperone system Hsp90/Cdc37 is an important target of CK2: both proteins are affected by CK2 which, by phosphorylating Cdc37 Ser13, regulates their activity in controlling the folding and the maintenance of a number of protein kinases in their active state [56]. Although this control by CK2 is quite unspecific and directed to several enzymes, it is evident that, being Akt among the major client proteins of Hsp90/Cdc37, a high activity of these chaperones has profound effects on Akt signaling and undoubtedly represents a profitable strategy by which CK2 potentiates the Akt survival message.

4.3 *The mTORC Connection*

Among the many substrates of CK2, Tel2 and Tti1 proteins have been identified [57]. These proteins interact with mTOR and are essential components of the mTORC1 and mTORC2 complexes. While mTORC1 controls cell growth and protein synthesis (by activating S6 kinase 1), mTORC2 is responsible of Akt Ser473 phosphorylation and regulates cell survival [58]. Fbxo9 is the ubiquitin ligase that mediates Tel2/Tti1 degradation. It has been demonstrated that, in multiple myeloma cells, Fbxo9 specifically targets Tel2 and Tti1 within mTORC1 only when they are

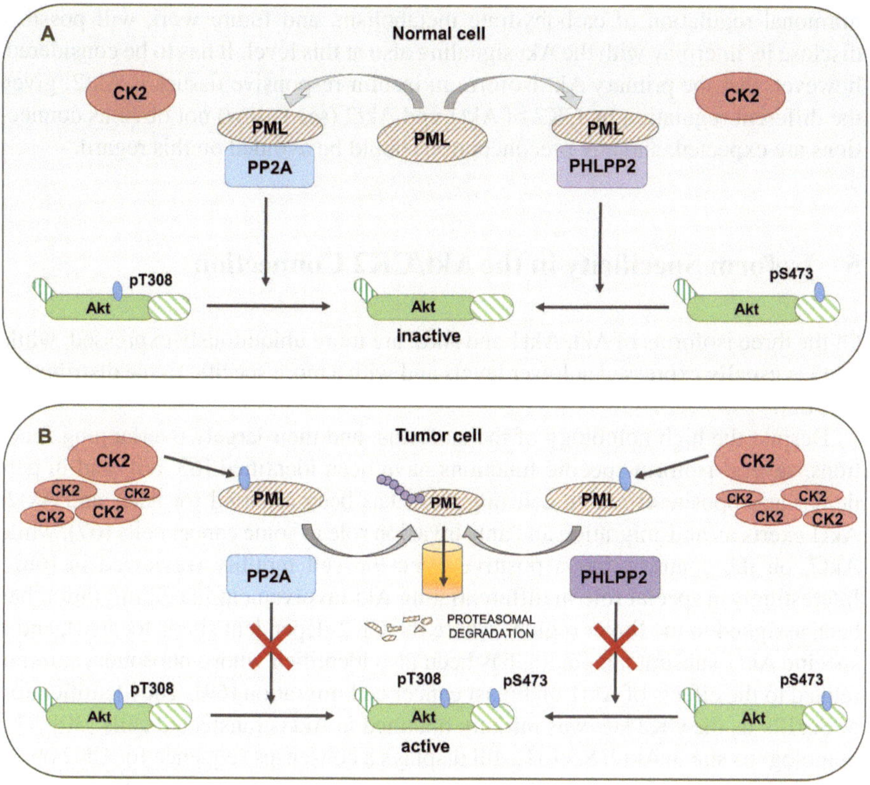

Fig. 3 Effects of CK2 on the PML-dependent regulation of Akt. Two conditions are represented of low (**a**) and high (**b**) CK2 expression level, corresponding to normal or tumor cells, respectively. When CK2 activity is too high, PML is phosphorylated and sent to proteasome. The consequent low amount of PML prevents its association to the phosphatases PP2A and PHLPP2 and their targeting to Akt, with the consequence of a more phosphorylated and more active Akt

phosphorylated by CK2 which, in response to growth factor deprivation, has moved to the cytosol and co-localized with them. As a consequence, mTORC1 is inactivated, while mTORC2 and Akt pathway are activated to sustain survival.

There could be several other CK2/Akt interconnecting points, suggested, for instance, by the identification of common substrates or signaling pathways affected by both the kinases, mostly not investigated for their relationships, yet. Among them, we would just like to mention the likely intersection of CK2 and Akt pathway at the level of insulin signaling: while the metabolic role of Akt and its regulation of cellular response to insulin have been described in detail and are well-known events (see [8] for review), the implication of CK2 in this field is still only poorly documented. It has been reported that insulin treatment of pancreatic β cells results in an elevated CK2 kinase activity, while, on the other hand, the inhibition of CK2 increases the insulin level [59]. As reviewed in [60], CK2 has therefore a role in hormonal regulation of carbohydrate metabolism, and future work will possibly disclose its interplay with the Akt signaling also at this level. It has to be considered, however, that the primary Akt isoform in insulin-responsive tissues is Akt2; given the different regulation by CK2 of Akt1 and Akt2 (see below) not obvious connections are expected, and any preconception should be avoided on this regard.

5 Isoform Specificity in the Akt/CK2 Connection

Of the three isoforms of Akt, Akt1 and Akt2 are more ubiquitously expressed, while Akt3 is usually expressed at lower levels and with a more specific tissue distribution [61, 62].

Despite the high homology of the isoforms, and their largely overlapping functions, several isoform-specific functions have been identified [63–66], and in particular an opposite effect on cell migration has been reported for Akt1 and Akt2: Akt1 exerts an anti-migration and anti-invasion role in some cancer cells [67], while Akt2, on the contrary, has a positive effect on cell motility (reviewed in [68]). Interestingly, a special role in differentiating Akt involvement in cell migration has been assigned to the linker region (where the CK2-dependent site is located), and a specific Akt1 substrate, palladin, has been also identified whose phosphorylation is related to the effects of Akt1 on breast cancer cell migration [69]. The identification of Ser129 as the CK2 site was initially obtained in Akt1-transfected cells [30]. The homologous site in Akt2, Ser131, still displays a consensus sequence for CK2 phosphorylation (Fig. 4a); however, we have recently demonstrated that Akt2 is not a CK2 substrate in cells [70]. We have also found that the phosphorylation of Ser129 by CK2 in Akt1 is essential for its recognition of palladin, providing an explanation of why Akt2, where the homologous site is not affected by CK2, is unable to phosphorylate this protein. This concept is schematized in Fig. 4b. The reasons of CK2 failure to phosphorylate Akt2 are under investigation. Presently, we can exclude that it is determined by different subcellular localization or by the presence of a disulfide bridge known to specifically occur in Akt2 in the vicinity of Ser131 [71].

We do not know whether other Akt isoform-specific substrates can be affected by Ser129 phosphorylation; however, the lesson learnt from the case of palladin suggests that this site can act as a sophisticated device, able to distinguish between isoforms and to dictate divergent cellular functions. It was already argued [72] that the pharmacological targeting of Akt has to be carefully considered in the light of the antimetastatic role of palladin phosphorylation by Akt1 in certain cancer cells; in the scenario of CK2/Akt connection, we suggest to add caution also in planning the employment of anti-CK2 drugs, which, blocking Akt1 Ser129 phosphorylation, in some circumstances could enhance cell motility and invasive migration of cancer cells.

As far as the isoform specificity is concerned in the frame of CK2/Akt relationship, it is worth to mention that differences have been observed also from the side

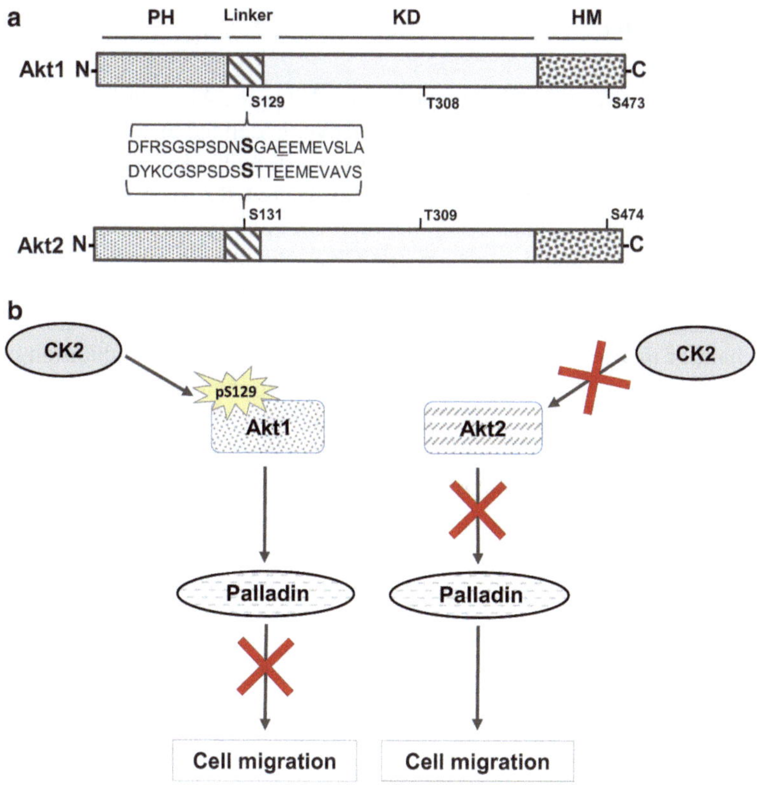

Fig. 4 Differential phosphorylation of Akt1 and Akt2 by CK2. (**a**) The structure organization of Akt1 and Akt2 is shown (*PH* pleckstrin homology domain, *KD* kinase domain, *HM* hydrophobic motif). The sequence surrounding Ser129 of Akt1 and the homologous site Ser131 of Akt2 are highlighted and aligned. Both sites display the minimal consensus for phosphorylation by CK2 (*underlined*), represented by an acidic residues at +3 position respect to the target site [74]. (**b**) The role of Akt1 phosphorylation at Ser129 in its targeting of palladin is represented (*left side*). As shown on the *right side*, the absence of CK2-dependent phosphorylation can explain why Akt2 does not phosphorylate palladin, permitting its pro-migration action

of CK2: we found that the monomeric catalytic subunit α of CK2 is slightly more active than the tetrameric enzyme $\alpha_2\beta_2$ on Akt1, at least when Akt1 is already phosphorylated at Thr308 and Ser473 [30]. Another intriguing observation is that the knockdown of the CK2 catalytic isoform α′ in pancreatic adenocarcinoma cells exerts more profound effects on the PI3K/Akt pathway than the knockdown of CK2 α, which is also less effective in reducing Akt Ser129 phosphorylation [73].

To which extent these findings could have general validity and physiological relevance is still to be determined.

6 Conclusions

Most of the above described interventions of CK2 in the Akt signaling are summarized in Fig. 5. CK2, being constitutively active, is defined as a lateral player [6], acting in a figurative horizontal way on a vertical, stimulus-dependent pathway; this is highlighted once more by the multiple sites where CK2 exerts its modulation of the PI3K/Akt pathway. Considering that both CK2 and Akt are frequently upregulated in cancer, one should conclude that CK2 inhibition is a potential antitumor therapy with the benefit of also blocking Akt and its survival message. Indeed, the first anti-CK2 drug is giving preliminary promising results in clinical studies [75]. In the light of their multifaceted cooperation, also the combined inhibition of both CK2 and Akt is expected to be highly successfully in inducing apoptosis of cancer cells. This approach has been already applied, showing, for example, that inhibition of the PI3K pathway potentiates the nocodazole-induced apoptotic response of cells treated with CK2 antisense oligodeoxynucleotides [52]; interestingly, Cheong and

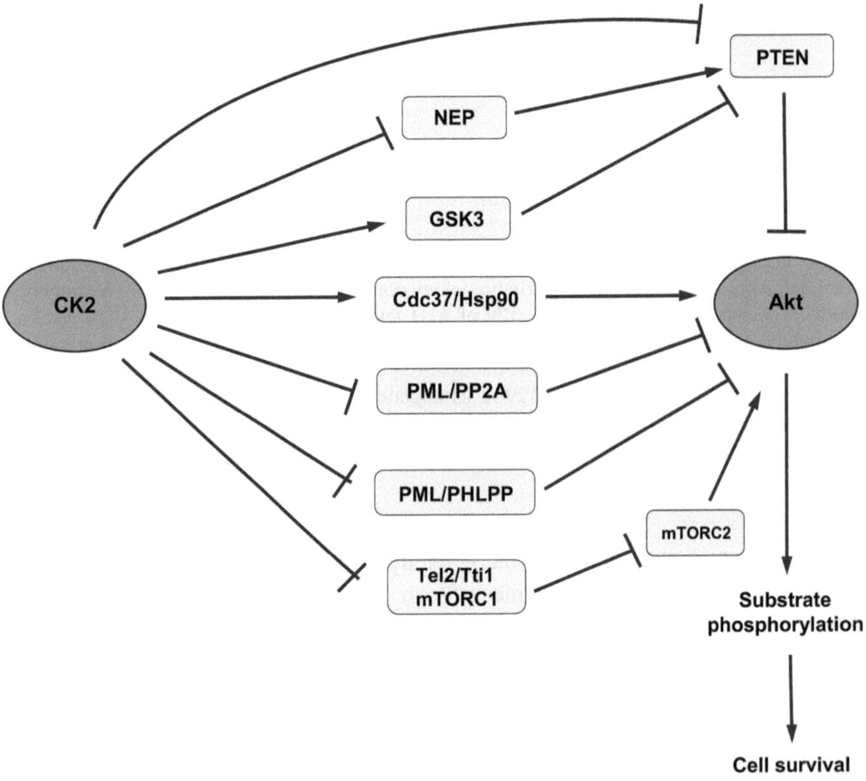

Fig. 5 Lateral player action of CK2 on the Akt signaling. The figure summarizes several levels of CK2 intervention, whose effects are positive on the stimulatory components of the Akt pathway, while negative on the inhibitory ones

colleagues found that PI3K/Akt and CK2 inhibitors at subtoxic concentrations synergistically induce apoptosis in AML cells, while healthy hematopoietic stem cells are instead resistant [32].

However, as often happens, "all that glitters is not gold": the divergent roles of Akt1 and Akt2, and our observation that CK2 can act by promoting an anti-migration role of Akt1, warn against cursory and hurried conclusions on the opportunity to inhibit CK2 as a general antitumor approach. We suggest that only a comprehensive picture of the different connections between CK2 and Akt in a specific tumor cell type would allow to understand if, for that tumor, a therapeutic strategy could be proposed based on dual targeting these kinases.

References

1. Pinna LA (2002) Protein kinase CK2: a challenge to canons. J Cell Sci 115:3873–3878
2. Meggio F, Pinna LA (2003) One-thousand-and-one substrates of protein kinase CK2? FASEB J 17:349–368
3. Ahmed K, Gerber DA, Cochet C (2002) Joining the cell survival squad: an emerging role for protein kinase CK2. Trends Cell Biol 12:226–230
4. Ahmad KA, Wang G, Unger G et al (2008) Protein kinase CK2 – a key suppressor of apoptosis. Adv Enzyme Regul 48:179–187
5. St-Denis NA, Litchfield DW (2009) Protein kinase CK2 in health and disease: from birth to death: the role of protein kinase CK2 in the regulation of cell proliferation and survival. Cell Mol Life Sci 66:1817–1829
6. Ruzzene M, Pinna LA (2010) Addiction to protein kinase CK2: a common denominator of diverse cancer cells? Biochim Biophys Acta 1804:499–504
7. Sale EM, Sale GJ (2008) Protein kinase B: signalling roles and therapeutic targeting. Cell Mol Life Sci 65:113–127
8. Manning BD, Cantley LC (2007) AKT/PKB signaling: navigating downstream. Cell 129: 1261–1274
9. Fayard E, Tintignac LA, Baudry A et al (2005) Protein kinase B/Akt at a glance. J Cell Sci 118:5675–5678
10. Bozulic L, Surucu B, Hynx D et al (2008) PKBalpha/Akt1 acts downstream of DNA-PK in the DNA double-strand break response and promotes survival. Mol Cell 30:203–213
11. Alessi DR, Cohen P (1998) Mechanism of activation and function of protein kinase B. Curr Opin Genet Dev 8:55–62
12. Brognard J, Sierecki E, Gao T et al (2007) PHLPP and a second isoform, PHLPP2, differentially attenuate the amplitude of Akt signaling by regulating distinct Akt isoforms. Mol Cell 25:917–931
13. Toker A (2012) Achieving specificity in Akt signaling in cancer. Adv Biol Regul 52:78–87
14. Cantley LC, Neel BG (1999) New insights into tumor suppression: PTEN suppresses tumor formation by restraining the phosphoinositide 3-kinase/AKT pathway. Proc Natl Acad Sci U S A 96:4240–4245
15. Di Cristofano A, Pandolfi PP (2000) The multiple roles of PTEN in tumor suppression. Cell 100:387–390
16. Torres J, Pulido R (2001) The tumor suppressor PTEN is phosphorylated by the protein kinase CK2 at its C terminus. Implications for PTEN stability to proteasome-mediated degradation. J Biol Chem 276:993–998
17. Miller SJ, Lou DY, Seldin DC et al (2002) Direct identification of PTEN phosphorylation sites. FEBS Lett 528:145–153

18. Vazquez F, Grossman SR, Takahashi Y et al (2001) Phosphorylation of the PTEN tail acts as an inhibitory switch by preventing its recruitment into a protein complex. J Biol Chem 276:48627–48630
19. Silva A, Yunes JA, Cardoso BA et al (2008) PTEN posttranslational inactivation and hyperactivation of the PI3K/Akt pathway sustain primary T cell leukemia viability. J Clin Invest 118:3762–3774
20. Gomes AM, Soares MV, Ribeiro P et al (2014) Adult B-cell acute lymphoblastic leukemia cells display decreased PTEN activity and constitutive hyperactivation of PI3K/Akt pathway despite high PTEN protein levels. Haematologica 99:1062–1068
21. He L, Fan C, Gillis A et al (2007) Co-existence of high levels of the PTEN protein with enhanced Akt activation in renal cell carcinoma. Biochim Biophys Acta 1772:1134–1142
22. Al-Khouri AM, Ma Y, Togo SH et al (2005) Cooperative phosphorylation of the tumor suppressor phosphatase and tensin homologue (PTEN) by casein kinases and glycogen synthase kinase 3beta. J Biol Chem 280:35195–35202
23. Xu D, Yao Y, Jiang X et al (2010) Regulation of PTEN stability and activity by Plk3. J Biol Chem 285:39935–39942
24. Siepmann M, Kumar S, Mayer G et al (2010) Casein kinase 2 dependent phosphorylation of neprilysin regulates receptor tyrosine kinase signaling to Akt. PLoS One 5:e13134
25. Patsoukis N, Li L, Sari D et al (2013) PD-1 increases PTEN phosphatase activity while decreasing PTEN protein stability by inhibiting casein kinase 2. Mol Cell Biol 33:3091–3098
26. Kang NI, Yoon HY, Kim HA et al (2011) Protein kinase CK2/PTEN pathway plays a key role in platelet-activating factor-mediated murine anaphylactic shock. J Immunol 186:6625–6632
27. Ning K, Miller LC, Laidlaw HA et al (2006) A novel leptin signalling pathway via PTEN inhibition in hypothalamic cell lines and pancreatic beta-cells. EMBO J 25:2377–2387
28. Heiker JT, Wottawah CM, Juhl C et al (2009) Protein kinase CK2 interacts with adiponectin receptor 1 and participates in adiponectin signaling. Cell Signal 21:936–942
29. Barata JT (2011) The impact of PTEN regulation by CK2 on PI3K-dependent signaling and leukemia cell survival. Adv Enzyme Regul 51:37–49
30. Di Maira G, Salvi M, Arrigoni G et al (2005) Protein kinase CK2 phosphorylates and upregulates Akt/PKB. Cell Death Differ 12:668–677
31. Kim JS, Eom JI, Cheong JW et al (2007) Protein kinase CK2alpha as an unfavorable prognostic marker and novel therapeutic target in acute myeloid leukemia. Clin Cancer Res 13:1019–1028
32. Cheong JW, Min YH, Eom JI et al (2010) Inhibition of CK2{alpha} and PI3K/Akt synergistically induces apoptosis of CD34+CD38− leukaemia cells while sparing haematopoietic stem cells. Anticancer Res 30:4625–4634
33. Shehata M, Schnabl S, Demirtas D et al (2010) Reconstitution of PTEN activity by CK2 inhibitors and interference with the PI3-K/Akt cascade counteract the antiapoptotic effect of human stromal cells in chronic lymphocytic leukemia. Blood 116:2513–2521
34. Borgo C, Cesaro L, Salizzato V et al (2013) Aberrant signalling by protein kinase CK2 in imatinib-resistant chronic myeloid leukaemia cells: biochemical evidence and therapeutic perspectives. Mol Oncol 7:1103–1115
35. Buontempo F, Orsini E, Martins LR et al (2014) Cytotoxic activity of the casein kinase 2 inhibitor CX-4945 against T-cell acute lymphoblastic leukemia: targeting the unfolded protein response signaling. Leukemia 28:543–553
36. Ryu BJ, Baek SH, Kim J et al (2012) Anti-androgen receptor activity of apoptotic CK2 inhibitor CX4945 in human prostate cancer LNCap cells. Bioorg Med Chem Lett 22:5470–5474
37. Chatterjee A, Chatterjee U, Ghosh MK (2013) Activation of protein kinase CK2 attenuates FOXO3a functioning in a PML-dependent manner: implications in human prostate cancer. Cell Death Dis 4:e543
38. Pierre F, Chua PC, O'Brien SE et al (2011) Pre-clinical characterization of CX-4945, a potent and selective small molecule inhibitor of CK2 for the treatment of cancer. Mol Cell Biochem 356:37–43

39. Noh EM, Lee YR, Chay KO et al (2011) Estrogen receptor alpha induces down-regulation of PTEN through PI3-kinase activation in breast cancer cells. Mol Med Rep 4:215–219
40. Stahl S, Branca RM, Efazat G et al (2011) Phosphoproteomic profiling of NSCLC cells reveals that ephrin B3 regulates pro-survival signaling through Akt1-mediated phosphorylation of the EphA2 receptor. J Proteome Res 10:2566–2578
41. Bliesath J, Huser N, Omori M et al (2012) Combined inhibition of EGFR and CK2 augments the attenuation of PI3K-Akt-mTOR signaling and the killing of cancer cells. Cancer Lett 322:113–118
42. Ku MJ, Park JW, Ryu BJ et al (2013) CK2 inhibitor CX4945 induces sequential inactivation of proteins in the signaling pathways related with cell migration and suppresses metastasis of A549 human lung cancer cells. Bioorg Med Chem Lett 23:5609–5613
43. Olsen BB, Svenstrup TH, Guerra B (2012) Downregulation of protein kinase CK2 induces autophagic cell death through modulation of the mTOR and MAPK signaling pathways in human glioblastoma cells. Int J Oncol 41:1967–1976
44. Zheng Y, McFarland BC, Drygin D et al (2013) Targeting protein kinase CK2 suppresses prosurvival signaling pathways and growth of glioblastoma. Clin Cancer Res 19:6484–6494
45. Parker R, Clifton-Bligh R, Molloy MP (2014) Phosphoproteomics of MAPK inhibition in BRAF mutated cells and a role for the lethal synergism of dual BRAF and CK2 inhibition. Mol Cancer Ther 13(7):1894–1906
46. Zanin S, Borgo C, Girardi C et al (2012) Effects of the CK2 inhibitors CX-4945 and CX-5011 on drug-resistant cells. PLoS One 7:e49193
47. Di Maira G, Brustolon F, Pinna LA et al (2009) Dephosphorylation and inactivation of Akt/PKB is counteracted by protein kinase CK2 in HEK 293T cells. Cell Mol Life Sci 66:3363–3373
48. Sato S, Fujita N, Tsuruo T (2000) Modulation of Akt kinase activity by binding to Hsp90. Proc Natl Acad Sci U S A 97:10832–10837
49. Ponce DP, Maturana JL, Cabello P et al (2011) Phosphorylation of AKT/PKB by CK2 is necessary for the AKT-dependent up-regulation of beta-catenin transcriptional activity. J Cell Physiol 226:1953–1959
50. Calleja V, Laguerre M, Parker PJ et al (2009) Role of a novel PH-kinase domain interface in PKB/Akt regulation: structural mechanism for allosteric inhibition. PLoS Biol 7:e17
51. Wu WI, Voegtli WC, Sturgis HL et al (2010) Crystal structure of human AKT1 with an allosteric inhibitor reveals a new mode of kinase inhibition. PLoS One 5:e12913
52. Guerra B (2006) Protein kinase CK2 subunits are positive regulators of AKT kinase. Int J Oncol 28:685–693
53. le Nguyen XT, Mitchell BS (2013) Akt activation enhances ribosomal RNA synthesis through casein kinase II and TIF-IA. Proc Natl Acad Sci U S A 110:20681–20686
54. Scaglioni PP, Yung TM, Cai LF et al (2006) A CK2-dependent mechanism for degradation of the PML tumor suppressor. Cell 126:269–283
55. Trotman LC, Alimonti A, Scaglioni PP et al (2006) Identification of a tumour suppressor network opposing nuclear Akt function. Nature 441:523–527
56. Miyata Y, Nishida E (2004) CK2 controls multiple protein kinases by phosphorylating a kinase-targeting molecular chaperone, Cdc37. Mol Cell Biol 24:4065–4074
57. Fernandez-Saiz V, Targosz BS, Lemeer S et al (2013) SCFFbxo9 and CK2 direct the cellular response to growth factor withdrawal via Tel2/Tti1 degradation and promote survival in multiple myeloma. Nat Cell Biol 15:72–81
58. Zoncu R, Efeyan A, Sabatini DM (2011) mTOR: from growth signal integration to cancer, diabetes and ageing. Nat Rev Mol Cell Biol 12:21–35
59. Wilhelm N, Kostelnik K, Gotz C et al (2012) Protein kinase CK2 is implicated in early steps of the differentiation of pre-adipocytes into adipocytes. Mol Cell Biochem 365:37–45
60. Al Quobaili F, Montenarh M (2012) CK2 and the regulation of the carbohydrate metabolism. Metabolism 61:1512–1517

61. Yang ZZ, Tschopp O, Hemmings-Mieszczak M et al (2003) Protein kinase B alpha/Akt1 regulates placental development and fetal growth. J Biol Chem 278:32124–32131
62. Santi SA, Lee H (2010) The Akt isoforms are present at distinct subcellular locations. Am J Physiol Cell Physiol 298:C580–C591
63. Bellacosa A, Testa JR, Moore R et al (2004) A portrait of AKT kinases: human cancer and animal models depict a family with strong individualities. Cancer Biol Ther 3:268–275
64. Stambolic V, Woodgett JR (2006) Functional distinctions of protein kinase B/Akt isoforms defined by their influence on cell migration. Trends Cell Biol 16:461–466
65. Dillon RL, Muller WJ (2010) Distinct biological roles for the akt family in mammary tumor progression. Cancer Res 70:4260–4264
66. Chin YM, Yuan X, Balk SP et al (2014) PTEN-deficient tumors depend on Akt2 for maintenance and survival. Cancer Discov 4(8):942–955
67. Yoeli-Lerner M, Yiu GK, Rabinovitz I et al (2005) Akt blocks breast cancer cell motility and invasion through the transcription factor NFAT. Mol Cell 20:539–550
68. Chin YR, Toker A (2009) Function of Akt/PKB signaling to cell motility, invasion and the tumor stroma in cancer. Cell Signal 21:470–476
69. Chin YR, Toker A (2010) The actin-bundling protein palladin is an Akt1-specific substrate that regulates breast cancer cell migration. Mol Cell 38:333–344
70. Girardi C, James P, Zanin S et al (2014) Differential phosphorylation of Akt1 and Akt2 by protein kinase CK2 may account for isoform specific functions. Biochim Biophys Acta 1843(9):1865–1874
71. Wani R, Bharathi NS, Field J et al (2011) Oxidation of Akt2 kinase promotes cell migration and regulates G1-S transition in the cell cycle. Cell Cycle 10:3263–3268
72. Toker A, Yoeli-Lerner M (2006) Akt signaling and cancer: surviving but not moving on. Cancer Res 66:3963–3966
73. Kreutzer JN, Ruzzene M, Guerra B (2010) Enhancing chemosensitivity to gemcitabine via RNA interference targeting the catalytic subunits of protein kinase CK2 in human pancreatic cancer cells. BMC Cancer 10:440. doi:10.1186/1471-2407-10-440
74. Pinna LA, Ruzzene M (1996) How do protein kinases recognize their substrates? Biochim Biophys Acta 1314:191–225
75. Siddiqui-Jain A, Drygin D, Streiner N et al (2010) CX-4945, an orally bioavailable selective inhibitor of protein kinase CK2, inhibits prosurvival and angiogenic signaling and exhibits antitumor efficacy. Cancer Res 70:10288–10298

Cellular Zinc Signalling Is Triggered by CK2

Thirayost Nimmanon and Kathryn M. Taylor

Abstract Zinc is an essential trace element involved in various biological processes. Too much or too little cellular free zinc can cause detrimental effects and cell death. The intracellular level of zinc is controlled by three groups of proteins: ZnT transporters (SLC30A, zinc exporters), ZIP channels (SLC39A, zinc importers) and zinc-binding proteins such as metallothioneins. Zinc channel, ZIP7, a hub of zinc release from the stores, has long been associated with aggressiveness and development of anti-endocrine resistance in breast cancer. However, little is known about its regulatory mechanism. Recently, we have demonstrated that this ZIP channel is post-translationally triggered by CK2 phosphorylation on two cytoplasmic residues resulting in zinc release from the stores and activation of downstream effectors which lead to cancer growth. This CK2-triggered zinc release acts to inhibit multiple protein tyrosine phosphatases, resulting in direct stimulation of normal growth, migration and, in increased amounts, also cancer growth. Interestingly, another ZIP channel, ZIP6, with a known role in EMT and cancer metastasis is also predicted to be phosphorylated by CK2. ZIP channels are therefore new substrates of CK2 that may help explain CK2-related cancer phenotypes, including cell proliferation, cell migration and metastasis.

Keywords Zinc signalling • SLC39A7 • ZIP7 • CK2 • Phosphorylation • Cancer

1 The Importance of Zinc in Cells

Zinc is an essential trace element known to be involved in various biological processes and required for normal cell growth and development. Even though there is only 30 nmol or 2 g of zinc distributed in all tissues and fluids in the human body [1], we cannot live without it. In fact, it is the second most abundant metal, after

T. Nimmanon • K.M. Taylor (✉)
Cardiff School of Pharmacy and Pharmaceutical Sciences, Cardiff University, Redwood Building, King Edward VII Avenue, Cardiff CF10 3NB, UK
e-mail: taylorkm@cardiff.ac.uk

K. Ahmed et al. (eds.), *Protein Kinase CK2 Cellular Function in Normal and Disease States*, Advances in Biochemistry in Health and Disease 12, DOI 10.1007/978-3-319-14544-0_9

iron, in the human body. Zinc has been associated with more than 3,000 enzymes and proteins, which constitute approximately 10 % of the proteome [2, 3], and shown to be the only metal detected in all classes of enzymes according to the IUPAC system [4]. Not surprisingly, therefore, it plays important roles in virtually all cellular processes, being involved in signal transduction, gene expression and apoptosis [5]. We need zinc even before we are born, since zinc plays essential roles in embryogenesis and organogenesis [6]. Zinc is crucial for growth and development [6], and proper zinc homeostasis may help prevent ageing-associated pathological conditions [7]. Zinc is required for proper functions of immune system [8] and free radical scavenging system [9], protection of the body from invaders and protection of the cell from free radical damage. Furthermore, zinc participates in cellular metabolism, proliferation, growth and differentiation [1] and has been shown to function as a neuromodulator and signalling molecule in the brain [10].

Importantly, zinc has been experimentally demonstrated in mast cells to be released from the cellular stores upon activation by an extracellular stimulus, producing a 'zinc wave' and activating multiple intracellular signalling cascades, thereby fulfilling all the criteria of an intracellular second messenger [11]. This role of zinc is not limited to mast cells, but it has also been found in other types of cells, including normal cells such as lymphocytes [12] and abnormal cells such as breast cancer cells [13]. For example, zinc release from the stores in breast cancer cells was shown to lead to activation of multiple downstream signalling cascades, resulting in cell growth and invasion [14]. Furthermore, this emerging role of zinc as a second messenger demonstrates how the cellular effects of zinc from an extracellular stimulus can be observed on the time scale of minutes in contrast to the transcription and DNA-binding properties of zinc, which take hours or days to accomplish. Therefore, zinc signalling may have a significant role in intracellular biological processes on a scale comparable to calcium signalling.

2 Cellular Zinc Homeostasis and ZIP Channels

Zinc does not freely traverse cellular membranes. The movement of zinc ions across cell membranes therefore requires specific transport proteins, consisting of ZnT transporters (SLC30A) and ZIP channels (SLC39A). ZnT transporters, or zinc exporters, decrease cytosolic zinc bioavailability, probably via Zn^{2+}/H^{+} exchange [15], by mobilising zinc from the cytoplasm to the cellular stores or eliminating zinc from the cell. In contrast, ZIP channels, or zinc importers, increase cytosolic zinc bioavailability by taking up zinc from the extracellular space or releasing zinc from the stores [16, 17]. In addition, zinc-binding proteins, particularly the cysteine-rich metallothioneins, orchestrate zinc homeostasis through a buffering reaction, controlling the free zinc level within picomolar levels under steady-state conditions, and a muffling reaction, modulating the transiently altered free zinc level under nonsteady-state conditions [18].

According to the phylogenetic tree, human ZIP channels are classified into four subfamilies: gufA subfamily (ZIP11), subfamily I (ZIP9), subfamily II (ZIP1-3) and LIV-1 subfamily (ZIP4-8, ZIP10 and ZIP12-14) [19] (Fig. 1). Using computer prediction analysis of the sequences, ZIP channels are proposed to have eight transmembrane (TM) domains with a long amino-terminus and a short carboxyl-terminus on the extracytoplasmic side, as well as a long histidine-rich cytoplasmic loop between TMs III and IV [20]. The cytoplasmic loop contains a histidine-rich repeat, defined as the presence of three to six consecutive HX motifs (H = histidine, X = any amino acid) [21], and is predicted to be the metal-binding site [19] (Fig. 2a). Exclusively for the LIV-1 subfamily, all the members, including ZIP7, are shown to have additional histidine-rich motifs at their amino-terminus and extracytoplasmic loop between TMs II and III [19] (Fig. 2b).

In contrast to all the other ZIP channels which are located on the plasma membrane, ZIP7 and ZIP13 are localised to intracellular membranes. ZIP7 has been found in the ER [22] and the Golgi [23], whereas ZIP13 is detected solely in the Golgi [24]. ZIP7 and ZIP13 are therefore believed to play a role in the control of zinc release from intracellular stores rather than the acquisition of zinc from outside the cell.

Based upon the muffler model, when zinc is taken up by a cell, the imported zinc is immediately muffled by the muffler metallothioneins, which then deliver zinc to the internal stores, such as the ER, via a ZnT transporter, such as ZnT5, ZnT6 and ZnT7 [18, 25], depicted in a schematic shown in Fig. 3. Upon the activation of ZIP7 by CK2 phosphorylation [13], zinc is released from these stores via the ZIP7 channel

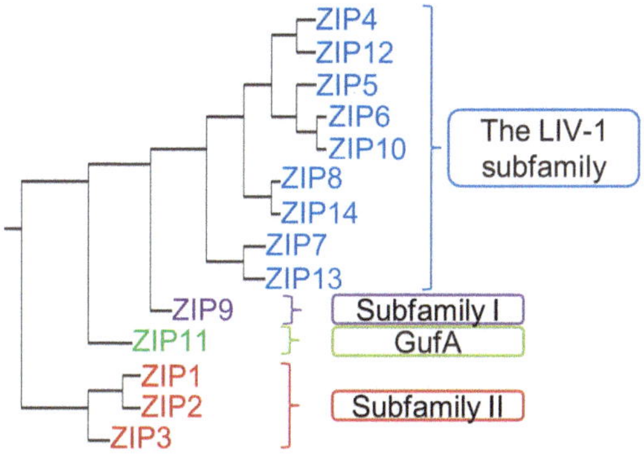

Fig. 1 Phylogenetic tree of human ZIP channels. Rooted phylogram was generated from the amino acid sequences of all the human ZIP channels retrieved from the NCBI database, using the ClustalW2 web service [66]; according to similarity of the sequences, human ZIP channels can be classified into four subfamilies: GufA (ZIP11), subfamily I (ZIP9), subfamily II (ZIP1-3) and LIV-1 subfamily (ZIP4–8, ZIP10 and ZIP12–14)

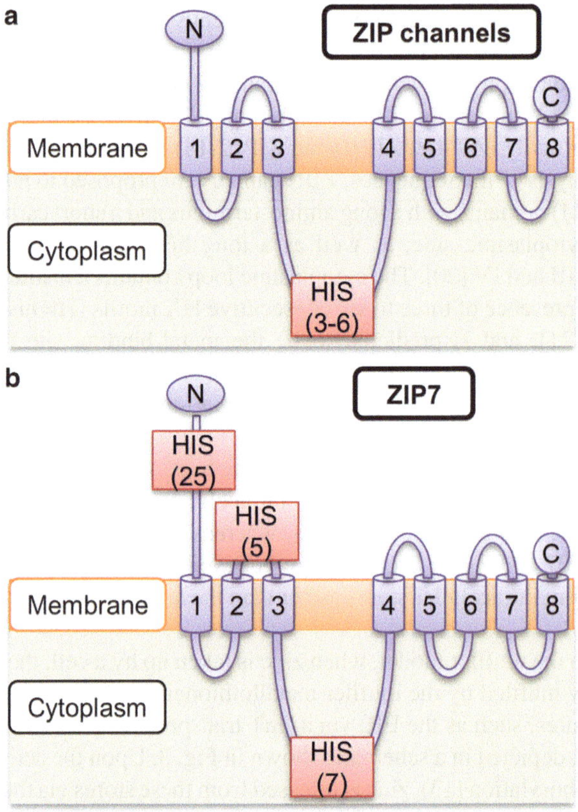

Fig. 2 Topological model of ZIP channels. The predicted structure of ZIP channels (**a**) and ZIP7 as a representative of the LIV-1 subfamily (**b**). In general, ZIP channels are proposed to have eight transmembrane domains, with a long extracytoplasmic amino-terminus, a long histidine-rich intracytoplasmic loop between TM III and TM IV and a short extracytoplasmic carboxyl-terminus. Specifically, a histidine-rich region is defined as an area with a $(HX)_n$ motif, where H signifies histidine and X signifies any amino acid residue and n ranges from 3 to 6. In contrast to other groups of ZIP channels, those in the LIV-1 subfamily have additional histidine-rich regions at their amino-terminus and extracytoplasmic loop between TMs II and III. HIS, histidine-rich repeats with number of histidine residues indicated

which resides on the ER membrane and is proposed to be a hub for zinc release from the stores [26]. The released zinc in turn exerts its cellular effects via widespread inhibition of protein tyrosine phosphatases [27], resulting in activation of various tyrosine kinases [26]. To control the extent of zinc action, the metallothioneins muffle any excess zinc and eventually deliver it again to the cellular stores or alternatively eliminate it from the cell [18] (Fig. 3, dotted lines) via ZnT1 which is the only zinc exporter located on the plasma membrane [28].

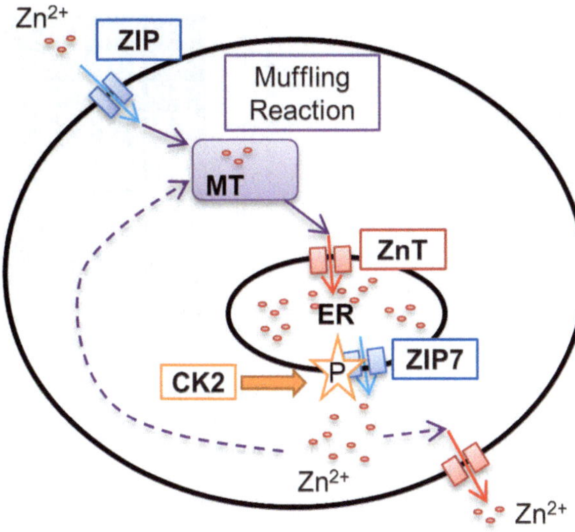

Fig. 3 Role of the muffling reaction in intracellular zinc homeostasis. According to the muffler model, after zinc is taken up from the extracellular space via a ZIP channel, the cytosolic free zinc is immediately muffled by the cysteine-rich metallothioneins (*purple straight thick arrow*) and the zinc is then shuttled directly to the stores, such as the ER. Upon activation by CK2 phosphorylation, zinc is released from stores via a ZIP7 channel allowing inhibition of tyrosine phosphatases. Excess zinc ions can be delivered back to the stores or removed from the cells via a ZnT transporter (*dashed arrows*). *MT* metallothioneins, *ER* endoplasmic reticulum

3 CK2 Triggers ZIP7-Mediated Zinc Release

To investigate if the ZIP channels could be activated by post-translational modification such as phosphorylation, the sequences were screened for potential phosphorylation sites using a variety of online databases. In order to optimise the success of this screening, we only took forward those predicted sites that had also been detected in mammalian cells by mass spectrometry from whole genome phosphor screens [29, 30]. This approach still found multiple sites for ZIP7, including two consecutive serine residues S^{275} and S^{276} [31], both of which reside in the intracytoplasmic loop between TMs III and IV. Interestingly, these residues are highly evolutionarily conserved in ZIP7 sequences of other species (Fig. 4) and match the CK2 consensus motif **S/T**-X-X-E, where S is serine, T is threonine, E is glutamine and X is any amino acid. Furthermore, most of these sites contain many additional acidic residues downstream of the phosphorylation site, which has previously been shown to increase the likelihood of phosphorylation by CK2 [32, 33].

Importantly, we have experimentally confirmed in a recent study that ZIP7 function is triggered by CK2 phosphorylation on serine residues S^{275} and S^{276} [13]. In this study, we showed that external stimulation, which activates zinc release from the stores, resulted in global tyrosine phosphorylation in the cells from 10 min after such stimulation [13]. This tyrosine phosphorylation was preceded by association

Fig. 4 Alignment of ZIP7 channels in mammals. The peptide sequences of ZIP7 were aligned with the TCoffee alignment tool [67]. The 50 % identical (*black*) and complementary (*grey*) residues are shaded. ZIP7 is found in various mammals, including human, Sumatran orang-utan, chimpanzee, rhesus monkey, dog, bovine, horse, pig, sheep, mouse and rat. The high conservation between the two serine residues in the intracytoplasmic loop between TM III and IV equivalent to S275 and S276 in human ZIP7 is indicated (*red box*)

of CK2α with ZIP7, which was demonstrated both by immunoprecipitation and the relatively new method of proximity ligation assay. The latter method produces red fluorescent dots when two molecules are within 40 nm, indicative of their physical interaction which can be fully quantified [13]. This technique demonstrated binding of CK2α with ZIP7 which also caused serine phosphorylation of ZIP7 at 2 min after external stimulation [13]. CK2α binding to ZIP7 resulted in ZIP7-dependent zinc release from the stores 5 min after external stimulation and consequential activation of the downstream effectors of zinc store release such as AKT and ERK [13]. Both the zinc release and the activation of AKT and ERK were successfully prevented by CK2 inhibition confirming that these changes were attributed to CK2 binding to ZIP7 and the subsequent zinc release [13].

To confirm the specific sites in ZIP7 to which CK2 bound, we performed site-directed mutagenesis, substituting the two serine residues S^{275} and S^{276} for alanine to prevent phosphorylation on these residues. The result showed that the mutation successfully prevented CK2 binding, prevented ZIP7 serine phosphorylation, prevented ZIP7-mediated zinc release, prevented activation of the downstream effectors of zinc release and prevented ZIP7-driven cell migration [13].

Taking all this data together, we have therefore demonstrated that CK2 triggers ZIP7-mediated zinc release from intracellular stores [13]. ZIP7 phosphorylation by CK2 on residues S^{275} and S^{276} was the trigger to open the ZIP7 zinc channel gate releasing zinc from the intracellular stores and increasing the zinc bioavailability in the cytoplasm. One major manifestation of this zinc release is the inhibition of widespread tyrosine phosphatases which will prevent deactivation of widespread tyrosine kinases and thus prolong the type of activation often associated with cancer [14]. This zinc release also causes activation of downstream effectors particularly mitogen-activated protein kinases (MAPKs) and AKT, collectively with activated tyrosine kinases resulting in promotion of cell growth and invasion [13] (Fig. 5).

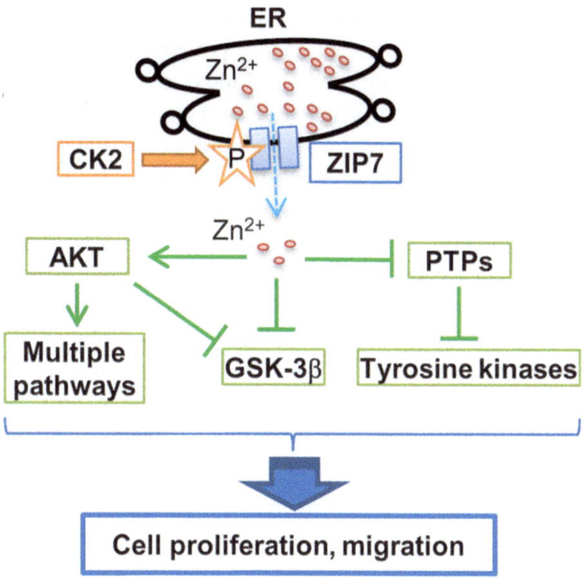

Fig. 5 CK2 triggers ZIP7-mediated zinc release from stores and downstream effects. ZIP7 phosphorylation by CK2 on residues S^{275} and S^{276} triggers ZIP7-mediated zinc release from the intracellular ER store [13]. The increased cytoplasmic zinc bioavailability directly causes activation of AKT [13, 14] as well as inhibition of both GSK-3β [34] and protein tyrosine phosphatases (PTPs) [27]. The activated AKT in turn activates multiple signalling pathways and causes inhibitory phosphorylation of GSK-3β [35], acting synergistically with zinc. The inhibition of PTPs results in widespread uncontrolled activation of tyrosine kinases. Given that GSK-3β is a known suppressor of EMT, AKT is a hub of multiple signalling pathways, and tyrosine kinases play various roles in carcinogenesis; CK2-triggered ZIP7-mediated zinc release from the ER ultimately promotes cell proliferation and migration [13, 14]

Additionally, both zinc [34] and AKT [35] have been shown to inhibitorily phosphorylate glycogen synthase kinase 3 beta (GSK-3β), a known suppressor of epithelial-mesenchymal transition (EMT) [36]. Zinc release from the stores therefore exerts its intensified inhibitory effect on GSK-3β, both directly and through activation of AKT, and thereby potently contributes to cell migration (Fig. 5). This profound finding has inevitably supported the importance of this ZIP channel, ZIP7, as a gatekeeper of zinc release from the stores and a hub of tyrosine kinase activation in the cell [13, 26].

4 ZIP7-Mediated Zinc Release Drives Aberrant Cancer Growth

Tissue-specific dysregulation of ZIP channel expression, with corresponding changes in cancerous tissue zinc level, has been implicated in many types of cancer [37, 38]. Among all the ZIP channels that have been implicated in cancers, ZIP7 has been closely associated with breast carcinogenesis [38]. To investigate the clinical importance of ZIP7 in breast cancer patients, we correlated ZIP7 gene expression in breast cancer samples with clinical outcome using data available from the Oncomine database [39]. ZIP7 was shown to be one of the 10 % of genes whose overexpression are associated with poor prognostic states in terms of relapse rate, mortality rate and tendency to develop lymph node metastasis and invasion [26]. We have also supported the association of ZIP7 with clinical outcome in a small clinical series, demonstrating positive correlation of ZIP7 mRNA expression with indicators of breast cancer progression, including the proliferation marker Ki-67, the epidermal growth factor receptor ErbB3 and lymph node spreading [40]. Furthermore, the plot generated using the Kaplan–Meier plotter survival analysis tool [41] demonstrated that patients with high ZIP7 gene expression are associated with a decreased relapse-free survival when compared to those with low ZIP7 expression (Fig. 6a). The difference between the two groups starts to be detected at 20 months after diagnosis and increases to 10 % at 120 months after diagnosis (Fig. 6a). In comparison to ZIP7 expression, patients with high CK2 expression have an even more remarkable decrease in their relapse-free survival, since both high CK2α and CK2β expressions are associated with a decrease in the relapse-free survival when compared to patients expressing lower levels of these genes. These differences between the groups are detected as early as 10 months after diagnosis and gradually increase up to 10 % at 40 months after diagnosis for both CK2α (Fig. 6b) and CK2β (Fig. 6c). Altogether, these findings suggest that ZIP7 plays an important role in breast cancer progression, thereby having great potential to be a clinical target for breast cancer treatment. However, given that ZIP7 gene expression does not have as much adverse impact on relapse-free survival as CK2 expression, ZIP7 is likely to be just one of many substrates of CK2 that are responsible for breast cancer progression.

In human breast cancer, ZIP7 has been shown to increase in expression in a breast cancer model that is resistant to either tamoxifen or fulvestrant [40]. Interestingly, ZIP7 is the only ZIP channel from the LIV-1 subfamily that is overexpressed in the tamoxifen-resistant model [40], which not only grows in the presence of the drug but also contains double the amount of zinc [40] and harbours a more aggressive behaviour, including increased cell mobility and invasiveness when compared to tamoxifen-responsive MCF-7 cells [42]. Therefore, the potential involvement of this channel in the development of anti-endocrine resistance as well as the observed aggressive behaviour of this cell model cannot be overstated.

We have experimentally proved that ZIP7 is essential for zinc redistribution from the stores in tamoxifen-resistant breast cancer cells, which causes activation of tyrosine kinase receptors and Src-dependent pathways [14]. These zinc-activated pathways in turn activate their downstream effectors, particularly mitogen-activated protein kinase (MAPK) and AKT, and thereby the overall result is increased cell growth and motility [14]. These findings therefore suggest that inhibition of ZIP7-mediated zinc release has high potential to be a novel strategy for the treatment of anti-hormonal resistant breast cancer, the complete mechanism of which needs further examination.

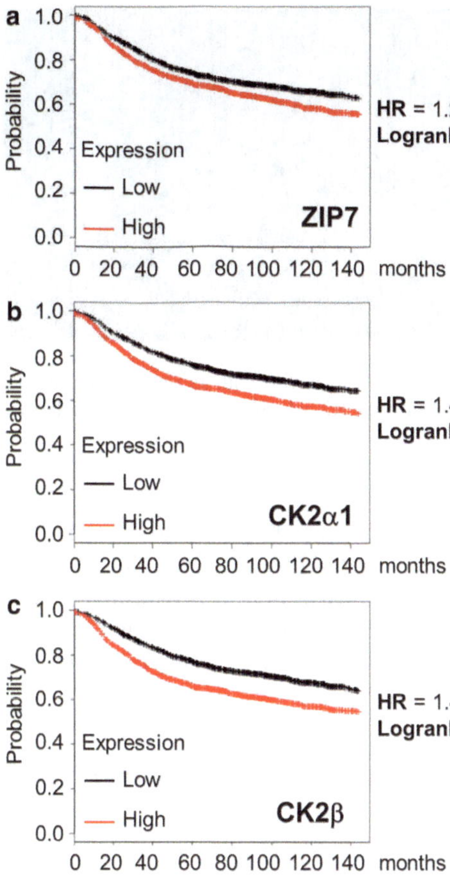

Fig. 6 Impact of ZIP7 and CK2 gene expressions on relapse-free survival in breast cancer patients. The Kaplan–Meier plots were generated using the Kaplan–Meier plotter survival analysis tool [41] for the percentage of breast cancer patients who survived up to 12 years without relapse ($n=3{,}455$). The group with a high gene expression (in *red*) of ZIP7 (**a**), CK2α (**b**) and CK2β (**c**) is compared to those with a low expression (in *black*) of the same gene. High expression of these three genes is associated with a significantly lower distant relapse-free rate when compared to their low-expression counterparts. The findings suggest that ZIP7 is one of many substrates of CK2 that are responsible for breast cancer aggressiveness

Our discovery that CK2 triggers ZIP7-mediated zinc release (Fig. 5) therefore provides a mechanism for the role of CK2 in cancer progression. Given that the protein kinase CK2 has important roles in various pathobiological features of cancers while having numerous potential substrates to which the effects can be attributed [43], it is highly possible that the phosphorylation of ZIP7 by CK2 mediates some of the well-characterised CK2-related cancer-promoting effects [44]. Additionally, CK2 cellular functions are closely associated with zinc, not only because the CK2 holoenzyme requires two zinc ions [45] (Fig. 7) but also because

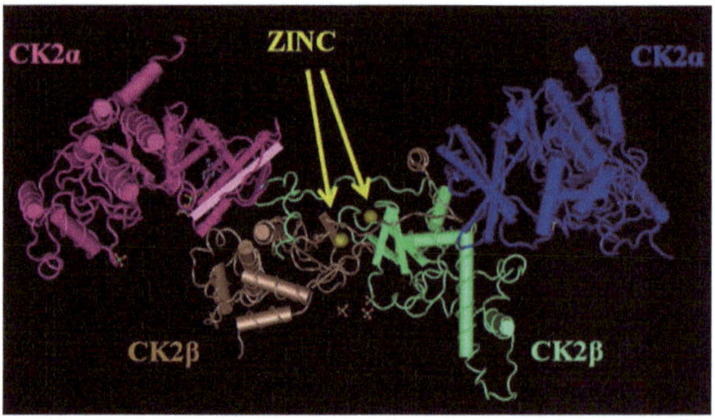

Fig. 7 Structure of the tetrameric CK2 holoenzyme with its zinc-binding site. This figure demonstrates the crystal structure of tetrameric CK2, which contains two zinc molecules within the beta-subunit of the enzyme [45]

it is a zinc-binding protein which requires the presence of a zinc ribbon motif to play a pivotal role in the formation of the CK2β homodimer [46, 47]. Given that CK2 has been believed to be constitutively active [48], it is plausible that, following CK2-triggered zinc release from cellular stores, zinc itself binds each regulatory beta-subunit of CK2, thereby functioning as a feedback regulator of CK2 activity.

5 Potential for CK2 Phosphorylation of Other ZIP Channels

Inasmuch as ZIP7 is functionally regulated by phosphorylation, it is entirely possible that other ZIP channels may also be regulated by the same mechanism. We have observed that phosphorylation sites that have been confirmed by mass spectrometry in mammalian cells were detected not only in ZIP7 but also in ZIP3, ZIP4, ZIP6, ZIP8 and ZIP10 [26]. We have therefore screened for potential phosphorylation sites in all ZIP channels using the predictive software that is available from a variety of online databases. Importantly, some sites in ZIP6 perfectly match the consensus motif for CK2 binding, **S/T**-X-X-E. These include residues S^{471} [49], S^{478} [49–52] and T^{479} [52–55] (Fig. 8). Importantly, when using online platforms PhosphoSitePlus [56] and KinomeXplorer [57], these sites are also predicted to be phosphorylated by protein kinase CK2. Additionally, residues T^{567}, T^{573} and S^{583} for ZIP10 have been predicted to be phosphorylated by CK2 according to the PhosphoNET database (Kinexus Bioinformatics Corporation), with only residue S^{583} matching the CK2 consensus motif. Moreover, none of the predicted sites for ZIP10 can be clearly assigned because of their failure to reach the maximum probability according to the PHOSIDA [58]. Thus, while ZIP6 has a high potential to be modified by CK2, the tendency for ZIP10 to be phosphorylated by CK2 is still questionable.

ZIP6 is known to be involved in EMT, both during the development process and cancer metastasis [38]. During zebrafish gastrulation, ZIP6 has been reported to play an important role in the EMT process in participation with signal transducer

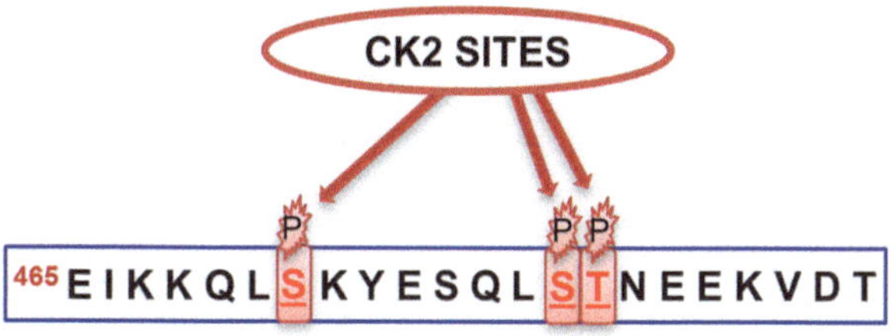

Fig. 8 Sites in ZIP6 predicted to be phosphorylated by CK2. This schematic demonstrates the amino acid sequence of ZIP6 between residues 265 and 286 which resides in the intracytoplasmic loop between TMs III and IV. We show the phosphorylation sites that have been experimentally confirmed in mammalian cells for ZIP6 by mass spectrometry and predicted to be phosphorylated by CK2, including residues S^{471} [49], S^{478} [49–52] and T^{479} [52–55]

and activator of transcription 3 (STAT3) and transcription factor Snail [59], both of which are involved in breast cancer metastasis [60, 61]. Based upon this finding, we further investigated the mechanistic role of ZIP6 in EMT in breast cancer metastasis [62], confirming the ZIP6 transcriptional transactivation by STAT3 [62]. We have also shown that the ZIP6 protein was produced as a pro-protein, stored in the ER [62] and required N-terminal cleavage to relocate it to the plasma membrane where it was activated and able to import zinc into the cell [62]. This ZIP6-mediated zinc influx then caused inhibitory phosphorylation of GSK-3β [62], either directly [34] or mediated by zinc-activated AKT [35], as previously mentioned. Inhibition of GSK-3β in turn resulted in nuclear retention of Snail and transcriptional repression of E-cadherin, leading to cell detachment [62] (Fig. 9). Furthermore, these detached cells were revealed to acquire anoikis resistance, thereby having increased ability to migrate and metastasise [62]. In summary, the role for ZIP6 in breast cancer EMT and its mechanistic link to GSK-3β inhibition and Snail-mediated E-cadherin repression have now been firmly established. However, it is still not known how ZIP6-mediated zinc transport is controlled which opens an opportunity to introduce CK2 phosphorylation as a candidate for a key ZIP6 functional regulator.

6 Zinc Signalling as a Target and/or Cancer Biomarker

The discovery that ZIP7 is functionally controlled by CK2 now provides ZIP7 as a potential therapeutic target as well as a potential biomarker of cancer status.

6.1 Targeting ZIP7 Using CK2 Inhibitors

ZIP7 is listed among the top 10 % of genes that are overexpressed in tumours with poor prognosis, and its mRNA expression is significantly correlated with cases of breast cancer that have high levels of poor outcome indicators such as Ki67 [40].

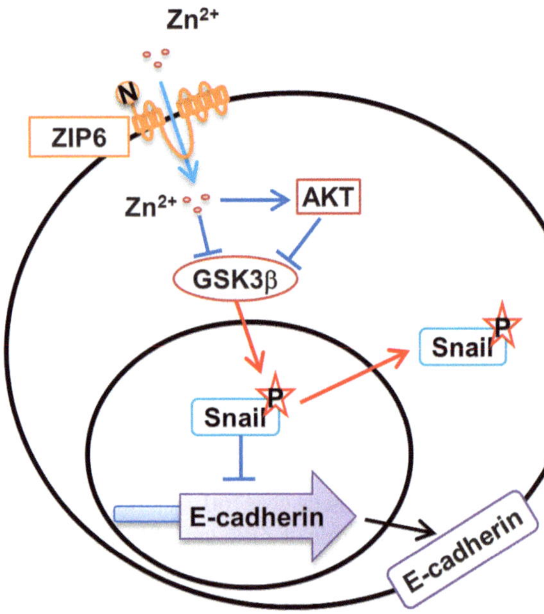

Fig. 9 Mechanistic role of ZIP6 in EMT. This figure demonstrates the ZIP6 signalling pathway associated with EMT. Upon activation by N-terminal cleavage, ZIP6 is relocated to the plasma membrane and mediates zinc influx into the cytoplasm [62]. The imported zinc inhibitorily phosphorylates GSK-3β both directly [34] and indirectly via its activation effect on AKT which also phosphorylates GSK-3β [35]. The inhibition of GSK-3β causes nuclear retention of the nuclear transcription factor Snail [68], resulting in transcriptional repression of E-cadherin, cell migration and metastasis [62]

ZIP7 may therefore share a common cancer-promoting role with CK2, which is thought to be essential for the cancer phenotype, acting as an oncogene upon overexpression [63]. This is further supported by the demonstration that a high expression of either ZIP7 or CK2 is associated with a significant decrease in the relapse-free survival when compared to a low-expression counterpart (Fig. 6).

We have demonstrated by manipulation of ZIP7 levels and activity that ZIP7 was responsible for the observed increase in cell growth, invasion and activation of downstream effectors of cellular zinc release [14]. The discovery that CK2 phosphorylation triggers ZIP7-mediated zinc release allows the application of the safe and well-tolerated CK2 inhibitors [64, 65] as a novel strategy to treat breast cancer and even to prevent the development of anti-endocrine resistance in breast cancer. Additionally, provided that zinc exerts its cellular effects by causing widespread sustained activation of tyrosine kinases via its inhibitory effects on protein tyrosine phosphatases [27], blocking of zinc mobilisation from the stores theoretically further benefits cancer patients by means of inhibiting a great variety of pathways associated with tyrosine kinases [14, 37] and potentially preventing the development of resistance to therapy.

6.2 Phosphorylated ZIP7 as a Cancer Biomarker

ZIP7 has been associated with breast cancer aggressiveness and anti-hormonal resistance [40]. ZIP7 activity therefore has high potential to be a good marker indicating the biological behaviour of breast cancer as well as its potential to become resistant to anti-hormonal therapy, a serious problem in the cancer clinic. The discovery that ZIP7 is activated by post-translational modification implies that the total ZIP7 level may not necessarily reflect the actual ZIP7 activity in a tumour, thus the need for specific determination of the level of phosphorylated or active form of ZIP7 instead. Furthermore, the level of phosphorylated ZIP7 may also be indirectly indicative of CK2 activity and helpful for predicting the responsiveness of a tumour to CK2 inhibitor treatment. Our group have developed a monoclonal antibody that recognises ZIP7 when its residues S^{275} and S^{276} are phosphorylated, and we are currently verifying the efficiency and clinical usefulness of this antibody.

References

1. World Health Organization and Food and Agriculture Organization of The United Nations (2002) Zinc. In: FAO/WHO expert consultation on human vitamin and mineral requirements. Rome, pp 257–270
2. Andreini C, Banci L, Bertini I, Rosato A (2006) Counting the zinc-proteins encoded in the human genome. J Proteome Res 5(1):196–201. doi:10.1021/pr050361j
3. Passerini A, Andreini C, Menchetti S, Rosato A, Frasconi P (2007) Predicting zinc binding at the proteome level. BMC Bioinformatics 8:39. doi:10.1186/1471-2105-8-39
4. Medicine FaNBoIo (2001) Zinc. In: Dietary reference intakes for vitamin A, vitamin K, arsenic, boron, chromium, copper, iodine, iron, manganese, molybdenum, nickel, silicon, vanadium, and zinc. National Academy Press, Washington, DC, pp 442–501
5. Hambidge KM, Krebs NF (2007) Zinc deficiency: a special challenge. J Nutr 137(4):1101–1105
6. Falchuk KH, Montorzi M (2001) Zinc physiology and biochemistry in oocytes and embryos. Biometals 14(3–4):385–395
7. Mocchegiani E, Costarelli L, Giacconi R, Cipriano C, Muti E, Rink L, Malavolta M (2006) Zinc homeostasis in aging: two elusive faces of the same "metal". Rejuvenation Res 9(2):351–354. doi:10.1089/rej.2006.9.351
8. John E, Laskow TC, Buchser WJ, Pitt BR, Basse PH, Butterfield LH, Kalinski P, Lotze MT (2010) Zinc in innate and adaptive tumor immunity. J Transl Med 8(1):118. doi:10.1186/1479-5876-8-118
9. Oteiza PI (2012) Zinc and the modulation of redox homeostasis. Free Radic Biol Med 53(9):1748–1759. doi:10.1016/j.freeradbiomed.2012.08.568
10. Colvin RA, Fontaine CP, Laskowski M, Thomas D (2003) Zn2+ transporters and Zn2+ homeostasis in neurons. Eur J Pharmacol 479(1–3):171–185
11. Yamasaki S, Sakata-Sogawa K, Hasegawa A, Suzuki T, Kabu K, Sato E, Kurosaki T, Yamashita S, Tokunaga M, Nishida K, Hirano T (2007) Zinc is a novel intracellular second messenger. J Cell Biol 177(4):637–645. doi:10.1083/jcb.200702081
12. Yu M, Lee WW, Tomar D, Pryshchep S, Czesnikiewicz-Guzik M, Lamar DL, Li G, Singh K, Tian L, Weyand CM, Goronzy JJ (2011) Regulation of T cell receptor signaling by activation-induced zinc influx. J Exp Med 208(4):775–785. doi:10.1084/jem.20100031

13. Taylor KM, Hiscox S, Nicholson RI, Hogstrand C, Kille P (2012) Protein kinase CK2 triggers cytosolic zinc signaling pathways by phosphorylation of zinc channel ZIP7. Sci Signal 5(210):ra11. doi:10.1126/scisignal.2002585
14. Taylor KM, Vichova P, Jordan N, Hiscox S, Hendley R, Nicholson RI (2008) ZIP7-mediated intracellular zinc transport contributes to aberrant growth factor signaling in antihormone-resistant breast cancer Cells. Endocrinology 149(10):4912–4920. doi:10.1210/en.2008-0351
15. Ohana E, Hoch E, Keasar C, Kambe T, Yifrach O, Hershfinkel M, Sekler I (2009) Identification of the Zn2+ binding site and mode of operation of a mammalian Zn2+ transporter. J Biol Chem 284(26):17677–17686. doi:10.1074/jbc.M109.007203
16. Lichten LA, Cousins RJ (2009) Mammalian zinc transporters: nutritional and physiologic regulation. Annu Rev Nutr 29:153–176. doi:10.1146/annurev-nutr-033009-083312
17. Fukada T, Kambe T (2011) Molecular and genetic features of zinc transporters in physiology and pathogenesis. Metallomics 3(7):662–674. doi:10.1039/c1mt00011j
18. Colvin RA, Holmes WR, Fontaine CP, Maret W (2010) Cytosolic zinc buffering and muffling: their role in intracellular zinc homeostasis. Metallomics 2(5):306–317. doi:10.1039/b926662c
19. Taylor KM, Nicholson RI (2003) The LZT proteins; the LIV-1 subfamily of zinc transporters. Biochim Biophys Acta 1611(1–2):16–30. doi:10.1016/S0005-2736(03)00048-8
20. Gaither LA, Eide DJ (2001) Eukaryotic zinc transporters and their regulation. Biometals 14(3–4):251–270
21. Guerinot ML (2000) The ZIP family of metal transporters. Biochim Biophys Acta 1465(1–2):190–198
22. Taylor KM, Morgan HE, Johnson A, Nicholson RI (2004) Structure-function analysis of HKE4, a member of the new LIV-1 subfamily of zinc transporters. Biochem J 377(Pt 1):131–139. doi:10.1042/BJ20031183
23. Huang L, Kirschke CP, Zhang Y, Yu YY (2005) The ZIP7 gene (Slc39a7) encodes a zinc transporter involved in zinc homeostasis of the Golgi apparatus. J Biol Chem 280(15):15456–15463. doi:10.1074/jbc.M412188200
24. Bin BH, Fukada T, Hosaka T, Yamasaki S, Ohashi W, Hojyo S, Miyai T, Nishida K, Yokoyama S, Hirano T (2011) Biochemical characterization of human ZIP13 protein: a homo-dimerized zinc transporter involved in the spondylocheiro dysplastic Ehlers-Danlos syndrome. J Biol Chem 286(46):40255–40265. doi:10.1074/jbc.M111.256784
25. Colvin RA, Bush AI, Volitakis I, Fontaine CP, Thomas D, Kikuchi K, Holmes WR (2008) Insights into Zn2+ homeostasis in neurons from experimental and modeling studies. Am J Physiol Cell Physiol 294(3):C726–C742. doi:10.1152/ajpcell.00541.2007
26. Hogstrand C, Kille P, Nicholson RI, Taylor KM (2009) Zinc transporters and cancer: a potential role for ZIP7 as a hub for tyrosine kinase activation. Trends Mol Med 15(3):101–111. doi:10.1016/j.molmed.2009.01.004
27. Haase H, Maret W (2005) Protein tyrosine phosphatases as targets of the combined insulinomimetic effects of zinc and oxidants. Biometals 18(4):333–338. doi:10.1007/s10534-005-3707-9
28. Sankavaram K, Freake HC (2012) The effects of transformation and ZnT-1 silencing on zinc homeostasis in cultured cells. J Nutr Biochem 23(6):629–634. doi:10.1016/j.jnutbio.2011.03.006
29. Puntervoll P, Linding R, Gemund C, Chabanis-Davidson S, Mattingsdal M, Cameron S, Martin DM, Ausiello G, Brannetti B, Costantini A, Ferre F, Maselli V, Via A, Cesareni G, Diella F, Superti-Furga G, Wyrwicz L, Ramu C, McGuigan C, Gudavalli R, Letunic I, Bork P, Rychlewski L, Kuster B, Helmer-Citterich M, Hunter WN, Aasland R, Gibson TJ (2003) ELM server: a new resource for investigating short functional sites in modular eukaryotic proteins. Nucleic Acids Res 31(13):3625–3630
30. Dinkel H, Van Roey K, Michael S, Davey NE, Weatheritt RJ, Born D, Speck T, Kruger D, Grebnev G, Kuban M, Strumillo M, Uyar B, Budd A, Altenberg B, Seiler M, Chemes LB, Glavina J, Sanchez IE, Diella F, Gibson TJ (2014) The eukaryotic linear motif resource ELM: 10 years and counting. Nucleic Acids Res 42(Database issue):D259–D266. doi:10.1093/nar/gkt1047
31. Kim JE, Tannenbaum SR, White FM (2005) Global phosphoproteome of HT-29 human colon adenocarcinoma cells. J Proteome Res 4(4):1339–1346. doi:10.1021/pr050048h

32. Meggio F, Pinna LA (2003) One-thousand-and-one substrates of protein kinase CK2? FASEB J 17(3):349–368. doi:10.1096/fj.02-0473rev
33. Pinna LA (1990) Casein kinase 2: an "eminence grise" in cellular regulation? Biochim Biophys Acta 1054(3):267–284
34. Ilouz R, Kaidanovich O, Gurwitz D, Eldar-Finkelman H (2002) Inhibition of glycogen synthase kinase-3beta by bivalent zinc ions: insight into the insulin-mimetic action of zinc. Biochem Biophys Res Commun 295(1):102–106, S0006-291X(02)00636-8 [pii]
35. Lee S, Chanoit G, McIntosh R, Zvara DA, Xu Z (2009) Molecular mechanism underlying Akt activation in zinc-induced cardioprotection. Am J Physiol Heart Circ Physiol 297(2):H569–H575. doi:10.1152/ajpheart.00293.2009
36. Kao SH, Wang WL, Chen CY, Chang YL, Wu YY, Wang YT, Wang SP, Nesvizhskii AI, Chen YJ, Hong TM, Yang PC (2013) GSK3β controls epithelial-mesenchymal transition and tumor metastasis by CHIP-mediated degradation of Slug. Oncogene. doi:10.1038/onc.2013.279
37. Taylor KM, Gee JMW, Kille P (2011) Zinc and cancer. In: Rink L (ed) Zinc in human health, vol 76. Ios Press, Amsterdam, pp 283–304. doi:10.3233/978-1-60750-816-8-283
38. Nimmanon T, Taylor K (2014) Zinc signalling and cancer. In: Fukada T, Kambe T (eds) Zinc signals in cellular functions and disorders. Springer, Tokyo, pp 285–313. doi:10.1007/978-4-431-55114-0_14
39. Rhodes DR, Yu J, Shanker K, Deshpande N, Varambally R, Ghosh D, Barrette T, Pandey A, Chinnaiyan AM (2004) ONCOMINE: a cancer microarray database and integrated data-mining platform. Neoplasia 6(1):1–6
40. Taylor KM, Morgan HE, Smart K, Zahari NM, Pumford S, Ellis IO, Robertson JF, Nicholson RI (2007) The emerging role of the LIV-1 subfamily of zinc transporters in breast cancer. Mol Med 13(7–8):396–406. doi:10.2119/2007-00040.Taylor
41. Gyorffy B, Lanczky A, Eklund AC, Denkert C, Budczies J, Li Q, Szallasi Z (2010) An online survival analysis tool to rapidly assess the effect of 22,277 genes on breast cancer prognosis using microarray data of 1,809 patients. Breast Cancer Res Treat 123(3):725–731. doi:10.1007/s10549-009-0674-9
42. Hiscox S, Morgan L, Barrow D, Dutkowskil C, Wakeling A, Nicholson RI (2004) Tamoxifen resistance in breast cancer cells is accompanied by an enhanced motile and invasive phenotype: inhibition by gefitinib ("Iressa", ZD1839). Clin Exp Metastasis 21(3):201–212
43. Tawfic S, Yu S, Wang H, Faust R, Davis A, Ahmed K (2001) Protein kinase CK2 signal in neoplasia. Histol Histopathol 16(2):573–582
44. Taylor KM, Kille P, Hogstrand C (2012) Protein kinase CK2 opens the gate for zinc signaling. Cell Cycle 11(10):1863–1864. doi:10.4161/cc.20414
45. Niefind K, Guerra B, Ermakowa I, Issinger OG (2001) Crystal structure of human protein kinase CK2: insights into basic properties of the CK2 holoenzyme. EMBO J 20(19):5320–5331. doi:10.1093/emboj/20.19.5320
46. Canton DA, Zhang C, Litchfield DW (2001) Assembly of protein kinase CK2: investigation of complex formation between catalytic and regulatory subunits using a zinc-finger-deficient mutant of CK2beta. Biochem J 358(Pt 1):87–94
47. Filhol O, Benitez MJ, Cochet C (2005) A zinc ribbon motif is essential for the formation of functional tetrameric protein kinase CK2. In: Iuchi S, Kuldell N (eds) Zinc finger proteins: from atomic contact to cellular function, Molecular biology intelligence unit. Landes Bioscience/Eurekah.com and Kluwer Academic/Plenum, Georgetown, TX, pp 121–127. doi:10.1007/0-387-27421-9_18
48. Pinna LA, Allende JE (2009) Protein kinase CK2 in health and disease: Protein kinase CK2: an ugly duckling in the kinome pond. Cell Mol Life Sci 66(11–12):1795–1799. doi:10.1007/s00018-009-9148-9
49. Zhou H, Di Palma S, Preisinger C, Peng M, Polat AN, Heck AJ, Mohammed S (2013) Toward a comprehensive characterization of a human cancer cell phosphoproteome. J Proteome Res 12(1):260–271. doi:10.1021/pr300630k
50. Wu X, Tian L, Li J, Zhang Y, Han V, Li Y, Xu X, Li H, Chen X, Chen J, Jin W, Xie Y, Han J, Zhong CQ (2012) Investigation of receptor interacting protein (RIP3)-dependent protein phos-

phorylation by quantitative phosphoproteomics. Mol Cell Proteomics 11(12):1640–1651. doi:10.1074/mcp.M112.019091

51. Franz-Wachtel M, Eisler SA, Krug K, Wahl S, Carpy A, Nordheim A, Pfizenmaier K, Hausser A, Macek B (2012) Global detection of protein kinase D-dependent phosphorylation events in nocodazole-treated human cells. Mol Cell Proteomics 11(5):160–170. doi:10.1074/mcp.M111.016014
52. Trinidad JC, Barkan DT, Gulledge BF, Thalhammer A, Sali A, Schoepfer R, Burlingame AL (2012) Global identification and characterization of both O-GlcNAcylation and phosphorylation at the murine synapse. Mol Cell Proteomics 11(8):215–229. doi:10.1074/mcp.O112.018366
53. Huttlin EL, Jedrychowski MP, Elias JE, Goswami T, Rad R, Beausoleil SA, Villen J, Haas W, Sowa ME, Gygi SP (2010) A tissue-specific atlas of mouse protein phosphorylation and expression. Cell 143(7):1174–1189. doi:10.1016/j.cell.2010.12.001
54. Wu F, Wang P, Zhang J, Young LC, Lai R, Li L (2010) Studies of phosphoproteomic changes induced by nucleophosmin-anaplastic lymphoma kinase (ALK) highlight deregulation of tumor necrosis factor (TNF)/Fas/TNF-related apoptosis-induced ligand signaling pathway in ALK-positive anaplastic large cell lymphoma. Mol Cell Proteomics 9(7):1616–1632. doi:10.1074/mcp. M000153-MCP201
55. Wisniewski JR, Nagaraj N, Zougman A, Gnad F, Mann M (2010) Brain phosphoproteome obtained by a FASP-based method reveals plasma membrane protein topology. J Proteome Res 9(6):3280–3289. doi:10.1021/pr1002214
56. Hornbeck PV, Kornhauser JM, Tkachev S, Zhang B, Skrzypek E, Murray B, Latham V, Sullivan M (2012) PhosphoSitePlus: a comprehensive resource for investigating the structure and function of experimentally determined post-translational modifications in man and mouse. Nucleic Acids Res 40(Database issue):D261–D270. doi:10.1093/nar/gkr1122
57. Horn H, Schoof EM, Kim J, Robin X, Miller ML, Diella F, Palma A, Cesareni G, Jensen LJ, Linding R (2014) KinomeXplorer: an integrated platform for kinome biology studies. Nat Methods 11(6):603–604. doi:10.1038/nmeth.2968
58. Gnad F, Ren S, Cox J, Olsen JV, Macek B, Oroshi M, Mann M (2007) PHOSIDA (phosphorylation site database): management, structural and evolutionary investigation, and prediction of phosphosites. Genome Biol 8(11):R250. doi:10.1186/gb-2007-8-11-r250
59. Yamashita S, Miyagi C, Fukada T, Kagara N, Che YS, Hirano T (2004) Zinc transporter LIVI controls epithelial-mesenchymal transition in zebrafish gastrula organizer. Nature 429(6989):298–302. doi:10.1038/nature02545
60. Pakala SB, Rayala SK, Wang RA, Ohshiro K, Mudvari P, Reddy SD, Zheng Y, Pires R, Casimiro S, Pillai MR, Costa L, Kumar R (2013) MTA1 promotes STAT3 transcription and pulmonary metastasis in breast cancer. Cancer Res 73(12):3761–3770. doi:10.1158/0008-5472.CAN-12-3998
61. Zhang A, Wang Q, Han Z, Hu W, Xi L, Gao Q, Wang S, Zhou J, Xu G, Meng L, Chen G, Ma D (2013) Reduced expression of Snail decreases breast cancer cell motility by downregulating the expression and inhibiting the activity of RhoA GTPase. Oncol Lett 6(2):339–346. doi:10.3892/ol.2013.1385
62. Hogstrand C, Kille P, Ackland ML, Hiscox S, Taylor KM (2013) A mechanism for epithelial-mesenchymal transition and anoikis resistance in breast cancer triggered by zinc channel ZIP6 and STAT3 (signal transducer and activator of transcription 3). Biochem J 455(2):229–237. doi:10.1042/BJ20130483
63. Duncan JS, Litchfield DW (2008) Too much of a good thing: the role of protein kinase CK2 in tumorigenesis and prospects for therapeutic inhibition of CK2. Biochim Biophys Acta 1784(1):33–47. doi:10.1016/j.bbapap.2007.08.017
64. Solares AM, Santana A, Baladron I, Valenzuela C, Gonzalez CA, Diaz A, Castillo D, Ramos T, Gomez R, Alonso DF, Herrera L, Sigman H, Perea SE, Acevedo BE, Lopez-Saura P (2009) Safety and preliminary efficacy data of a novel casein kinase 2 (CK2) peptide inhibitor administered intralesionally at four dose levels in patients with cervical malignancies. BMC Cancer 9:146. doi:10.1186/1471-2407-9-146

65. Siddiqui-Jain A, Drygin D, Streiner N, Chua P, Pierre F, O'Brien SE, Bliesath J, Omori M, Huser N, Ho C, Proffitt C, Schwaebe MK, Ryckman DM, Rice WG, Anderes K (2010) CX-4945, an orally bioavailable selective inhibitor of protein kinase CK2, inhibits prosurvival and angiogenic signaling and exhibits antitumor efficacy. Cancer Res 70(24):10288–10298. doi:10.1158/0008-5472.CAN-10-1893
66. Larkin MA, Blackshields G, Brown NP, Chenna R, McGettigan PA, McWilliam H, Valentin F, Wallace IM, Wilm A, Lopez R, Thompson JD, Gibson TJ, Higgins DG (2007) Clustal W and Clustal X version 2.0. Bioinformatics 23(21):2947–2948. doi:10.1093/bioinformatics/btm404
67. Notredame C, Higgins DG, Heringa J (2000) T-coffee: a novel method for fast and accurate multiple sequence alignment. J Mol Biol 302(1):205–217. doi:10.1006/jmbi.2000.4042
68. Zhou BP, Deng J, Xia W, Xu J, Li YM, Gunduz M, Hung MC (2004) Dual regulation of Snail by GSK-3beta-mediated phosphorylation in control of epithelial-mesenchymal transition. Nat Cell Biol 6(10):931–940. doi:10.1038/ncb1173

CK2: A Global Regulator of Cell Death

Janeen H. Trembley, Fatima Qaiser, Betsy T. Kren, and Khalil Ahmed

Abstract Protein kinase CK2 has emerged as a major signal involved in diverse cellular functions of health and disease. The nature of its broad of range of functions is underscored by the large number of potential substrates of CK2 present in various locales in the cell. CK2 has gained much attention for its role in cancer biology, which is attributed to its functions both in cell growth and proliferation as well as in the regulation of cell death. Indeed, it appears that CK2 impact on cell death may be one of its most important functions, especially in the context of cancer biology where both cell proliferation and cell death are dysregulated and elevated CK2 in cancer would have an effect on both of these activities. Just as CK2 has been proposed to have a global role in cell growth-related activities, it appears that it may have an analogous global role in the suppression of apoptosis when it is elevated and induce cell death when it is downregulated. In this review, we have highlighted the current

J.H. Trembley • B.T. Kren
Cellular and Molecular Biochemistry Research Laboratory (151), Minneapolis Veterans Affairs Health Care System, Minneapolis, MN 55417, USA

Department of Laboratory Medicine and Pathology, University of Minnesota, Minneapolis, MN 55455, USA

Masonic Cancer Center, University of Minnesota, Minneapolis, MN 55455, USA

F. Qaiser
Cellular and Molecular Biochemistry Research Laboratory (151), Minneapolis Veterans Affairs Health Care System, Minneapolis, MN 55417, USA

Department of Laboratory Medicine and Pathology, University of Minnesota, Minneapolis, MN 55455, USA

K. Ahmed (✉)
Department of Laboratory Medicine and Pathology, University of Minnesota Medical School and Minneapolis VA Health Care System, Minneapolis, MN USA
e-mail: ahmedk@umn.edu

K. Ahmed et al. (eds.), *Protein Kinase CK2 Cellular Function in Normal and Disease States*, Advances in Biochemistry in Health and Disease 12, DOI 10.1007/978-3-319-14544-0_10

status of CK2 involvement in the processes related to cell death with a focus on apoptosis. It is proposed that a newly identified mechanism of CK2 regulation of cell death relates to its impact on early intracellular dynamics of Ca^{2+} signaling which profoundly alter mitochondrial function and lead to cell death.

Keywords Protein kinase CK2 • Apoptosis • Cell death • Suppression of apoptosis • Mitochondria • Endoplasmic reticulum • Ca^{2+} • ROS • Bcl-2 • Bcl-xL • Stress • IAPs • Survivin • Mitochondrial membrane potential • Mitochondrial permeability transition

Abbreviations

CK2	The adopted acronym for previous name casein kinase
CK2α	42 kDa catalytic subunit of CK2
CK2 α′	38 kDa catalytic subunit of CK2
CK2β	28 kDa regulatory subunit of CK2
DR	Death receptor
ER	Endoplasmic reticulum
IAPs	Inhibitor of apoptosis proteins (including cIAP1, cIAP2, XIAP, and survivin)
MNA	1,8-Dihydroxy-4-nitro-anthracene-9,10-dione
OGN	Oligonucleotides
NM	Nuclear matrix
RIP1	Receptor-interacting protein-1
ROS	Reactive oxygen species
TBB	4,5,6,7-Tetrabrombenzotriazole
TBCA	Tetrabromocinnamic acid
TNFα	Tumor necrosis factor α
TRADD	TNF receptor type I-associated death domain protein
TRAF2	TNF receptor-associated factor-2
TRAIL/Apo2-L	Tumor necrosis factor-related apoptosis-inducing ligand

1 Introduction

The acronym "CK2" was adopted over 20 years ago to replace the previously used misnomer "casein kinase" since casein is not a genuine physiological substrate for protein kinase CK2. Indeed, the issue of distinguishing CK2 from other protein kinases belonging to the group of casein kinases has been addressed in considerable detail recently [1] and is discussed further in this book (Chap. 13). Over time, CK2 has emerged as an important player in an ever-increasing number of cellular activities in health and disease states, acquiring a status of "master regulator" of cell functions.

Among its numerous activities, CK2 regulation of cell death (such as apoptosis), originally documented in 2001, has emerged as a key role [2–4] and, as was commented in a recent review, may possibly be the most important function of CK2 [1]. Notwithstanding the critical roles of CK2 in normal cell functions including development (e.g., [5–16]), the study of CK2 in the context of neoplastic diseases has been a particularly fertile area of research ever since it was originally observed that this kinase was elevated in proliferating normal and neoplastic cells [13, 17]; over time it has been extensively documented that CK2 is consistently elevated in all the cancers that have been studied (for reviews, see, e.g., [18, 19]). CK2 is elevated (albeit transiently) during normal and benign cell proliferation; however, this is in contrast to the consistent and stable increase compared to the levels in normal counterpart cells observed during cancer cell proliferation (e.g., [20]). The observed involvement of CK2 in normal and neoplastic cell proliferation raised the issue of the basis of its function in cancer cells. Thus, the discovery of the ability of elevated CK2 to potently suppress cell death represented a key feature of this kinase that is distinct in cancer vs normal cells, revealing, for the first time, a fundamental link of this signal to the cancer cell phenotype [4, 21, 22]. It may be recalled that two of the well-known characteristics of cancer cells that distinguish them from normal cells are unrestricted cell proliferation and even more importantly a universal ability to evade cell death; these features are among the hallmark characteristics of cancer cells discussed in an elegant recent review on the subject [23]. In this chapter, we present an account of the emerging broad involvement of CK2 in the regulation of cell death with a focus on our contributions to this area of research.

2 A Brief Overview of Modes of Cell Death

Several hallmark characteristics of cancer cells have been documented [23]. The process of cell death, an integral part of cell biology, has been recognized to be hijacked by cancer cells resulting in their tremendous capability to evade death, thus becoming a highly significant hallmark of cancer [23]. Studies on cell death have identified several mechanisms involved in the execution of this process. Among them, the process of apoptosis or programmed cell death appears to represent a major mode of cell death. Both proliferation and death are deregulated in cancer, but the latter is now recognized to play a particularly prominent role in the pathogenesis of cancer (see, e.g., [24]). Indeed, it appears that in some cancers the deregulation of cell death (apoptosis) plays an even more important role than deregulation of cell proliferation, a notable example being that of prostate cancer [25].

The manner of cell death characterized as necrosis (often referred to as homicide) has been known for a long time as a major mechanism in response to pathological events (see, e.g., [26]); however, over time it has been recognized that cells can undergo various types of death. The term "programmed cell death" was originally proposed to describe physiological activity during organ developmental [27]. Subsequently, a major form of programmed cell death was defined by the term "apoptosis" [28]. In recent years, apoptosis has gained much attention as a key mechanism of cell death responsive to a large variety of inducers. However, additional pathways such as autophagy, necroptosis, and autosis continue to gain attention, as has been discussed extensively in numerous review articles (see, e.g., [24, 29–41]). It is noteworthy that diverse mechanisms of cell death as indicated above do have several common features including involvement of certain signaling pathways; further, multiple mechanisms of cell death may occur simultaneously. Here we provide a very brief account of the topic of apoptosis to recapitulate some salient features that will be addressed subsequently in the context of the involvement of CK2 as a modulator of cell death activity.

Apoptotic cell death has been identified to occur via intrinsic or extrinsic pathways. The intrinsic pathway is followed in response to intracellular damage which could be initiated by various types of agents. The extrinsic pathway is involved when membrane-associated death receptors are activated through the action of their cognate ligands. The intrinsic pathway (also referred to as type I apoptosis) can be activated by a wide range of chemical and physical agents co-opting mitochondria as the focal point with activation of Bcl-2 family proteins (including BH3-only members) eventually resulting in the release of cytochrome *c*. Cytochrome *c* combines with APAF1 (apoptotic protease activating factor-1) to form the apoptosome, and upstream caspases are recruited resulting in their activation and action on the executioner caspase-3 and caspase-7 [40, 42]. The release of cytochrome *c* reflects the damage to mitochondria indicated by a change in mitochondrial permeability transition. The entire process of type I apoptosis is under the regulatory control of antiapoptotic (e.g., Bcl-2) and proapoptotic (e.g., Bax/Bak) members of the Bcl-2 family. Several members of the caspase family are known to be involved in various stages of apoptosis and respond to specific death signals (e.g., [40]). For example, while caspase-3-mediated apoptosis is a frequent pathway in type I apoptosis, other caspases such as caspase-2, caspase-4, and caspase-12 have been documented to serve as the prime mediators of execution in response to stress-mediated apoptosis (e.g., [42–45]).

A second apoptotic pathway refers to the extrinsic pathway which involves death receptors (DRs) activated by proapoptotic and pro-inflammatory cytokines including, e.g., Fas ligand (FasL) and TNFα (for details see, e.g., [40]). In this context, the discovery of Apo2-L/TRAIL is of particular note; the complexity of death receptor-mediated apoptosis in various systems has been discussed in detail [46, 47]. In general, this pathway follows the formation of death-inducing signaling complex (DISC) leading to the activation of caspase-8. In analysis of FADD- and TRADD-dependent DR signaling, it has been observed that while FADD signaling follows the caspase-8 activation pathway to apoptosis, the TRADD signaling leads to the assembly of an apical complex that contains RIP1, TRAF2 or TRAF5, and inhibitor of apoptosis proteins cIAPs 1 and 2, causing activation of diverse gene expression programs through the NF-κB pathway. Interestingly, caspase-8 is also known to act on Bid causing its cleavage and entry of t-Bid into mitochondria, thereby resulting in expansion of the apoptotic machinery [40, 46]. It is clear that interplay of members of the apoptotic machinery (including all the caspases) contributes to dynamic regulatory functions in the outcome of the cell fate in response to various proapoptotic signals [40, 48].

As mentioned earlier, a number of non-apoptotic mechanisms of cell death have been elaborated on in recent years. These include, e.g., necrosis (an oldest known mechanism of pathological cell death involving several types of necrosis), necroptosis (considered as regulated necrosis or possibly a combination of both necrosis and apoptosis), and autophagy; however, as noted above, additional names denoting extra aspects of cell death process continue to emerge. The term necroptosis indicates that the process of necrosis follows a regulated program and that the morphological features of necroptosis are distinct from that of necrosis and apoptosis [49]. DNA damage-mediated cell death involving, e.g., poly-ADP-ribose polymerase 1 (PARP1), leads to necrotic cell death or necroptosis [50]. Several of the pathways involving cell death indicate the involvement of Ca^{2+}; in this context, it is noteworthy that cellular Ca^{2+} shifts were among the earliest changes observed and preceded the appearance of necrosis which emphasizes the time scale of observations subsequent to the induction of injury [26, 51]. Additional distinct features of this type of cell death include the involvement of RIP1 kinase activity [52] and the possible function of small molecule inhibitors (called necrostatins) in blocking necroptosis [40]. Autophagy is another non-apoptotic form of cell death and appears to be regulated by a group of autophagy-related genes (*ATG*). It is well known that under nutrient deprivation, autophagy provides a mechanism of cell survival; however, under many conditions where apoptosis is unavailable, it assumes the function of inducing cell death [40, 53]. A cross talk between functions of antiapoptotic Bcl-2 and autophagy protein Beclin-1 is strongly suggested [54]. Other roles of apoptosis are also known as, e.g., in host defense mechanisms involving NOD-like receptors (see, e.g., [40]). One may speculate a role of CK2 in this aspect since many of the viral proteins are phosphorylated by CK2 [14].

The brief discussion of some but not all of the pathways of death indicates the breadth of the impact of this function on cell biology in normal and disease states and in particular the importance of these pathways in the pathobiology of cancer cells pertaining to the deregulation of cell death as among the fundamental characteristics of the cancer cell phenotype. It has been proposed that CK2 has a global function in activities related to cell growth [55, 56] and emerging supporting evidence continues to mount in the literature (e.g., [57]). Likewise, it appears that CK2 may have a global function in the activities related to cell death, and thus it would be an intriguing aspect of studies on CK2 to determine the potential sites of action of CK2 in modulating various modes of cell death. In the ensuing discussion, we have attempted to present the current status of this field with emphasis on the role of CK2 in the regulation of the apoptotic machinery.

3 CK2 and Cell Death Related to Loss of Growth Factors

One of the earliest hints of the function of CK2 in cell growth and cell death was derived from studies on androgenic regulation of rat ventral prostate [58, 59]. This organ is strictly dependent on the availability of androgens for continued growth and undergoes rapid cell death leading to involution upon loss of androgen expression, which served as an original model to describe apoptosis [60]. It was observed that loss of androgens in the rat resulted in a rapid induction of cell death in the ventral prostate which was preceded by rapid decline in phosphorylation of nonhistone proteins which related to the loss of nuclear-associated protein kinase activity (subsequently identified as CK2) [58, 59, 61, 62]. Studies revealed that CK2 was distributed in the nuclear and cytoplasmic fractions and that within the nucleus it appeared to be distributed in the nucleoplasm and chromatin and nuclear matrix structures [62, 63]. These studies further established that the loss of CK2 from the nuclear compartments preceded cell death in rat ventral prostate in animals subjected to androgen deprivation revealing a novel observation on dynamic rapid shuttling of CK2 within the cellular compartments depending on the status of growth signaling [62, 64, 65]; subsequently, this concept was further expanded in other studies [66]. Analogous observations to those made in the rat ventral prostate were made in cultured prostate cancer cells that responded to growth factors and/or androgen [67]. To sum, these investigations from the authors' laboratory provided two fundamental observations which hinted at the role of CK2 in the regulation of cell death, namely, CK2 dynamic shuttling in and out of the nuclear compartment on institution of growth signal or its loss, respectively, and the regulation of cell survival decision based on the nuclear CK2 status. Subsequently, analogous observations were reported relating to the function of parathyroid hormone-related protein (PTHrP) which is known to protect cells from apoptotic stimuli, an activity that correlates with its nuclear translocation; it was found that inhibition of apoptosis by this protein was associated with increased and sustained expression of nuclear CK2 level [68]. Involvement of various cellular organelles has been discussed as playing a role in cell death signaling [69]; the above-described observations exemplify the role of the nucleus in apoptotic activity by serving as the dynamic locale of a regulatory signal such as CK2.

4 CK2 Regulation of Apoptosis Induced by Chemicals, Drugs, and Physical Agents

While the studies described in the above section strongly hinted at the involvement of CK2 in cell death, a more definitive function of CK2 in regulation of cell death was originally provided in work dealing with drug-induced cell death where it was demonstrated that increased expression of CK2 resulted in blocking of cell death promoted by etoposide and diethylstilbestrol [2]. These compounds induced cell death in a dose- and time-dependent manner in prostate cancer cells which was associated with an initial shuttling of CK2 from the cytosolic fraction to the nucleus with a high association with the nuclear matrix. Further, it was found that when CK2 subunits were transiently overexpressed in various cell lines, there was a strong protection against cell death following CK2α and CK2α plus CK2β overexpression but not following CK2β overexpression alone. The results were analogous for both etoposide and diethylstilbestrol drugs and for different prostate cancer cell lines, suggesting that the protective effect of CK2α reflected a general characteristic of CK2 function relating to the activity of the CK2α catalytic subunit [2]. As discussed subsequently, downregulation of CK2 impacts the mitochondrial apoptotic circuitry, and in that context it is interesting to note that downregulation of CK2α was found to cause a reduction in the expression of the message for Bcl-xL [70].

In line with the above data are the observations that the presence of chemical inhibitors of CK2 such as apigenin or TBB caused a dose- and time-dependent inhibition of cell viability in a variety of cell lines. The effect of these agents related to the inhibition of CK2 catalytic activity since transfection of cells with kinase-dead CK2α (K68M) expression plasmid or treatment of cells with CK2α siRNA or antisense oligonucleotides (OGN) also demonstrated a potent loss of cell viability [71]. Likewise, it was originally documented that treatment of cells with antisense or siRNA to CK2α induces apoptosis in cell culture and in mouse xenograft cancer models [22, 70–76]. Analogous observations have been made in numerous reported studies and have been the subject of several recent reviews (see, e.g., [18, 77, 78]).

Chemopreventive agents generally induce moderate apoptotic activity in cells and are known to affect a diverse range of cellular signaling pathways [79, 80]. It was therefore of interest to determine the effect of two of the chemopreventive agents (epigallocatechin-3-gallate or EGCG and resveratrol) on CK2 signaling in certain prostate cancer cells [81]. These studies demonstrated that apoptosis induced by EGCG and resveratrol was associated with significant downregulation of CK2 activity and expression in cancer cells. Overexpression of CK2α protected cells against the effects of these agents suggesting that targeting of the CK2 signal may, in part, play a role in the chemopreventive activity of polyphenolic chemopreventive agents such as EGCG and resveratrol in cancer cells. It was reported that apigenin inhibited 22Rv1 xenograft tumor growth by upregulating IGFBP3 protein [82]; a possible mechanism of this observation may relate to IGFBP3 being a substrate for CK2 which is inhibited by apigenin [83].

Physical agents such as heat and radiation are known to induce cell death. It has been reported that heat shock and UV radiation in different cells induce rapid translocation of CK2 to the nuclear compartment, with a significant increase in the nuclear matrix [3, 84, 85]. Likewise, it was subsequently shown that CK2 undergoes nuclear translocation and inhibits apoptosis caused by ionizing radiation [86]. In another study, it was demonstrated that UV-induced NF-κB activation depended on phosphorylation of IκBα at a cluster of C-terminal sites with the CK2 consensus phosphorylation sequence. This response also involved p38 MAP kinase so that inhibition of this pathway increased UV-induced cell death; thus, these investigators proposed that the p38-CK2-NF-κB axis represented an important component of the mammalian UV response [87]. Several of the above examples accord with the influence of CK2 on the intrinsic mode of apoptosis.

5 Regulation of Death Receptor-Mediated Apoptosis by CK2

As described above, death receptors play an important role in the mediation of cell death (extrinsic apoptosis), and induction of apoptosis in cancer cells has been documented via the death receptor pathway [30, 40, 46]. Several observations have suggested a strong link of CK2 to the regulation of extrinsic mode of apoptosis in prostate, colon, and rhabdomyosarcoma cell lines [21, 88–90]. However, it may be noted that under certain conditions, when type II apoptosis occurs, there is involvement of the mitochondrial pathway also, as exemplified by TRAIL/Apo2-L where, depending on the dose employed, TRAIL/Apo2-L can induce both type I and type II apoptosis. To explore the involvement of CK2 under various conditions, an analysis of CK2 regulation of death receptor-mediated apoptosis was undertaken by treating prostate cancer cells with TNFα, FasL, or TRAIL/Apo2-L. In all of these cases, cells underwent a rapid death which was blocked by prior overexpression of CK2α. Further, it was observed that cells treated with low doses of a chemical inhibitor of CK2 (such as apigenin or TBB) were sensitized to the action of TNFα. For example, cells treated with 20 μM apigenin demonstrated about 25 % apoptosis at 24 h, while those treated with 40 ng/ml of TNFα (in the absence of cycloheximide) were unaffected; however, when the two treatments were combined, there was a significant enhancement in the apoptotic activity, increasing to about 60 %. Likewise, TBB and TNFα employed at low suboptimal concentrations were ineffective in producing cell death individually, but when the two agents were combined at these dose levels, there was a potent induction of apoptosis in the treated cells [21].

In a subsequent study, it was documented in prostate cancer cell lines that treatment with doses of TBB and TRAIL/Apo2-L that were minimally effective for inducing cell death added individually caused a potent induction of apoptosis when added together at the suboptimal concentrations. This was evidenced by significant activation of caspase-3, caspase-8, and caspase-9, with detection of cleaved caspase-3 and caspase-8. Under these conditions, significant loss of c-FLIP$_L$ was observed when cells were first treated with TBB followed by treatment with TRAIL/Apo2-L, suggesting moderate inhibition of CK2 sensitized cells to suboptimal concentrations of TRAIL/Apo2-L for induction of apoptosis. Apoptotic cell death was confirmed by analysis of DNA fragmentation and lamin A cleavage. An examination of the mitochondrial circuitry in this experimental model demonstrated that in prostate cancer cell lines treated with TBB prior to TRAIL/Apo2-L, there was a significant augmentation of the proapoptotic protein Bax with concomitant loss of antiapoptotic proteins Bcl-2 and Bcl-X$_L$ and a profound increase in cytochrome *c* release. These events were completely reversed by prior transient overexpression of CK2α in prostate cancer cells, as was evidenced by complete blocking of the engagement of mitochondrial apoptotic proteins activated by TRAIL/Apo2-L, blocking of activation of caspases, and reversal of loss of c-FLIP$_L$ [91]. An interesting observation on the regulation of apoptosis by CK2 suggested the involvement of ER stress response on inhibition of CK2 activity such that the mitochondrial pathway induced in prostate cancer cell lines (LNCaP and PC-3) was not accompanied by activation of caspase-9, but in the case of LNCaP cells, there was an associated upregulation of DR5 and activation of caspase-8 resulting in activation of caspase-3 [92]. It may also be recalled that Bid, which plays an important role in type II apoptosis, is cleaved by activated caspase-8 resulting in t-Bid which enters mitochondria and causes release of apoptosis-inducing factors—importantly, the phosphorylation of Bid by CK2 renders it insensitive to action of caspase-8 [93]. The various observations discussed above exemplify the strong link of CK2 in regulation of intrinsic and extrinsic modes of apoptosis.

6 Regulation of Caspases by CK2

Given the importance of the functions of caspases in the operation of apoptosis, it has been of interest to determine the potential role of CK2 in regulation of caspase activities. In this context, an early study on the regulation of caspase-inhibiting protein ARC demonstrated that its phosphorylation by CK2 at Thr^{149} promoted its targeting to mitochondria where its binding to caspase-8 took place; this suggested a mechanism where ARC required phosphorylation for its function [94]. A potential role of CK2 in regulation of caspases was further indicated in studies examining the activation of caspase-9 in TNF-α-treated murine cells where it was observed that caspase-9 was phosphorylated by CK2 at a Ser^{348} which rendered it refractory to caspase-8 cleavage. Studies also suggested that TNF-α cross-linking resulted in dephosphorylation of caspase-9, thereby making it susceptible to cleavage by activated caspase-8, suggesting a pro-survival mechanism of action of CK2 [95]. Interestingly, it has been observed that caspase-9 is unable to be activated when phosphorylated by AKT [96]. Since AKT phosphorylation on Ser^{129} has been reported to result in upregulation of its activity [97], it may be surmised that such a mechanism may accord with suppression of apoptosis on upregulation of CK2. Response of various caspase activities to downregulation of CK2 (and induction of apoptosis) has been documented in several studies (e.g., [70, 98]). A detailed investigation of potential regulation of caspases by CK2 was undertaken recently [99, 100]. In these studies, an elegant approach was developed to screen CK2 targets in caspase signaling by designing a peptide match program to identify sequences from the human proteome that contained overlapping recognition motifs for caspases and kinases. It was documented that protein kinase CK2 had the most extensive consensus sites for phosphorylation that overlapped with caspase recognition motifs. Further studies identified numerous shared candidate targets of CK2 and caspases, including procaspase-3. Thus, these studies implicate the global involvement of CK2 in the regulation of caspase signaling pathways [100] and accord with the aforementioned response of caspases to sensitization of cells using CK2 inhibitors to TNF-α- and TRAIL/Apo2-L-mediated apoptosis, thus reinforcing the originally proposed notion of a global impact of CK2 on cell death processes [21, 91].

7 Impact of CK2 on Inhibitor of Apoptosis Proteins

Inhibitor of apoptosis proteins (cIAP1, cIAP2, XIAP) and survivin have attracted considerable interest as modulators of apoptosis, and because of their elevation in cancer, cells have also attracted much attention as potential targets for cancer therapy [101–106]. A study of the effect of CK2 on survivin expression demonstrated that treatment of cells with CK2 inhibitors TBB or DMAT causing apoptosis in colon and several other cancer cells demonstrated a concomitant decrease in survivin mRNA and protein. Further, it was found that augmented CK2α expression resulted in β-catenin-Tcf/Lef-dependent upregulation of survivin; these effects were abolished by TBB. Similarly, molecular downregulation of CK2α was associated with reduced β-catenin and survivin levels leading to the conclusion that CK2 activity promoted survival by increasing survivin expression via β-catenin-Tcf/Lef-mediated transcription [107]. Subsequently, it was also documented that CK2α-specific enhancement of β-catenin transcriptional activity and survivin may depend on AKT hyperactivation by CK2-mediated phosphorylation of AKT at Ser^{129} [108]. Our unpublished work also confirmed the role of transcriptional regulation of survivin in response to CK2 modulation. However, it should be noted that posttranslational regulation of survivin is also plausible since Thr^{48} in the BIR domain of survivin is critical to its mitotic and antiapoptotic activities and can be phosphorylated by CK2 in vitro [109].

In studies on the impact of CK2 on IAPs in prostate cancer cells, a number of strategies were followed to examine the response of IAPs (including survivin) to altered CK2 activity [110]. It was observed that various inhibitors of CK2 affected the levels of IAPs and survivin in a dose-dependent manner. Suboptimal levels of apoptosis-inducing agents including CK2 inhibitors (TBB and apigenin), TNFα, and etoposide (as discussed above) individually caused moderate reductions in immunoreactive levels of IAPs; however, when the CK2 inhibitors were combined with etoposide or TNFα, there was a marked increase in the loss of IAPs (including survivin). These results suggested that one mechanism by which moderate inhibition of CK2 activity sensitized the cells to the effects of apoptotic agents even at suboptimal levels involves modulation of IAPs. Further, prior overexpression of CK2α in the presence of the expression plasmid pcDNA6-CK2α resulted in complete blocking of the effect of these apoptosis-inducing agents (used at optimal levels) on cIAP2 (used as a representative IAP) [110].

Since IAPs are known to be localized to different compartments in the cell [104, 111], it was interesting to note that their response to altered CK2 activity varied in different cellular compartments. For example, an analysis of the effect of TBB at optimal apoptosis-inducing concentration (80 μM) on survivin showed that it was heavily localized in the nuclear matrix (NM) fraction compared with cytoplasm in prostate cancer cells, and in the presence of TBB, the cytoplasmic signal was completely lost, while the effect on the NM-associated survivin was moderate. In the case of XIAP, most of it was localized in the cytoplasm with a minor amount in the NM fraction, and CK2 was reduced in cytosol in the presence of TBB. cIAP2 was found to be present mostly in the NM fraction, and the NM CK2 subpopulation was dramatically reduced in the presence of TBB. In the presence of a second inhibitor of CK2 (MNA at 80 μM which induced apoptotic response in prostate cancer cells), it was confirmed that the cIAP2 subpopulation localized predominantly in the NM fraction was greatly reduced in the presence of the inhibitor, while cIAP1 and XIAP subpopulations localized in the cytoplasmic fraction similarly demonstrated a strong reduction response to MNA inhibitor. In the case of survivin (present mostly in the NM with a small amount in the cytoplasm), the response was observed largely in the cytoplasmic fraction. Thus, it is clear that CK2 activity has a potent regulatory effect on various IAPs and that the effects are of a complex nature; studies in our laboratory are being directed to further analyze the mechanism of CK2 regulation of IAPs. It is of interest to note that IAPs have been proposed as targets for cancer therapy; however, because of their complex regulation by CK2, such strategies may be compromised by the elevated levels of CK2 in cancer cells [110].

8 Some Examples of Other Loci of CK2 Activity Relating to Cell Death Regulation

CK2 is known to have numerous potential substrates as has been cataloged [14], and since the time of that report, it is most likely that many additional potential substrates have been identified; an example of a recent addition was mentioned above [31]. An examination of the various substrates of CK2 indicates their broad distribution in the cell compartments and association with diverse functions including cell death. Here, a few examples of some of the loci of CK2 activity relating to regulation of cell death are presented.

The potential link of CK2 to the regulation of caspases was discussed in an earlier section. It is therefore interesting to note that CK2 was previously found to regulate the activity of caspase-inhibiting protein ARC via its phosphorylation at Thr^{149} which controls its subcellular localization to exert its antiapoptotic action [94]. CK2-mediated phosphorylation can produce its effect in ways other than producing a direct response. For example, CK2-mediated phosphorylation may affect apoptotic activity of Fas-associated protein 1 (FAF1) by influencing its nuclear localization rather than directly affecting its activity [112]. Similarly, c-Myc which is known to promote cell proliferation or induce apoptosis under different conditions was found to be stabilized by CK2-mediated phosphorylation [113], whereas phosphorylation of Max at Ser^{11} by CK2 renders it insensitive to cleavage by caspase-5 [114]. It was documented that a direct effect of CK2-mediated phosphorylation of the scaffold protein XRCC1 was to enable the assembly and activity of DNA single-strand break repair protein complexes in vitro and at sites of chromosomal breakage; various data suggested that inhibition of CK2 resulted in loss of repair of cellular DNA single-strand breaks by XRCC1 [115]. AKT has been considered an important kinase whose activation is associated with blocking of cell death, and it is of particular note that AKT phosphorylation at Ser^{129} results in its hyperactivation [97]. Effects on AKT phosphorylation by inhibition of CK2 have been documented [98, 116]. Likewise, CK2 inhibition is also apparent in the phosphorylation of PTEN Ser^{370} and Ser^{380} [98]. Other potentially important substrates of CK2 are p53 and MDM-2; phosphorylation of MDM-2 promotes its association with p53, but at present further studies are needed to determine the precise role of the phosphorylation of these proteins in the context of their functions [117, 118]. Along this line, a recent observation is noteworthy which demonstrated that knockdown of CK2α in xenograft tumor cells resulted in increased expression of p53 and Tap63 which would accord with their role in growth arrest and apoptosis [70]. Another proapoptotic protein whose function is modulated by CK2 is IGFBP-3 where it was observed that downregulation of CK2 enhanced the apoptotic potential of IGFBP-3 [83]; again, it may be speculated that under conditions where CK2 is elevated (as in cancers), it might result in blocking the apoptotic potential of IGFBP-3 through its phosphorylation. PML tumor suppressor regulates pathways involved in growth suppression, and induction of apoptosis was observed to show an inverse relationship in its expression relative to CK2, which accords with the observation that phosphorylation of PML at Ser^{517} by CK2 results in its ubiquitin-mediated degradation [119, 120]. On the other hand, it has been shown that PTEN, an antagonist of PI3K signaling with an important role in cell death and survival, is stabilized by phosphorylation by CK2 [121].

NFκB plays diverse roles in cell function (including those related to cell death); phosphorylation of p65 at Ser^{529} has been shown to be mediated by CK2 [122, 123]. Further, the NFκB inhibitor IκB has also been shown to be a substrate for CK2 so that its phosphorylation promotes its degradation [124]. In studies on the role of aberrant activation of NFκB in breast cancer, it was documented that inhibition of IKK and CK2 activity decreased NFκB activity in different breast cancer cells, whereas aberrant expression of IKK or CK2 resulted in increased nuclear levels of NFκB/Rel. Further, it was demonstrated that CK2 played a role in Her-2/neu-mediated signaling and activation of NFκB. In these studies, ectopic expression of CK2 appeared to be sufficient to induce NFκB activation [125, 126]. Studies by these authors have led to the recognition of Wnt and NFκB pathways as two key signal transduction pathways that are regulated by CK2 activity in embryonic development and in cancer [127]. Studies in cultured cells or in vivo xenograft models have shown that p65 Ser^{529} phosphorylation is very responsive to the downregulation of CK2, and indeed this response can serve as a surrogate marker for altered CK2 function (see, e.g., [70, 75, 76, 128–131]). The JAK/STAT pathway has also been found to be a target of CK2 function [132]. Protein B23 (nucleophosmin/numatrin), a ubiquitous nucleolar protein which is essential for rRNA synthesis, is also a substrate of CK2 and has been used as a biomarker of CK2 activity in the cell [133, 134]. Thus, it would appear that modulations in CK2 activity and function may be monitored by changes in the phosphorylation of some of the above-described proteins which may in turn serve as markers of cellular response to altered CK2 status in the cell.

9 Temporal Events in Triggering of Apoptosis Following CK2 Downregulation

In the foregoing discussion, several examples of CK2 involvement in the modulation of cell death-related activities were presented. These examples illustrated the involvement of various cellular compartments (such as nuclei, plasma membrane, ER, and mitochondria) in the regulation of cell death [69], as illustrated in Fig. 1.

Mitochondria represent a major locus of regulation of apoptotic activity, and the regulation of apoptotic circuitry by CK2 was clearly indicated by upregulation of Bax and downregulation of Bcl-2 and Bcl-xL with release of cytochrome *c* on downregulation of CK2. Equally importantly, these events were completely blocked on upregulation of CK2 [21, 91]. An early effect on production of ROS was suggested to be an early event in the activation of the mitochondrial apoptotic machinery in response to downregulation of CK2 [135, 136]. Further investigation of the events that trigger the mitochondrial apoptotic circuitry on downregulation of CK2 was reported recently [137]. In these studies, inhibitors of CK2 activity (TBB and TBCA) were employed to study their temporal effect on cell viability of prostate cancer cells. Remarkably, the CellTiter 96® Aqueous One assay demonstrated that PC3-LN4 cells began to indicate a change in cell viability within 2 h following the treatment with CK2 inhibitor, and further analysis by crystal violet clonal survival assays demonstrated the initiation of loss of cell viability as early as 6 h following inhibition of CK2. These observations suggested the presence of preceding signals that trigger the initiation of events leading to loss of cell viability and death. Further analysis to address this issue indicated that there was a significant loss of mitochondrial membrane potential ($\Delta\psi_m$) in cancer cells which was somewhat lesser in a benign prostate epithelial cell line; these observations accorded with previous observation showing greater sensitivity of cancer cell lines to downregulation of CK2 compared with normal or benign cell lines [73].

It was originally proposed that induction of apoptosis induced by downregulation of CK2 was via the upstream production of H_2O_2 [135]. However, further analysis revealed that treatment of cells with catalase (a ROS scavenger) did not block the changes in $\Delta\psi_m$ induced by CK2 inhibition; these results suggested that changes

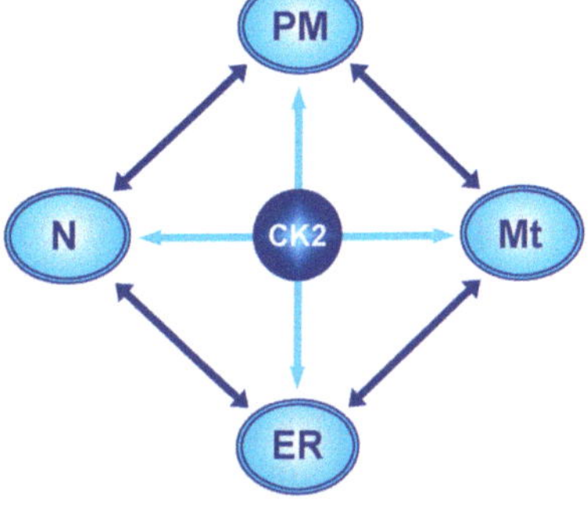

Fig. 1 The cartoon illustrates CK2 involvement and function in various subcellular loci. This might occur by intracellular modulation of CK2 activity at these loci as well as via mechanisms such as dynamic shuttling as has been proposed [62, 66, 145]

in mitochondrial membrane potential preceded production of ROS on inhibition of CK2 [137]. Considering that mitochondrial membrane potential and permeability are under intimate control of cellular Ca^{2+} (see, e.g., [138]), an analysis of the effect of BAPTA (a Ca^{2+} chelator) revealed that pretreatment with this agent prior to treatment of cells to TBB resulted in the prevention of loss of $\Delta\psi_m$ indicating the involvement of Ca^{2+} as an early signal in response to CK2 inhibition; this was supported by the observation of rapid intracellular Ca^{2+} release on the treatment of PC3-LN4 cells with TBB. The above study results underscore the importance of CK2 in the process of regulation of cell death since mitochondrial membrane permeability transition (MPT), $\Delta\psi_m$, and Ca^{2+} shuttling are known to be intimately linked to the process of cell death [29, 139–141]. At present it is not entirely clear whether the effects on mitochondrial changes upon CK2 inhibition relate to the effects on the CK2 cytoplasmic pool in general or reflect effects on the CK2 pool localized to mitochondria or both compartments. Some support for the involvement of mitochondrial CK2 inhibition in this process has been suggested in studies on the effects of CK2 inhibitors on highly purified isolated mitochondria which demonstrated that inhibitors of CK2 produced significant changes in permeability transition and swelling within 5 min which were rescued by addition of excess Ca^{2+} chelator EGTA, thus further supporting the notion that early mitochondrial function changes are critical to initiation of apoptotic signaling and may likely be in response to inhibition of mitochondrial CK2 [137]. Based on these and other studies in the authors' laboratory, a temporal sequence of events that trigger and propagate the apoptotic signaling can be envisioned as shown in Fig. 2. We hypothesize that one of the earliest responses to loss of CK2 activity which may involve different cellular loci (as indicated in Fig. 1) is a shift in intracellular Ca^{2+} flux that affects mitochondrial membrane permeability transition and mitochondrial membrane potential. The resultant concatenation of events involves production of ROS and initiation of negative effects on cell viability and proliferative capacity, such as through downregulation of AKT and NFκB signaling. The ensuing events that occur over time from this point on are the activation of apoptotic signaling (including the Bcl-2 family of proteins), release of cytochrome *c*, activation of caspases, cleavage of lamin A/C, loss of IAP expression, and resultant cell death. We anticipate that future work will further elaborate on this scheme for the impact of CK2 in early events associated with induction of cell death.

10 Concluding Remarks

An important outcome of the discovery of CK2 impact on cell death in cancer cells was the consideration of its targeting for cancer therapy [22] which has now gained much impetus [142]. Several strategies of targeting CK2 for cancer therapy have emerged including a chemical inhibitor of CK2 (referred to as CX4945, administered orally) which was found to be effective in inducing cell death in a variety of cancer cells and affecting the activity of several downstream pro-survival pathways [116, 130, 131]. A peptide inhibitor of CK2 phosphorylation sites (called

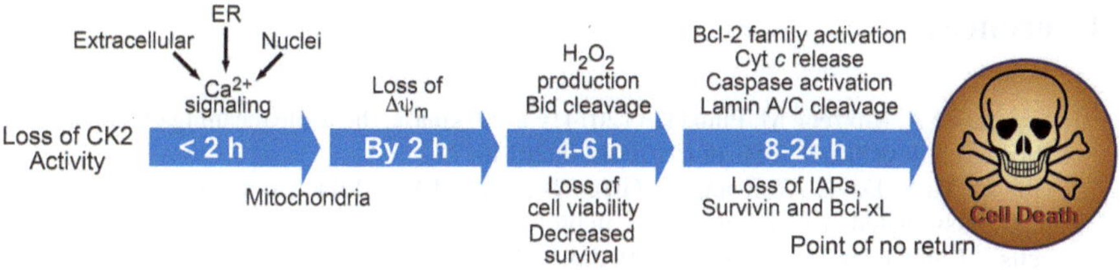

Fig. 2 A proposed model showing the hypothesis on earliest events that trigger apoptotic activity in response to inhibition of CK2 activity or in response to apoptotic signals that affect CK2 status at various cellular loci. It is suggested that rapid inhibition of CK2 at various loci triggers early intracellular Ca^{2+} shuttling that results in altered mitochondrial membrane permeability transition associated with loss of mitochondrial membrane potential occurring within 2 h. These events precede the formation of ROS which occurs between 4 and 6 h resulting in the initiation of decreased cell viability and survival. This is followed by the expression of the terminal events such as activation of Bcl-2 family, loss of IAPs, release of cytochrome *c*, and activation of caspases, thus heralding the "point of no return" leading to cell death [137]. Conceivably, the model could also be applicable to altered CK2 in response to other apoptotic inducers

CIGB-300) has been tested in preclinical and clinical studies on cervical cancer [143]. As discussed in detail elsewhere in this book, another approach to cancer therapy targeting CK2 has been to undertake its molecular downregulation employing siRNA or antisense to CK2αα′ delivered in an encapsulated format to achieve downregulation of both catalytic subunits of CK2 [70, 74, 75, 129, 144]. In this context, it is also tempting to note that the above-described observations on sensitization of cancer cells to TRAIL/Apo2-L by moderate inhibition or downregulation of CK2 could have strong translational implications for cancer therapy. With these current developments, targeting CK2 for cancer therapy has the potential of becoming a reality in the not too distant future.

The foregoing discussion highlights the importance of identifying the mechanisms involved in the regulation of cell death processes by CK2. The regulation of cell death via modulation of CK2 has far-reaching span and involves diverse cellular loci. Several of these mechanisms need to be investigated to ascertain the details of the intervening processes as that could contribute to novel information for identifying additional combinatorial targets in cancer therapy. At present our observations hint that rapid intracellular shifts of Ca^{2+} may be one of the earliest responses in the mediation of cell death in response to downregulation of CK2 and may be the integrating mechanism linking cell fate decisions in response to events at various intracellular locales including the nucleus, membrane, endoplasmic reticulum, and mitochondria. Future studies should provide new insights into the mechanisms involved in the global function of CK2 as a regulator of cell death.

Acknowledgments The original work in the authors' laboratory was supported in part by the US Department of Veterans Affairs Merit Review Program, grant number I01BX001731 (K.A.), Department of Health and Human Services, and National Cancer Institute, grant numbers R01CA150182, R01CA15062 (K.A.), and R21CA158730 (B.T.K.). One of us (K.A.) would like to dedicate this work to the memory of AHU and GS.

References

1. Venerando A, Ruzzene M, Pinna LA (2014) Casein kinase: the triple meaning of a misnomer. Biochem J 460(2):141–156. doi:10.1042/BJ20140178
2. Guo C, Yu S, Davis AT, Wang H, Green JE, Ahmed K (2001) A potential role of nuclear matrix-associated protein kinase CK2 in protection against drug-induced apoptosis in cancer cells. J Biol Chem 276(8):5992–5999. doi:10.1074/jbc.M004862200
3. Ahmed K, Gerber DA, Cochet C (2002) Joining the cell survival squad: an emerging role for protein kinase CK2. Trends Cell Biol 12(5):226–230
4. Ahmad KA, Wang G, Unger G, Slaton J, Ahmed K (2008) Protein kinase CK2—a key suppressor of apoptosis. Adv Enzyme Regul 48:179–187. doi:10.1016/j.advenzreg.2008.04.002
5. Blanquet PR (2000) Casein kinase 2 as a potentially important enzyme in the nervous system. Prog Neurobiol 60(3):211–246, doi:S0301-0082(99)00026-X [pii]
6. Dominguez I, Mizuno J, Wu H, Song DH, Symes K, Seldin DC (2004) Protein kinase CK2 is required for dorsal axis formation in Xenopus embryos. Dev Biol 274(1):110–124. doi:10.1016/j.ydbio.2004.06.021, S0012-1606(04)00438-5 [pii]
7. Blond O, Jensen HH, Buchou T, Cochet C, Issinger OG, Boldyreff B (2005) Knocking out the regulatory beta subunit of protein kinase CK2 in mice: gene dosage effects in ES cells and embryos. Mol Cell Biochem 274(1–2):31–37
8. Buchou T, Vernet M, Blond O, Jensen HH, Pointu H, Olsen BB, Cochet C, Issinger OG, Boldyreff B (2003) Disruption of the regulatory β subunit of protein kinase CK2 in mice leads to a cell-autonomous defect and early embryonic lethality. Mol Cell Biol 23(3): 908–915
9. Huillard E, Ziercher L, Blond O, Wong M, Deloulme JC, Souchelnytskyi S, Baudier J, Cochet C, Buchou T (2010) Disruption of CK2beta in embryonic neural stem cells compromises proliferation and oligodendrogenesis in the mouse telencephalon. Mol Cell Biol 30(11):2737–2749. doi:10.1128/MCB.01566-09
10. Mestres P, Boldyreff B, Ebensperger C, Hameister H, Issinger OG (1994) Expression of casein kinase 2 during mouse embryogenesis. Acta Anat (Basel) 149(1):13–20
11. Ahmed K (1999) Nuclear matrix and protein kinase CK2 signaling. Crit Rev Eukaryot Gene Expr 9(3–4):329–336
12. Trembley JH, Wang G, Unger G, Slaton J, Ahmed K (2009) Protein kinase CK2 in health and disease: CK2: a key player in cancer biology. Cell Mol Life Sci 66(11–12):1858–1867. doi:10.1007/s00018-009-9154-y
13. Rayan A, Goueli SA, Lange P, Ahmed K (1985) Chromatin-associated protein kinases in human normal and benign hyperplastic prostate. Cancer Res 45(5):2277–2282
14. Meggio F, Pinna LA (2003) One-thousand-and-one substrates of protein kinase CK2? FASEB J 17(3):349–368. doi:10.1096/fj.02-0473rev
15. Rebholz H, Nishi A, Liebscher S, Nairn AC, Flajolet M, Greengard P (2009) CK2 negatively regulates Galphas signaling. Proc Natl Acad Sci U S A 106(33):14096–14101. doi:10.1073/pnas.0906857106, 0906857106 [pii]
16. Rebholz H, Zhou M, Nairn AC, Greengard P, Flajolet M (2013) Selective knockout of the casein kinase 2 in d1 medium spiny neurons controls dopaminergic function. Biol Psychiatry 74(2):113–121. doi:10.1016/j.biopsych.2012.11.013, S0006-3223(12)00994-8 [pii]
17. Munstermann U, Fritz G, Seitz G, Lu YP, Schneider HR, Issinger OG (1990) Casein kinase II is elevated in solid human tumours and rapidly proliferating non-neoplastic tissue. Eur J Biochem 189(2):251–257
18. Guerra B, Issinger OG (2008) Protein kinase CK2 in human diseases. Curr Med Chem 15(19):1870–1886. doi:10.1007/s00018-009-9148-9
19. Tawfic S, Yu S, Wang H, Faust R, Davis A, Ahmed K (2001) Protein kinase CK2 signal in neoplasia. Histol Histopathol 16(2):573–582
20. Yenice S, Davis AT, Goueli SA, Akdas A, Limas C, Ahmed K (1994) Nuclear casein kinase 2 (CK-2) activity in human normal, benign hyperplastic, and cancerous prostate. Prostate 24(1):11–16

21. Wang G, Ahmad KA, Ahmed K (2005) Modulation of death receptor-mediated apoptosis by CK2. Mol Cell Biochem 274(1–2):201–205
22. Wang H, Davis A, Yu S, Ahmed K (2001) Response of cancer cells to molecular interruption of the CK2 signal. Mol Cell Biochem 227(1–2):167–174. doi:10.1023/A:1013112908734
23. Hanahan D, Weinberg RA (2011) Hallmarks of cancer: the next generation. Cell 144(5):646–674. doi:10.1016/j.cell.2011.02.013
24. Evan GI, Vousden KH (2001) Proliferation, cell cycle and apoptosis in cancer. Nature 411(6835):342–348. doi:10.1038/35077213
25. McKenzie S, Kyprianou N (2006) Apoptosis evasion: the role of survival pathways in prostate cancer progression and therapeutic resistance. J Cell Biochem 97(1):18–32. doi:10.1002/jcb.20634
26. Judah JD, Ahmed K, McLean AE (1965) Pathogenesis of cell necrosis. Fed Proc 24(5):1217–1221
27. Lockshin RA, Williams CM (1965) Programmed cell death—I. Cytology of degeneration in the intersegmental muscles of the Pernyi silkmoth. J Insect Physiol 11:123–133
28. Kerr JF, Wyllie AH, Currie AR (1972) Apoptosis: a basic biological phenomenon with wide-ranging implications in tissue kinetics. Br J Cancer 26(4):239–257
29. Lemasters JJ, Nieminen AL, Qian T, Trost LC, Elmore SP, Nishimura Y, Crowe RA, Cascio WE, Bradham CA, Brenner DA, Herman B (1998) The mitochondrial permeability transition in cell death: a common mechanism in necrosis, apoptosis and autophagy. Biochim Biophys Acta 1366(1–2):177–196
30. Ashkenazi A, Dixit VM (1999) Apoptosis control by death and decoy receptors. Curr Opin Cell Biol 11(2):255–260
31. Liu Y, Shoji-Kawata S, Sumpter RM Jr, Wei Y, Ginet V, Zhang L, Posner B, Tran KA, Green DR, Xavier RJ, Shaw SY, Clarke PG, Puyal J, Levine B (2013) Autosis is a Na+, K+-ATPase-regulated form of cell death triggered by autophagy-inducing peptides, starvation, and hypoxia-ischemia. Proc Natl Acad Sci U S A 110(51):20364–20371. doi:10.1073/pnas.1319661110
32. Green DR, Levine B (2014) To be or not to be? How selective autophagy and cell death govern cell fate. Cell 157(1):65–75. doi:10.1016/j.cell.2014.02.049
33. Zornig M, Hueber A, Baum W, Evan G (2001) Apoptosis regulators and their role in tumorigenesis. Biochim Biophys Acta 1551(2):F1–F37, doi:S0304419X01000312 [pii]
34. Linkermann A, Green DR (2014) Necroptosis. N Engl J Med 370(5):455–465. doi:10.1056/NEJMra1310050
35. Lockshin RA, Zakeri Z (2004) Apoptosis, autophagy, and more. Int J Biochem Cell Biol 36(12):2405–2419. doi:10.1016/j.biocel.2004.04.011
36. Kroemer G, Martin SJ (2005) Caspase-independent cell death. Nat Med 11(7):725–730. doi:10.1038/nm1263
37. Nikoletopoulou V, Markaki M, Palikaras K, Tavernarakis N (2013) Crosstalk between apoptosis, necrosis and autophagy. Biochim Biophys Acta 1833(12):3448–3459. doi:10.1016/j.bbamcr.2013.06.001
38. Danial NN, Korsmeyer SJ (2004) Cell death: critical control points. Cell 116(2):205–219
39. Elmore S (2007) Apoptosis: a review of programmed cell death. Toxicol Pathol 35(4):495–516. doi:10.1080/01926230701320337
40. Degterev A, Yuan J (2008) Expansion and evolution of cell death programmes. Nat Rev Mol Cell Biol 9(5):378–390. doi:10.1038/nrm2393
41. Hitomi J, Christofferson DE, Ng A, Yao J, Degterev A, Xavier RJ, Yuan J (2008) Identification of a molecular signaling network that regulates a cellular necrotic cell death pathway. Cell 135(7):1311–1323. doi:10.1016/j.cell.2008.10.044
42. Zou H, Li Y, Liu X, Wang X (1999) An APAF-1.cytochrome c multimeric complex is a functional apoptosome that activates procaspase-9. J Biol Chem 274(17):11549–11556
43. Lassus P, Opitz-Araya X, Lazebnik Y (2002) Requirement for caspase-2 in stress-induced apoptosis before mitochondrial permeabilization. Science 297(5585):1352–1354. doi:10.1126/science.1074721, 297/5585/1352 [pii]

44. Hitomi J, Katayama T, Eguchi Y, Kudo T, Taniguchi M, Koyama Y, Manabe T, Yamagishi S, Bando Y, Imaizumi K, Tsujimoto Y, Tohyama M (2004) Involvement of caspase-4 in endoplasmic reticulum stress-induced apoptosis and Abeta-induced cell death. J Cell Biol 165(3):347–356. doi:10.1083/jcb.200310015, jcb.200310015 [pii]
45. Nakagawa T, Zhu H, Morishima N, Li E, Xu J, Yankner BA, Yuan J (2000) Caspase-12 mediates endoplasmic-reticulum-specific apoptosis and cytotoxicity by amyloid-beta. Nature 403(6765):98–103. doi:10.1038/47513
46. Wilson NS, Dixit V, Ashkenazi A (2009) Death receptor signal transducers: nodes of coordination in immune signaling networks. Nat Immunol 10(4):348–355. doi:10.1038/ni.1714
47. Ashkenazi A, Dixit VM (1998) Death receptors: signaling and modulation. Science 281(5381):1305–1308
48. Chen L, Willis SN, Wei A, Smith BJ, Fletcher JI, Hinds MG, Colman PM, Day CL, Adams JM, Huang DC (2005) Differential targeting of prosurvival Bcl-2 proteins by their BH3-only ligands allows complementary apoptotic function. Mol Cell 17(3):393–403. doi:10.1016/j.molcel.2004.12.030, S1097276505010403 [pii]
49. Festjens N, Vanden Berghe T, Vandenabeele P (2006) Necrosis, a well-orchestrated form of cell demise: signalling cascades, important mediators and concomitant immune response. Biochim Biophys Acta 1757(9–10):1371–1387. doi:10.1016/j.bbabio.2006.06.014, S0005-2728(06)00212-X [pii]
50. Zong WX, Ditsworth D, Bauer DE, Wang ZQ, Thompson CB (2004) Alkylating DNA damage stimulates a regulated form of necrotic cell death. Genes Dev 18(11):1272–1282. doi:10.1101/gad.1199904, 1199904 [pii]
51. Judah JD, Ahmed K, McLean AE (1966) Cation changes following liver injury due to carbon tetrachloride. Ann Med Exp Biol Fenn 44(2):338–342
52. Holler N, Zaru R, Micheau O, Thome M, Attinger A, Valitutti S, Bodmer JL, Schneider P, Seed B, Tschopp J (2000) Fas triggers an alternative, caspase-8-independent cell death pathway using the kinase RIP as effector molecule. Nat Immunol 1(6):489–495. doi:10.1038/82732
53. Shimizu S, Kanaseki T, Mizushima N, Mizuta T, Arakawa-Kobayashi S, Thompson CB, Tsujimoto Y (2004) Role of Bcl-2 family proteins in a non-apoptotic programmed cell death dependent on autophagy genes. Nat Cell Biol 6(12):1221–1228. doi:10.1038/ncb1192, ncb1192 [pii]
54. Pattingre S, Tassa A, Qu X, Garuti R, Liang XH, Mizushima N, Packer M, Schneider MD, Levine B (2005) Bcl-2 antiapoptotic proteins inhibit Beclin 1-dependent autophagy. Cell 122(6):927–939. doi:10.1016/j.cell.2005.07.002, S0092-8674(05)00692-6 [pii]
55. Guo C, Davis AT, Ahmed K (1998) Dynamics of protein kinase CK2 association with nucleosomes in relation to transcriptional activity. J Biol Chem 273(22):13675–13680. doi:10.1074/jbc.273.22.13675
56. Barz T, Ackermann K, Dubois G, Eils R, Pyerin W (2003) Genome-wide expression screens indicate a global role for protein kinase CK2 in chromatin remodeling. J Cell Sci 116(Pt 8):1563–1577
57. Basnet H, Su XB, Tan Y, Meisenhelder J, Merkurjev D, Ohgi KA, Hunter T, Pillus L, Rosenfeld MG (2014) Tyrosine phosphorylation of histone H2A by CK2 regulates transcriptional elongation. Nature. doi:10.1038/nature13736, nature13736 [pii]
58. Ahmed K, Ishida H (1971) Effect of testosterone on nuclear phosphoproteins of rat ventral prostate. Mol Pharmacol 7(3):323–327
59. Goueli SA, Ahmed K (1991) Nature of the intrinsic protein kinases involved in phosphorylation of non-histone proteins in intact prostatic nuclei: further identification of androgen-sensitive protein kinase reactions. Mol Cell Biochem 101(2):145–155
60. Wyllie AH, Kerr JF, Currie AR (1980) Cell death: the significance of apoptosis. Int Rev Cytol 68:251–306
61. Ahmed K, Wilson MJ (1975) Chromatin-associated protein phosphokinases of rat ventral prostate. Characteristics and effects of androgenic status. J Biol Chem 250(6):2370–2375
62. Ahmed K, Yenice S, Davis A, Goueli SA (1993) Association of casein kinase 2 with nuclear chromatin in relation to androgenic regulation of rat prostate. Proc Natl Acad Sci U S A 90(10):4426–4430

63. Tawfic S, Ahmed K (1994) Association of casein kinase 2 with nuclear matrix. Possible role in nuclear matrix protein phosphorylation. J Biol Chem 269(10):7489–7493
64. Tawfic S, Ahmed K (1994) Growth stimulus-mediated differential translocation of casein kinase 2 to the nuclear matrix. Evidence based on androgen action in the prostate. J Biol Chem 269(40):24615–24620
65. Tawfic S, Faust RA, Gapany M, Ahmed K (1996) Nuclear matrix as an anchor for protein kinase CK2 nuclear signalling. J Cell Biochem 62(2):165–171, doi:10.1002/(SICI)1097-4644(199608)62:2<165::AID-JCB4>3.0.CO;2-Q
66. Faust M, Montenarh M (2000) Subcellular localization of protein kinase CK2. A key to its function? Cell Tissue Res 301(3):329–340
67. Guo C, Yu S, Davis AT, Ahmed K (1999) Nuclear matrix targeting of the protein kinase CK2 signal as a common downstream response to androgen or growth factor stimulation of prostate cancer cells. Cancer Res 59(5):1146–1151
68. Okoumassoun LE, Russo C, Denizeau F, Averill-Bates D, Henderson JE (2007) Parathyroid hormone-related protein (PTHrP) inhibits mitochondrial-dependent apoptosis through CK2. J Cell Physiol 212(3):591–599. doi:10.1002/jcp.21055
69. Galluzzi L, Bravo-San Pedro JM, Kroemer G (2014) Organelle-specific initiation of cell death. Nat Cell Biol 16(8):728–736. doi:10.1038/ncb3005, ncb3005 [pii]
70. Brown MS, Diallo OT, Hu M, Ehsanian R, Yang X, Arun P, Lu H, Korman V, Unger G, Ahmed K, Van Waes C, Chen Z (2010) CK2 modulation of NF-κB, TP53, and the malignant phenotype in head and neck cancer by anti-CK2 oligonucleotides in vitro or in vivo via sub-50-nm nanocapsules. Clin Cancer Res 16(8):2295–2307. doi:10.1158/1078-0432.ccr-09-3200
71. Wang G, Unger G, Ahmad KA, Slaton JW, Ahmed K (2005) Downregulation of CK2 induces apoptosis in cancer cells—a potential approach to cancer therapy. Mol Cell Biochem 274(1–2):77–84
72. Faust RA, Tawfic S, Davis AT, Bubash LA, Ahmed K (2000) Antisense oligonucleotides against protein kinase CK2-α inhibit growth of squamous cell carcinoma of the head and neck in vitro. Head Neck 22(4):341–346, doi:10.1002/1097-0347(200007)22:4<341::AID-HED5>3.0.CO;2–3 [pii]
73. Slaton JW, Unger GM, Sloper DT, Davis AT, Ahmed K (2004) Induction of apoptosis by antisense CK2 in human prostate cancer xenograft model. Mol Cancer Res 2(12):712–721
74. Ahmad KA, Wang G, Slaton J, Unger G, Ahmed K (2005) Targeting CK2 for cancer therapy. Anticancer Drugs 16(10):1037–1043. doi:10.1097/00001813-200511000-00001
75. Unger GM, Kren BT, Korman VL, Kimbrough TG, Vogel RI, Ondrey FG, Trembley JH, Ahmed K (2014) Mechanism and efficacy of sub-50-nm tenfibgen nanocapsules for cancer cell-directed delivery of anti-CK2 RNAi to primary and metastatic squamous cell carcinoma. Mol Cancer Ther 13(8):2018–2029. doi:10.1158/1535-7163.MCT-14-0166, 1535-7163.MCT-14-0166 [pii]
76. Trembley JH, Unger GM, Korman VL, Tobolt DK, Kazimierczuk Z, Pinna LA, Kren BT, Ahmed K (2012) Nanoencapsulated anti-CK2 small molecule drug or siRNA specifically targets malignant cancer but not benign cells. Cancer Lett 315(1):48–58. doi:10.1016/j.canlet.2011.10.007
77. Trembley JH, Chen Z, Unger G, Slaton J, Kren BT, Van Waes C, Ahmed K (2010) Emergence of protein kinase CK2 as a key target in cancer therapy. Biofactors 36(3):187–195. doi:10.1002/biof.96
78. St-Denis NA, Litchfield DW (2009) Protein kinase CK2 in health and disease: From birth to death: the role of protein kinase CK2 in the regulation of cell proliferation and survival. Cell Mol Life Sci 66(11–12):1817–1829
79. Khan N, Afaq F, Saleem M, Ahmad N, Mukhtar H (2006) Targeting multiple signaling pathways by green tea polyphenol (–)-epigallocatechin-3-gallate. Cancer Res 66(5):2500–2505. doi:10.1158/0008-5472.CAN-05-3636, 66/5/2500 [pii]
80. Siddiqui IA, Adhami VM, Bharali DJ, Hafeez BB, Asim M, Khwaja SI, Ahmad N, Cui H, Mousa SA, Mukhtar H (2009) Introducing nanochemoprevention as a novel approach for cancer control: proof of principle with green tea polyphenol epigallocatechin-3-gallate. Cancer Res 69(5):1712–1716. doi:10.1158/0008-5472.CAN-08-3978, 0008-5472.CAN-08-3978 [pii]

81. Ahmad KA, Harris NH, Johnson AD, Lindvall HC, Wang G, Ahmed K (2007) Protein kinase CK2 modulates apoptosis induced by resveratrol and epigallocatechin-3-gallate in prostate cancer cells. Mol Cancer Ther 6(3):1006–1012. doi:10.1158/1535-7163.MCT-06-0491
82. Shukla S, Mishra A, Fu P, MacLennan GT, Resnick MI, Gupta S (2005) Up-regulation of insulin-like growth factor binding protein-3 by apigenin leads to growth inhibition and apoptosis of 22Rv1 xenograft in athymic nude mice. FASEB J 19(14):2042–2044. doi:10.1096/fj.05-3740fje
83. Cobb LJ, Mehta H, Cohen P (2009) Enhancing the apoptotic potential of insulin-like growth factor-binding protein-3 in prostate cancer by modulation of CK2 phosphorylation. Mol Endocrinol 23(10):1624–1633. doi:10.1210/me.2008-0365
84. Gerber DA, Souquere-Besse S, Puvion F, Dubois MF, Bensaude O, Cochet C (2000) Heat-induced relocalization of protein kinase CK2. Implication of CK2 in the context of cellular stress. J Biol Chem 275(31):23919–23926. doi:10.1074/jbc.M002697200
85. Davis AT, Wang H, Zhang P, Ahmed K (2002) Heat shock mediated modulation of protein kinase CK2 in the nuclear matrix. J Cell Biochem 85(3):583–591. doi:10.1002/jcb.10158
86. Yamane K, Kinsella TJ (2005) CK2 inhibits apoptosis and changes its cellular localization following ionizing radiation. Cancer Res 65(10):4362–4367. doi:10.1158/0008-5472.CAN-04-3941
87. Kato T Jr, Delhase M, Hoffmann A, Karin M (2003) CK2 is a C-terminal IkappaB kinase responsible for NF-kappaB activation during the UV response. Mol Cell 12(4):829–839, doi:S1097276503003587 [pii]
88. Ravi R, Bedi A (2002) Sensitization of tumor cells to Apo2 ligand/TRAIL-induced apoptosis by inhibition of casein kinase II. Cancer Res 62(15):4180–4185
89. Izeradjene K, Douglas L, Delaney A, Houghton JA (2004) Influence of casein kinase II in tumor necrosis factor-related apoptosis-inducing ligand-induced apoptosis in human rhabdomyosarcoma cells. Clin Cancer Res 10(19):6650–6660. doi:10.1158/1078-0432.CCR-04-0576, 10/19/6650 [pii]
90. Izeradjene K, Douglas L, Delaney A, Houghton JA (2005) Casein kinase II (CK2) enhances death-inducing signaling complex (DISC) activity in TRAIL-induced apoptosis in human colon carcinoma cell lines. Oncogene 24(12):2050–2058. doi:10.1038/sj.onc.1208397, 1208397 [pii]
91. Wang G, Ahmad KA, Ahmed K (2006) Role of protein kinase CK2 in the regulation of tumor necrosis factor-related apoptosis inducing ligand-induced apoptosis in prostate cancer cells. Cancer Res 66(4):2242–2249. doi:10.1158/0008-5472.CAN-05-2772, 66/4/2242 [pii]
92. Hessenauer A, Schneider CC, Götz C, Montenarh M (2011) CK2 inhibition induces apoptosis via the ER stress response. Cell Signal 23(1):145–151. doi:10.1016/j.cellsig.2010.08.014
93. Desagher S, Osen-Sand A, Montessuit S, Magnenat E, Vilbois F, Hochmann A, Journot L, Antonsson B, Martinou JC (2001) Phosphorylation of bid by casein kinases I and II regulates its cleavage by caspase 8. Mol Cell 8(3):601–611, doi:S1097-2765(01)00335-5 [pii]
94. Li PF, Li J, Muller EC, Otto A, Dietz R, von Harsdorf R (2002) Phosphorylation by protein kinase CK2: a signaling switch for the caspase-inhibiting protein ARC. Mol Cell 10(2):247–258, doi:S1097276502006007 [pii]
95. McDonnell MA, Abedin MJ, Melendez M, Platikanova TN, Ecklund JR, Ahmed K, Kelekar A (2008) Phosphorylation of murine caspase-9 by the protein kinase casein kinase 2 regulates its cleavage by caspase-8. J Biol Chem 283(29):20149–20158. doi:10.1074/jbc.M802846200
96. Cardone MH, Roy N, Stennicke HR, Salvesen GS, Franke TF, Stanbridge E, Frisch S, Reed JC (1998) Regulation of cell death protease caspase-9 by phosphorylation. Science 282(5392):1318–1321
97. Di Maira G, Salvi M, Arrigoni G, Marin O, Sarno S, Brustolon F, Pinna LA, Ruzzene M (2005) Protein kinase CK2 phosphorylates and upregulates Akt/PKB. Cell Death Differ 12(6):668–677. doi:10.1038/sj.cdd.4401604, 4401604 [pii]
98. Siddiqui-Jain A, Drygin D, Streiner N, Chua P, Pierre F, O'Brien SE, Bliesath J, Omori M, Huser N, Ho C, Proffitt C, Schwaebe MK, Ryckman DM, Rice WG, Anderes K (2010) CX-4945, an orally bioavailable selective inhibitor of protein kinase CK2, inhibits prosurvival

and angiogenic signaling and exhibits antitumor efficacy. Cancer Res 70(24):10288–10298. doi:10.1158/0008-5472.can-10-1893
99. Turowec JP, Duncan JS, Gloor GB, Litchfield DW (2011) Regulation of caspase pathways by protein kinase CK2: identification of proteins with overlapping CK2 and caspase consensus motifs. Mol Cell Biochem 356(1–2):159–167. doi:10.1007/s11010-011-0972-5
100. Duncan JS, Turowec JP, Duncan KE, Vilk G, Wu C, Luscher B, Li SS, Gloor GB, Litchfield DW (2011) A peptide-based target screen implicates the protein kinase CK2 in the global regulation of caspase signaling. Sci Signal 4(172):ra30. doi:10.1126/scisignal.2001682
101. Reed JC (2002) Apoptosis-based therapies. Nat Rev Drug Discov 1(2):111–121. doi:10.1038/nrd726
102. Salvesen GS, Duckett CS (2002) IAP proteins: blocking the road to death's door. Nat Rev Mol Cell Biol 3(6):401–410. doi:10.1038/nrm830
103. Altieri DC (2010) Survivin and IAP proteins in cell-death mechanisms. Biochem J 430(2):199–205. doi:10.1042/bj20100814
104. McEleny KR, Watson RW, Coffey RN, O'Neill AJ, Fitzpatrick JM (2002) Inhibitors of apoptosis proteins in prostate cancer cell lines. Prostate 51(2):133–140. doi:10.1002/pros.10061
105. Finlay D, Vamos M, Gonzalez-Lopez M, Ardecky RJ, Ganji SR, Yuan H, Su Y, Cooley TR, Hauser CT, Welsh K, Reed JC, Cosford ND, Vuori K (2014) Small-molecule IAP antagonists sensitize cancer cells to TRAIL-induced apoptosis: roles of XIAP and cIAPs. Mol Cancer Ther 13(1):5–15. doi:10.1158/1535-7163.MCT-13-0153, 1535-7163.MCT-13-0153 [pii]
106. Rodriguez-Berriguete G, Fraile B, de Bethencourt FR, Prieto-Folgado A, Bartolome N, Nunez C, Prati B, Martinez-Onsurbe P, Olmedilla G, Paniagua R, Royuela M (2010) Role of IAPs in prostate cancer progression: immunohistochemical study in normal and pathological (benign hyperplastic, prostatic intraepithelial neoplasia and cancer) human prostate. BMC Cancer 10:18. doi:10.1186/1471-2407-10-18, 1471-2407-10-18 [pii]
107. Tapia JC, Torres VA, Rodriguez DA, Leyton L, Quest AF (2006) Casein kinase 2 (CK2) increases survivin expression via enhanced beta-catenin-T cell factor/lymphoid enhancer binding factor-dependent transcription. Proc Natl Acad Sci U S A 103(41):15079–15084. doi:10.1073/pnas.0606845103, 0606845103 [pii]
108. Ponce DP, Yefi R, Cabello P, Maturana JL, Niechi I, Silva E, Galindo M, Antonelli M, Marcelain K, Armisen R, Tapia JC (2011) CK2 functionally interacts with AKT/PKB to promote the beta-catenin-dependent expression of survivin and enhance cell survival. Mol Cell Biochem 356(1–2):127–132. doi:10.1007/s11010-011-0965-4
109. Barrett RM, Colnaghi R, Wheatley SP (2011) Threonine 48 in the BIR domain of survivin is critical to its mitotic and anti-apoptotic activities and can be phosphorylated by CK2 in vitro. Cell Cycle 10(3):538–548
110. Wang G, Ahmad KA, Harris NH, Ahmed K (2008) Impact of protein kinase CK2 on inhibitor of apoptosis proteins in prostate cancer cells. Mol Cell Biochem 316(1–2):91–97. doi:10.1007/s11010-008-9810-9
111. Stauber RH, Mann W, Knauer SK (2007) Nuclear and cytoplasmic survivin: molecular mechanism, prognostic, and therapeutic potential. Cancer Res 67(13):5999–6002. doi:10.1158/0008-5472.CAN-07-0494, 67/13/5999 [pii]
112. Olsen BB, Jessen V, Hojrup P, Issinger OG, Boldyreff B (2003) Protein kinase CK2 phosphorylates the Fas-associated factor FAF1 in vivo and influences its transport into the nucleus. FEBS Lett 546(2–3):218–222, doi:S0014579303005751 [pii]
113. Channavajhala P, Seldin DC (2002) Functional interaction of protein kinase CK2 and c-*Myc* in lymphomagenesis. Oncogene 21(34):5280–5288. doi:10.1038/sj.onc.1205640
114. Krippner-Heidenreich A, Talanian RV, Sekul R, Kraft R, Thole H, Ottleben H, Luscher B (2001) Targeting of the transcription factor Max during apoptosis: phosphorylation-regulated cleavage by caspase-5 at an unusual glutamic acid residue in position P1. Biochem J 358(Pt 3):705–715
115. Loizou JI, El-Khamisy SF, Zlatanou A, Moore DJ, Chan DW, Qin J, Sarno S, Meggio F, Pinna LA, Caldecott KW (2004) The protein kinase CK2 facilitates repair of chromosomal DNA single-strand breaks. Cell 117(1):17–28, doi:S0092867404002065 [pii]

116. Pierre F, Chua PC, O'Brien SE, Siddiqui-Jain A, Bourbon P, Haddach M, Michaux J, Nagasawa J, Schwaebe MK, Stefan E, Vialettes A, Whitten JP, Chen TK, Darjania L, Stansfield R, Bliesath J, Drygin D, Ho C, Omori M, Proffitt C, Streiner N, Rice WG, Ryckman DM, Anderes K (2011) Pre-clinical characterization of CX-4945, a potent and selective small molecule inhibitor of CK2 for the treatment of cancer. Mol Cell Biochem 356(1–2):37–43. doi:10.1007/s11010-011-0956-5
117. Meek DW, Cox M (2011) Induction and activation of the p53 pathway: a role for the protein kinase CK2? Mol Cell Biochem. doi:10.1007/s11010-011-0966-3
118. Allende-Vega N, McKenzie L, Meek D (2008) Transcription factor TAFII250 phosphorylates the acidic domain of Mdm2 through recruitment of protein kinase CK2. Mol Cell Biochem 316(1–2):99–106. doi:10.1007/s11010-008-9816-3
119. Scaglioni PP, Yung TM, Cai LF, Erdjument-Bromage H, Kaufman AJ, Singh B, Teruya-Feldstein J, Tempst P, Pandolfi PP (2006) A CK2-dependent mechanism for degradation of the PML tumor suppressor. Cell 126(2):269–283. doi:10.1016/j.cell.2006.05.041, S0092-8674(06)00814-2 [pii]
120. Scaglioni PP, Yung TM, Choi S, Baldini C, Konstantinidou G, Pandolfi PP (2008) CK2 mediates phosphorylation and ubiquitin-mediated degradation of the PML tumor suppressor. Mol Cell Biochem 316(1–2):149–154. doi:10.1007/s11010-008-9812-7
121. Vazquez F, Grossman SR, Takahashi Y, Rokas MV, Nakamura N, Sellers WR (2001) Phosphorylation of the PTEN tail acts as an inhibitory switch by preventing its recruitment into a protein complex. J Biol Chem 276(52):48627–48630. doi:10.1074/jbc.C100556200
122. Wang D, Westerheide SD, Hanson JL, Baldwin AS Jr (2000) Tumor necrosis factor alpha-induced phosphorylation of RelA/p65 on Ser529 is controlled by casein kinase II. J Biol Chem 275(42):32592–32597. doi:10.1074/jbc.M001358200, M001358200 [pii]
123. Pando MP, Verma IM (2000) Signal-dependent and -independent degradation of free and NF-κB bound IκB-α. J Biol Chem 275(28):21278–21286. doi:10.1074/jbc.M002532200, M002532200 [pii]
124. McElhinny JA, Trushin SA, Bren GD, Chester N, Paya CV (1996) Casein kinase II phosphorylates I kappa B alpha at S-283, S-289, S-293, and T-291 and is required for its degradation. Mol Cell Biol 16(3):899–906
125. Romieu-Mourez R, Landesman-Bollag E, Seldin DC, Traish AM, Mercurio F, Sonenshein GE (2001) Roles of IKK kinases and protein kinase CK2 in activation of nuclear factor-κB in breast cancer. Cancer Res 61(9):3810–3818
126. Romieu-Mourez R, Landesman-Bollag E, Seldin DC, Sonenshein GE (2002) Protein kinase CK2 promotes aberrant activation of nuclear factor-{kappa}B, transformed phenotype, and survival of breast cancer cells. Cancer Res 62(22):6770–6778
127. Dominguez I, Sonenshein GE, Seldin DC (2009) Protein kinase CK2 in health and disease: CK2 and its role in Wnt and NF-kappaB signaling: linking development and cancer. Cell Mol Life Sci 66(11–12):1850–1857. doi:10.1007/s00018-009-9153-z
128. Trembley JH, Unger GM, Tobolt DK, Korman VL, Wang G, Ahmad KA, Slaton JW, Kren BT, Ahmed K (2011) Systemic administration of antisense oligonucleotides simultaneously targeting CK2alpha and alpha′ subunits reduces orthotopic xenograft prostate tumors in mice. Mol Cell Biochem 356(1–2):21–35. doi:10.1007/s11010-011-0943-x
129. Trembley JH, Unger GM, Korman VL, Abedin MJ, Nacusi LP, Vogel RI, Slaton JW, Kren BT, Ahmed K (2014) Tenfibgen ligand nanoencapsulation delivers bi-functional anti-CK2 RNAi oligomer to key sites for prostate cancer targeting using human xenograft tumors in mice. PLoS One 9(10):e109970. doi:10.1371/journal.pone.0109970
130. Zheng Y, McFarland BC, Drygin D, Yu H, Bellis SL, Kim H, Bredel M, Benveniste EN (2013) Targeting protein kinase CK2 suppresses pro-survival signaling pathways and growth of glioblastoma. Clin Cancer Res. doi:10.1158/1078-0432.CCR-13-0265
131. Gray GK, McFarland BC, Rowse AL, Gibson SA, Benveniste EN (2014) Therapeutic CK2 inhibition attenuates diverse prosurvival signaling cascades and decreases cell viability in human breast cancer cells. Oncotarget 5(15):6484–6496, doi:2248 [pii]

132. Zheng Y, Qin H, Frank SJ, Deng L, Litchfield DW, Tefferi A, Pardanani A, Lin FT, Li J, Sha B, Benveniste EN (2011) A CK2-dependent mechanism for activation of the JAK-STAT signaling pathway. Blood 118(1):156–166. doi:10.1182/blood-2010-01-266320
133. Perera Y, Farina HG, Gil J, Rodriguez A, Benavent F, Castellanos L, Gomez RE, Acevedo BE, Alonso DF, Perea SE (2009) Anticancer peptide CIGB-300 binds to nucleophosmin/B23, impairs its CK2-mediated phosphorylation, and leads to apoptosis through its nucleolar disassembly activity. Mol Cancer Ther 8(5):1189–1196. doi:10.1158/1535-7163.MCT-08-1056
134. Wang G, Pan Y, Ahmad KA, Ahmed K (2010) Protein B23/nucleophosmin/numatrin nuclear dynamics in relation to protein kinase CK2 and apoptotic activity in prostate cells. Biochemistry (Mosc) 49(18):3842–3852. doi:10.1021/bi9021928
135. Ahmad KA, Wang G, Ahmed K (2006) Intracellular hydrogen peroxide production is an upstream event in apoptosis induced by down-regulation of casein kinase 2 in prostate cancer cells. Mol Cancer Res 4(5):331–338. doi:10.1158/1541-7786.MCR-06-0073
136. Hanif IM, Ahmad KA, Ahmed K, Pervaiz S (2009) Involvement of reactive oxygen species in apoptosis induced by pharmacological inhibition of protein kinase CK2. Ann N Y Acad Sci 1171:591–599. doi:10.1111/j.1749-6632.2009.04916.x
137. Qaiser F, Trembley JH, Kren BT, Wu JJ, Naveed AK, Ahmed K (2014) Protein kinase CK2 inhibition induces cell death via early impact on mitochondrial function. J Cell Biochem 115(12):2103–2115. doi:10.1002/jcb.24887
138. Perry SW, Norman JP, Barbieri J, Brown EB, Gelbard HA (2011) Mitochondrial membrane potential probes and the proton gradient: a practical usage guide. Biotechniques 50(2):98–115. doi:10.2144/000113610
139. Lemasters JJ, Theruvath TP, Zhong Z, Nieminen AL (2009) Mitochondrial calcium and the permeability transition in cell death. Biochim Biophys Acta 1787(11):1395–1401. doi:10.1016/j.bbabio.2009.06.009
140. Isenberg JS, Klaunig JE (2000) Role of the mitochondrial membrane permeability transition (MPT) in rotenone-induced apoptosis in liver cells. Toxicol Sci 53(2):340–351
141. Stavrovskaya IG, Kristal BS (2005) The powerhouse takes control of the cell: is the mitochondrial permeability transition a viable therapeutic target against neuronal dysfunction and death? Free Radic Biol Med 38(6):687–697. doi:10.1016/j.freeradbiomed.2004.11.032
142. Pinna LA, Allende JE (2009) Protein kinase CK2 in health and disease: Protein kinase CK2: an ugly duckling in the kinome pond. Cell Mol Life Sci 66(11–12):1795–1799. doi:10.1007/s00018-009-9148-9
143. Perea SE, Baladron I, Garcia Y, Perera Y, Lopez A, Soriano JL et al (2011) CIGB-300, a synthetic peptide-based drug that targets the CK2 phosphoacceptor domain. Translational and clinical research. Mol Cell Biochem 356:45–50. doi:10.1007/s11010-011-0950-y
144. Unger G, Trembley J, Kren B, Ahmed K (2012) Nanoparticles in cancer therapy. Encyclopedia of cancer: SpringerReference (www.springerreference.com). Springer-Verlag Berlin Heidelberg, Heidelberg. doi:10.1007/SpringerReference_175620 2012-03-05 10:28:01 UTC
145. Filhol O, Cochet C (2009) Protein kinase CK2 in health and disease: Cellular functions of protein kinase CK2: a dynamic affair. Cell Mol Life Sci 66(11–12):1830–1839. doi:10.1007/s00018-009-9151-1

Protein Kinase CK2: Systematic Relationships with Other Posttranslational Modifications

David W. Litchfield and Laszlo Gyenis

Abstract A wealth of biochemical and genetic evidence has demonstrated that protein kinase CK2 has critical roles in the regulation and execution of numerous biological processes. Large-scale proteomic and phosphoproteomics studies have further reinforced the widespread impact of CK2 on cellular events through interactions with many cellular proteins or protein complexes and through phosphorylation of a vast number of cellular proteins. Given its global participation in many fundamental processes, it is not surprising that CK2 has been implicated in numerous human diseases, a factor that has spurred interest in CK2 as a candidate for molecular-targeted therapy. Despite this growing profile, many questions regarding its precise mechanisms of regulation remain. In fact, several lines of evidence suggest that CK2 is constitutively active, leading to a speculation that CK2 is an unregulated enzyme. Accordingly, there is an apparent paradox that leads to the question of how an unregulated enzyme such as CK2 can be a participant in regulatory processes. In an effort to resolve this paradox, studies in our lab and others have focused on an investigation of the relationships between CK2 and other cellular pathways. Using a combination of computational predictions and database mining together with proteomic strategies and biochemical assays, we have been elucidating systematic relationships between CK2 and regulatory pathways where CK2 phosphorylation sites overlap other posttranslational modifications. Overall, these studies suggest intriguing mechanisms by which CK2 can participate in regulatory events and also how alterations in CK2 levels that accompany disease may promote pathological rewiring of regulatory pathways.

Keywords Casein kinase 2 • CK2 • Phosphorylation • Apoptosis • Caspase • Hierarchical phosphorylation • Kinase inhibitor

D.W. Litchfield (✉) • L. Gyenis (✉)
Department of Biochemistry, Schulich School of Medicine & Dentistry, University of Western Ontario, London, ON, Canada, N6A 5C1

Department of Oncology, Schulich School of Medicine & Dentistry, University of Western Ontario, London, ON, Canada, N6A 5C1
e-mail: litchfi@uwo.ca; lgyenis@uwo.ca

K. Ahmed et al. (eds.), *Protein Kinase CK2 Cellular Function in Normal and Disease States*, Advances in Biochemistry in Health and Disease 12, DOI 10.1007/978-3-319-14544-0_11

1 Introduction

Despite its comparatively long history, in many ways, protein kinase CK2 remains an enigmatic participant in regulatory processes. More than half a century after its enzymatic activity was first described in the literature [1], CK2 has been associated with an extensive repertoire of biological processes with an ever-expanding list of potential cellular substrates and interaction partners [2–4]. The CK2 family is a small branch of the protein kinase superfamily that is comprised of two members in humans designated as CK2α and CK2α′ encoded by distinct genes on different chromosomes [5, 6]. Although classified as a protein serine/threonine kinase on the basis of its sequence features, CK2 has been shown to have the capacity to phosphorylate tyrosine residues in vitro as well as in yeast and mammalian cells [7–9]. While either of its catalytic subunits displays enzymatic activity in monomeric form, CK2 has typically been described as a tetrameric enzyme where two catalytic subunits (either CK2α or CK2α′) form complexes with a dimer of regulatory CK2β subunits [10]. There are several examples where alterations in the levels of CK2 have been implicated in human disease prompting interest in CK2 as a potential therapeutic target (reviewed in [11–14]). In fact, CK2 inhibitors have recently entered clinical trials with numerous other inhibitors undergoing preclinical investigation [15–24]. To evaluate the effects of CK2 inhibitors in experimental models and ultimately in clinical samples should CK2 inhibitors continue to progress in clinical trials, it is important that the regulation and functions of CK2 be thoroughly understood. This chapter will summarize some of the current understanding of the regulation and functions of CK2 and will then highlight recent efforts to place CK2 within the intricate regulatory networks that govern cellular processes related to proliferation and survival.

2 CK2: A Pleiotropic Regulator of Key Cellular Processes

Genetic and biochemical studies in a variety of experimental models have implicated CK2 as a regulatory participant in a broad array of biological processes [6, 25, 26]. In a similar respect, the number of candidate CK2 substrates continues to mount providing further indications for the role(s) of CK2 in a plethora of regulatory processes [27, 28]. In fact, phosphopeptides with phosphorylated residues conforming to the consensus for CK2 phosphorylation are among the most abundant within the human proteome with >10 % of identified phosphopeptides typically conforming to the CK2 consensus [3]. Taken together with the recent identification of nearly 40,000 phosphorylation sites from a single cell line, it is therefore conceivable that CK2 could phosphorylate thousands of individual sites within cells resulting in a tremendous impact on cellular processes [29]. While a comprehensive listing of all the biological processes in which CK2 could have role(s) is well beyond the scope of this discussion, processes where CK2 has been implicated include fundamental processes such as

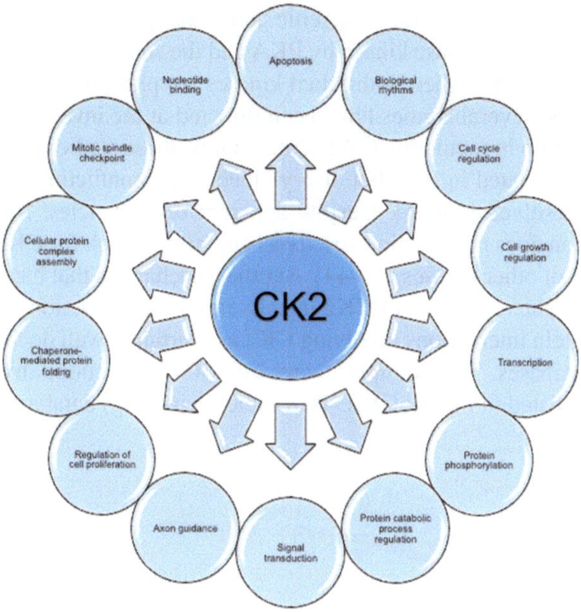

Fig. 1 Summary of Gene Ontology processes reported for CK2. This summary was compiled from EMBL-EBI (http://www.ebi.ac.uk/) and UniProt (http://www.uniprot.org) databases

transcription and translation, regulation of protein stability and turnover, circadian rhythms, cell cycle progression, and cell survival [2, 6, 30] (Fig. 1). This broad impact of CK2 on fundamental processes may provide an explanation, at least in part, for why CK2 is exploited by viruses during infection or in cancer cells that gain a selective survival advantage over their normal counterparts [26, 31, 32].

3 Regulated or Constitutively Active?

Since its discovery, a number of potential mechanisms for the regulation of CK2 have been proposed. CK2 was originally defined as a messenger-independent kinase to distinguish it from cAMP-dependent protein kinase (PKA) and other kinases regulated by second messengers such as Ca2+ or cyclic nucleotides [33]. Although unaffected by known second messengers, biochemical studies demonstrated that the catalytic activity of CK2 could be stimulated by positively charged compounds such as polyamines and inhibited by negatively charged compounds such as heparin [34, 35]. Even more than a quarter century after these findings were initially reported, the physiological relevance of these findings remains unclear. More recently, other cellular metabolites including inositol phosphates such as IP6 and inositol pyrophosphates such as IP7 have been reported to modulate CK2 activity at least in relation to the phosphorylation of specific substrates [36, 37]. Based on the phosphorylation of phosphorylase kinase by PKA and the subsequent discovery of other protein kinase cascades where individual kinases are phosphorylated and activated by other kinases, several studies have been directed at the investigation of the role of phosphorylation in regulating CK2. Collectively, these studies demonstrated that CK2 is phosphorylated in cells but also yielded some conflicting results that have not yet been resolved [38–41]. Despite apparent discrepancies, it is evident that phosphorylation does not function strictly as an on/off switch for CK2 as is the case with a number of other kinases [42–44]. Another mechanism that has been proposed to contribute to the regulation of CK2 in cells involves interactions with other proteins [45]. Protein interactions involving CK2 are certainly well documented within interaction databases. For example, there are >300 interactions involving human CK2α (Fig. 2) listed in the BioGRID (www.thebiogrid.org) database [46]. With the

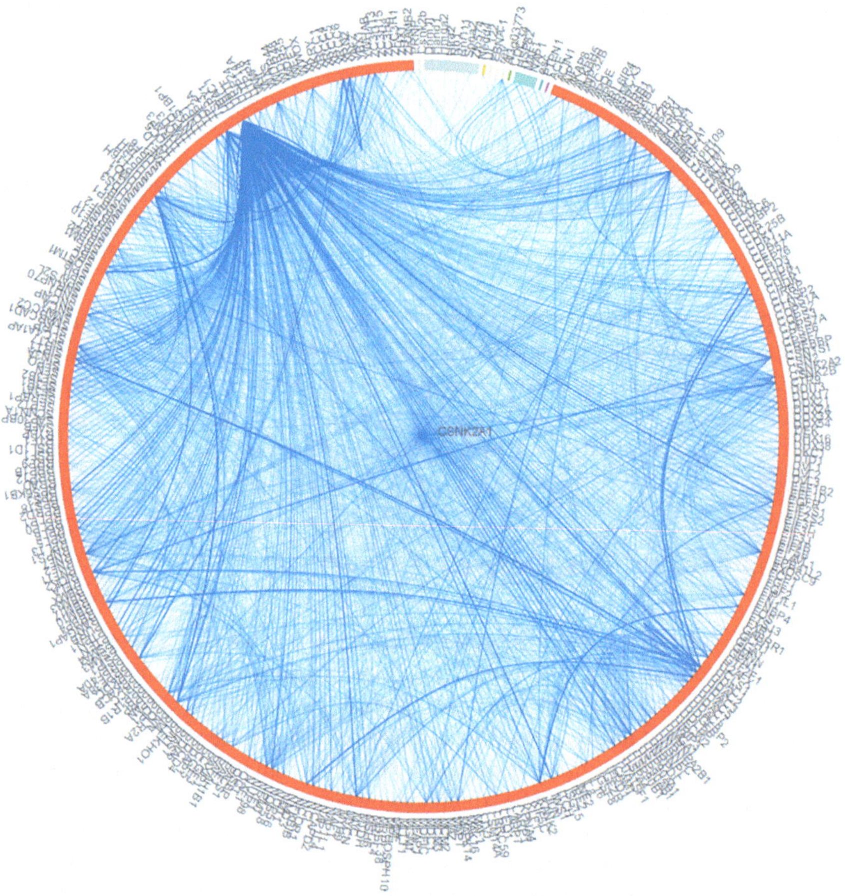

Fig. 2 Interaction network for human CK2α (gene name CSNK2A1) derived with the BioGRID database (http://thebiogrid.org). The BioGRID Graphical Network Viewer was used to generate the illustration on the basis of 368 physical or genetic interactions (BioGRID version 3.2.116) for CK2α that have a total of 6,978 interactions within the network

exception of CK2β, a remarkably conserved protein that has long been described as a CK2 subunit within tetrameric CK2 holoenzyme complexes, these interaction partners are not universal components of CK2 complexes. It has been speculated that CK2 interactors could include substrates or indirect regulators of CK2 that control the subcellular localization of CK2 and/or coordinate its interactions with substrates [4, 6, 47]. One additional mechanism that warrants consideration is the demonstration that CK2 activity can be modulated when it forms higher-order structures or complexes of tetramers in vitro [48–50] and potentially in cells [51]. Overall, while it is apparent that a number of distinct mechanisms could contribute in some way to modulating the ability of CK2 to phosphorylate its substrates in cells, it is clear that there is no singular mechanism for regulating the catalytic activity of CK2.

Although there have been several indications that CK2 could be regulated in cells through a variety of distinct mechanisms described above, there are other findings that raise questions about whether or not the catalytic activity of CK2 is actually regulated. For example, in contrast to several other protein kinases such as cyclin-dependent protein kinases (CDKs) and mitogen-activated protein kinases (MAPKs), the catalytic subunits of CK2 display high specific activity when expressed as recombinant proteins in bacteria, indicating that other eukaryotic proteins are not required to convert the catalytic CK2 subunits into active enzymes. It is also noteworthy that, although it is evident that CK2β can have a dramatic influence on the phosphorylation of some CK2 substrates, the catalytic subunits of CK2 typically express similar enzymatic activity toward the majority of substrates in the presence or absence of the regulatory CK2β subunits [52–54]. This observation does not necessarily diminish the importance of CK2β as a key regulator, but does suggest that the role of CK2β is not restricted to activation of CK2 akin to the activation of a CDK by cyclin. Examination of the phosphorylation of CK2 substrates such as the eukaryotic translation elongation factor EEF1D also suggests that phosphorylation at CK2 sites readily occurs to high stoichiometry in cells [55]. Once EEF1D is phosphorylated, its CK2 site appears to be stably phosphorylated without dramatic alterations in phosphate occupancy. Collectively, these observations suggest that the enzymatic activity of CK2 is not subject to regulation in a manner seen with many other protein kinases such as the CDKs or MAPKs. This latter suggestion is supported by structural insights that reveal that the activation loop of CK2 lacks regulatory hallmarks such as phosphorylation sites that have prominent roles in the regulation of many other kinases [56, 57].

4 Resolving the Paradox: An Unregulated Enzyme as a Participant in Regulatory Processes

At first glance, the prospect that CK2 could be a constitutively active (i.e., potentially unregulated) enzyme is at odds with its involvement as a regulatory participant in fundamental biological processes or with its capacity to promote pathogenesis (Fig. 1). However, it is important to recognize that there are many discrete populations of CK2 in cells that can be distinguished on the basis of features such as localization or composition of protein complexes. With the emergence of systematic proteomics profiling, it has also become apparent that many proteins—not just histones that are subject to an intricate combination of posttranslational modifications to generate the "histone code"—undergo multiple posttranslational modifications. Based on the existence of discrete CK2 populations, we postulated that individual populations of CK2 could be subject to distinct mechanisms of regulation rather than being subject to a universal mode of regulation that simultaneously impacts all of the CK2 within the cell. Furthermore, based on the widespread prevalence of multiple posttranslational modifications on many individual proteins, we were especially interested in the prospect that CK2 becomes a regulatory participant in cellular processes through systematic interactions with other regulatory modifications. Two examples of systematic relationships involving CK2 phosphorylation and other posttranslational modifications that will be discussed are the involvement of CK2 in hierarchical phosphorylation and sites with overlapping CK2 and caspase consensus recognition motifs.

5 Hierarchical Phosphorylation: Priming and Primed

The involvement of CK2 in hierarchical phosphorylation was established more than 20 years ago with the demonstration that CK2 can prime substrates such as glycogen synthase and inhibitor 2 of protein phosphatase I to subsequent phosphorylation by GSK-3 [58–60]. The prospect that CK2 could also be the primed kinase was implied by the demonstration that phosphorylated residues, including phosphoserine or phosphotyrosine, can substitute for glutamic acid or aspartic acid as specificity determinants for CK2 phosphorylation [61–63]. While a limited number of examples of CK2 being primed by other kinases have been identified to date, it is noteworthy that two cell cycle-regulated kinases, Plk1 and Cdk1, have been shown to prime substrates for subsequent phosphorylation by CK2 [64, 65]. These observations offer a possible explanation for the involvement of CK2 as a regulatory participant in pathways that control cell cycle progression despite the fact that it displays similar activity at all stages of the cell cycle. In this respect, cell cycle-dependent phosphorylation events catalyzed by CK2 would arise as a consequence of the CK2-catalyzed phosphorylation being dependent on prior phosphorylation by cell cycle-regulated kinases such as Plk1 or Cdk1. Detailed investigation of hierarchical phosphorylation relationships on individual proteins using traditional biochemical approaches can be a painstaking process that undoubtedly limits the number of characterized examples. By comparison, phosphoproteomics studies reveal a large number of multiple phosphorylated peptides and/or peptides with clusters of phosphorylated residues [66]. Consequently, it can be readily envisaged that kinases, particularly enzymes such as CK2 where phosphorylated residues can serve as specificity determinants, could have widespread involvement in multisite phosphorylation. Therefore, to systematically investigate the role of CK2 in hierarchical phosphorylation, we have devised a workflow that combines the use of peptide libraries with phosphorylated residues at specific positions in relation to the phosphorylatable residue together with database mining. While the results from this workflow will be presented in detail elsewhere [67], these studies reinforce the prediction that CK2 has widespread involvement in hierarchical phosphorylation. In addition to explanations for the regulated phosphorylation of substrates by an unregulated kinase, these studies also offer the promise of further extending the repertoire of cellular CK2 substrates. To enable identification of CK2 substrates that feature hierarchical phosphorylation, a key objective of these studies is elucidation of the spatial determinants for hierarchical phosphorylation.

6 Convergence of CK2 with Caspase Pathways

As described in the preceding section, hierarchical phosphorylation represents a mechanism that would enable CK2 to be a regulatory participant in cellular processes through its ability to catalyze regulated phosphorylation events such as the phosphorylation of Wee1 that occurs in a cell cycle-dependent manner [64] even in the absence of changes in its own catalytic activity. In essence, an unregulated kinase acquires regulatory involvement through its relationship with other regulated enzymes. This section will highlight another systematic relationship between CK2 and other posttranslational modifications, namely, the intersection of CK2 with caspase signaling [68, 69]. Our effort to study the convergence of CK2 with caspase signaling was motivated by the remarkable resemblance of the consensus recognition motifs of CK2 and caspases [70]. A systematic workflow combining computational predictions and database mining with peptide arrays and biochemical assays revealed many key regulatory proteins with overlapping CK2 and caspase recognition motifs [69]. Proteins with predicted CK2 phosphorylation sites proximal to sites of caspase cleavage include many known caspase substrates as well as caspases and pro-caspases. These observations suggest that CK2 could influence caspase action at both the levels of caspases and their substrates. Given that increased levels of CK2 have been observed in a number of different cancers and in virally infected cells, these results suggest an intriguing mechanism by which the constitutive activity of CK2 can be redirected to enhance cancer cell survival or to enable viruses to modulate caspase action in infected host cells. Another important consideration relevant to the overlap between CK2 and caspase recognition motifs relates to the nature of the substrates that may be phosphorylated when CK2 is elevated. Particularly intriguing is the prospect that some of the substrates that are phosphorylated when CK2 is elevated to abnormally high levels are pathological substrates that would not normally be phosphorylated by CK2 in normal cells. In these situations, the constitutive activity could be exploited for the pathological rewiring of pathways to promote outcomes such as survival that will confer a selective advantage for cancer cells or be favorable for the infecting virus.

Our initial effort to investigate the relationship between CK2 and caspase signaling was restricted to the identification and analysis of candidate proteins with CK2 sites immediately adjacent to the caspase cleavage site (i.e., P2 or P1′). For all of the candidates that were examined in this context, phosphorylation blocked cleavage [69]. As a logical extension of that work, we then devised a quantitative proteomics strategy based on the TAILS (terminal amine isotopic labeling of substrates) workflow to investigate the impact of phosphorylation on cleavage of cellular proteins with endogenous phosphorylation at sites that are phosphorylated in cells [71]. This strategy revealed that phosphorylation had both a positive and negative impact on cleavage corroborating findings obtained by Dix et al. [72] using an alternative proteomics strategy known as PROTOMAP [73]. We also performed positional scanning with peptide libraries to further elucidate the relationship between phosphorylation and cleavage. As seen in our earlier studies, phosphorylation adjacent to the cleavage site at either P2 or P1′ blocked cleavage. Phosphorylated residues at other positions on the C-terminal side of the cleavage site also inhibited cleavage, with the inhibitory effects diminishing with increasing distance of the phosphorylation site from the cleavage site. Similar trends were observed for phosphorylated residues on the N-terminal side of the cleavage site with the exception of the P3 position located two residues away from the cleavage site. Similar to results reported by Dix et al. [72], phosphorylation at P3 did not inhibit cleavage. In fact, with sequences evaluated by Dix et al. [72], phosphorylation actually promoted cleavage by caspase 8. Collectively, these results demonstrate that the relationship between phosphorylation and cleavage is complex with complete inhibition of cleavage only observed when phosphorylated residues are immediately adjacent to the cleavage site. Although CK2 is certainly not the only kinase involved in the regulation of caspase pathways [74], the prevalence of acidic residues within sites that are cleaved by caspases results in a higher likelihood that CK2, as compared to the majority of other kinases that do not have acidic specificity determinants, has sites that overlap caspase cleavage sites. Again, this observation illustrates the utility of CK2 as a vehicle to enable cancer cells or virally infected cells to circumvent caspase pathways.

7 Closing Thoughts and Implications

The realization that the actions of CK2 are coordinated with other regulatory modifications has a number of important implications. The involvement of CK2 in hierarchical phosphorylation, either as a priming kinase or as the primed kinase, adds an additional level of control to regulatory events. Furthermore, when CK2, often considered to be an unregulated enzyme, is the primed kinase, it would adopt the regulatory characteristics of the priming kinase. With respect to the convergence of CK2 with caspases, perhaps the most intriguing aspect of this relationship is the potential for CK2 to attenuate caspase action to promote cell survival. By exploiting the constitutive activity of CK2, cancer cells or virally infected cells with elevated levels of CK2 could gain enhanced survival through increased phosphorylation of caspases and caspase substrates. It is even conceivable that when CK2 is overexpressed or directed to new locations in these cells, it could phosphorylate pathological substrates (i.e., proteins in that it does not phosphorylate in normal cells) with a resultant pathological rewiring of caspase pathways (Fig. 3). With the emergence of precision medicine and molecular-targeted agents, it is also important to recognize that drugs directed at a single target could have unintended, or at least unexpected, consequences related to relationships such as those that we have described here for CK2. Under some circumstances, it could be envisaged that these relationships could be advantageous. For example, it may be possible to consider the use of CK2 inhibitors to enhance the susceptibility of cancer cells to apoptotic stimuli. Looking to the future, we can expect that our knowledge of the sites that undergo posttranslational modification in cells will continue to expand with ongoing improvements in proteomic technologies. This information will undoubtedly reveal more systematic relationships between CK2 and other regulatory modifications—and perhaps ultimately a logical code that enables us to decipher the roles of CK2 within the intricate regulatory networks that control biological processes.

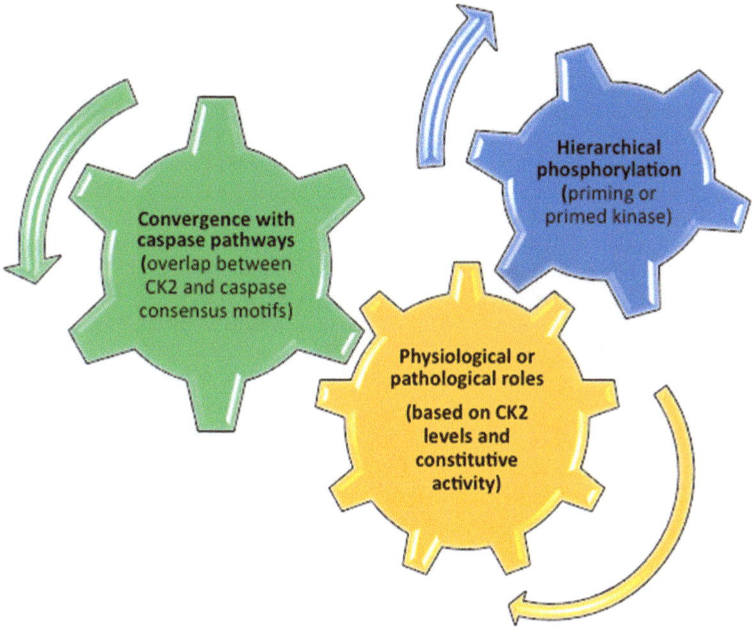

Fig. 3 Schematic illustration to conceptualize the involvement of CK2 as a central constituent within regulatory networks. Based on its constitutive activity and alterations in expression that have been observed in cancer cells or in virally infected cells, CK2 could have both physiological and pathological substrates that intersect other regulatory pathways with the potential for perturbing processes that promote the cell survival (see text for additional discussion)

Acknowledgments Our experimental studies on CK2 have been supported by operating grants from the Canadian Institutes of Health Research and the Canadian Cancer Society Research Institute. We also acknowledge past and present members of the Litchfield lab including James Duncan, Sam Fess, Michelle Gabriel, Adam Rabalski, Nicole St. Denis, Jacob Turowec, Greg Vilk, and Stephanie Zukowski whose experimental contributions have contributed to the concepts highlighted in this chapter. Finally, we apologize to our colleagues for omission of important citations that were incurred in the interests of being concise—and encourage readers to view other chapters in this book for more insights on protein kinase CK2.

References

1. Burnett G, Kennedy EP (1954) The enzymatic phosphorylation of proteins. J Biol Chem 211(2):969–980
2. Meggio F, Pinna LA (2003) One-thousand-and-one substrates of protein kinase CK2? FASEB J 17(3):349–368. doi:10.1096/fj.02-0473rev, 17/3/349 [pii]
3. Salvi M, Sarno S, Cesaro L, Nakamura H, Pinna LA (2009) Extraordinary pleiotropy of protein kinase CK2 revealed by weblogo phosphoproteome analysis. Biochim Biophys Acta 1793(5):847–859. doi:10.1016/j.bbamcr.2009.01.013, S0167-4889(09)00029-9 [pii]
4. Gyenis L, Litchfield DW (2008) The emerging CK2 interactome: insights into the regulation and functions of CK2. Mol Cell Biochem 316(1–2):5–14. doi:10.1007/s11010-008-9830-5
5. Manning G, Whyte DB, Martinez R, Hunter T, Sudarsanam S (2002) The protein kinase complement of the human genome. Science 298(5600):1912–1934. doi:10.1126/science.1075762, 298/5600/1912 [pii]
6. Litchfield DW (2003) Protein kinase CK2: structure, regulation and role in cellular decisions of life and death. Biochem J 369(Pt 1):1–15. doi:10.1042/BJ20021469, BJ20021469 [pii]
7. Vilk G, Weber JE, Turowec JP, Duncan JS, Wu C, Derksen DR, Zien P, Sarno S, Donella-Deana A, Lajoie G, Pinna LA, Li SS, Litchfield DW (2008) Protein kinase CK2 catalyzes tyrosine phosphorylation in mammalian cells. Cell Signal 20(11):1942–1951. doi:10.1016/j.cellsig.2008.07.002, S0898-6568(08)00197-6 [pii]
8. Wilson LK, Dhillon N, Thorner J, Martin GS (1997) Casein kinase II catalyzes tyrosine phosphorylation of the yeast nucleolar immunophilin Fpr3. J Biol Chem 272(20):12961–12967
9. Basnet H, Su XB, Tan Y, Meisenhelder J, Merkurjev D, Ohgi KA, Hunter T, Pillus L, Rosenfeld MG (2014) Tyrosine phosphorylation of histone H2A by CK2 regulates transcriptional elongation. Nature. doi:10.1038/nature13736, nature13736 [pii]
10. Turowec JP, Duncan JS, French AC, Gyenis L, St Denis NA, Vilk G, Litchfield DW (2010) Protein kinase CK2 is a constitutively active enzyme that promotes cell survival: strategies to identify CK2 substrates and manipulate its activity in mammalian cells. Methods Enzymol 484:471–493. doi:10.1016/B978-0-12-381298-8.00023-X, B978-0-12-381298-8.00023-X [pii]
11. Duncan JS, Litchfield DW (2008) Too much of a good thing: the role of protein kinase CK2 in tumorigenesis and prospects for therapeutic inhibition of CK2. Biochim Biophys Acta 1784(1):33–47. doi:10.1016/j.bbapap.2007.08.017, S1570-9639(07)00210-5 [pii]
12. Trembley JH, Wang G, Unger G, Slaton J, Ahmed K (2009) Protein kinase CK2 in health and disease: CK2: a key player in cancer biology. Cell Mol Life Sci 66(11–12):1858–1867. doi:10.1007/s00018-009-9154-y
13. Ruzzene M, Pinna LA (2010) Addiction to protein kinase CK2: a common denominator of diverse cancer cells? Biochim Biophys Acta 1804(3):499–504. doi:10.1016/j.bbapap.2009.07.018, S1570-9639(09)00185-X [pii]
14. Guerra B, Issinger OG (2008) Protein kinase CK2 in human diseases. Curr Med Chem 15(19):1870–1886
15. Battistutta R, Cozza G, Pierre F, Papinutto E, Lolli G, Sarno S, O'Brien SE, Siddiqui-Jain A, Haddach M, Anderes K, Ryckman DM, Meggio F, Pinna LA (2011) Unprecedented selectivity

and structural determinants of a new class of protein kinase CK2 inhibitors in clinical trials for the treatment of cancer. Biochemistry 50(39):8478–8488. doi:10.1021/bi2008382
16. Pierre F, Chua PC, O'Brien SE, Siddiqui-Jain A, Bourbon P, Haddach M, Michaux J, Nagasawa J, Schwaebe MK, Stefan E, Vialettes A, Whitten JP, Chen TK, Darjania L, Stansfield R, Anderes K, Bliesath J, Drygin D, Ho C, Omori M, Proffitt C, Streiner N, Trent K, Rice WG, Ryckman DM (2011) Discovery and SAR of 5-(3-chlorophenylamino)benzo[c][2,6] naphthyridine-8-carboxylic acid (CX-4945), the first clinical stage inhibitor of protein kinase CK2 for the treatment of cancer. J Med Chem 54(2):635–654. doi:10.1021/jm101251q
17. Siddiqui-Jain A, Bliesath J, Macalino D, Omori M, Huser N, Streiner N, Ho CB, Anderes K, Proffitt C, O'Brien SE, Lim JK, Von Hoff DD, Ryckman DM, Rice WG, Drygin D (2012) CK2 inhibitor CX-4945 suppresses DNA repair response triggered by DNA-targeted anticancer drugs and augments efficacy: mechanistic rationale for drug combination therapy. Mol Cancer Ther 11(4):994–1005. doi:10.1158/1535-7163.MCT-11-0613, 1535-7163.MCT-11-0613 [pii]
18. Prins RC, Burke RT, Tyner JW, Druker BJ, Loriaux MM, Spurgeon SE (2013) CX-4945, a selective inhibitor of casein kinase-2 (CK2), exhibits anti-tumor activity in hematologic malignancies including enhanced activity in chronic lymphocytic leukemia when combined with fludarabine and inhibitors of the B-cell receptor pathway. Leukemia 27(10):2094–2096. doi:10.1038/leu.2013.228, leu2013228 [pii]
19. Zheng Y, McFarland BC, Drygin D, Yu H, Bellis SL, Kim H, Bredel M, Benveniste EN (2013) Targeting protein kinase CK2 suppresses prosurvival signaling pathways and growth of glioblastoma. Clin Cancer Res 19(23):6484–6494. doi:10.1158/1078-0432.CCR-13-0265, 1078-0432.CCR-13-0265 [pii]
20. Martins LR, Lucio P, Melao A, Antunes I, Cardoso BA, Stansfield R, Bertilaccio MT, Ghia P, Drygin D, Silva MG, Barata JT (2014) Activity of the clinical-stage CK2-specific inhibitor CX-4945 against chronic lymphocytic leukemia. Leukemia 28(1):179–182. doi:10.1038/leu.2013.232, leu2013232 [pii]
21. Moucadel V, Prudent R, Sautel CF, Teillet F, Barette C, Lafanechere L, Receveur-Brechot V, Cochet C (2011) Antitumoral activity of allosteric inhibitors of protein kinase CK2. Oncotarget 2(12):997–1010, 361 [pii]
22. Prudent R, Moucadel V, Nguyen CH, Barette C, Schmidt F, Florent JC, Lafanechere L, Sautel CF, Duchemin-Pelletier E, Spreux E, Filhol O, Reiser JB, Cochet C (2010) Antitumor activity of pyridocarbazole and benzopyridoindole derivatives that inhibit protein kinase CK2. Cancer Res 70(23):9865–9874. doi:10.1158/0008-5472.CAN-10-0917, 0008-5472.CAN-10-0917 [pii]
23. Sarno S, Papinutto E, Franchin C, Bain J, Elliott M, Meggio F, Kazimierczuk Z, Orzeszko A, Zanotti G, Battistutta R, Pinna LA (2011) ATP site-directed inhibitors of protein kinase CK2: an update. Curr Top Med Chem 11(11):1340–1351, BSP/CTMC/E-Pub/-00083-11-11 [pii]
24. Cozza G, Pinna LA, Moro S (2012) Protein kinase CK2 inhibitors: a patent review. Expert Opin Ther Pat 22(9):1081–1097. doi:10.1517/13543776.2012.717615
25. Glover CV 3rd (1998) On the physiological role of casein kinase II in Saccharomyces cerevisiae. Prog Nucleic Acid Res Mol Biol 59:95–133
26. St-Denis NA, Litchfield DW (2009) Protein kinase CK2 in health and disease: from birth to death: the role of protein kinase CK2 in the regulation of cell proliferation and survival. Cell Mol Life Sci 66(11–12):1817–1829. doi:10.1007/s00018-009-9150-2
27. Franchin C, Cesaro L, Salvi M, Millioni R, Iori E, Cifani P, James P, Arrigoni G, Pinna L (2014) Quantitative analysis of a phosphoproteome readily altered by the protein kinase CK2 inhibitor quinalizarin in HEK-293T cells. Biochim Biophys Acta. doi:S1570-9639(14)00252-0 [pii], 10.1016/j.bbapap.2014.09.017
28. Wang C, Ye M, Bian Y, Liu F, Cheng K, Dong M, Dong J, Zou H (2013) Determination of CK2 specificity and substrates by proteome-derived peptide libraries. J Proteome Res 12(8):3813–3821. doi:10.1021/pr4002965
29. Sharma K, D'Souza RC, Tyanova S, Schaab C, Wisniewski JR, Cox J, Mann M (2014) Ultradeep human phosphoproteome reveals a distinct regulatory nature of tyr and ser/thr-based signaling. Cell Rep 8(5):1583–1594. doi:10.1016/j.celrep.2014.07.036, S2211-1247(14)00620-2 [pii]

30. Tsuchiya Y, Akashi M, Matsuda M, Goto K, Miyata Y, Node K, Nishida E (2009) Involvement of the protein kinase CK2 in the regulation of mammalian circadian rhythms. Sci Signal 2(73):ra26. doi:10.1126/scisignal.2000305, 2/73/ra26 [pii]
31. Ahmed K, Gerber DA, Cochet C (2002) Joining the cell survival squad: an emerging role for protein kinase CK2. Trends Cell Biol 12(5):226–230, S0962892402022791 [pii]
32. Cao JY, Shire K, Landry C, Gish GD, Pawson T, Frappier L (2014) Identification of a novel protein interaction motif in the regulatory subunit of casein kinase 2. Mol Cell Biol 34(2): 246–258. doi:10.1128/MCB.00968-13, MCB.00968-13 [pii]
33. Tuazon PT, Traugh JA (1991) Casein kinase I and II–multipotential serine protein kinases: structure, function, and regulation. Adv Second Messenger Phosphoprotein Res 23:123–164
34. Hathaway GM, Traugh JA (1984) Kinetics of activation of casein kinase II by polyamines and reversal of 2,3-bisphosphoglycerate inhibition. J Biol Chem 259(11):7011–7015
35. Hathaway GM, Lubben TH, Traugh JA (1980) Inhibition of casein kinase II by heparin. J Biol Chem 255(17):8038–8041
36. Solyakov L, Cain K, Tracey BM, Jukes R, Riley AM, Potter BV, Tobin AB (2004) Regulation of casein kinase-2 (CK2) activity by inositol phosphates. J Biol Chem 279(42):43403–43410. doi:10.1074/jbc.M403239200, M403239200 [pii]
37. Rao F, Cha J, Xu J, Xu R, Vandiver MS, Tyagi R, Tokhunts R, Koldobskiy MA, Fu C, Barrow R, Wu M, Fiedler D, Barrow JC, Snyder SH (2014) Inositol pyrophosphates mediate the DNA-PK/ATM-p53 cell death pathway by regulating CK2 phosphorylation of Tti1/Tel2. Mol Cell 54(1):119–132. doi:10.1016/j.molcel.2014.02.020, S1097-2765(14)00165-8 [pii]
38. Litchfield DW, Dobrowolska G, Krebs EG (1994) Regulation of casein kinase II by growth factors: a reevaluation. Cell Mol Biol Res 40(5–6):373–381
39. Litchfield DW, Luscher B, Lozeman FJ, Eisenman RN, Krebs EG (1992) Phosphorylation of casein kinase II by p34cdc2 in vitro and at mitosis. J Biol Chem 267(20):13943–13951
40. Litchfield DW, Lozeman FJ, Cicirelli MF, Harrylock M, Ericsson LH, Piening CJ, Krebs EG (1991) Phosphorylation of the beta subunit of casein kinase II in human A431 cells. Identification of the autophosphorylation site and a site phosphorylated by p34cdc2. J Biol Chem 266(30):20380–20389
41. Ackerman P, Glover CV, Osheroff N (1990) Stimulation of casein kinase II by epidermal growth factor: relationship between the physiological activity of the kinase and the phosphorylation state of its beta subunit. Proc Natl Acad Sci U S A 87(2):821–825
42. Messenger MM, Saulnier RB, Gilchrist AD, Diamond P, Gorbsky GJ, Litchfield DW (2002) Interactions between protein kinase CK2 and Pin1. Evidence for phosphorylation-dependent interactions. J Biol Chem 277(25):23054–23064. doi:10.1074/jbc.M200111200, M200111200 [pii]
43. St-Denis NA, Derksen DR, Litchfield DW (2009) Evidence for regulation of mitotic progression through temporal phosphorylation and dephosphorylation of CK2alpha. Mol Cell Biol 29(8):2068–2081. doi:10.1128/MCB.01563-08, MCB.01563-08 [pii]
44. St-Denis NA, Bailey ML, Parker EL, Vilk G, Litchfield DW (2011) Localization of phosphorylated CK2alpha to the mitotic spindle requires the peptidyl-prolyl isomerase Pin1. J Cell Sci 124(Pt 14):2341–2348. doi:10.1242/jcs.077446, jcs.077446 [pii]
45. Olsten ME, Weber JE, Litchfield DW (2005) CK2 interacting proteins: emerging paradigms for CK2 regulation? Mol Cell Biochem 274(1–2):115–124
46. Chatr-Aryamontri A, Breitkreutz BJ, Heinicke S, Boucher L, Winter A, Stark C, Nixon J, Ramage L, Kolas N, O'Donnell L, Reguly T, Breitkreutz A, Sellam A, Chen D, Chang C, Rust J, Livstone M, Oughtred R, Dolinski K, Tyers M (2013) The BioGRID interaction database: 2013 update. Nucleic Acids Res 41(Database issue):D816–D823. doi:10.1093/nar/gks1158, gks1158 [pii]
47. Olsten ME, Litchfield DW (2004) Order or chaos? An evaluation of the regulation of protein kinase CK2. Biochem Cell Biol 82(6):681–693. doi:10.1139/o04-116, o04-116 [pii]
48. Glover CV (1986) A filamentous form of Drosophila casein kinase II. J Biol Chem 261(30):14349–14354

49. Lolli G, Pinna LA, Battistutta R (2012) Structural determinants of protein kinase CK2 regulation by autoinhibitory polymerization. ACS Chem Biol 7(7):1158–1163. doi:10.1021/cb300054n
50. Leroy D, Valero E, Filhol O, Heriche JK, Goldberg Y, Chambaz EM, Cochet C (1994) Modulation of the molecular organization and activity of casein kinase 2 by naturally occurring polyamines. Cell Mol Biol Res 40(5–6):441–453
51. Hubner GM, Larsen JN, Guerra B, Niefind K, Vrecl M, Issinger OG (2014) Evidence for aggregation of protein kinase CK2 in the cell: a novel strategy for studying CK2 holoenzyme interaction by BRET. Mol Cell Biochem. doi:10.1007/s11010-014-2196-y
52. Arrigoni G, Marin O, Pagano MA, Settimo L, Paolin B, Meggio F, Pinna LA (2004) Phosphorylation of calmodulin fragments by protein kinase CK2. Mechanistic aspects and structural consequences. Biochemistry 43(40):12788–12798. doi:10.1021/bi049365c
53. Turowec JP, Vilk G, Gabriel M, Litchfield DW (2013) Characterizing the convergence of protein kinase CK2 and caspase-3 reveals isoform-specific phosphorylation of caspase-3 by CK2alpha′: implications for pathological roles of CK2 in promoting cancer cell survival. Oncotarget 4(4):560–571, 948 [pii]
54. Grankowski N, Boldyreff B, Issinger OG (1991) Isolation and characterization of recombinant human casein kinase II subunits alpha and beta from bacteria. Eur J Biochem 198(1):25–30
55. Gyenis L, Duncan JS, Turowec JP, Bretner M, Litchfield DW (2011) Unbiased functional proteomics strategy for protein kinase inhibitor validation and identification of bona fide protein kinase substrates: application to identification of EEF1D as a substrate for CK2. J Proteome Res 10(11):4887–4901. doi:10.1021/pr2008994
56. Niefind K, Guerra B, Pinna LA, Issinger OG, Schomburg D (1998) Crystal structure of the catalytic subunit of protein kinase CK2 from Zea mays at 2.1 A resolution. EMBO J 17(9):2451–2462. doi:10.1093/emboj/17.9.2451
57. Niefind K, Guerra B, Ermakowa I, Issinger OG (2001) Crystal structure of human protein kinase CK2: insights into basic properties of the CK2 holoenzyme. EMBO J 20(19):5320–5331. doi:10.1093/emboj/20.19.5320
58. Roach PJ (1990) Control of glycogen synthase by hierarchal protein phosphorylation. FASEB J 4(12):2961–2968
59. Roach PJ (1991) Multisite and hierarchal protein phosphorylation. J Biol Chem 266(22):14139–14142
60. Cohen P (1992) Signal integration at the level of protein kinases, protein phosphatases and their substrates. Trends Biochem Sci 17(10):408–413, 0968-0004(92)90010-7 [pii]
61. Litchfield DW, Arendt A, Lozeman FJ, Krebs EG, Hargrave PA, Palczewski K (1990) Synthetic phosphopeptides are substrates for casein kinase II. FEBS Lett 261(1):117–120
62. Meggio F, Perich JW, Reynolds EC, Pinna LA (1991) Phosphotyrosine as a specificity determinant for casein kinase-2, a growth related Ser/Thr-specific protein kinase. FEBS Lett 279(2):307–309, 0014-5793(91)80174-2 [pii]
63. Hrubey TW, Roach PJ (1990) Phosphoserine in peptide substrates can specify casein kinase II action. Biochem Biophys Res Commun 172(1):190–196, S0006-291X(05)80192-5 [pii]
64. Watanabe N, Arai H, Iwasaki J, Shiina M, Ogata K, Hunter T, Osada H (2005) Cyclin-dependent kinase (CDK) phosphorylation destabilizes somatic Wee1 via multiple pathways. Proc Natl Acad Sci U S A 102(33):11663–11668. doi:10.1073/pnas.0500410102, 0500410102 [pii]
65. Yata K, Lloyd J, Maslen S, Bleuyard JY, Skehel M, Smerdon SJ, Esashi F (2012) Plk1 and CK2 act in concert to regulate Rad51 during DNA double strand break repair. Mol Cell 45(3):371–383. doi:10.1016/j.molcel.2011.12.028, S1097-2765(12)00034-2 [pii]
66. Schweiger R, Linial M (2010) Cooperativity within proximal phosphorylation sites is revealed from large-scale proteomics data. Biol Direct 5:6. doi:10.1186/1745-6150-5-6, 1745-6150-5-6 [pii]
67. St. Denis N, Gabriel M, Turowec JP, Gloor GB, Li SS-C, Gingras A-C, Litchfield DW (2014) Systematic investigation of hierarchical phosphorylation by protein kinase CK2. J Proteomics (epub ahead of print). doi:10.1016/j.jprot.2014.10.020

68. McDonnell MA, Abedin MJ, Melendez M, Platikanova TN, Ecklund JR, Ahmed K, Kelekar A (2008) Phosphorylation of murine caspase-9 by the protein kinase casein kinase 2 regulates its cleavage by caspase-8. J Biol Chem 283(29):20149–20158. doi:10.1074/jbc.M802846200, M802846200 [pii]
69. Duncan JS, Turowec JP, Duncan KE, Vilk G, Wu C, Luscher B, Li SS, Gloor GB, Litchfield DW (2011) A peptide-based target screen implicates the protein kinase CK2 in the global regulation of caspase signaling. Sci Signal 4(172):ra30. doi:10.1126/scisignal.2001682, 4/172/ra30 [pii]
70. Turowec JP, Duncan JS, Gloor GB, Litchfield DW (2011) Regulation of caspase pathways by protein kinase CK2: identification of proteins with overlapping CK2 and caspase consensus motifs. Mol Cell Biochem 356(1–2):159–167. doi:10.1007/s11010-011-0972-5
71. Turowec JP, Zukowski SA, Knight JD, Smalley DM, Graves LM, Johnson GL, Li SS, Lajoie GA, Litchfield DW (2014) An unbiased proteomic screen reveals caspase cleavage is positively and negatively regulated by substrate phosphorylation. Mol Cell Proteomics 13(5):1184–1197. doi:10.1074/mcp.M113.037374, M113.037374 [pii]
72. Dix MM, Simon GM, Wang C, Okerberg E, Patricelli MP, Cravatt BF (2012) Functional interplay between caspase cleavage and phosphorylation sculpts the apoptotic proteome. Cell 150(2):426–440. doi:10.1016/j.cell.2012.05.040, S0092-8674(12)00766-0 [pii]
73. Dix MM, Simon GM, Cravatt BF (2014) Global identification of caspase substrates using PROTOMAP (protein topography and migration analysis platform). Methods Mol Biol 1133:61–70. doi:10.1007/978-1-4939-0357-3_3
74. Kurokawa M, Kornbluth S (2009) Caspases and kinases in a death grip. Cell 138(5):838–854. doi:10.1016/j.cell.2009.08.021, S0092-8674(09)01038-1 [pii]

Role for Protein Kinase CK2 on Cell Proliferation: Assessing CK2 Complex Components in the Nucleus During the Cell Cycle Progression

Miwako Kato Homma, Takeshi Shibata, Toshiyuki Suzuki, Masato Ogura, Hiroko Kozuka-Hata, Masaaki Oyama, and Yoshimi Homma

Abstract CK2 is one of the serine threonine kinases known to be essential for basic cell viability and survival. Until recently, genetic, biochemical, and cell biological studies have indicated the involvement of this enzyme in the control of cell proliferation and in signal transduction. It has been reported that more than 300 proteins have been identified as CK2 substrates in cells; however, the clarification of the functional relationship between those substrates and CK2 in the aspect of cell proliferation and survival is still required. The identification of the cellular factors involved in CK2 function is important to delineate its molecular mechanism in the cells. We previously demonstrated a significant increase in CK2 activity by growth factor stimulation of quiescent cells and identified eIF5 in the CK2 complex as being a downstream molecule. Also, we described the cell cycle-dependent association of CK2 with an endogenous tumor suppressor adenomatous polyposis coli (APC) protein, which was further investigated to uncover the negative regulation of CK2 activity by the APC-C terminal domain that is lost in more than 60 % of FAP-derived cancer cells. These results established the importance of kinase activity that seems to be non-constitutively active and controlled by unknown cellular mechanisms for properly regulating cell proliferation. We therefore examined the CK2 protein complex in the nucleus in distinct phases of cell proliferation by employing synchronized cells. Cell lysates were made after the stimulation with FBS, and nuclear fractions were extracted and immunoprecipitated with anti-CK2 polyclonal antibodies.

M.K. Homma (✉) • T. Suzuki • M. Ogura • Y. Homma
Department Biomolecular Science, Fukushima Medical University School of Medicine, 1-Hikarigaoka, Fukushima 960-1295, Japan
e-mail: mkhomma@fmu.ac.jp

T. Shibata
ABSciex. K.K., Tokyo, Japan

H. Kozuka-Hata • M. Oyama
Medical Proteomics Laboratory, The Institute of Medical Sciences, The University of Tokyo, Tokyo 108-8639, Japan

K. Ahmed et al. (eds.), *Protein Kinase CK2 Cellular Function in Normal and Disease States*, Advances in Biochemistry in Health and Disease 12, DOI 10.1007/978-3-319-14544-0_12

Using two-dimensional gel electrophoresis followed by mass spectrometry analysis, we identified 22 proteins, including hnRNP, histone-binding proteins, and lamin-B, which were localized in the nuclear fraction preferentially associated with CK2 in the early G_1 phase. In further studies, using nuclear CK2 immunoprecipitates followed by nanoLC mass spectrometry analysis, we identified 140 proteins as CK2-interacting proteins in the nucleus. Intriguingly, more than 20 % of these proteins were constituted by DNA-binding and RNA-binding proteins, suggesting the involvement of CK2 function as a dynamic regulator for gene function associated with cell cycle progression. Our approach can be extended to other cell states and cellular compartments and provide the broad biochemical framework for understanding the role of the kinase.

Keywords Adenomatous polyposis coli (APC) • Cell cycle • CK2 • Eukaryotic translation initiation factor 5 (eIF5) • Mass spectrometry (MS) • Nuclear functions • Phosphorylation • Posttranslational modifications • Protein complex • Proteomics

Abbreviations

2D	Two dimensional
Ala	Alanine
APC	Adenomatous polyposis coli
CK2	Casein kinase 2/II
DMEM	Dulbecco's Modified Eagle's Medium
eIF5	Eukaryotic translation initiation factor 5
FBS	Fetal bovine serum
FDR	False discovery rate
Glu	Glutamic acid
nanoLC	Nanoscale liquid chromatography
SDS-PAGE	Sodium dodecyl sulfate polyacrylamide gel electrophoresis

1 CK2 Activity and Growth State of the Cells

CK2 is one of the old kinases identified more than 60 years ago [1], and its important role on numerous aspects for cellular function has been elucidated [2–4]. As shown in Fig. 1, CK2 is a heterotetrameric enzyme consisting of catalytic subunit α or α' and two regulatory β subunits, which catalyzes phosphate transfer using both GTP and ATP as phosphate donors. Genetic studies have shown that CK2 is indispensable to cell proliferation, survival, and development, since a double knockout of the two catalytic subunits, α and α', is lethal in yeast and mouse [2, 5, 6]. Contrarily, the catalytic subunits α and α' are considered to function

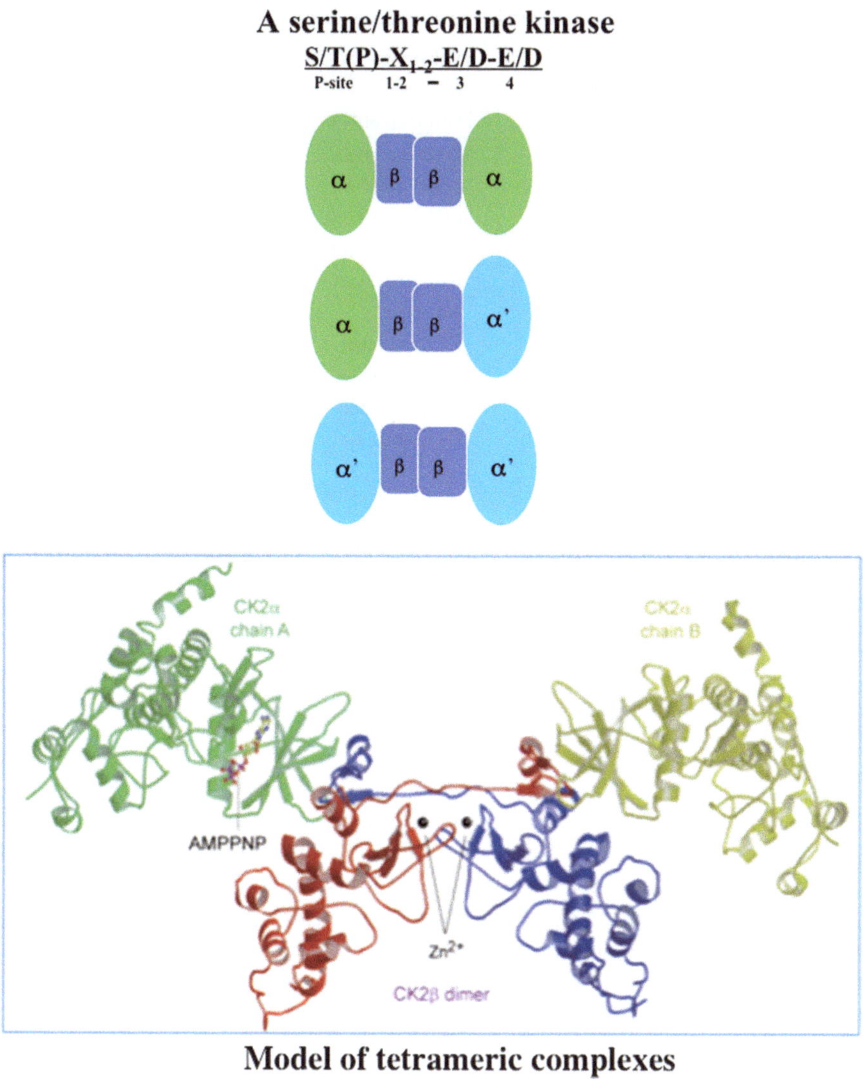

Fig. 1 Ribbon representation of the CK2 holoenzyme from the crystal structure (PBD no. 1JWH)

complementary, and a single knockout of either α or α′ remains viable in yeast. The knockout effect of CK2α′ was observed to be defective in spermatogenesis causing globozoospermia in male mice [3], whereas a knockout of CK2β in mice showed embryonic lethality [7]. Importantly, CK2 is implicated in tumorigenesis and transformation by promoting cell survival signaling pathways through upregulating proto-oncogene products such as c-Jun and c-Myc and transcriptional activators like β-catenin in the cell [8–10]. An increase in cellular contents of CK2 subunits has also been reported to be associated with carcinogenesis. A two- to threefold increase

Table 1 Summary of CK2 activities observed by extracellular stimuli

Stimulus	Cell type	Substrates and effect	References
Serum	WI138 cells	Increase	Carroll and Marshak
Serum	WI138 cells	DSD, no effect	Litchfield
EGF	WI138 cells	DSD, no effect	Litchfield
FGF	WI138 cells	DSD, no effect	Litchfield
PDGF	WI138 cells	DSD, no effect	Litchfield
Serum	WI138 cells	DSD, no effect	Litchfield
Insulin	3T3-L1	ETE, increase	Sommercorn
Insulin	Rat adipocytes	No effect	Haystead
IGF-1	Balb/c 3T3	ETE, increase	Klarlund
EGF	A431	ETE, increase	Ackerman
EGF	Swiss 3T3	ETE, no effect	Ahn
Serum	TIG	ETE, increase	Homma

in cellular CK2 content was observed in leukemic cells, and a statistically significant increase of CK2 activity was detected in kidney, head and neck, and bladder carcinomas [11, 12].

1.1 Changes in CK2 Activity in Response to Extracellular Stimuli

Numerous studies have been carried out to clarify the molecular mechanisms for the activation of CK2 by using a variety of external stimuli in different kinds of cultured cells, as shown in Table 1. These studies were mostly based on established biochemical methods; however, the results on changes in enzymatic activity in response to various stimuli have shown diverse effects and not to be consistent (*for review* [13–15]). In addition to the fact that the regulatory mechanisms of the CK2 catalytic subunit by itself have not been well defined until now, CK2 has been considered to be a constitutively active protein kinase. The structural transition of the CK2 holoenzyme with regard to holoenzyme formation and aggregation [16, 17] has recently been resolved. We therefore examined changes in CK2 activity during cell cycle progression and demonstrated a significant increase in CK2 activity in normal human fibroblasts synchronously induced to enter G_1 by growth factor stimulation [18]. For these experiments, normal human fetal lung fibroblast TIG-7 cells were grown in Dulbecco's Modified Eagle's Medium (DMEM) supplemented with 10 % fetal bovine serum (FBS). On the other hand, for synchronization experiments, logarithmically growing cells were starved in 0.1 % FBS for 48 h and then cultured in fresh media containing 10 % FBS for several additional hours as indicated in reference [18]. CK2 immunoprecipitates were analyzed for CK2 activity in vitro by the p81 filter method. As we have demonstrated, activation of CK2 appears to be transient following stimulation of the quiescent cells with FBS, reaching the highest level in its activity in the first 3 h and decreasing thereafter. These results suggest a potential role for CK2 in cell cycle progression and also that CK2 activity is controlled in relation to the progression of cell cycle. In order to delineate the dynamics of CK2 signaling, we screened for interacting molecules that associate with CK2 holoenzyme. We eventually identified eIF5 by mass spectrometric analysis in the CK2 complex when the highest level of activity was exhibited during cell cycle progression.

2 CK2 Phosphorylation Levels of Eukaryotic Translational Initiation Factor 5 Correlate with the Growth State in the Cell

To determine whether eIF5 is subject to phosphorylation modification in vivo, we performed ^{32}P-labeling of TIG cells and immunoprecipitated eIF5, which was found to be phosphorylated in vivo (Fig. 2a). Its phosphorylation level was low in G_0, dramatically increased at 3 h, and further increased in the S and G_2/M phases, although the expression levels of the eIF5 protein were consistent during the time course of the experiment. On the other hand, when growing cells were serum starved and gated into G_0/G_1, eIF5 phosphorylation level decreased to the basal level over a 48 h period. The interaction between eIF5 and CK2 was observable 3 h after FBS stimulation only in G_1 but not in G_0 or G_2/M. These results suggest that the phosphorylated state of eIF5 is closely correlated with the growing state of the cells. We also demonstrated that CK2 is the kinase responsible for phosphorylation of eIF5 and identified the exact phosphorylation sites in eIF5 by peptide mass and sequencing analysis (Fig. 2b, *upper*). Two serine residues 389 and 390 of eIF5 were major sites of phosphorylation by CK2 both in vivo and in vitro, and phosphorylation of two threonines, 207 and 208, was also detected (Fig 2b, *lower*).

2.1 *Phenotypes of Phosphorylation-Site Mutants in eIF5*

We constructed mutant eIF5 cDNA plasmids in which Ser or Thr CK2 sites were substituted by alanines (Ala) and expressed in COS cells, in order to test their ability to be phosphorylated. We found that cell proliferation was significantly abrogated by all three eIF5 mutants, which were severely impaired in phosphorylation compared to wt eIF5-expressing cells [18]. In contrast, cell proliferation was enhanced by 30–50 % in eIF5 mutants substituting Ser and Thr residues with glutamic acid (Glu) that mimic phosphorylated states in eIF5. Flow cytometric analysis indicated that the expression of mutant eIF5 with Ala induced decreases in the percentage of cells in the G_2/M phases and increases in the G_0/G_1 and S phases of the cell cycle. These results point out the possibility that cell growth was strictly correlated with the ability of eIF5 to be phosphorylated in vivo. Actually, cyclin B1-associated

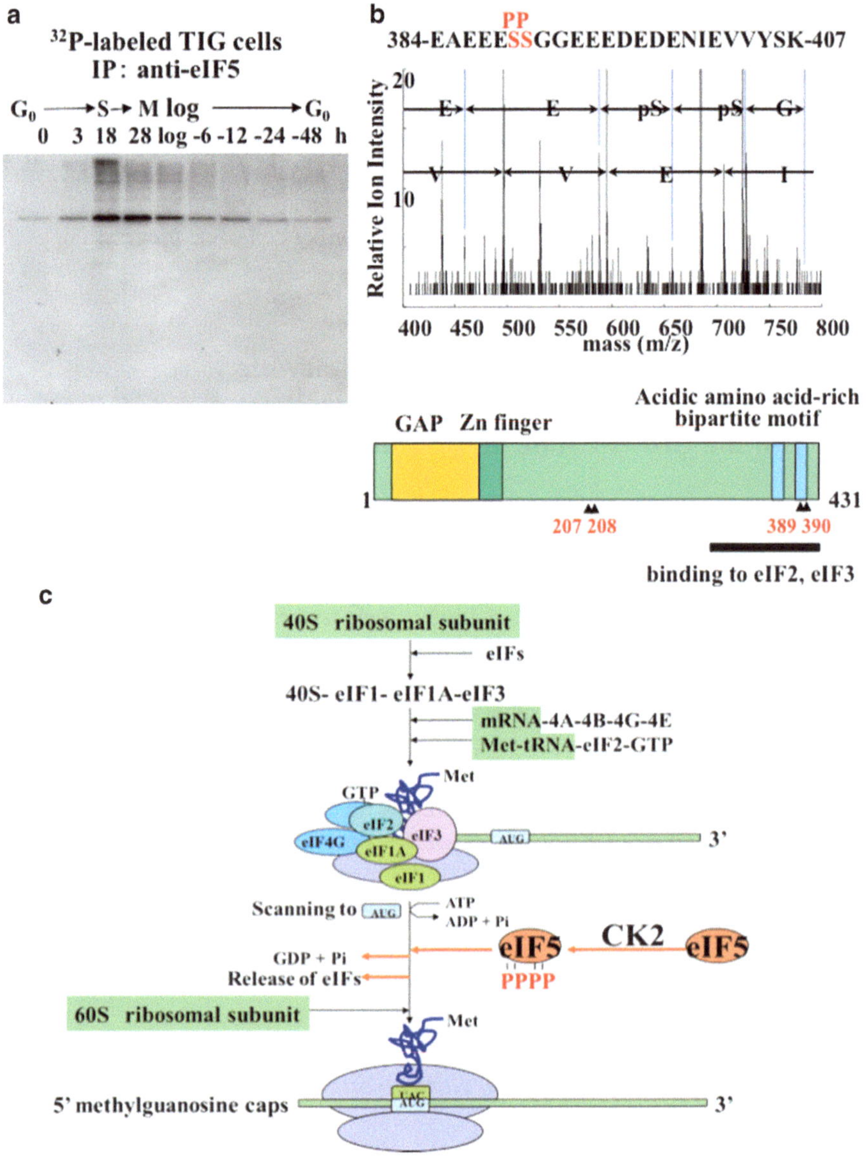

Fig. 2 Cell cycle-dependent phosphorylation of eIF5 in vivo. (**a**) Phosphorylated state of endogenous eIF5 correlates with the growing state of the cells. Logarithmically growing normal human fibroblast TIG cells were serum starved for 48 h (G_0) and cultured synchronously to progress through the cell cycle by the addition of FBS (G_1), and then logarithmically growing cells were starved again in 0.1 % FBS for indicated hours to be gated into G_0 phase. The length of these treatments is indicated. Cells were labeled with ^{32}P-Pi for 1 h at 37 °C before harvesting and were immunoprecipitated with anti-eIF5 monoclonal antibody and separated by 10 % SDS-PAGE. An image from autoradiography indicates elevated phosphorylation of eIF5 associating with the progression of cell cycle [18]. (**b**) Results of mass spectrometry analysis obtained by peptides derived from eIF5 proteolytic digests to identify its phosphorylation sites by CK2 at four sites. (**b**, *lower*) Schematic structure of human eIF5, indicating phosphorylation sites by *red-colored numbers*, is also shown. (**c**) A model of the functional role for phosphorylated eIF5 on the formation of mature initiation complex, with newly synthesized mRNA and initiation factors, engaged on the ribosomes to initiate protein translations

kinase activity was significantly inactivated by these mutants. From these results, it was suggested that blocking phosphorylation of eIF5 by CK2 inhibited cell cycle progression. By analyzing the formation of multifactor complex comprising translation-initiation factors, including eIF1, eIF2, eIF3, and eIF4 on sucrose gradient velocity sedimentation, we found that phosphorylation site mutation impairs stable association of eIF5 with pre-initiation complex. These results may imply that eIF5 phosphorylation by CK2 is important for the proper association of eIF5 with the mature translation-initiation complex that is engaged in translation in vivo (Fig. 2c). Collectively, these data point toward CK2 involvement in regulating translation through the association and phosphorylation of eIF5.

3 Downregulation of CK2 Activity by Association with APC Protein

We observed CK2 being associated with the adenomatous polyposis coli (APC) protein during the later phase in the cell cycle [19]. The tumor suppressor APC is linked to FAP (familial adenomatous polyposis) and is inactivated by mutations frequently observed in most colorectal tumors (Fig. 3a). To analyze the biochemical significance of APC on cell proliferation, we established a method to isolate APC proteins with polyclonal antibodies which were raised against the N-terminus of APC protein. Figure 3b demonstrates the results of immunoprecipitation followed by Western blot indicating that the protocol is quite efficient to immunoprecipitate not only the full-length APC from normal human cells, TIG-3 and TIG-7 cells, but also the variable masses of truncated forms of APC in colorectal carcinomas, DLD-1, SW480, KMS-4, and KMS-8 cells; from the latter, two lines were established from different individuals of FAP background. The arrowheads in Fig. 3a indicate a nonsense mutation site in each colorectal carcinoma cell line, and those protein sizes of endogenous APC were consistent with previous mutational analyses. Notably, the expression of APC is upregulated in carcinomas. In colorectal carcinomas, activation of endogenous enzymes such as phospholipase C, PI3-kinase, and protein kinases has been reported. Following our extensive approach to prove potential associations of APC with kinase activities, we identified CK2 as an APC-associating kinase in the normal cells and also colorectal carcinoma cells, and demonstrated that CK2α and CK2 holoenzyme bind directly to the N-terminal region of the APC protein. By analyzing APC immunoprecipitates for association with CK2 in synchronized cells, we demonstrated that APC and CK2 interactions are cell cycle regulated (Fig. 3c).

To further examine the significance of APC-CK2 interactions, we analyzed the catalytic activity of CK2 in the presence of APC. Also, we observed the inhibition of CK2 activity in the presence of APC immunoprecipitates from normal cells consisting of full-length APC. We therefore prepared recombinant APC fragments overlapping the 300 kDa to verify the effects of APC itself. As a result, C-terminal fragments of APC exhibited strong inhibition toward CK2 activity depending on the fragment concentrations [19]. This result indicates large differences in the ability of full-length

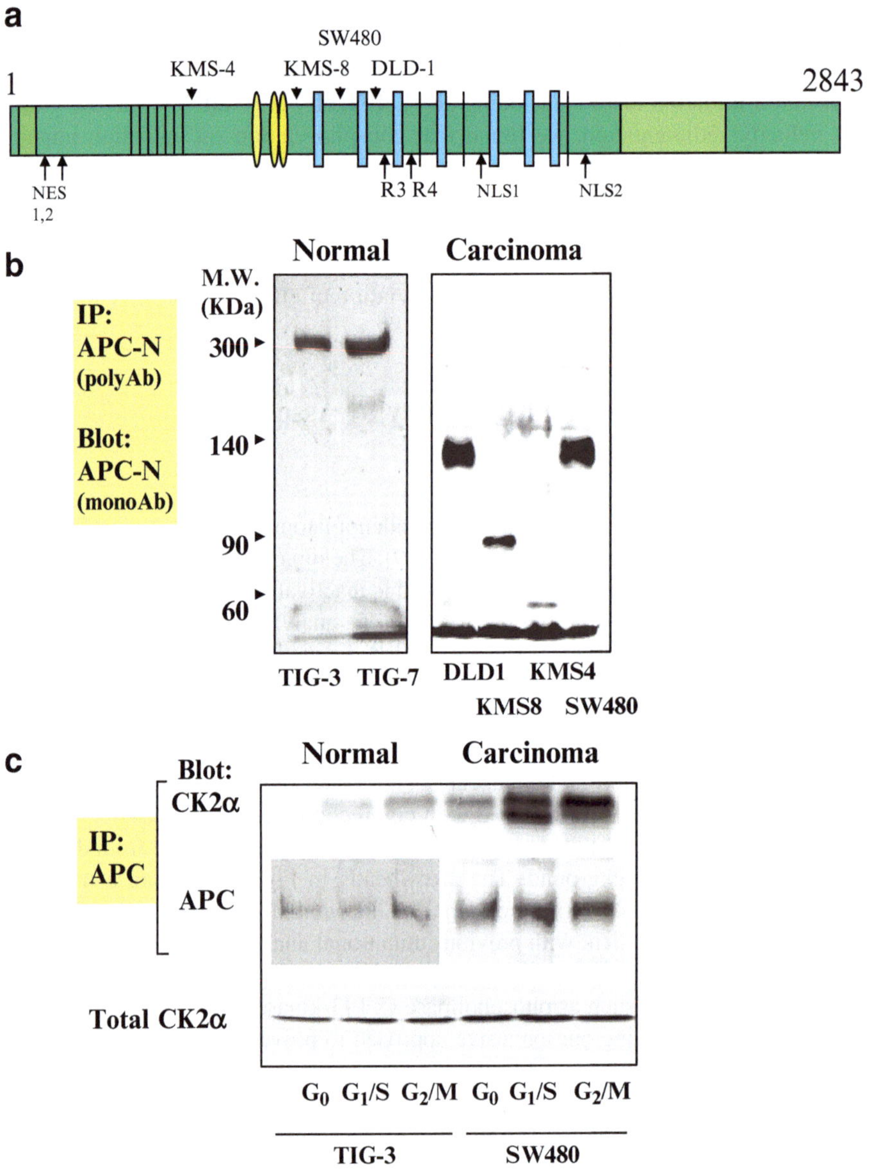

Fig. 3 Adenomatous polyposis coli protein associates with CK2 in the cells. (**a**) Key domains of the APC protein: a causative gene for familial adenomatous polyposis. (**b**) Detection of full-length and truncated forms of APC protein by Western blot following immunoprecipitation by polyclonal anti-APC antibodies raised against N-terminal amino acids in APC. Normal human fibroblast TIG-3, TIG-7 cells, colorectal adenoma cell lines DLD-1, SW480 cells, and colorectal carcinoma cells KMS-4, KMS-8. (**c**) A cell cycle-dependent association of APC with CK2 observed both in normal TIG cells and SW480 cells. Two cell lines were synchronized separately, and cell populations arrested in each phase of the cell cycle as indicated were immunoprecipitated by anti-APC antibodies. CK2 proteins in each immunoprecipitates were detected by Western blot using anti-CK2 antibodies [19]

and truncated forms of APC to regulate CK2 activity. In other words, the inhibitory effect of APC toward CK2 activity may be absent in carcinoma cells which lack the APC-C-ter, although the two molecules could interact in vivo. To further investigate the effect of the interaction between APC and CK2, cells expressing this region of APC were monitored in growth and transformation assays. HEK293 cells expressing the C-terminal fragment of APC strongly suppressed growth rates, increasing the doubling times by twofold compared to control cells [20]. The potential significance of the interaction between APC and CK2 was then verified for regulating cell proliferation. We suggest a model in which interactions of APC with CK2 perturb cell proliferation as well as cellular transformation as described [20].

4 CK2-Associating Proteins in the Nuclear Fraction Following the Serum Stimulation

4.1 Spatiotemporal Organization of CK2 in Response to the External Stimuli

Recent analysis of the spatiotemporal organization of individual CK2 subunits in living cells has indicated that they are dynamic and that they are involved in different multimolecular assemblies [21]. For the first step, in order to understand the cellular function of CK2 as a tetrameric complex in the nuclear fraction, we conducted immunoprecipitation of the CK2 complex using the immobilized polyclonal antibodies against CK2α which recognize both monomeric and tetrameric enzyme followed by conventional two-dimensional gel electrophoresis and mass spectrometry analysis.

4.2 Conventional Methods by 2D Gel Development Combined with Mass Spectrometry Analysis

For synchronization experiments, logarithmically growing TIG cells were starved in 0.1 % FBS for 48 h (G_0) and then cultured in fresh media containing 10 % FBS for an additional 3 h to obtain cell populations enriched in the G_1 phase of the cell cycle. The two sets of cell populations were harvested, fractionated into the cytosol and nuclear fraction according to the methods [22], and then immunoprecipitated using anti-CK2 polyclonal antibodies. Briefly, after the removal of the cytosolic fraction, nuclear proteins were solubilized by 0.6 M NaCl in 20 mM Hepes buffer (pH 7.4), containing 25mM β-glycerophosphate, 20 % glycerol, 1 mM DTT, 2 mM EDTA, 1 μg/ml aprotinin, 1 μg/ml leupeptin, and 1mM PMSF. About 30 μg proteins in the CK2 immunoprecipitates was processed for isoelectric focusing using IPTG gel strips (*p*I 4–7, 18 cm, GE Healthcare) followed by 10 % SDS-PAGE. 2D gels were silver stained following methanol fixation (Silver Stain MS Kit, Wako, Japan). As shown in Fig. 4a, gels were analyzed to detect changes in protein spot intensities

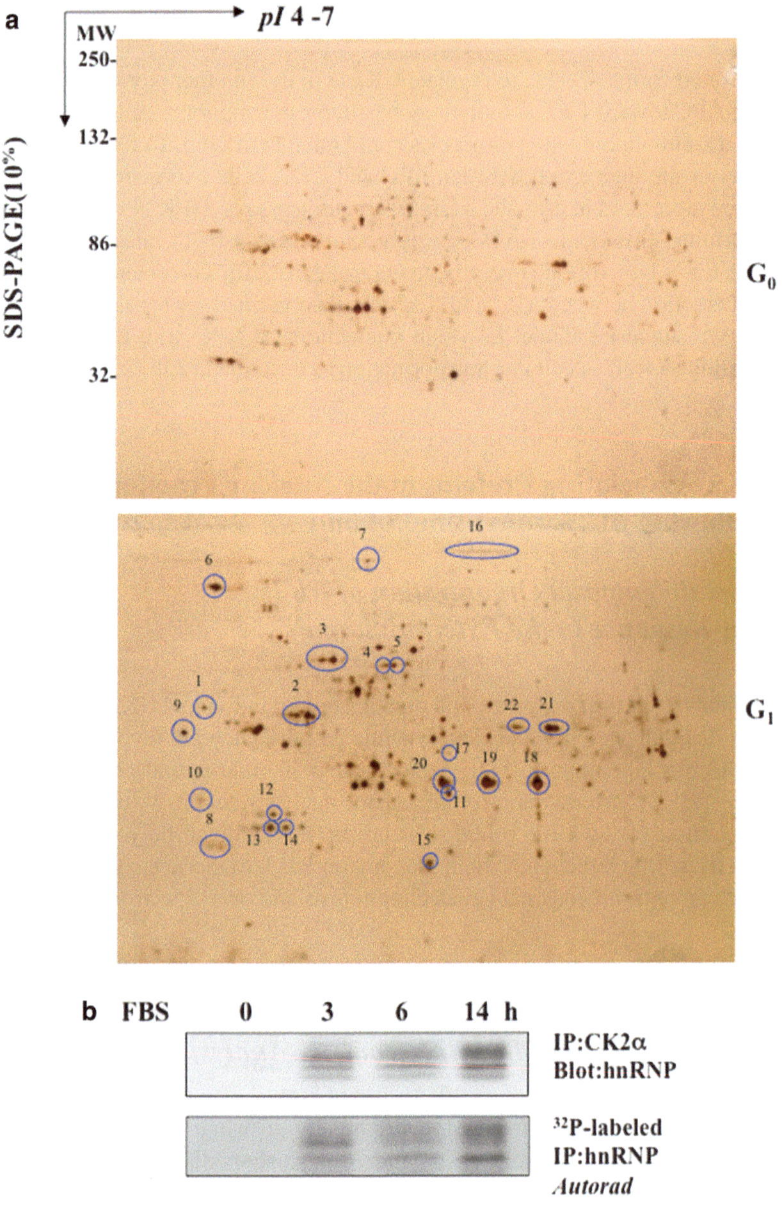

Fig. 4 Cell cycle-dependent association of nuclear proteins with CK2. (**a**) 2D gel images showing CK2-associating nuclear proteins in synchronized TIG cells. Serum-starved TIG cells (G_0, *upper*) were stimulated with FBS for 3 h (G_1, *lower*). The nuclear fraction was extracted and immunoprecipitation was done by anti-CK2α antibodies. Proteins were equilibrated with Pharmalite (pH 4–7) and then applied on two-dimensional gel electrophoresis. Proteins were visualized by silver stain reagent. The primary sequence of proteins marked in circles was identified by mass spectrometry as shown in Table 2. (**b**) hnRNP was associated with CK2 in the nuclear fraction in TIG cells, as shown by immunoprecipitation followed by blotting with anti-hnRNP antibody (*upper*, stimulated without or with FBS for 3, 6, and 18 h). hnRNPs in CK2 complex were shown to be phosphorproteins as determined by ^{32}P-labeling (*lower*) followed by immunoprecipitation with anti-hnRNP. (**c**) NanoLC mass spectrometry analysis identified phosphopeptides from hnRNPH1 and hnRNPA1. Product ion mass spectrum corresponds to the peptide sequence HTGPNSPDTANDGFVR derived from hnRNPH1 (*upper*) and SESPKEPEQLRK from hnRNPA1 (*lower*), respectively

c

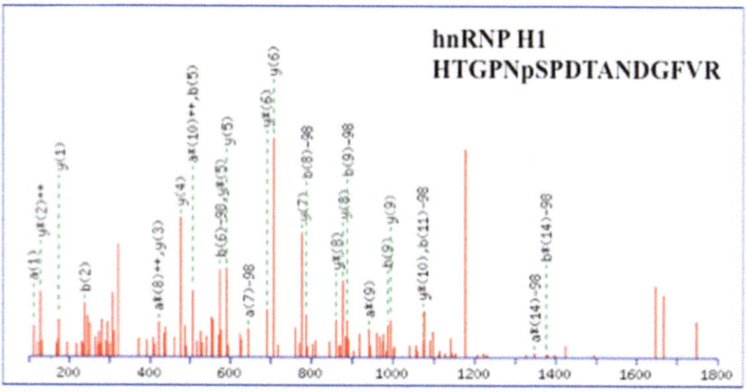

B	B Ions	B+2H	B-NH3	B-H2O	AA	Y Ions	Y+2H	Y-NH3	Y-H2O	Y
1	138.1	69.5			H	1,764.7	882.9	1,747.7	1,746.7	16
2	239.1	120.1		221.1	T	1,627.7	814.3	1,610.7	1,609.7	15
3	296.1	148.6		278.1	G	1,526.6	763.8	1,509.6	1,508.6	14
4	393.2	197.1		375.2	P	1,469.6	735.3	1,452.6	1,451.6	13
5	507.2	254.1	490.2	489.2	N	1,372.6	686.8	1,355.5	1,354.5	12
6	674.2	337.6	657.2	656.2	S+80	1,258.5	629.8	1,241.5	1,240.5	11
7	771.3	386.1	754.3	753.3	P	1,091.5	546.3	1,074.5	1,073.5	10
8	886.3	443.7	869.3	868.3	D	994.5	497.7	977.4	976.5	9
9	987.4	494.2	970.3	969.4	T	879.4	440.2	862.4	861.4	8
10	1,058.4	529.7	1,041.4	1,040.4	A	778.4	389.7	761.4	760.4	7
11	1,172.4	586.7	1,155.4	1,154.4	N	707.4	354.2	690.3	689.3	6
12	1,287.5	644.2	1,270.4	1,269.5	D	593.3	297.2	576.3	575.3	5
13	1,344.5	672.8	1,327.5	1,326.5	G	478.3	239.6	461.3		4
14	1,491.6	746.3	1,474.5	1,473.5	F	421.3	211.1	404.2		3
15	1,590.6	795.8	1,573.6	1,572.6	V	274.2	137.6	257.2		2
16	1,764.7	882.9	1,747.7	1,746.7	R	175.1	88.1	158.1		1

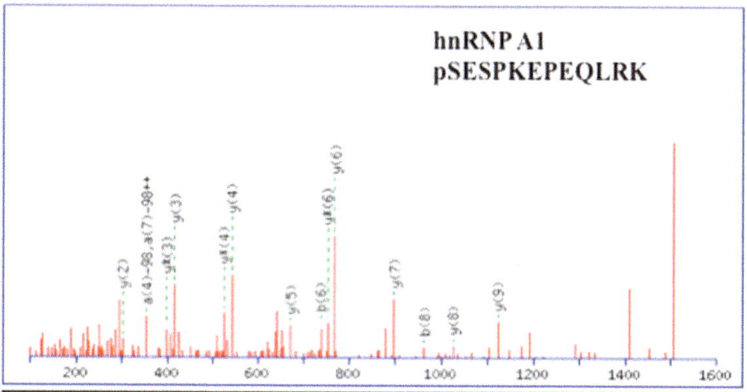

B	B Ions	B+2H	B-NH3	B-H2O	AA	Y Ions	Y+2H	Y-NH3	Y-H2O	Y
1	168.0	84.5		150.0	S+80	1,507.7	754.4	1,490.7	1,489.7	12
2	297.1	149.0		279.1	E	1,340.7	670.9	1,323.7	1,322.7	11
3	384.1	192.5		366.1	S	1,211.6	606.3	1,194.6	1,193.6	10
4	481.1	241.1		463.1	P	1,124.6	562.8	1,107.6	1,106.6	9
5	609.2	305.1	592.2	591.2	K	1,027.6	514.3	1,010.6	1,009.6	8
6	738.3	369.6	721.2	720.3	E	899.5	450.3	882.5	881.5	7
7	835.3	418.2	818.3	817.3	P	770.5	385.7	753.4	752.4	6
8	964.4	482.7	947.3	946.4	E	673.4	337.2	656.4	655.4	5
9	1,092.4	546.7	1,075.4	1,074.4	Q	544.4	272.7	527.3		4
10	1,205.5	603.3	1,188.5	1,187.5	L	416.3	208.7	399.3		3
11	1,361.6	681.3	1,344.6	1,343.6	R	303.2	152.1	286.2		2
12	1,507.7	754.4	1,490.7	1,489.7	K	147.1	74.1	130.1		1

Fig. 4 (continued)

on the gel images using a Melanie III Viewer. In total, 22 protein spots with changing intensities of more than twofold in each of the three biological replicate samples. These proteins excised from wet gels were destained with a solution containing $Na_2S_2O_3$ according to the manufacturer's protocol, digested with porcine trypsin (100 ng, Promega), and desalted on C18 ZipTips (Millipore). The resulting peptides were analyzed by tandem mass spectrometer (Q-TOF electrospray ionization mass spectrometer, QSTAR, Applied Biosystems).

Nuclear protein analysis derived from G_0 and G_1 cells revealed major changes in matrix proteins involved in multipotent nuclear functions such as vimentin, tubulin, nuclear membrane component lamin-B, histone-binding protein RBBP4 for the exchange of H3-H4 complex and the assembly of new nucleosomes, hnRNPC1/C2 and hnRNPH1 involved in RNA metabolism, and TAR DNA-binding protein, which were shown to be associated with CK2 and appeared more in G_1 cells compared with G_0 during the progression of the cell cycle. These proteins are listed in Table 2 according to their scores obtained from mass spectrometry analysis. As one of the CK2-associated proteins, we analyzed heterogeneous RNA-binding proteins, known as RNA splicing factors involved in RNA metabolism, for its in vivo binding to CK2 and for its cell cycle-associated phosphorylation state. As shown in Fig. 4b, hnRNP initiated the interaction with CK2 in the nuclear fraction when G_0 cells were stimulated with FBS to progress into G_1. Phosphorylation levels of hnRNP associating with CK2α were also increased in response to growth stimulation by FBS. According to a separate experiment using nanoLC mass spectrometry, performed as described in the following section, Fig. 4c shows a typical mass spectrum generated after TiO_2-enrichment of phosphopeptides from trypsin-digested CK2-associating proteins, hnRNPA1 and hnRNPH1, and their in vivo phosphorylation. The spectrum in Fig. 4c *lower panel* significantly matched the trypsin-generated peptide of hnRNPA1 by 99 % confidence, showing the addition of 79.96633 at the first serine position of a peptide SESPKEPEQLRK corresponding to Ser4 of hnRNPA1. This phosphorylation site completely matches the predicted phosphorylation site by CK2, suggesting that CK2α may be involved in nuclear function as an upstream kinase or as an indispensable binding protein for RNA transport and RNA processing, for the progression of the cell cycle in mammalian cells.

4.3 Integrated Proteomics Technology

Recent mass spectrometry has emerged as a key technology for screening protein posttranslational modifications including phosphorylation and also clarifying each of the components in a multiprotein complex in the cells [23–27]. In order to characterize CK2-dependent pathways associated with the cell cycle, we performed large-scale studies of complex samples using a nanoLC mass spectrometry approach. CK2-associating proteins from nuclear fractions were prepared from logarithmically growing TIG cells which were serum starved or arrested in G_0 phase and then cultured synchronously to progress through the cell cycle by the addition

Table 2 Proteins associating with CK2 identified by 2D gel

Spot no.	NCBI	Access no.	Names	MW	p*I*	% Cov.	Total no. peptides
1	NP_003371	P08670	Vimentin	53.6	5.05	27	13
2	NP_001257328	Q71U36	Tubulin alpha-1A chain	50.1	4.94	29	11
3	NP_005564	P20700	Lamin-B1	66.1	5.11	22	11
4	NP_116126	Q03252	Lamin-B2	67.3	5.29	13	7
5	NP_116126	Q03252	Lamin-B2	67.3	5.29	13	7
6	NP_003290	P14625	Heat-shock protein 90	92.5	4.76	20	14
7	NP_001839	P12109	Collagen alpha-1(VI) chain	108.5	5.26	8	7
8	NP_000357	P09493	Tropomyosin alpha-1 chain	32.7	4.69	16	4
9	NP_001128727	Q09028	Histone-binding protein RBBP4	47.6	4.74	12	4
10	NP_001012321	P08865	40 S ribosomal protein SA	32.8	4.79	12	3
11	NP_001003790	O94905	Erlin-2	37.8	5.47	10	3
12	NP_001070910	P07910	HNRNP C1/C2	33.6	4.95	12	3
13	NP_001070910	P07910	HNRNP C1/C2	33.6	4.95	12	3
14	NP_001070910	P07910	HNRNP C1/C2	33.6	4.95	12	3
15	NP_001193469	P47756	F-actin-capping protein	30.6	5.69	8	2
16	NP_001181883	P43243	Matrin-3	94.5	5.87	4	3
17	NP_003356	P31930	Cytochrome b-c1 complex	52.6	5.94	4	2
18	NP_031401	Q13148	TAR DNA-binding protein 43	44.7	5.85	28	10
19	NP_031401	Q13148	TAR DNA-binding protein 43	44.7	5.85	25	10
20	NP_031401	Q13148	TAR DNA-binding protein 43	44.7	5.85	13	6
21	NP_005511	P31943	HNRNP H	49.2	5.89	28	7
22	NP_005511	P31943	HNRNP H	49.2	5.89	18	6

of FBS for 3 h. Immunoprecipitations were performed using immobilized-normal IgG or anti-CK2 antibodies that were conjugated to agarose beads, and proteins in the complex were eluted by glycine-HCl, pH 2.5, followed by neutralization by 2 M Tris solution to pH 8.0. These proteins were digested each with 10 ng/μl trypsin in 20 mM ammonium bicarbonate, added to the mixture three times during incubation at 37 °C for overnight. These digests were followed by peptide extraction with 50 % acetonitrile containing 0.1 % acetic acid (Fig. 5). The peptide solutions were analyzed by TripleTOF 5600 mass spectrometry (ABSciex). The raw data files were analyzed by ABSciex MS Data Converter to create peak lists on the basis of the fragmentation spectra. Peptides and proteins were identified by Mascot (Matrix Science, UK) against IPI human database with a precursor mass tolerance of 20 ppm, a fragment ion mass tolerance of 0.02 Da, and strict trypsin specificity allowing for up to two missed cleavages. We identified 140 interacting proteins by 95 % confidence as components of the nuclear CK2 as listed in Table 3. Those proteins cover diverse nuclear functions including nuclear matrix proteins, vimentin, centrin, calnexin, and KIAA; RNA-binding protein, RBM15, hnRNP, splicing factor TRA2B, and RNA helicase DDX21; nucleosome function, H1, H4, and H2B; and other functions, SRRM2, blc2-associated protein, and BAZ1A. Comparing these results with those obtained from the 2D gel analysis as shown in Table 2, we demonstrate three common proteins, vimentin, lamin-B1, and hnRNP, which were observed in both analyses. In addition, as described in the previous section, we validated the interaction of CK2 with one of the newly identified associating protein hnRNP in more detail, demonstrating phosphorylation at Ser 104 of hnRNPH1 and at Ser4 of hnRNPA1, respectively (Fig. 4c). Our analysis revealed numerous novel interactions and thus provides a valuable resource for understanding how this kinase signaling pathway affects different cellular functions. Intriguingly, more than 20 % of CK2-interacting proteins were constituted by DNA- and RNA-binding proteins, suggesting the involvement of CK2 function as a dynamic regulator for gene function in the nucleus which may be associated with cell cycle progression. These proteomics technologies could clearly identify the molecular composite constitution surrounding the kinase in the cell. We recently demonstrated an association of nuclear and cytosolic proteins with CK2 occurring in a cell cycle-dependent manner, suggesting that possible binding partner switching might occur following the progression of the cell cycle (collaborative work with Dr. Oyama M, manuscript in preparation, [28]).

4.4 *Phosphoproteins Associating with CK2 in the Nucleus*

The identification and characterization of the in vivo substrates of an individual protein kinase is important for understanding its cellular functions in physiological states such as proliferation and pluripotency [29]. To clarify the possibility that CK2-binding proteins may be regulated by phosphorylation, we performed phosphopeptide enrichment by TiO_2 of proteolyzed immunoprecipitates from TIG cells.

a

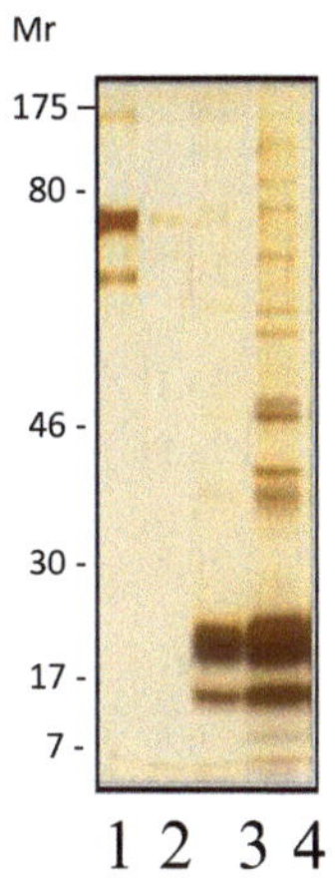

b

Normal human fibroblast TIG cells, 44 plates
Arrested in G_0 and synchronously progressed in G_1

↓

Lysed cells

↓

Nuclear fraction

↓

Immunoprecipitated with anti-CK2α antibodies

↓

Eluted CK2 complex by glycine-HCl
Adjusted pH to 8.8

↓

Proteolysed with trypsin and desalted by TiO_2 column

↓

Applied on nanoLC Triple TOF^{TM} 5600 system

c

ID statistics (Protein-Thresfolded):
2101 total spectra, 2101 non-empty spectra; 178740 proteins searched

Unused (Confidence) Cutoff	Proteins detected	Proteins before grouping	Distinct peptides	Spectra identified	% Total spectra
>1.3 (95)	140	629	437	686	32.7
Cutoff applied: >0.05	157	679	441	693	33.0

The number of spectrum obtained as phosphorylated peptides

99 % confidence	46	**FDR 1 % Phosphopeptide**	306
95 % confidence	283	**FDR 1 % Confident Peptide**	356

Fig. 5 Assessing the components of the CK2 complex and their phosphorylation statues. (**a**) Purification of CK2-associating proteins from nuclear fractions. TIG cells synchronously progressed through the cell cycle in early G_1 phase were fractionated to obtain nuclear fractions. Immunoprecipitations were performed using normal IgG (*lane* 3) or anti-CK2 antibodies (*lane* 4). Proteins in the complex were eluted by glycine-HCl, pH 2.5 followed by neutralization by 2 M Tris solution, and an aliquot of each eluate was applied on a 10 % SDS-PAGE, visualized by silver-staining kit. BSA was applied on *lane* 1(100 ng) and *2* (10 ng). (**b**) Analytical conditions of preparing sample for mass spectrometry are indicated on the *right panel*. (**c**) Mass spectrometry identification of CK2-associating proteins is listed in Tables 3 and 4. ID statistics and the number of spectrum derived from mass analysis data are indicated as separate inserts

Table 3 Proteins associating with CK2 identified by nanoLC mass spectrometry

No.	Access no.	Names	Unused cutoff	% Cov.	Peptides
1	782992.3	Serine/arginine repetitive matrix protein 2	61.53	14.8	38
2	886854.1	BCL2-associated transcription factor 1 isoform 2	28	12.1	15
3	418471.6	VIM vimentin	24.3	34.1	14
4	647720.1	Serine/arginine repetitive matrix protein 1	23.94	22.9	15
5	930226.1	Highly similar to actin, cytoplasmic 2	16.23	29.8	12
6	217467.3	HIST1H1E histone H1.4	13.93	31.1	8
7	883896.1	LIM domain and actin-binding protein 1 isoform a	11.35	9.6	10
8	104050.3	Thyroid hormone receptor-associated protein 3	10.2	4.9	5
9	12199.1	Coiled-coil domain-containing protein 86	9.97	19.4	5
10	969501.1	TCOF1, treacle protein isoform f	8	4.3	5
11	453473.6	HIST2H4A	8	48.5	4
12	220717.2	RNA-binding protein 15	8	6.5	4
13	215928.4	Centrin-2	8	10.5	4
14	18370.4	Isoform 2 of supervillin	8	3.4	5
16	909288.1	HNRNP A1	6	5.5	4
17	607584.1	Myb-binding protein 1A	6	2.6	5
18	515061.3	HIST1H2BJ histone H2B type 1-J	6	23	3
19	220158.1	Alpha-adducin	4.26	4.8	2
20	556514.1	RNA-binding motif protein 14 variant (Fragment)	4.03	4.9	2
21	412415.2	BAZ1A	4	1.4	2
22	515115.3	ABLIM1, putative uncharacterized protein ABLIM1	4	9	2
23	914995.1	A-kinase anchor protein 2	4	5.2	2
24	555647.4	TRA2B splicing factor variant	4	17.7	2
25	13933.2	Desmoplakin	4	1.6	3
26	954159.1	AHCTF1P1	4	2.1	2
27	953077.1	CTNNA1, putative uncharacterized protein	4	4.2	2
28	942853.1	RNA-binding motif protein 39	4	6.8	2
29	872909.1	Ribosomal RNA methyltransferase NOP2	4	5.1	2
30	796848.1	SRSF2, arginine-/serine-rich splicing factor	4	20.6	2
31	290952.7	RRP1B, ribosomal RNA processing protein 1	4	5.1	2
32	941747.1	Calnexin	4	5.7	2

(continued)

Table 3 (continued)

No.	Access no.	Names	Unused cutoff	% Cov.	Peptides
33	895905.3	TP53 cellular tumor antigen p53	4	3.7	2
34	789281.2	KIAA1429	4	1.7	2
35	477179.1	DDX21, nucleolar RNA helicase 2	4	3.9	2
36	215879.1	SRSF6, serine-/arginine-rich splicing factor 6	4	6	4
37	5614.6	SPTBN1, spectrin beta chain	4	1.6	3
38	965597.1	DBN1	3.7	5	3
39	45550.4	PPP1R9B neurabin-2	3.66	3.4	2
40	329547.3	Zinc finger domain-containing protein 13	3.12	3.5	2
41	966829.1	ALB 69 kDa protein	2.39	13.4	4
42	879810.1	SPTAN1	2.37	1.4	1
43	922722.1	Periphilin-1	2.21	6.1	2
44	642213.1	RNA-binding protein	2.02	11.8	1
45	908725.1	ZC3H18	2	4.9	2
46	514411.2	RSF1, spacing factor 1	2	3.1	1
47	937545.1	PDS5B, sister chromatid cohesion protein PDS5	2	2.6	1
48	893035.1	CAD	2	1.2	1
49	872028.2	NUMA1	2	2.2	1
50	845355.1	ATRX	2	1.4	1
51	646374.1	PRPF38B, 52 kDa protein	2	9.4	1
52	644087.1	LMNA progerin	2	5.2	1
53	640210.2	MLLT4 afadin isoform 2	2	1.8	1
54	420096.4	Plectin	2	0.7	1
55	291607.2	ITPR3, inositol 1,4,5-trisphosphate receptor type 3	2	1.2	1
56	291032.7	Cytospin-B	2	3.8	1
57	32342.4	TRIP12 protein	2	1.8	1
58	6103.1	CD2BP2	2	13.8	1
59	953543.1	PELP1	2	2.1	1
60	953142.1	ZC3H14	2	3.6	1
61	946415.1	ACIN1, apoptotic chromatin condensation inducer	2	1.7	1
62	910135.1	AFAP1, actin filament-associated protein 1 isoform a	2	3.2	1
63	844406.1	Arginine- and glutamate-rich protein 1	2	9.9	1
64	830039.1	Splicing factor U2AF 65 kDa subunit	2	3	1
64	655922.1	Chromodomain Y-like protein	2	3.3	1
65	642195.2	Zinc finger MYM-type protein 4	2	1.4	1
66	477330.3	Rho GTPase-activating protein 11A	2	2.2	1
67	436632.1	Nipped-B-like protein	2	0.7	1

(continued)

Table 3 (continued)

No.	Access no.	Names	Unused cutoff	% Cov.	Peptides
68	306301.2	Pyruvate dehydrogenase E1 alpha 1 isoform 2	2	5.8	2
69	292817.6	KIAA1462	2	1.8	1
70	218500.6	Histone-lysine N-methyltransferase MLL	2	0.7	1
71	217975.4	Lamin-B1	2	2.7	1
72	216631.1	Beta-2-syntrophin	2	10.5	1
73	175136.5	RBM15B, RNA-binding protein 15B	2	2.5	1
74	31627.4	DNA-directed RNA polymerase II subunit RPB1	2	2.1	1
75	8868.3	MAP1B	2	1.2	1
76	8557.6	Insulin-like growth factor 2 mRNA-binding protein 1	2	3.8	2
77	968132.1	HNRNPH1	2	9.4	1
78	968092.1	HNRNPD	2	19.6	1
79	967721.1	SNHG4	2	1.8	1
80	965804.1	CAP1 47 kDa protein	2	4.2	1
81	954039.1	Disabled homolog 2	2	1.6	1
82	953619.1	CLASRP	2	1.6	1
83	953308.1	PCM1	2	0.8	1
84	945115.1	Rho guanine nucleotide exchange factor 7	2	1.4	2
85	943523.1	Centrosomal protein of 57 kDa	2	3.1	1
86	942431.1	Epidermal growth factor receptor substrate 15-like 1	2	2.3	1
87	941257.2	Serine/threonine-protein kinase D2	2	1.6	1
88	940614.1	PDCD4 50 kDa protein	2	2.2	1
89	940224.1	UPF0606 protein C11orf41	2	0.5	1
90	940183.1	PKP4	2	2.6	1
91	939811.1	Ribosomal protein S3	2	7	1
92	929732.1	FCHO2	2	1.2	1
93	927445.1	UAP56-interacting factor	2	10.1	1
94	927389.1	RCC1	2	4.8	1
95	926274.2	CALD1	2	2.8	1
96	924436.1	HSPB1	2	5.4	2
97	918003.1	SSFA2	2	0.6	1
98	915466.1	EPN2	2	2.8	1
99	915253.1	RNA-binding protein 10	2	1.4	1
100	910164.1	SLC16A1	2	4.4	1
101	909184.1	WDR77	2	11.6	1
102	909030.1	KTN1	2	1.3	1

(continued)

Table 3 (continued)

No.	Access no.	Names	Unused cutoff	% Cov.	Peptides
103	892704.1	PUM2	2	3.4	1
104	879287.1	CDCA5	2	3.3	1
105	873776.1	MORC2	2	0.9	1
106	872788.1	DPYSL3, dihydropyrimidinase-related protein 3	2	1.9	1
107	871240.2	RBM16	2	1.5	1
108	868816.1	RPL12P38	2	7.7	1
109	855962.1	FK506-binding protein 15	2	1.7	1
110	796446.1	KDM2B	2	3.2	1
111	795990.1	DKFZp686A1195	2	0.9	1
112	794894.1	RPL23A protein	2	5.1	1
113	747462.2	EPB41L2	2	4.1	1
114	646643.1	SRSF10	2	3.8	1
115	643027.1	LARP1	2	1.5	1
116	514439.5	POGZ	2	1.4	1
117	477661.1	E3 ubiquitin-protein ligase ZFP91	2	3.5	1
118	465457.4	Cyclin-Y	2	3.2	1
119	411937.4	NOP56	2	3	1
120	410716.1	Bromodomain-containing protein 3	2	2	1
121	396104.2	Male-specific lethal 3 homolog	2	2.4	1
122	386908.3	Alpha-parvin	2	8.8	1
123	376222.1	YTH domain-containing protein 1	2	1.3	1
124	304932.2	Ribosomal RNA-processing protein 8	2	3.5	1
125	302829.5	Retinoblastoma-associated protein	2	1.2	1
126	293746.2	Multiple myeloma tumor-associated protein 2	2	6.1	1
127	290204.1	U1 small nuclear ribonucleoprotein 70 kDa	2	3	1
128	218850.5	Secretory carrier-associated membrane protein 2	2	4	1
129	216047.3	SWI/SNF complex subunit SMARCC2	2	1.2	1
130	170744.1	Transcription factor Elf-4	2	2	1
131	106955.3	C11orf84	2	3.4	1
132	30968.4	C9orf142	2	5.9	1
133	29081.3	DNA ligase 3	2	1.2	1
134	21924.1	H1FX, histone H1x	2	5.2	1
135	11857.1	Chromatin assembly factor 1 subunit B	2	2.9	1
136	9146.4	TRAF-type zinc finger domain-containing protein 1	2	2.2	2

(continued)

Table 3 (continued)

No.	Access no.	Names	Unused cutoff	% Cov.	Peptides
137	8200.2	YEATS domain-containing protein 2	2	1.3	1
138	6937.2	Ubiquitin-conjugating enzyme E2 J1	2	4.7	1
139	2564.3	DNA repair protein XRCC1	2	3.5	3
140	816.1	14-3-3 protein epsilon	2	4.3	1

For the measurements on the mass spectrometer, the peptides were separated using Eksigent nanoLC System and an HiQSil C18 analytical column (particle size 3 μm, bed length 15 cm, 100 Å, KYA Tech. USA). The peptides were eluted at a flow rate of 300 nl/min with formation of a solvent gradient of buffer A and B (1–2 % buffer B in 1 min, 2–25 % B in 59 min, 25–40 % B in 10 min, 40–100 % B in 5 min). The MS was operated in data-dependent mode to automatically switch between Triple Q and LTQ-MS/MS acquisition. Survey full scan MS spectra (from *m/z* 400 to 1,500) were acquired with resolution R=60,000. Ion selection threshold was 1,000 counts for MS/MS; an activation Q-value of 0.25 and activation time of 30 ms were also applied for MS/MS. Protein identification was performed by automated database searches using the Sequest algorithm rel. 27.11 (Sorcerer built 4.04, Sage-N Research Inc., Milpitas, CA) and the Swiss-Prot database rel. 57.15 (forward-reverse, limited to 32,448 mouse entries). The considered enzyme specificity was fully tryptic allowing two missed cleavages. Parent and fragment mass tolerance (MS) were set to 10 ppm and 1 Da, respectively. Methionine oxidation and carbamidomethylation (on C) were considered as optional modification. Protein identification was based on (a) peptide thresholds, with a 95.0 % confidence minimum (Sequest: ΔCn scores of greater than 0.4 and XCorr scores of greater than 2.0 or 3.5 for doubly and triply charged peptides, respectively), and (b) protein thresholds, with 95.0 % confidence

Table 4 shows the top 102 phosphopeptides according to their peptide scores obtained from CK2-associating proteins. In total, a 172 phosphopeptides from 42 proteins were identified by this study at 99% confidence level. Three hundred and six phosphopeptides and 317 phospho-amino acids were identified as 1 % false discovery rate (FDR) including 291 sites at serine and 26 sites at threonine. This indicates that the sites detected from our nuclear preparation were exclusively serine or threonine residues. Proline-directed phosphorylation sites, Pro at the +1 position, and acidophilic sites, a Glu/Asp at position +3, accounted for 72.5 % of all detected phosphorylation. This proteomic analysis disclosed many novel phosphorylation sites of CK2-associating proteins in the early G_1 phase nucleus including matrix proteins SRRM2, vimentin, and supervillin; bcl2-associated transcription factor 1; nucleosomal protein histone H1.4, centrin-2; RNA-binding proteins hnRNPA1, hnRNPH1, RBM15, RBM14, RBM39, and DDX21; ribosomal RNA processing protein RRP1B; retinoblastoma-associating protein RB1; zinc finger containing protein ZC3H13; and adaptor protein AKAP2.

These proteins identified by our study were classified by the gene ontology database DAVID, and the top twenty proteins are listed in Table 5 according to their P-values. These functional annotation data clearly suggest the role of CK2 for dynamic gene regulation: RNA processing, RNA metabolism, nucleosome assembly,

Table 4 Phosphorylated peptides identified by nanoLC mass spectrometry

No.	Access no.	Names	Unused cutoff	Sequence	Modifications	Conf
1	782992.3	Serine/arginine repetitive matrix protein 2	61.53	AQTPPGPSLSGSK	Phospho(T)@3	99
1			61.53	AQTPPGPSLSGSKSPCPQEK	Phospho(S)@14	99
1			61.53	DRSPPKSPEKLPQSSSSESSPPSPQPTK	Phospho(S)@3; Phospho(S)@17	99
1			61.53	GEGDAPFSEPGTTSTQRPSSPETATK	Phospho(S)@20	99
1			61.53	HASSSPESPKPAPAPGSHR	Phospho(S)@8	99
1			61.53	MALPPQEDATASPPRQK	Phospho(S)@12	99
1			61.53	NHSGSRTPPVALNSSR	Phospho(S)@5; Phospho(T)@7	99
1			61.53	SGSSQELDVKPSASPQER	Phospho(S)@4	99
1			61.53	SLSGSSPCPK	Phospho(S)@1	99
1			61.53	SPSVSSPEPAEK	Phospho(S)@3	99
1			61.53	SRSGSSQELDVKPSASPQER	Phospho(S)@3	99
1			61.53	SRTPLLPR	Phospho(S)@1	99
1			61.53	SSGHSSSELSPDAVEK	Phospho(S)@5	99
1			61.53	SSGHSSSELSPDAVEK	Phospho(S)@6	99
1			61.53	SSSPVTELASR	Phospho(S)@3	99
1			61.53	SSSPVTELASRSPIR	Phospho(S)@12	99
1			61.53	SVSPCSNVESR	Phospho(S)@3	99
1			61.53	AQTPPGPSLSGSKSPCPQEK	Phospho(T)@3; Phospho(S)@14	99
1			61.53	CRSPGMLEPLGSSR	Phospho(S)@3	99
1			61.53	DRSPPKSPEKLPQSSSSESSPPSPQPTK	Phospho(S)@3; Phospho(S)@7	99

(continued)

Table 4 (continued)

No.	Access no.	Names	Unused cutoff	Sequence	Modifications	Conf
1			61.53	GEFSASPMLK	Phospho(S)@6	99
1			61.53	HASSSPESPKPAPAPGSHR	Phospho(S)@4	99
1			61.53	HSCSGSSPPR	Phospho(S)@7	99
1			61.53	NHSGSRTPPVALNSSR	Phospho(S)@5	99
1			61.53	QGSITSPQANEQSVTPQRR	Phospho(T)@15	99
1			61.53	RGEGDAPFSEPGTTSTQRPSSPETATK	Phospho(S)@21	99
1			61.53	RSLSGSSPCPK	Phospho(S)@2	99
1			61.53	SGTPPRQGSITSPQANEQSVTPQRR	Phospho(S)@1; Phospho(S)@19	99
1			61.53	SLTRSPPAIR	Phospho(S)@1; Phospho(S)@5	99
1			61.53	SRSGSSQELDVKPSASPQER	Phospho(S)@5	99
1			61.53	SRSPLAIR	Phospho(S)@1	99
1			61.53	TPPVALNSSR	Phospho(T)@1	99
1			61.53	VKAQTPPGPSLSGSK	Phospho(T)@5	99
1			61.53	VSGRTSPPLLDR	Phospho(S)@2	99
2	886854.1	Bcl-2-associated TF1	28	IDISPSTLR	Phospho(S)@4	99
2			28	KAEGEPQEESPLKSK	Phospho(S)@10	99
2			28	QKSPEIHR	Phospho(S)@3	99
2			28	RIDISPSTLR	Phospho(S)@5	99
2			28	RIDISPSTLR	Phospho(S)@7	99
2			28	STFREESPLR	Phospho(S)@7	99
2			28	EKSTFREESPLR	Phospho(S)@9	99
2			28	FDYSPPLHK	Phospho(S)@4	99
2			28	NTPSQHSHSIQHSPER	Phospho(S)@9	99
2			28	TPSQHSHSIQHSPER	Phospho(S)@12	99

2			28	YSPSQNSPIHHIPSR	Phospho(S)@7	99
3	418471.6	VIM vimentin	24.3	LRSSVPGVR	Phospho(S)@4	99
3			24.3	SLYASSPGGVYATR	Phospho(S)@5	99
3			24.3	SLYASSPGGVYATR	Phospho(S)@6	99
3			24.3	YASSPGGVYATR	Phospho(S)@4	99
4	647720.1	Serine/arginine repetitive matrix protein 1	23.94	EKTPELPEPSVK	Phospho(T)@3	99
4			23.94	HRPSPPATPPPK	Phospho(S)@4; Phospho(T)@8	99
4			23.94	HRPSPPATPPPK	Phospho(T)@8	99
4			23.94	KVELSESEEDKGGK	Phospho(S)@5	99
4			23.94	RLSPSASPPR	Phospho(S)@3; Phospho(S)@7	99
4			23.94	RLSPSASPPR	Phospho(S)@7	99
4			23.94	RQSPSPSTRPIR	Phospho(S)@3	99
4			23.94	RYSPPIQR	Phospho(S)@3	99
4			23.94	TASPPPPPK	Phospho(S)@3	99
4			23.94	VPKPEPIPEPKEPSPEK	Phospho(S)@14	99
4			23.94	APQTSSSPPPVR	Phospho(S)@7	99
4			23.94	APQTSSSPPPVRR	Phospho(S)@7	99
4			23.94	KVELSESEEDKGGK	Phospho(S)@7	99
4			23.94	QSPSPSTRPIR	Phospho(S)@6	99
5	217467.3	HIST1H1E histone H1.4	13.93	SETAPAAPAAPAPAEK	Phospho(T)@3	99
6	883896.1	LIM domain protein 1	11.35	ASSLSESSPPK	Phospho(S)@3	99
6			11.35	ASSLSESSPPK	Phospho(S)@5	99

(continued)

Table 4 (continued)

No.	Access no.	Names	Unused cutoff	Sequence	Modifications	Conf
6			11.35	ASSLSESSPPK	Phospho(S)@8	99
6			11.35	ETPHSPGVEDAPIAK	Phospho(S)@5	99
6			11.35	ETPHSPGVEDAPIAK	Phospho(T)@2	99
6			11.35	LRSPPEALVQGR	Phospho(S)@3	99
6			11.35	RETPHSPGVEDAPIAK	Phospho(S)@6	99
6			11.35	SEVQQPVHPKPLSPDSR	Phospho(S)@13	99
6			11.35	TPHSPGVEDAPIAK	Phospho(S)@4	99
7	104050.3	Thyroid hormone receptor-associated protein	10.2	ASAVSELSPR	Phospho(S)@8	99
7			10.2	IDISPSTFR	Phospho(S)@4	99
7			10.2	RIDISPSTFRK	Phospho(S)@5	99
7			10.2	SPVGKSPPSTGSTYGSSQK	Phospho(S)@6	99
8	12199.1	Coiled-coil domain-containing protein	9.97	FESNPEETREPGSPPSVQR	Phospho(S)@13	99
8			9.97	LQQGAGLESPQGQPEPGAASPQR	Phospho(S)@20	99
8			9.97	ESNPEETREPGSPPSVQR	Phospho(S)@12	99
8			9.97	AGLGSPERPPK	Phospho(S)@5	99
8			9.97	AGLGSPERPPKTSPGSPR	Phospho(S)@13	99
9	969501.1	Treacle protein isoform f	8	KLGAGEGGEASVSPEKTSTTSK	Phospho(S)@13	99
9			8	LDSSPSVSSTLAAK	Phospho(S)@4	99
9			8	LGAGEGGEASVSPEK	Phospho(S)@12	99
10	220717.2	RNA-binding protein 15	8	DRTPPLLYR	Phospho(T)@3	99
10			8	SLSPGGAALGYR	Phospho(S)@3	99

10			8	HCAPSPDRSPELSSSR	Phospho(S)@5	99
10			8	TTAPTEGKSPLKKEDR	Phospho(S)@9	99
11	215928.4	Centrin-2	8	KRMSPKPELTEEQK	Phospho(S)@4	99
11			8	SPKPELTEEQKQEIR	Phospho(S)@1	99
12	18370.4	Isoform 2 of supervillin	8	SPSFGDPQLSPEARPR	Phospho(S)@10	99
12			8	SPSFGDPQLSPEARPR	Phospho(S)@3	99
12			8	CTSHSETPTVDDEEKVDER	Phospho(S)@5	99
12			8	DSSFTEVPRSPK	Phospho(S)@10	99
12			8	FSSSIENSDSPVR	Phospho(S)@3	99
13	909288.1	HNRNP A1	6	SESPKEPEQLR	Phospho(S)@3	99
13			6	SESPKEPEQLRK	Phospho(S)@1	99
13			6	SKSESPKEPEQLR	Phospho(S)@5	99
14	607584.1	Myb-binding protein 1A	6	KNQKPSQVNGAPGSPTEPAGQK	Phospho(S)@14	99
14			6	EIPSATQSPISK	Phospho(S)@8	99
14			6	KNQKPSQVNGAPGSPTEPACQK	Phospho(T)@16	99

Table 5 Functional annotation chart of CK2-associating proteins in the nucleus

No.	Names	*P*-Value
1	RNA splicing	1.00E-05
2	RNA processing	1.20E-05
3	Cytoskeleton organization	1.90E-05
4	mRNA processing	2.10E-04
5	mRNA metabolic process	4.60E-04
6	Regulation of actin polymerization or depolymerization	6.10E-04
7	Regulation of cellular component size	8.00E-04
8	Nucleocytoplasmic transport	8.80E-03
9	Nuclear transport	9.10E-03
10	Regulation of protein polymerization	1.50E-02
11	Nucleosome assembly	2.20E-02
12	Chromatin assembly	2.40E-02
13	Nuclear import	2.40E-02
14	Protein-DNA complex assembly	2.60E-02
15	Nucleosome organization	2.70E-02
16	DNA packaging	4.10E-02
17	Chromatin assembly and disassembly	4.70E-02
18	rRNA transcription	4.80E-02
19	Nuclear mRNA splicing via spliceosome	6.50E-02
20	Chromatin organization	8.40E-02

chromatin organization, and nuclear cytoskeletal organization. Our results revealed that RNA- and DNA-binding proteins were enriched in nuclear CK2-associating proteins indicating that CK2 may play important roles for their cellular functions during the progression of cell cycle. Recently, an integrated phosphoproteomics workflow identified spliceosomal complexes as the CK2 substrates in HeLa cells [30], suggesting an involvement of CK2 in the RNA splicing machinery. Other findings demonstrated CK2 phosphorylation of splicing factor Prp3p [31], Rad 51[32], and other proteins closely related to gene regulation [33–35]. Applying the above data sets to an algorithm such as SCANSITE could predict substrates recognized by specific kinases in relation to the nuclear CK_2 complex. Knowledge of these sites will likely provide a framework for proposing a regulatory mechanism for complex phosphorylation that occurred in the cells.

5 Conclusion and Outlook

Protein kinases in eukaryotes mediate cellular protein phosphorylation events and comprise the largest catalytic superfamilies. The living body maintains a basal metabolic system indispensable for survival and energy homeostasis.

External stimulation such as that from a hormone or a growth factor transmits a specified signal through the cell membrane which is integrated to respond to the original stimuli. When a signal transmission system plays on the dynamic reply, the most important molecules controlling the specificity of the transduction will be kinases.

CK2 has been considered a constitutively nonregulated kinase. However, we showed that CK2 activation occurred during the progression of cell cycle in response to growth stimuli of G_0-arrested human normal fibroblasts. We demonstrated that when CK2 exhibited the highest enzymatic activity in the cell during cell cycle progression, in G_1 phase, CK2 directly phosphorylated one of the translational initiation factors, eIF5, at Ser 389 and Ser390 and participated in the initiation of normal protein synthesis. We identified CK2 as an upstream kinase toward eIF5. Results obtained by the expression of eIF5 mutants that lack CK2 phosphorylation sites suggest that CK2 may be involved in the regulation of cell cycle progression through the phosphorylation of a key molecule for translation. Moreover, we revealed that CK2 interacted with APC following the progression of the cell cycle, and its activity was negatively regulated by this association. Interestingly, it has been known that a pool of CK2 molecules translocate from cytosolic fraction, where most of them are localized in a normal state, to the nucleus after the stimulation by growth factors [28, 36]. In the nuclear fraction, expression of CK2 was evident in human cancer cells. We then elucidated the molecules that associate with and may regulate CK2 enzymatic activity in the cells, as well as downstream target molecules for CK2. A nuclear fraction of human normal cells from synchronously growing cells according to the time course of the signal transmission was prepared, at G_0 and early G_1 cells, and we identified an individual component of the kinase complex by mass analysis. Notably, the data provided important information suggesting a diverse function of CK2 in the nucleus. In conclusion, by exploiting the strategy described here, a conventional approach to use 2D gel and a mostly integrated method with nanoLC mass spectrometry, a number of nuclear proteins and phosphorylated peptides have been characterized that may have brought into sharper focus the mechanisms and pathways associating with CK2 uncovered.

It is important to address the molecular mechanisms for determining the specific activation or downregulation of each kinase, which functions as a molecular switch of the signal transduction systems. From this point of view, both proteomics technologies to clarify the molecular constitution associating with each kinase and also structural chemistry technology to understand the regulatory mechanisms for kinase activities are useful approaches. Determining each of the exact phosphorylation residues is often essential for elucidating specific kinase-substrate relationships, which provides an insight for understanding essential signaling pathways and ultimately providing a novel molecular target to understand numerous pathologies. For example, in order to improve the metabolism in the living cells and also to understand disorders associated with energy homeostasis such as type II diabetes, lipid metabolism, and inflammation, studies clarifying regulatory mechanisms by biochemical and structural view points are critical. Therefore, future studies to unveil a new mechanism for the control of CK2 catalytic activity, in the aspects of its conformational changes by chemical modifications including phosphorylation, acetylation, and glycosylation [37], will be required. We demonstrated that autophosphorylation occurred in α subunit of CK2 on more than ten sites including a couple of tyrosine residues at Tyr12, Tyr23, and Tyr26 in the N-terminal region, which contributes to its full-enzymatic activity (collaborative work with Dr. Oyama-M and Dr. Koshiba-S, manuscript in preparation [38]). Collectively, these data will provide additional evidence for its multipotent ability and a functional specialization of the individual CK2 subunits to influence most of the key physiological processes.

Acknowledgments The authors acknowledge Drs. Dongxia Li and Tamiko Sueoka for discussions on the manuscript, Ms Junko Yamaki for her technical assistance, and Pete McCann, Roy Cameron, and Naoko Shiota for their critical reading on the manuscript. This work was supported by grants #23570170 from the Ministry of Education, Science, Sports, and Culture of Japan, Research Seeds Quest Program from the Japan Science and Technology Agency, and the Grant for Joint Research Project of the Institute of Medical Science, the University of Tokyo.

References

1. Burnett G, Kennedy E (1954) The enzymatic phosphorylation of proteins. J Biol Chem 211:969–980
2. Hanna DE, Rethinaswamy A, Glover CV (1995) Casein Kinase II is required for cell cycle progression during G1 and G2/M in S Cerevisiae. J Biol Chem 270:25905–25914
3. Filhol O, Martiel J-L, Cochet C (2004) Protein kinase CK2: a new view of an old molecular complex. EMBO Rep 5:351–355
4. Meggio F, Pinna LA (2003) One-thousand-and-one substrates of protein kinase CK2. FASEB J 17:349–368
5. Lou DY, Dominquez I, Toselli P, Landesman-Bollaq E, O'Brien C, Seldin DC (2008) The alpha catalytic subunit of protein kinase CK2 is required for mouse embryonic development. Mol Cell Biol 28:131–139
6. Xu X, Toselli PA, Russell LD, Seldin DC (1999) Globozoospermia in mice lacking the casein kinase II alpha' catalytic subunit. Nat Genet 23:118–121
7. Buchou T, Vernet M, Blond O, Jensen HH, Pointu H, Olsen BB, Cochet C, Issinger OG, Boldyreff B (2003) Disruption of the regulatory beta subunit of protein kinase CK2 in mice leads to a cell-autonomous defect and early embryonic lethality. Mol Cell Biol 23:908–915
8. Ahmed K, Gerber DA, Cochet C (2002) Joining the cell survival squad: an emerging role for protein kinase CK2. Trends Cell Biol 12:226–230
9. Litchfield DW (2003) Protein kinase CK2: Structure, regulation and role in cellular decision of life and death. Biochem J 369:1–15
10. Haitao J, Wang J, Nika H et al (2009) EGF-induced ERK activation promotes CK2 mediated disassociation of a-catenin from b-catenin and transactivation of b-catenin. Mol Cell 36:547–559
11. Duncan JS, Litchfield DW (2008) Too much of good things:the role of protein kinase CK2 in tumorigenesis and prospects for therapeutic inhibition of CK2. Biochim Biophys Acta 1784:33–47
12. Trembely JH, Wang G, Unger G, Slaton J, Ahmed K (2009) Protein kinase CK2 in health and disease: CK2 a player in cancer biology. Cell Mol Life Sci 66:1858–1867
13. Litchfield DW, Dobrowolska G, Krebs EG (1994) Regulation of casein kinase II by growth factors: a reevaluation. Cell Mol Biol Res 4:373–381

14. Sommercorn J, Mulligan JA, Lozeman FJ, Krebs EG (1987) Activation of casein kinase II in response to insulin and to epidermal growth factor. Proc Natl Acad Sci U S A 84:8834–8838
15. Ahn NG, Weiel JE, Chan CP, Krebs EG (1990) Identification of multiple epidermal growth factor-stimulated protein serine/threonine kinases from Swiss 3T3 cells. J Biol Chem 265:11487–11494
16. Niefind K, Issinger O-G (2010) Conformational plasticity of the catalytic subunit of protein kinase CK2 and its consequences for regulation and drug design. Biochim Biophys Acta 1804:484–492
17. Schnitzler A, Olsen BB, Issinger O-G, Niefind K (2014) The protein kinase CK2 [Andante] holoenzyme structure supports.J. Mol Biol 426:1871–1882
18. Homma MK, Wada I, Suzuki T, Yamaki J, Krebs EG, Homma Y (2005) CK2 phosphorylation of eukaryotic translation initiation factor 5 potentiates cell cycle progression. Proc Natl Acad Sci U S A 102:15688–15693
19. Homma MK, Li D, Krebs EG, Homma Y (2002) Association and regulation of casein kinase 2 activity by adenomatous polyposis coli protein. Proc Natl Acad Sci U S A 99:5959–5964
20. Homma MK, Homma Y (2005) Regulatory role of CK2 during the progression of cell cycle. Mol Cell Biochem 274:47–52
21. Montenarh M, Gots C (2013) The interactome of protein kinase CK2. In: Pinna EA (ed) CK2 Protein kinase. Wiley, New York, pp 76–116
22. Dignam JD (1990) Preparation of extracts from higher eukaryotes. In: Deutscher MP (ed) Methods in Enzymology, vol 182, Guide to Protein Purification. Academic, San Diego, pp 203–224
23. Ozlu N, Montigatti F, Renard BY, Field CM et al (2010) Binding partner switching on microtubules and Aurora-B in the mitosis to cytokinesis transition. Mol Cell Proteomics 9:336–350
24. Lessard J, Wu JI, Ranish JA et al (2007) An essential switch in subunit composition of a chromatin remodeling complex during neural development. Neuron 55:201–205
25. Old WM, Shabb JB, Houel S et al (2009) Functional proteomics identifies targets of phosphorylation by B-Raf signaling in melanoma. Mol Cell 34:115–131
26. Eric SW, Old WM, Resing KA, Ahn NG (2007) Mapping protein post-translational modifications with mass spectrometry. Nat Methods 3:798–806
27. Oyama M, Kozuka-Hata H et al (2009) Temporal perturbation of tyrosine phosphoproteome dynamics reveals the system-wide regulatory networks. Mol Cell Proteomics 8:226–231
28. Homma MK, Oyama M, Kozuka-Hata H, Ogura M and Homma. Y. Nuclear localization and phosphorylation status of protein kinase CK2 during cell cycle progression: switching in the CK2 complex, Manuscript in preparation for publication.
29. Phanstiel DH, Brumbaugh J, Wenger CD et al (2011) Proteomic and phosphoproteomic comparison of human ES and iPS cells. Nat Methods 8:821–827
30. Bian Y, Ye M, Wang C et al (2013) Global screening of CK2 kinase substrates by an integrated phosphoproteomics workflow. Sci Rep 3:3460. doi:10.1038/srep0346
31. Lehnert S, Gotz C, Kartarius S, Schafer B, Montenarh M (2008) Protein kinase CK2 interacts with the splicing factor hPrp3P. Oncogene 27:2390–2400
32. Yata K, Lloyd J et al (2012) Plk1 and CK2 act in concert to regulate Rad51 during DNA double strand break repair. Mol Cell 45:371–383
33. Khan D, He S, Yu J, Winter S, Cao W et al (2013) Protein kinase CK2 regulates the dimerization of histone deactylase 1(HDAC1) and HDAC2 during mitosis. J Biol Chem 288:16518–16528
34. Montaqnoli A, Valsasina B, Brotherton D, Troiani S (2006) Identification of Mcm2 phosphorylation sites by S-phase-regulating kinases. J Biol Chem 281:10281–10290
35. Panova TB, Panov KI, Russell J, Zomerdijk JCBM (2006) Casein kinase 2 associates with initiation-competent RNA polymerase I and has multiple roles in ribosomal DNA transcription. Mol Cell Biol 26:5957–5968

36. Homma MK, Homma Y (2008) Cell cycle and activation of CK2. Mol Cell Biochem 316:49–55
37. Tarrant MK, Rho H-S, Xie Z et al (2012) Regulation of CK2 by phosphorylation and O-GlcNAcylation revealed by semisynthesis. Nat Chem Biol 8:262–269
38. Homma MK, Oyama M, Kozuka-Hata H, Yamaki J, Koshiba S, Homma Y. Essential role for autophosphorylation of CK2α in the N-terminal region at tyrosine residues. Manuscript in preparation for publication

“Genuine” Casein Kinase: The False Sister of CK2 That Phosphorylates Secreted Proteins at S-x-E/pS Motifs

Elena Tibaldi, Giorgio Arrigoni, Giorgio Cozza, Luca Cesaro, and Lorenzo A. Pinna

Abstract The early discovery of protein kinase CK2 (an acronym derived from “casein kinase 2”) in 1954 was made possible by its ability to readily phosphorylate in vitro casein. For a while CK2 was suspected to be itself *a bona fide* casein kinase, because it proved able to re-phosphorylate the same residues which are phosphorylated in native casein. Later, however it was shown that, albeit similar and sometimes overlapping, the consensus sequences of CK2 and the genuine casein kinase isolated from the Golgi apparatus of the lactating mammary gland (G-CK) were definitely distinct (S/T-x-x-E/D/pS vs. S-x-E/pS). While CK2 was recognized as one of the most pleiotropic members of the big family of eukaryotic protein kinases (the so-called kinome) and has been implicated in several global diseases, with special reference to neoplasia, G-CK remained for decades an orphan enzyme, believed to play a dedicated but unexciting role in the biosynthesis of dairy proteins. Even though it became later evident that G-CK is also present in the Golgi apparatus of many tissues and is responsible for the phosphorylation of secreted proteins at S-x-E/pS motifs, its identity remained a mystery until 2012, when it was shown to be indistinguishable from Fam20C, a member of the four-jointed (FJ) atypical kinase family, causative of the Raine syndrome and other biomineralization disorders. Now we know that G-CK/Fam20C is as pleiotropic as CK2, being responsible for the generation of the largest proportion of the phosphosecretome, and, similar to CK2, it may represent a valuable druggable target.

E. Tibaldi
Department of Molecular Medicine, University of Padova, Padova, Italy

G. Arrigoni
Department of Biomedical Sciences and CNR, Institute of Neuroscience, University of Padova, Padova, Italy

Proteomic Center of Padova University, Padova, Italy

G. Cozza • L. Cesaro • L.A. Pinna (✉)
Department of Biomedical Sciences and CNR, Institute of Neuroscience, University of Padova, Padova, Italy
e-mail: lorenzo.pinna@unipd.it

K. Ahmed et al. (eds.), *Protein Kinase CK2 Cellular Function in Normal and Disease States*, Advances in Biochemistry in Health and Disease 12, DOI 10.1007/978-3-319-14544-0_13

Keywords CK2 • Fam20C • G-CK • Dentin matrix protein 4 • Casein kinases • Phosphorylation • Golgi apparatus • Raine syndrome • Biomineralization • Phosphosecretome

1 Introduction: A Historical Account

Protein kinase CK2 (together with CK1) was the first protein kinase to be detected, as early as in 1954, in liver extracts [1] and shortly later in yeast, brain and other tissues [2]. This was made possible by using as phosphoacceptor substrate casein, i.e., one of the rare phosphoproteins already known at that time, a circumstance still implicit in the acronyms, derived from "casein kinases" 1 and 2.

For a while it was unclear whether the same enzymes were also responsible for the biosynthetic phosphorylation of casein in the tissue where this protein is produced, before being secreted with milk, i.e., the lactating mammary gland. CK2, in particular, was suspected to be the "genuine" casein kinase because, unlike CK1, it was shown to be able to re-phosphorylate seryl residues which are phosphorylated in native casein fractions. At the end of the 1980s, however, by comparing preparations of genuine casein kinase from the Golgi-enriched fraction of the lactating mammary gland with CK2, it was possible to show that the consensus sequences for the two kinases, albeit similar, are definitely distinct (S-x-E/pS and S/T-x-x-E/D/pS, respectively) [3, 4]. In particular, as summarized in Fig. 1, peptides reproducing the

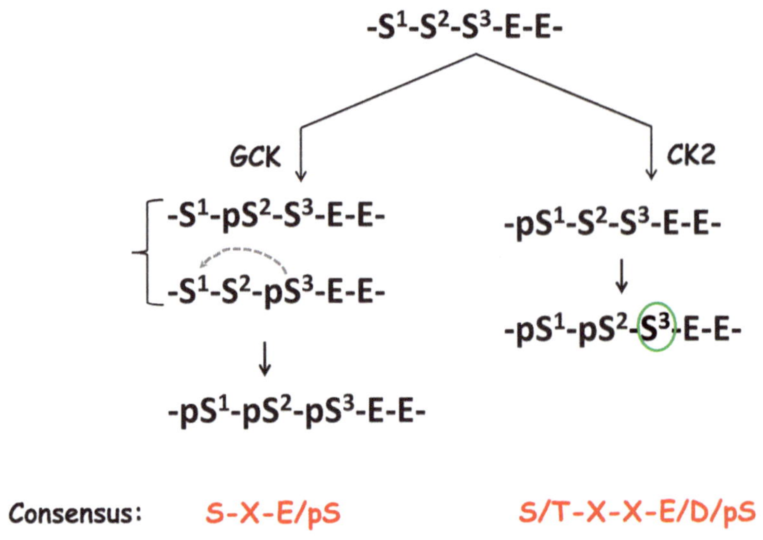

Fig. 1 Complete vs. incomplete phosphorylation of the SSSEE motif by genuine casein kinase from the Golgi-enriched portion of the lactating mammary gland (GEF-CK) and by "casein kinase 2" (CK2) respectively. Drawn from ref. 4. Consensuses were elucidated in ref. 3

S-S-S-E-E motifs recurrent in casein fractions are phosphorylated with different orders and to different extent by the two "casein kinases": CK2 phosphorylates only the first and second serines of the row but not the third one; in contrast, the casein kinase from the Golgi-enriched fraction of the lactating mammary gland (GEF-CK, later shortened to G-CK) started phosphorylating the third and second serines, and, once Ser 3 was phosphorylated, it was able to incorporate phosphate also in the first serine.

At about the same time, the primary sequence of CK2 subunits was elucidated [5], and its numerous endogenous targets started being detected (reviewed in ref. 6), revealing its striking pleiotropicity and suggesting its implication in a variety of global diseases.

By sharp contrast, the "genuine" casein kinase was erroneously believed to be a dedicated enzyme, just committed to the phosphorylation of casein, a storage protein nearly devoid of biological interest. Consequently, G-CK was looked at as a mere object of curiosity-driven investigation. This may at least partially explain why G-CK remained for such a long time an "orphan" enzyme, whose genes were unknown and antibodies not available.

Substantial steps forward in the understanding of G-CK biochemical properties and biological functions were made possible by the development of an exceptionally selective peptide substrate derived from β-casein β (28–40) which is readily phosphorylated by G-CK, but not by CK2 and CK1, nor, to the best of our knowledge, by any other protein kinase [7]. Exploiting this peptide substrate, it was possible to purify G-CK from the Golgi apparatus and to show that it is not a dedicated "casein kinase," being detectable also in the Golgi apparatus of liver and of many other tissues [8] and being able to phosphorylate several other secreted proteins at naturally occurring pS-x-E/pS motifs [9–11]. This kind of information, in conjunction with the availability of repertoires of phosphopeptides generated from secreted proteins (serum and cerebrospinal liquid), led to the proposal that G-CK is responsible for the generation of the largest proportion of the secreted phosphoproteome [12] (see Fig. 2). Somewhat paradoxically, all these inferences were drawn from studies performed with an orphan enzyme: despite many efforts, in fact the molecular and genetic identity of G-CK escaped detection till 2012.

2 The "Genuine" Casein Kinase Is Not a Member of the Kinome

Looking at our lab notes in retrospective, already in 2005 mass spectrometry (MS) analysis of proteins co-migrating with in-gel G-CK casein kinase activity revealed a protein whose accession number (IPI00363789.1) corresponded to a protein named "similar to cDNA sequence BC004044" and 2 years later was again identified as "dentin matrix protein 4" (DMP4) of unknown function and identical to Fam20C, a member of the four-jointed (FJ) family of calcium-binding proteins. Although DMP4/Fam20C was repeatedly identified in our G-CK preparations (see Fig. 3),

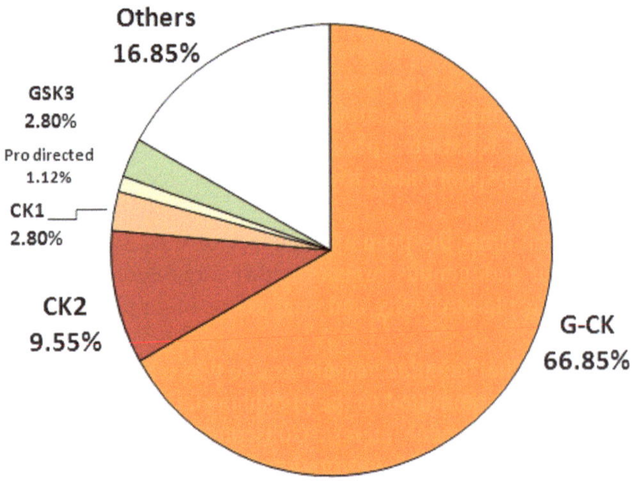

Fig. 2 Major contribution of G-CK to the generation of the phosphosecretome. Constructed with data drawn from ref. 12

we overlooked the possibility that it could be responsible for G-CK activity, based on the wrong presumption that such an activity had to be necessarily due to a typical protein kinase. In hindsight, this was hard to reconcile with the observation that G-CK is refractory to up to 0.5 mM staurosporine, a potent and promiscuous inhibitor of nearly all the members of the kinome. Only later, having become aware that FJ may represent a family of atypical (protein) kinases [13], we reevaluated those old data and started considering the possibility that Fam20C and/or Fam20A might be responsible for the casein phosphorylation activity of our G-CK preparations, and we were able to show that on PAGE-SDS, the G-CK band immunoreacted with Fam20C antibodies [14] (see Fig. 4).

At about the same time, two other labs were approaching the solution of the same enigma: the Irvine group who was trying to find a function for Fam20C, whose gene alteration was already known to be responsible for the Raine syndrome, a deadly osteoclerotic bone dysplasia [15], and Vincent Tagliabracci, who in Jack Dixon's lab was addressing the problem from a bioinformatics standpoint, looking for a protein having a signal peptide committing it to secretion and an atypical kinase domain but devoid of transmembrane sequence. This latter approach turned out to be the most successful, leading to the incontrovertible evidence that Fam20C is indistinguishable from G-CK, displaying identical properties *in primis* the same consensus and the ability to phosphorylate the specific G-CK β (28–40) peptide substrate as well as casein itself and a wide spectrum of secreted proteins at S-x-E/pS motifs [16]. The same conclusion was independently reached in our and Irvine's labs at about the same time [14, 17] and provided the demonstration that the enzyme responsible for the in vivo phosphorylation of casein, at variance with CK1 and CK2, whose acronyms refer to their ability to phosphorylate casein in vitro, is not a member of the kinome but an atypical protein kinase.

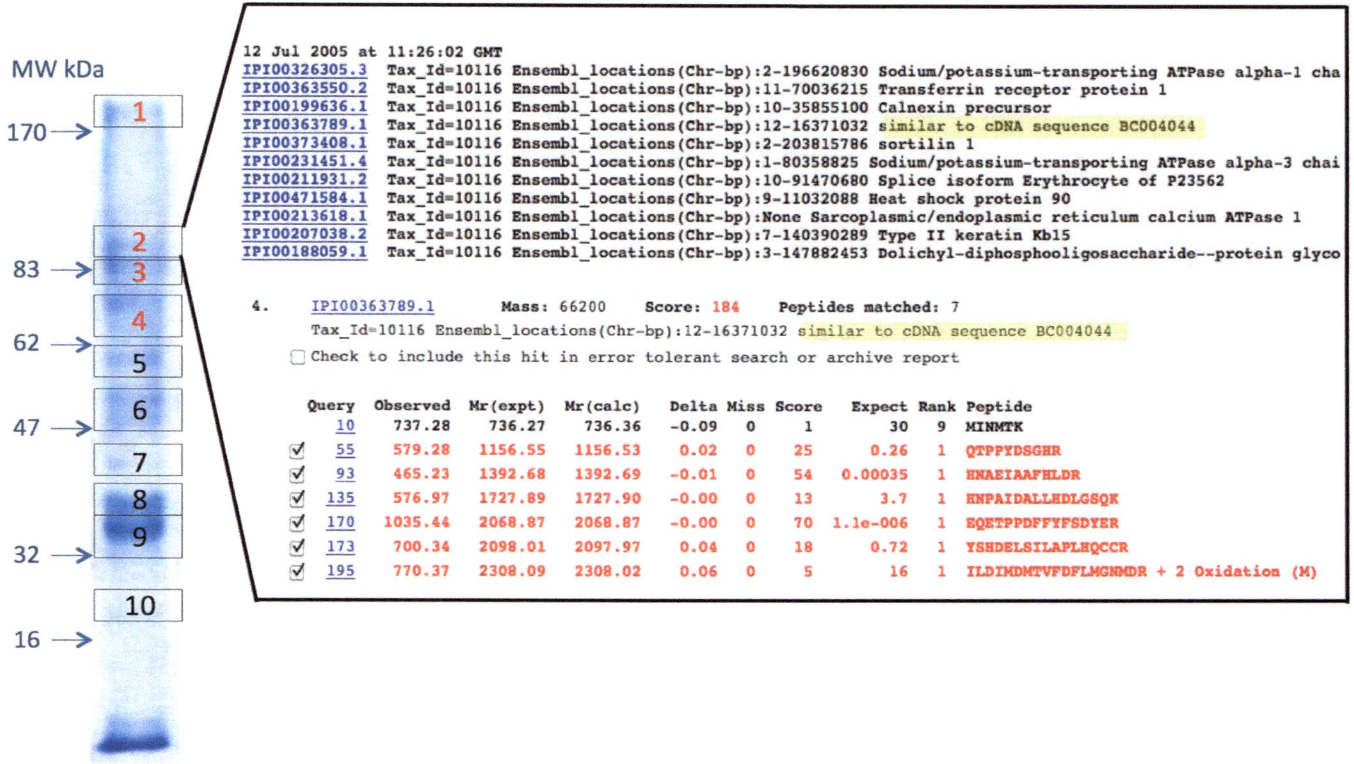

Fig. 3 Early identification of Fam20C/DMP4 by MS analysis of proteins co-purifying with G-CK activity. A partially purified preparation of G-CK obtained as in ref. 23 was subjected to SDS-PAGE, and the proteins were stained with colloidal Coomassie Brilliant Blue. The gel was cut into segments as indicated, and each segment assayed for G-CK activity, which was found to peak in segment 2. MS analysis of the same segment (conducted as described in ref. 27) led to the identification of 11 proteins, listed on the *right side* of the figure. One of these, highlighted in *yellow*, corresponded to "similar to cDNA sequence BC004044," later renamed DMP4 or Fam20C. The identified peptides are listed below. For additional information see main text

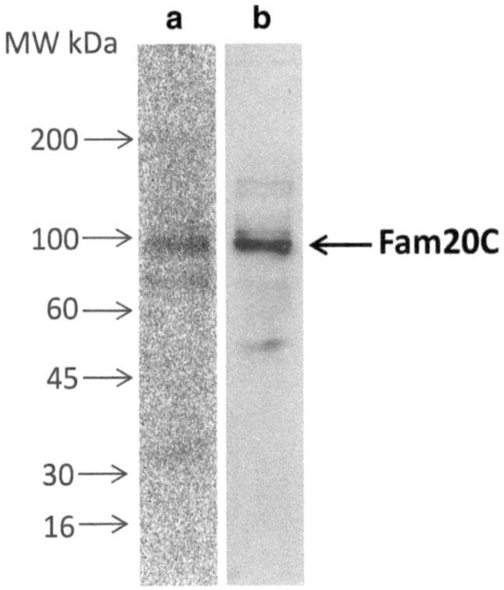

Fig. 4 Fam20C co-migrates with G-CK activity on SDS-PAGE. (**a**) G-CK, purified from the Golgi apparatus of the lactating mammary gland as described in ref. 23, was subjected to SDS-PAGE. The position at which G-CK migrated was detected by in-gel kinase assay using casein as substrate and [γ-^{32}P]-ATP as phosphate donor. (**b**) An aliquot of G-CK run in parallel with the in-gel kinase assay was subjected to Western blot analysis with anti-FAM20C antibody. Data drawn from ref. 14

3 Distinctive Feature of Fam20C/G-CK and CK2

A number of distinctive properties of Fam20C/G-CK and CK2 are listed in Table 1. It can be seen that these two kinases are sharply different in nearly all respects, sharing only the ability to recognize seryl residues specified by negatively charged side chains, at close but distinct positions, n+3 in the case of CK2 and n+2 in the case of Fam20C/G-CK. Also to note is the fact that while CK2 tolerates threonine instead of serine as phosphoacceptor residue and aspartic acid as specificity determinant, this is not the case of Fam20C/G-CK. The circumstance however that quite often additional acidic residues (often Glu) besides the one at position n+3 are present at CK2 sites [18] and that in some cases more than one phospho-serine are clustered together, raises the possibility that a number of phosphosites retrieved in currently available databases fulfill both the consensus of CK2 and that of Fam20C/G-CK. As shown in Fig. 5, in fact, a significant number of phosphosites commonly ascribed to CK2 also display the consensus for being phosphorylated by Fam20C/G-CK. In these cases the physiological implication of either CK2 or Fam20C/G-CK remains an open question, whose clear-cut answer will require additional information and/or experimentation.

Also worthy to mention in this respect is the global analysis of SSSEE stretches which are among the most typical Fam20C/G-CK targets in casein fractions (see above, Fig. 1). This sequence of 5 amino acids is recurrent 142-fold in the

Table 1 Distinctive properties of CK2 and Fam20C/G-CK

	CK2	Fam20C/G-CK
Belongings (protein family)	ePKs ("kinome")	FJ atypical PKs
Structure	Monomer α(α') and tetramer $\alpha(\alpha')_2\beta_2$	Monomer
Consensus	S/T-x-x-E/D/pS	S-x-E/pS
Phosphate donor	ATP and GTP	Only ATP
Preferred cation	$Mg^{2+} > Mn^{2+}$	$Mn^{2+} >> Mg^{2+}$
Cellular localization	Ubiquitous	Golgi and secreted
Inhibition of staurosporine	$IC_{50} \cong 16$ μM	Unaffected up to 500 μM
Drastic inhibition by flavonoids ($IC_{50} < 1$ μM)	Quercetin Fisetin Keamferol Luteolin Apigenin Myricetin	None
Inhibition by heparin (5 μg/ml)	>90 %	Unaffected
Stimulation by sphingosine (25 μM)	No	Yes

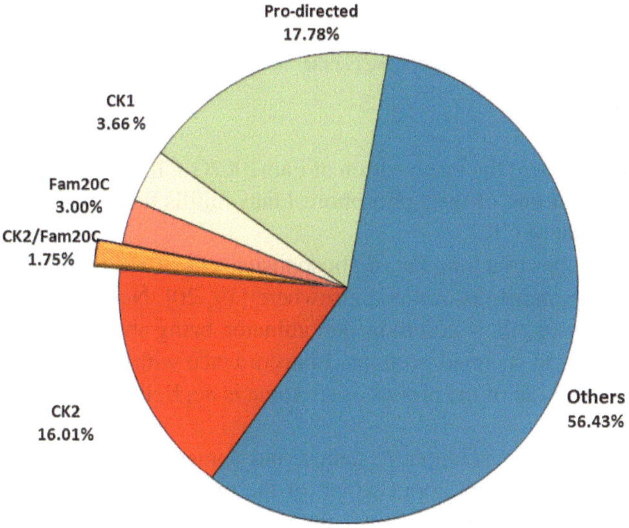

Fig. 5 Major contribution of "casein kinases" to the generation of the human phosphoproteome. Phosphosites assigned to CK2 but displaying also the Fam20C/G-CK consensus are highlighted. Constructed with data reported in ref. 28

human proteome, and it has been reported as fully or partially phosphorylated in at least 28 nonredundant phosphopeptides retrieved in the PhosphoSitePlus database. Considering their sequences also beyond their C- and N-termini and the position of phosphate(s) in their SSS triplets, it turns out that 9 out of these 28 phosphopeptides

Table 2 SSSEE phosphosites whose occupancy implies the intervention of Fam20C/G-CK

ID	Sequence	Name	Localization
P38398	DSCSSsEEIKK	Breast cancer type 1 susceptibility protein	Nucleus; chromosome
P47710	sIssssEEMSL	Alpha-S1-casein	Secreted
P05814	EsLsssEESIT	Beta-casein	Secreted
Q9Y6X4	AVDsssEEIEV	Soluble lamin-associated protein of 75	Nucleus envelope. Nucleus inner membrane; peripheral membrane protein; nucleoplasmic side
Q6ZNB6	QFssssEEGDE	NF-X1-type zinc finger protein	Membrane; single-pass membrane protein
P30414	KEAsssEEPRN	NK-tumor recognition protein	Membrane; peripheral membrane protein
Q96T83	RTKsSsEEVLE	Sodium/hydrogen exchanger	Golgi apparatus > trans-Golgi network membrane; multi-pass membrane protein. Recycling endosome membrane; multi-pass membrane protein
P18583	AsEsssEEKDD	Protein SON	Nucleus speckle
B3KS81	QSRsssEERDH	Serine/arginine repetitive matrix protein 5	
Q5SRQ6	VKMsssEEVsW	Casein kinase II subunit beta	

Drawn from PhosphoSitePlus. Lower case denotes serines which are phosphorylated

(listed in Table 2) need the intervention of Fam20C/G-CK for becoming phosphorylated since at least one of their phosphoresidues fulfills the consensus of Fam20C/G-CK but not that of CK2.

The same applies to a long list of phosphosites elucidated in the infancy of studies on phosphoproteins, as outlined elsewhere [19, 20]. None of them conform to the CK2 consensus, their common denominator being the motif pS-x-E/pS and mostly belonging to secreted proteins. In accordance with this, the contribution of CK2 to the generation of the phosphosecretome is negligible as compared to that of G-CK (see Fig. 2).

While the common acidophilic nature and sometimes overlapping consensus sequences may shed doubt about which of the two kinases is responsible for the generation of a subset of phosphosites, the possibility that CK2 and Fam20C/G-CK might cooperate in the targeting of the same proteins seems unlikely due to their neatly different localization. While in fact CK2 is ubiquitous in most subcellular compartments, being particularly abundant in nuclei, it is undetectable in the Golgi apparatus, where instead Fam20C is exclusively localized [8], being also detectable in secreted fluids like milk [21] and cell culture-conditioned medium [16]. This feature, dictated by an N-terminal signal peptide which is cleaved in the Golgi at the end of its biosynthesis, commits Fam20C to secretion, making it the first-choice kinase available for the phosphorylation of secreted proteins [19]. This also provides the rationale for the observation mentioned above that many secreted proteins found to be phosphorylated at S-x-E motifs in early studies on protein phosphorylation, including not only caseins but also lactalbumin, fibrinogen, pepsin, and several hormonal neuropeptides, are targeted by Fam20C/G-CK [19, 20].

Apart from their common acidophilic nature, accounting for the confusing nomenclature of "casein kinases," CK2 and Fam20C/G-CK neatly differentiate in every other respect, not surprisingly if it is considered that they belong to two distinct and unrelated families of proteins (Table 1). Noteworthy is the insensitivity of Fam20C/G-CK to all known inhibitors of protein kinases tested on it. Insensitivity to staurosporine is especially remarkable as this compound inhibits nearly all protein kinases in the nano-molar or low micromolar range. Even CK2, which is considered refractory to staurosporine, is inhibited with an IC_{50} value around 15 μM [22]. By sharp contrast, no detectable inhibition of Fam20C/G-CK could be observed, even when raising the staurosporine concentration up to 0.5 mM [23]. Also notable is its insensitivity to flavonoids, a class of compounds including many protein kinase inhibitors. At least six flavonoids have been reported to inhibit CK2 with IC_{50} values in the sub-micromolar range [14], but none of these are appreciably affecting Fam20C/G-CK (see Table 1).

Staurosporine and flavonoids are directed to the ATP-binding site, which is highly conserved across bona fide protein kinases, including CK2, but deeply altered in Fam20C [24]. Heparin, in contrast, is a well-known CK2 inhibitor which competes against the acidic phosphoacceptor substrate by interacting with the same basic residues [25, 26]. Heparin however also fails to inhibit Fam20C/G-CK, when its activity is assayed with substrates displaying the canonical acidic consensus (S-x-E).

For the time being, the lack of any pharmacological tool able to inhibit Fam20C/G-CK hampers cell studies aimed at shedding light on the biological functions of this atypical pleiotropic kinase. On the other hand, it can be argued that from a practical, potentially therapeutic perspective, an upregulation of Fam20C/G-CK rather than its downregulation would be desirable. Its known or suspected pathological implications, in fact, in Raine syndrome and possibly in other biomineralization disorders are accounted for by a loss of function and not by a gain of function/unscheduled activity as in the popular cases of many "onco-kinases" (CK2 included).

In connection with this, a promising hint comes from the observation that the activity of G-CK partially purified from the Golgi apparatus of the lactating mammary gland is enhanced severalfold by sphingosine [10]. Such a behavior has been recently confirmed using recombinant Fam20C and a variety of its natural targets and shown to become especially apparent under physiological conditions, where Mn^{2+} is replaced by 1 mM Mg^{2+} as activator cation [Cozza G, Tibaldi E, Tagliabracci VS, Dixon JE, Pinna LA unpublished work]. These data on the one side disclose the possibility that Fam20C/G-CK is a mediator of sphingosine signaling and on the other pave the road toward the development of new pharmacological strategies for the treatment of biomineralization disorders caused by the defective activity of Fam20C/G-CK.

Acknowledgment This work was supported by grants from AIRC (Associazione Italiana per la Ricerca sul Cancro) Grant IG10312.

References

1. Burnett G, Kennedy EP (1954) The enzymatic phosphorylation of proteins. J Biol Chem 211:969–980
2. Rabinowitz M, Lipmann F (1960) Reversible phosphate transfer between yolk phosphoprotein and adenosine triphosphate. J Biol Chem 235:1043–1050
3. Meggio F, Boulton AP, Marchiori F, Borin G, Lennon DP et al (1988) Substrate-specificity determinants for a membrane-bound casein kinase of lactating mammary gland. A study with synthetic peptides. Eur J Biochem 177:281–284
4. Meggio F, Perich JW, Meyer HE, Hoffmann-Posorske E, Lennon DP et al (1989) Synthetic fragments of beta-casein as model substrates for liver and mammary gland casein kinases. Eur J Biochem 186:459–464
5. Saxena A, Padmanabha R, Glover CV (1987) Isolation and sequencing of cDNA clones encoding alpha and beta subunits of Drosophila melanogaster casein kinase II. Mol Cell Biol 7:3409–3417
6. Pinna LA (1990) Casein kinase 2: an "eminence grise" in cellular regulation? Biochim Biophys Acta 1054:267–284
7. Lasa-Benito M, Marin O, Meggio F, Pinna LA (1996) Golgi apparatus mammary gland casein kinase: monitoring by a specific peptide substrate and definition of specificity determinants. FEBS Lett 382:149–152
8. Lasa M, Marin O, Pinna LA (1997) Rat liver Golgi apparatus contains a protein kinase similar to the casein kinase of lactating mammary gland. Eur J Biochem 243:719–725
9. Lasa M, Chang PL, Prince CW, Pinna LA (1997) Phosphorylation of osteopontin by Golgi apparatus casein kinase. Biochem Biophys Res Commun 240:602–605
10. Brunati AM, Marin O, Bisinella A, Salviati A, Pinna LA (2000) Novel consensus sequence for the Golgi apparatus casein kinase, revealed using proline-rich protein-1 (PRP1)-derived peptide substrates. Biochem J 351(Pt 3):765–768
11. Tibaldi E, Arrigoni G, Martinez HM, Inagaki K, Shimasaki S et al (2010) Golgi apparatus casein kinase phosphorylates bioactive Ser-6 of bone morphogenetic protein 15 and growth and differentiation factor 9. FEBS Lett 584:801–805
12. Salvi M, Cesaro L, Tibaldi E, Pinna LA (2010) Motif analysis of phosphosites discloses a potential prominent role of the Golgi casein kinase (GCK) in the generation of human plasma phospho-proteome. J Proteome Res 9:3335–3338
13. Ishikawa HO, Takeuchi H, Haltiwanger RS, Irvine KD (2008) Four-jointed is a Golgi kinase that phosphorylates a subset of cadherin domains. Science 321:401–404
14. Lolli G, Cozza G, Mazzorana M, Tibaldi E, Cesaro L et al (2012) Inhibition of protein kinase CK2 by flavonoids and tyrphostins. A structural insight. Biochemistry 51(31):6097–6107
15. Raine J, Winter RM, Davey A, Tucker SM (1989) Unknown syndrome: microcephaly, hypoplastic nose, exophthalmos, gum hyperplasia, cleft palate, low set ears, and osteosclerosis. J Med Genet 26:786–788
16. Tagliabracci VS, Engel JL, Wen J, Wiley SE, Worby CA et al (2012) Secreted kinase phosphorylates extracellular proteins that regulate biomineralization. Science 336:1150–1153
17. Ishikawa HO, Xu A, Ogura E, Manning G, Irvine KD (2012) The Raine syndrome protein FAM20C is a Golgi kinase that phosphorylates bio-mineralization proteins. PLoS One 7:e42988
18. Meggio F, Pinna LA (2003) One-thousand-and-one substrates of protein kinase CK2? FASEB J 17:349–368
19. Tagliabracci VS, Pinna LA, Dixon JE (2013) Secreted protein kinases. Trends Biochem Sci 38:121–130
20. Venerando A, Ruzzene M, Pinna LA (2014) Casein kinase: the triple meaning of a misnomer. Biochem J 460:141–156
21. Duncan JS, Wilkinson MC, Burgoyne RD (2000) Purification of Golgi casein kinase from bovine milk. Biochem J 350(Pt 2):463–468

22. Meggio F, Donella Deana A, Ruzzene M, Brunati AM, Cesaro L et al (1995) Different susceptibility of protein kinases to staurosporine inhibition. Kinetic studies and molecular bases for the resistance of protein kinase CK2. Eur J Biochem 234:317–322
23. Tibaldi E, Arrigoni G, Brunati AM, James P, Pinna LA (2006) Analysis of a sub-proteome which co-purifies with and is phosphorylated by the Golgi casein kinase. Cell Mol Life Sci 63:378–389
24. Xiao J, Tagliabracci VS, Wen J, Kim SA, Dixon JE (2013) Crystal structure of the Golgi casein kinase. Proc Natl Acad Sci U S A 110:10574–10579
25. Vaglio P, Sarno S, Marin O, Meggio F, Issinger OG et al (1996) Mapping the residues of protein kinase CK2 alpha subunit responsible for responsiveness to polyanionic inhibitors. FEBS Lett 380:25–28
26. Sarno S, Vaglio P, Meggio F, Issinger OG, Pinna LA (1996) Protein kinase CK2 mutants defective in substrate recognition. Purification and kinetic analysis. J Biol Chem 271:10595–10601
27. Berggard T, Arrigoni G, Olsson O, Fex M, Linse S et al (2006) 140 mouse brain proteins identified by Ca^{2+}-calmodulin affinity chromatography and tandem mass spectrometry. J Proteome Res 5:669–687
28. Ruzzene M, Tosoni K, Zanin S, Cesaro L, Pinna LA (2011) Protein kinase CK2 accumulation in "oncophilic" cells: causes and effects. Mol Cell Biochem 356:5–10

Part III
Function of CK2 in Cancer and Its Therapeutic Targeting

critical for the activation of multiple pro-survival signaling pathways, and its catalytic activity is invariably elevated in various types of tumors. However, the precise role of CK2 has never been addressed in RCCs. In this study, we have analyzed the activity of CK2 and the expression of its subunits in a small cohort of RCC tumors. This analysis revealed, in the majority of tumor samples, an upregulation of the CK2 catalytic subunits that was not correlated with mRNA abundance in the majority of tumor samples. Moreover, relative levels of the three CK2 subunits varied significantly between tumor samples, and a positive correlation was observed between low CK2β expression and an upregulation of the ZEB2 mesenchymal marker in a subset of tumor samples. Using the CK2 inhibitor CX-4945 to downregulate the CK2 catalytic activity in 786-O cells as a model of VHL-deficient renal cancer cell line, we showed that CK2 represents a potential promising therapeutic target in RCCs.

Keywords Renal cancer • VHL • 786-O cells • Protein kinase CK2 • CK2 subunits • Posttranscriptional regulation • Signaling pathways • EMT • CK2-targeted therapy • CX-4945

1 Introduction

Renal cell carcinomas account for 3.5 % of all cancers, with an incidence increasing steadily over the years. Despite an early detection, the prognosis for patients with metastatic renal cancer remains unpredictable. The prognosis of these carcinomas is often poor; one third of patients have metastatic disease at presentation, and 40 % of patients undergoing a surgical resection will develop metastases. The main histological subtype of RCCs is represented by clear cell renal cell carcinomas (ccRCCs) (85 % of RCCs). Ninety percent of ccRCCs harbor a biallelic inactivation of the tumor suppressor gene *VHL* (von Hippel-Lindau). The absence of the VHL protein leads (1) to the upregulation of hypoxia-related target genes and (2) to a decrease in the apoptotic process, thus playing a key role in renal tumorigenesis.

Aberrant activation of protein kinases through dysregulated gene expression and enzymatic activity is a key oncogenic force underlying human tumorigenesis including that of ccRCCs. Although less common than in other tumor types, several chromosomal amplifications are present in RCC tumors that are associated with dysregulated kinase expression [1]. Consequently, the development of molecular targeted therapies involving kinase inhibitors has changed the clinical management of RCCs [2]. Protein kinase CK2 is a highly ubiquitous and multifaceted serine/threonine kinase described as a multisubunit holoenzyme generated by the tight association of two α or α′ catalytic subunits with a dimer of β regulatory subunits. CK2 activity is invariably elevated in various types of tumors as well as in cancer cell lines [3, 4]. Moreover, CK2 expression/activity is critical for activation of multiple pro-survival signaling pathways. These characteristics identify CK2 as an attractive therapeutic target. CX-4945 is the first selective orally bioavailable CK2 inhibitor to advance into human clinical trials against several types of cancers as hematopoietic and breast cancer [5, 6]. In RCCs, *VHL* inactivation leads to cellular phenotypic changes, characteristic of an epithelial-to-mesenchymal transition

(EMT), a cellular process in which CK2 has been shown to play an important role ultimately leading to the acquisition of cancer stem cell properties [7]. However, the precise role of CK2 has, up to now, never been addressed in RCCs.

In this context, we have studied the enzymatic activity, the expression by Western blotting, and the mRNA levels by RT-qPCR of the catalytic (α, α′) and the regulatory (β) CK2 subunits, as well as of known EMT markers, in surgical biopsies of normal and tumor tissues from 15 patients with RCC. We observed an increase in CK2 enzymatic activity, correlated with an accumulation of the catalytic CK2 α/α′ subunits in more than 75 % of tumor samples, compared with the normal tissue of the same patients. However, this upregulation at the protein level was not correlated with any mRNA upregulation, suggesting an important posttranscriptional dysregulation in RCCs.

In addition, using the highly aggressive 786-O cell line as a model, we showed that renal cancer cells can be responsive to CK2-targeted therapeutic agents. Our results show that CK2 represents a promising therapeutic target in RCCs and that the detection of its upregulated expression could help in the molecular classification of these tumors.

2 Material and Methods

2.1 Patient Characteristics

We obtained paraffin-embedded samples and the corresponding frozen samples of 15 RCCs that were diagnosed on the basis of histological and clinical findings at the Grenoble Hospital. Prior patient consent and approval from the local research ethics committee was obtained before we used these clinical materials for research purposes. The clinical characteristics of the patients with RCC are summarized in Table 1.

2.2 Reagent, Drugs, and Antibodies

Antibodies P-FAK (Y576/577) (#3281), P-p130Cas (Y410) (#4011), PARP (#9542), AKT (#9272), STAT-3 (#4904), P-STAT-3 (#9145), p38MAPK (#9212), P-P38MAPK (#9211), and HSP90 (#4874) were from Cell Signaling Technology (Cambridge, MA). Anti-paxillin (610052), FAK (610088), and p130Cas (610271) were from BD Biosciences (San Jose, CA). Antibodies against p21 (sc-397), P53 DO-1 (sc-126), and ZEB (sc-25388) were purchased from Santa Cruz, and P-AKT (S129) (#AP3020a) was from Abgent (San Diego, CA). Anti-α-catenin (ab49105), anti-actin (ab8226), and anti-P-p21 (ab47300) were from Abcam. Anti-P-p42-p44 MAPK (#9100) was from New England Biolabs, anti-P-α-catenin (#11330) was from SAB, anti-survivin (NB500-201) was from Novus Biologicals, and anti-GAPDH (#AM4300) was from Life Technologies. Dimethyl sulfoxide, protease, and phosphatase inhibitor cocktails (respectively #P8340 and #P5726) were from Sigma-Aldrich (St. Louis, MO).

Table 1 Clinical characteristics of the 15 patients of this study

Characteristics	$N=15$	$CK2\alpha+\alpha'$ (%)	
		High expression	Low expression
Gender			
Male	8 (53 %)	6	2
Female	7	6	1
Age<65	11 (73 %)	8	3
Age>65	4	4	0
Risk factors			
Smoking	8	8	0
HTA	4	3	1
Obesity	3	2	1
Genetic syndrome (VHL)	0		
Dialysis	0		
Pathology classification			
Clear cell renal carcinoma	14	11 (78.5)	3 (22.5)
Tubulopapillary	1	1	0
pTNM staging[a]			
T classification			
pT1–T2	9	8	1
pT3–T4	5	3	2
N classification			
pNx	12	10 (83)	2 (17)
pN2	1	0	1
Distant metastasis			1
At the diagnosis	2	1	0
After the diagnosis	1	1	3
Vascular invasion	4	1	2
Lymphatic invasion	4	2	
Fuhrman grade			
I–II	5	4 (80)	1 (20)
III–IV	10	3 (30)	7 (70)
Expression of CK2			
Western blot	15	12 (80)	3 (20)
Kinase activity	15	12 (80)	3 (20)

[a]One undetermined

2.3 Cell Lines and Culture

ccRCC cell lines from ATCC® (786-O, ACHN, RCC4, CAKI) were grown in 10-cm-diameter dishes in a humidified incubator (37 °C, 5 % CO_2) with RPMI 1640 medium (Gibco) containing 10 % fetal calf serum, penicillin (100 U/mL), and streptomycin (100 μg/mL). The VHL(−) cell lines consist of the parental renal carcinoma cell line, 786-O WT, and its derivative lines containing either the empty expression vector HA-pBABE (786-O pB) or a functional VHL construct HA-VHL (786-O VHL). Stable transfected cells were maintained in medium supplemented with 2 μg/ml puromycin.

2.4 *Western Blot Analysis*

Cell pellets were suspended in 150 μL of RIPA buffer (Tris–HCl pH 7.4 10 mM, NaCl 150 mM, SDS 0.1 %, Na deoxycholate 0.5 %, EDTA 1 mM, Triton X100 1 %) containing protease and phosphatase inhibitor cocktails. Homogenates were quantified using BCA protein assay kit (Thermo Scientific). SDS-PAGE was performed using precast 4–12 % gradient gel (Bio-Rad) and electrophoresed in NuPAGE buffer at 150 V for 75 min. Separated proteins at 20 μg/lane were transferred to PVDF membranes (60 min at 100 V). Blotted membranes were blocked during 1 h at room temperature with saturation buffer (1 % BSA in TBST) and then incubated with primary antibody diluted in saturation buffer, during 90 min at room temperature or overnight at 4 °C. Secondary antibodies were added for 1 h. Luminata Forte Western HRP substrate (Millipore) was added to make revelation with Fusion FX (PerkinElmer). Anti-actin, GAPDH, and HSP90 were used for loading control.

2.5 *Viability Assay*

Cytotoxicity was measured using PrestoBlue assay (Invitrogen, Carlsbad, CA). Cell lines were seeded in a 96-well microtiter plate at a concentration of 1×10^5 cells/ml. Cells were allowed to grow for 24 h at 37 °C and 5 % CO_2. The cells were exposed to CX-4945 or vehicle (DMSO) with concentrations ranging between 1 and 6 μM. The microtiter plate was incubated for further 48 h and thereafter 10 μl PrestoBlue was added. The plates were incubated for 1 h and the fluorescence was read at 580 nm, using a FluoStar Optima plate reader (BMG LabTech, Ortenberg, Germany).

2.6 *Migration and Invasion Assay*

Cell migration and invasion assays were performed with 786-O WT, 786-O pB, and 786-O VHL cells. Cells were trypsinized, counted and resuspended in serum-depleted media (0.5 % FBS), and plated at 2.10^5 cells/well in 24-well Boyden chamber (BD Biosciences) or 5.10^4 cells/well in 24-well Matrigel Invasion Chamber plates (BD Biosciences) using 10 % FBS as chemoattractant. After 6 h for migration assay or 24 h for invasion assay, migratory or invasive cells were fixed with PFA, stained with hematoxylin/eosin. The number of cells within an entire field that migrated through the membrane was counted with a microscope.

Table 2 Primer sequences for real-time qPCR

Gene (Acc N°)	Forward primer (5′–3′)	Reverse primer (5′–3′)	Amplicon (pb)
HPRT1 (NM_000194)	atggacaggactgaacgtcttgct	ttgagcacacagagggctacaatg	80
CK2β (NM_001320)	ttggacctggagcctgatgaagaa	tagcgggcgtggatcaatccataa	101
CK2α (NM_001895)	tgtccgagttgcttcccgatactt	ttgccagcatacaacccaaactcc	104
CK2α′ (NM_001896)	agcccaccaccgtatatcaaacct	atgctttctgggtcgggaagaagt	92
ZEB2 (NM_014795.2)	cgcttgacatcactgaagga	aatgcacagagtgtggcaag	203
U6snRNA (X07425)	ctcgcttcggcagcaca	aacgcttcacgaatttgcgt	94
GAPDH (NM_002046)	atggggaaggtgaaggtcg	ggggtcattgatggcaacaata	108
36B4 (RPLPO) (NM_001002)	gaaatcctgggtgtccgcaatgtt	agacaaggccaggactcgtttgta	113

2.7 RT-qPCR

RNA was isolated using TRIzol® (*Invitrogen*™). One microgram of total RNA was reverse transcribed (kit *AffinityScript*, Agilent™) and cDNA samples were amplified by using gene-specific primers (forward and reverse) that are listed in Table 2. Real-time PCR reactions were done using the SYBR Green Master Mix and ABI PRISM® 7900HT detection system (Applied Biosystems). The thermal cycling conditions were 95 °C for 5 min immediately followed by 45 cycles of 10 s at 95 °C and 30 s at 60 °C. At the end, a dissociation program was perform that begins with 1 min incubation at 95 °C and then 30 s incubation at 55 °C followed by a ramp up to 95 °C. The housekeeping genes HPRT, 36B4, GAPDH, and U6 were used to standardize the mRNA levels of the target genes. The relative quantification in gene expression was determined using the *delta-delta Ct* method [8]. The calibrator used was a pool of different ccRCC cell lines (786-O, ACHN, RCC4, CAKI).

2.8 CK2 Activity

Measure of CK2 activity was as previously described [9] following the P-cellulose filter procedure. Briefly, reaction was started by addition of the sample (3 µl) to the reaction mixture (20 mM Tris–HCl, pH 7.5, 20 mM $MgCl_2$, 150 mM NaCl, 20 µM ATP, 15 µM of the synthetic peptide RRREDEESDDEE and 1 µ Ci of [γ-32P] ATP/reaction) in a final volume of 20 µl. Reaction was performed at 22 °C for 5 min.

2.9 Statistical Analysis

The results were subjected to statistical analysis by the Student's *t* test; the level of significance was set as * $p<.05$ and ** $p<.01$.

3 Results

3.1 CK2 Subunit Expression in Renal Cell Carcinomas

Western blot analyses, CK2 kinase activity, and quantitative polymerase chain reaction (RT-qPCR) were used to assess CK2 protein and mRNA expression in different RCCs and adjacent normal tissues. We observed that in 79.9 % (12/15) of renal cancer samples, CK2α or CK2α′ subunits were highly expressed at the protein level. In comparison, only 13.3 % (2/15) of normal renal samples had similar CK2α or CK2α′ expression compared to tumor samples, and one exhibited a significantly higher expression (Fig. 1a, b).

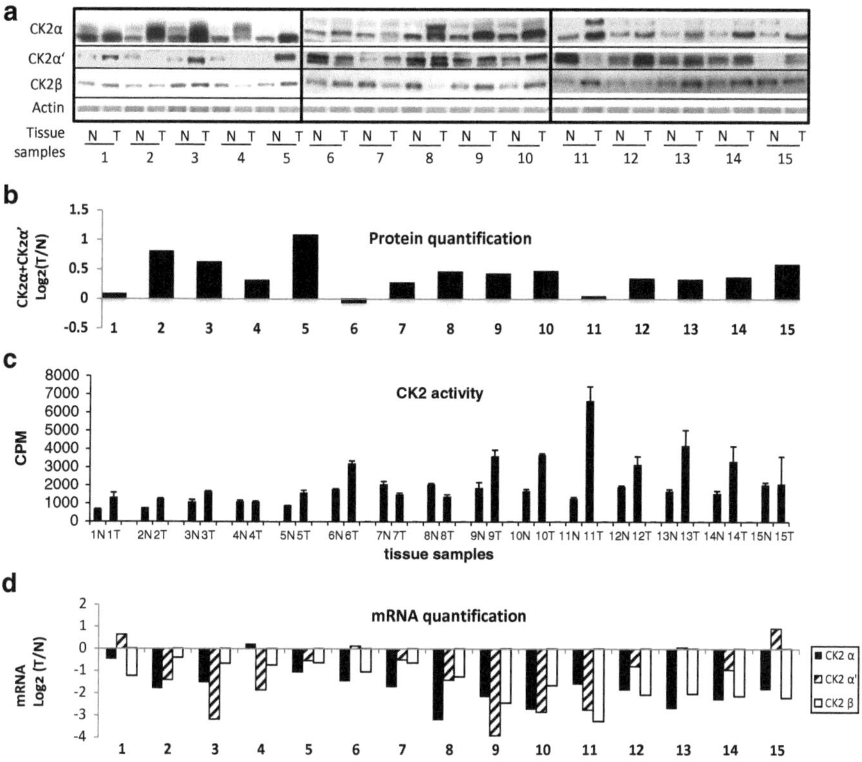

Fig. 1 Expression of CK2 subunits in RCC samples. (**a**) Western blot analysis of CK2 subunit expression (CK2α, CK2α′, and CK2β) on a selected subset of 15 RCC samples and adjacent normal tissues. Actin was used as a loading control. (**b**) CK2 subunit signals were quantified using ImageJ software. The log2 of the ratio between (CK2α+CK2α′) in tumor and normal adjacent tissues was calculated after normalization for actin expression. (**c**) CK2 activity was determined in the same samples. Results are the mean of assays run in triplicates. (**d**) RT-qPCR analysis of mRNA levels for CK2α, CK2α′, and CK2β in the same samples. The results are represented as the log2 of the ratio between mRNA for CK2α+CK2α′ or CK2β in tumor and normal adjacent tissues

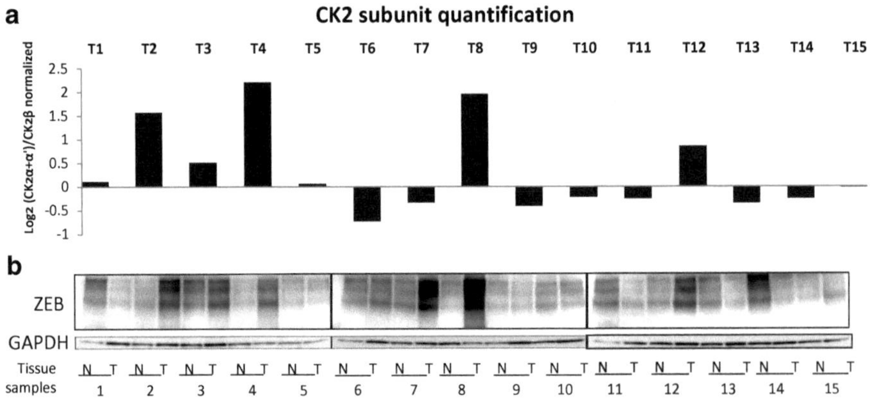

Fig. 2 Unbalanced expression of CK2 subunits in RCC samples. (**a**) CK2 subunit expression in the 15 RCC samples was quantified as described in Fig.1 and the CK2 subunit signals were represented as the log2 of the ratio between CK2α+CK2α′ and CK2β in tumor samples after normalization for their expression in normal adjacent tissue. (**b**) Western blot analysis of ZEB2 expression on the 15 RCC samples and adjacent normal tissues. GAPDH was used as a loading control

This overexpression was correlated with an enhanced CK2 kinase activity. However, this correlation was not absolute (Fig. 1c). Surprisingly, an RT-qPCR analysis of the relative expression of CK2α/α′ mRNA in the RCC samples and adjacent normal tissues showed a strong downregulation of these transcripts in tumor samples compared with normal tissues. These data reveal a negative correlation between mRNA abundance and protein expression, suggesting an important post-transcriptional dysregulation in RCCs (Fig. 1d).

Relative levels of the three CK2 subunits varied significantly between tumor samples.

In particular, the ratio between catalytic and regulatory subunits was significantly altered in a subset of samples (Fig. 2a; #2, #3, #4, #8, #12 compared with #6, #7, #9, #10, #11, #13, #15).

Since we previously reported that an unbalanced expression of CK2 subunits in breast tumor samples is correlated with induction of markers of the epithelial-to-mesenchymal transition (EMT), we analyzed the ZEB2 protein and mRNA expression in the different RCCs and adjacent normal tissues by Western blot and RT-qPCR. Interestingly, in a subset of tumor samples expressing low CK2β levels, the ZEB2 mesenchymal marker was significantly upregulated (Fig. 2b; #2, #3, #4, #8, #12 compared with #5, #9, #11, #13, #14, #15). However, this correlation was not absolute, for example, with sample #7 which exhibited low CK2β and enhanced ZEB2 expression.

3.2 CK2α Overexpression Is Correlated with Classification in Renal Cancer

Next, we investigated the association between CK2α/α′ expression and the clinicopathological characteristics of the RCC cases. We did not find a significant association of CK2α/α′ expression or CK2 activity with patient's age, sex, risk factors, and vascular or lymphatic invasion in 15 cases. However, we observed a trend of correlation between the expression level of CK2α/α′ and the status of clinical stage ($p=.1$), Fuhrman grade (I–II vs. III–IV) ($\chi^2=2.73$), and pathological classification ($p=.05$) (pT1-T2 vs. pT3-T4) ($\chi^2=4.48$) in renal cancer patients (Table 1).

3.3 CK2α Regulates Proliferation and Apoptosis of Renal Cancer Cells

CK2 subunit expression was analyzed in five RCC cell lines (ACHN, CAKI, RCC4 VHL$^+$, RCC4 VHL$^-$, 786-O) with different CK2α, α′/CK2β ratio (data not shown). To evaluate the importance of CK2 activity on the proliferation of renal cancer cells, 786-O RCC cells (786-O WT) or isogenic 786-O cells either reconstituted with functional VHL (786-O VHL) or expressing the empty expression vector pBABE (786-O pB) were exposed to increasing concentrations of CX-4945 for 48 h and analyzed for cell viability by PrestoBlue assay. Growth of 786-O WT, 786-O pB, and 786-O VHL cells was inhibited in a dose-dependent fashion (Fig. 3a) showing a similar sensitivity to the drug ($IC_{25}=17$ μM). We further analyzed the potential of CX-4945 to induce renal cancer cell apoptosis. 786-O WT cells were treated with CX-4945 and cell apoptosis was determined after 48 h of treatment. Effect on apoptotic cell death was evaluated by the examination of PARP cleavage, a key substrate of activated caspases and an early indicator of apoptosis. CX-4945 induced a substantial increase in cleaved PARP accumulation which is indicative of a pro-apoptotic effect (Fig. 3b). Survivin is an anti-apoptotic protein which is highly expressed in RCC patients [11]. We found that incubation of 786-O WT cells with CX-4945 resulted in a marked decreased in survivin expression (Fig. 3c). This effect is consistent with the observation that CK2 interacts with AKT to promote the β-catenin-dependent expression of survivin and enhance cell survival [12].

3.4 CK2 Inhibition Suppresses Constitutive AKT and Wnt Signaling Activation in Renal Cancer Cells

CK2 has been shown to positively regulate the PI3K pathway by affecting PTEN stability [13] and by phosphorylating serine 129 of AKT, which promotes the catalytic activity of AKT [14]. CK2-targeted CX-4945 has been shown to inhibit the

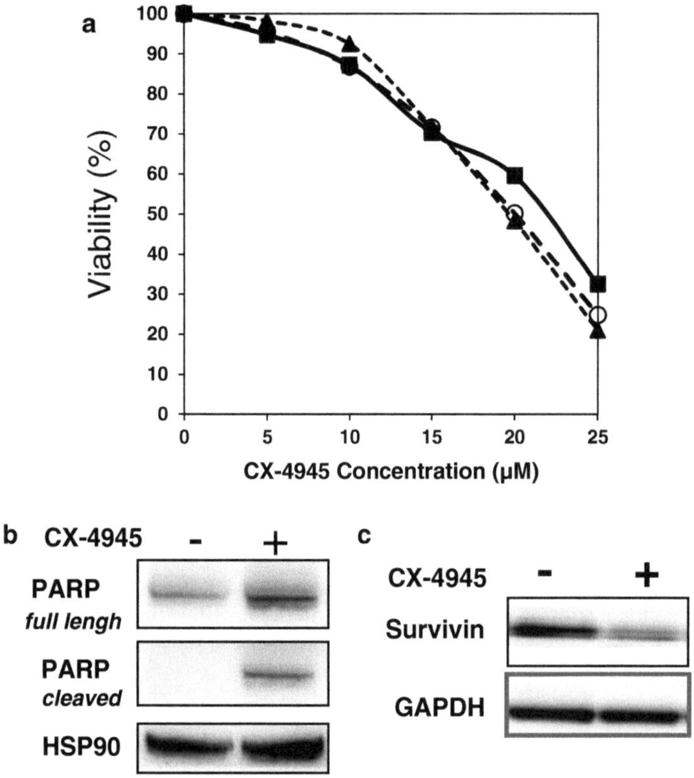

Fig. 3 CX-4945 inhibited cell survival of 786-O cells. (**a**) 786-O WT (*black square*), 786-O pB (*open circle*), or 786-O VHL (*black triangle*) were cultured with dose-titrated CX-4945 for 48 h and cell viability was assessed by the PrestoBlue assay. 786-O WT cells cultured for 48 h in the absence or presence of 6 μM CX-4945 were analyzed by Western blot for PARP cleavage (**b**) or for surviving expression (**c**). HSP90 or GAPDH were used as loading controls

activation of CK2α and its downstream molecules, AKT and p21 [5]. A strong decrease in the constitutive phosphorylation of AKT-S129 was observed in CX-4945-treated cells (Fig. 4a). Activated AKT has been reported to phosphorylate and thereby decrease the stability of the p21 cell-cycle inhibitor protein [15]. Treatment of 786-O WT cells with CX-4945 resulted in reduced phosphorylation of p21 (T145) (Fig. 4a). CX-4945-induced dephosphorylation of the downstream AKT effector p21 (T145) has been reported in human breast cancer cells [5]. It has been shown that the inhibitory effect of α-catenin on β-catenin transactivation is abrogated via CK2α-dependent phosphorylation of α-catenin at S641 which promotes β-catenin transactivation and tumor cell invasion [16].

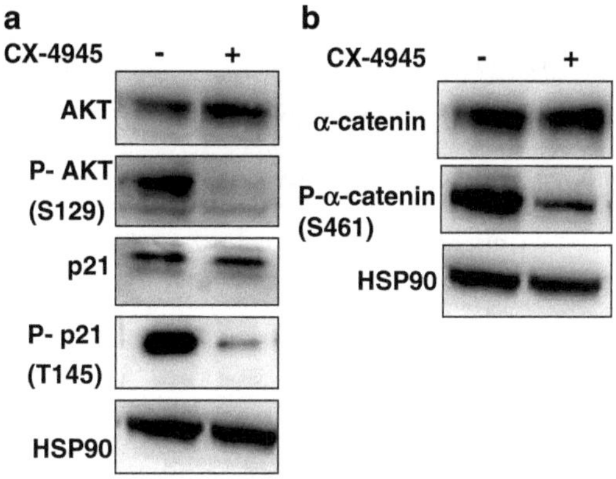

Fig. 4 CX-4945 inhibited phosphorylation of AKT, p21, and α-catenin. 786-O WT cells cultured for 48 h in the absence or presence of 6 μM CX-4945 were analyzed by immunoblot with indicated antibodies. HSP90 was used as loading control

As expected, phosphorylation of α-catenin at serine 641 was inhibited by CX-4945 in 786-O WT cells (Fig. 4b).

3.5 CK2 Inhibition Triggers Activation of the ERK and p38 MAPK Pathways in Renal Cancer Cells

Various genotoxic stresses activate the p38MAPK signaling and it has been shown that in response to TNFα stimulation of HeLa cells, a direct interaction of activated p38MAPK with CK2 is functionally responsible for phosphorylation-dependent p53 activation [17]. Moreover, it has been observed that CK2 inhibitors promote activation of the p38MAPK, thereby affecting cell shape and cytoskeleton of astrocytes and endothelial cells [18]. Here, the addition of CX-4945 to 786-O WT cells induced a strong activation of the p38MAPK (Fig.5a). Of note, p38MAPK activation associated with growth inhibition and cellular senescence has already been reported in human RCC cell lines treated with interleukin-4 [19]. Moreover, the p38MAPK pathway induces and stabilizes the p53 protein level [20, 21]. Accordingly, we found that in 786-O WT cells, CX-4945-induced p38MAPK activation was correlated with increased p53 expression (Fig. 5b). Although activation of ERK is generally related to growth stimulating actions of many growth factors, several reports suggest that ERKs may also mediate apoptotic signaling induced by anti-neoplastic agents [22]. P-ERK was also stimulated by the addition of CX-4945 to 786-O cells (Fig. 5c). This agrees with the observation that downregulation of CK2 induces autophagic cell death through the activation of the ERK signaling pathway in glioblastoma cells [23].

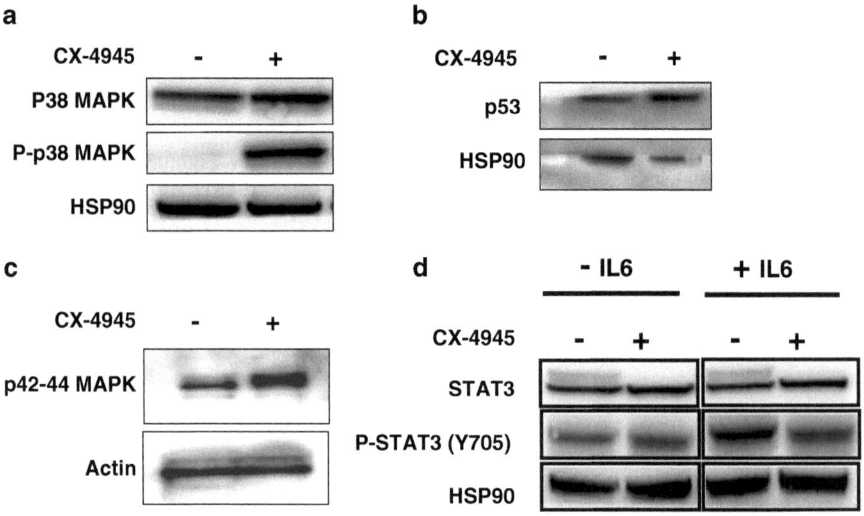

Fig. 5 CK2 inhibition leads to ERK/p38MAPK activation and to decreased IL6-induced STAT-3 activation. 786-O WT cells cultured for 48 h in the absence or presence of 6 μM CX-4945 were analyzed by immunoblot with indicated antibodies (**a**–**c**). The same treatment was followed by stimulation with IL-6 for 10 min (**d**). Lysates were immunoblotted with indicated antibodies. HSP90 or actin were used as loading controls

3.6 CK2 Inhibition Attenuates IL6-Induced Activation of JAK/STAT-3 in Renal Cancer Cells

Chemokine receptor signaling pathways are implicated in the pathobiology of RCC [24].

In particular, P-STAT-3 S727 is an independent prognostic factor for RCC [25]. Since a cross talk between CK2 and JAK/STAT-3 signaling has been reported [10, 26, 27], we tested whether inhibition of CK2 affects STAT-3 activation in renal cancer cells. 786-O WT cells were incubated for 48 h in the absence or presence of CX-4945, stimulated with IL-6 for 10 min to induce STAT-3 activation and then examined for phospho-tyrosine STAT-3 levels. Downregulation of CK2 activity in 786-O WT cells led to reduced IL-6-induced STAT-3 activation (Fig. 5d). It was reported that inhibition of the JAK/STAT-3 signaling pathway in RCC leads to reduced levels of survivin expression [28]. This is consistent with our observation that CX-4945 decreased survivin expression in 786-O WT cells (Fig. 3c).

3.7 Effects of CX-4945 on Cell Migration and Invasion

Migration and invasion of 786-O cells were evaluated using the Boyden chamber assay. 786-O WT, 786-O pB, and 786-O VHL cells exhibited similar behavior in migration assay (Fig. 6a) and CX-4945 had minor effects on their mobility (Fig. 6b).

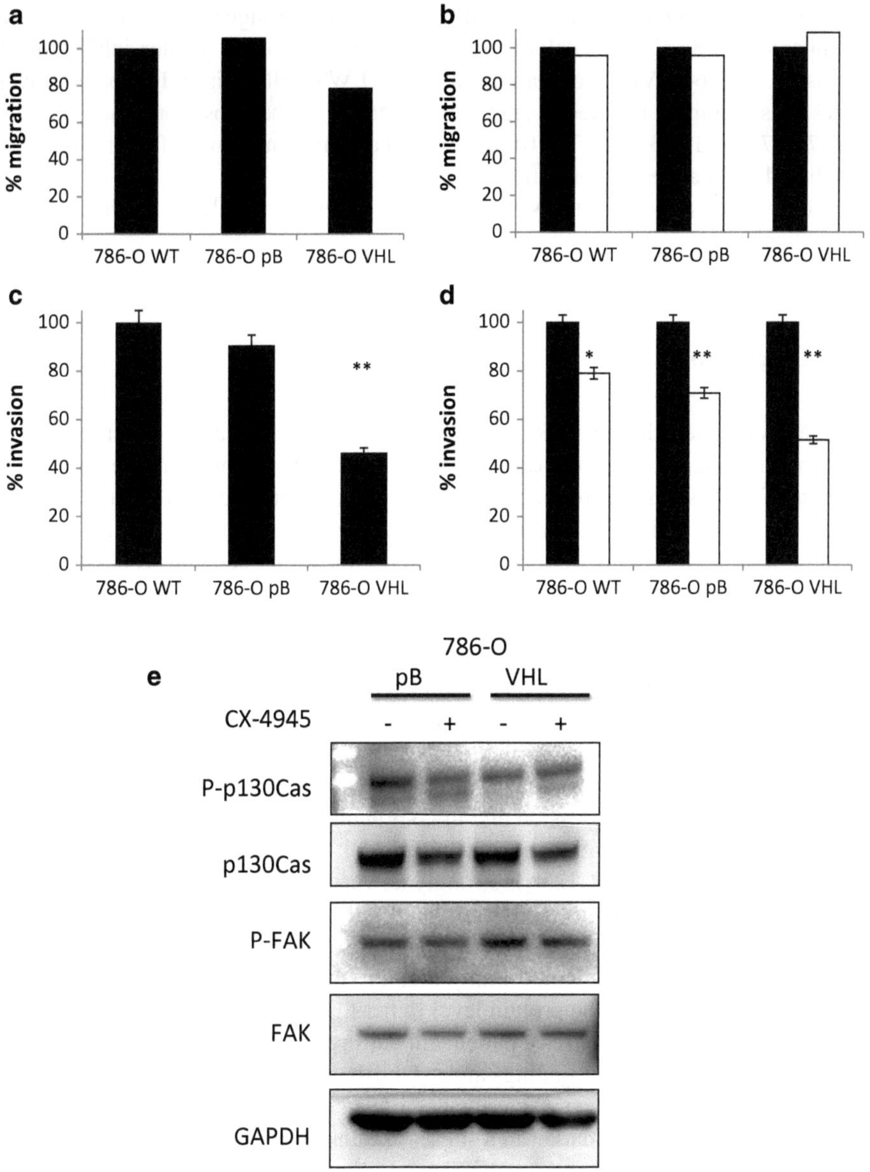

Fig. 6 Effects of CK2 inhibition on the migration/invasion of 786-O cells. 786-O WT, 786-O pB, or 786-O VHL cells were plated in 24-well Boyden chamber plates (**a**, **b**) or Matrigel Invasion Chamber plates (**c**, **d**) using 10 % FBS as chemoattractant. Cells were also incubated in the absence (*black bar*) or presence (*white bar*) of 6 μM CX-4945. After 6 h for migration assay or 24 h for invasion assay, cells that migrated through the membrane were counted. (**e**) 786-O pB or 786-O VHL cells cultured for 24 h in the absence or presence of 6 μM CX-4945 were analyzed by immunoblot with indicated antibodies. GAPDH was used as loading control

In contrast, 786-O WT and 786-O pB cells showed stronger capability to invade compared to 786-O VHL cells (Fig. 6c). CX-4945 showed greater inhibition of invasion in 786-O VHL compared with 786-O WT cells (Fig. 6d). For both cell lines, this treatment resulted in significant decrease in the phosphorylation of FAK (Y576/577) and p130Cas (Y410), which are key components for efficient integrin-mediated focal adhesion formation (Fig. 6e).

These results show that CX-4945 has the potential to inhibit the invasion of 786-O cells.

4 Discussion

Challenges remain in the selection of biomarkers that will allow stratification of patients and prediction of the effectiveness of a kinase inhibitor in one individual patient. CK2 has gained prominence as a nodal kinase and a cancer-related protein, due to its ability to drive cell proliferation and survival. Consequently, CK2 upregulation, described in many tumors, has a strong impact on cellular processes indispensable for cancer cell survival.

In this investigation, we analyzed the protein and mRNA expressions of CK2 subunits in renal carcinoma and normal renal tissues by Western blot, CK2 kinase activity, and RT-qPCR.

We present evidence that CK2 subunits are overexpressed at the protein level in renal carcinoma compared to normal renal tissues, suggesting that CK2 participates in the pathogenesis of renal cancer. Surprisingly, there was no correlation between aberrant CK2 subunit expression and mRNA abundance, reflecting distinctive protein expression and stabilization mechanisms. It is noteworthy that it has been recently reported that miR-125b is strongly downregulated in breast tumors and that the expression of CK2α is inversely correlated with miR-125b expression, suggesting a key posttranscriptional regulation [29]. This posttranscriptional nature of aberrant CK2 expression in RCCs would imply that any predictive assays used in patients need to be performed at the protein level rather than at the mRNA level. However, due to the limited sample size of patients in our investigation, further studies would be needed to confirm these findings and establish the role of CK2 as a reliable clinical predictor for renal cancer. Nevertheless, our analysis revealed an upregulation of the ZEB2 mesenchymal marker in a subset of tumor samples expressing low CK2β levels, suggesting a link between CK2β expression and the EMT process. In breast cancer, a correlation between an altered CK2α/CK2β expression ratio and EMT has been previously established [7]. Moreover, a higher amount of CK2α subunit compared with CK2β subunit has been found in an MDR (multidrug-resistant) leukemia cell line, where it seems to contribute to the apoptosis-resistant phenotype [14]. The kidney is mesenchymal in origin and develops through MET (mesenchymal-to-epithelial transition) to differentiate into mature nephrons [30]. In RCCs, this transition is reversed leading to EMT and dedifferentiation through activation of transcriptional regulators such as ZEB2. Therefore, a link may also exist in RCCs between dysregulated expression of CK2 subunits and EMT-associated features.

To understand the biological functions of CK2 in renal carcinoma, we used the CK2 inhibitor CX-4945 to downregulate the CK2 catalytic activity in 786-O cells as a model of VHL-deficient renal cancer cell line, a situation observed in the majority of RCCs.

We found that CX-4945 inhibited the growth of 786-O WT as well as 786-O VHL cells, suggesting that this drug blocks the growth of renal cancer cells regardless of their VHL status.

As already observed in different cancers, CK2 expression/activity is required for activation of pro-survival pathways [26].

Our findings show that in renal cancer cells, CX-4945 inhibited the activated AKT and Wnt signaling pathways, the IL6-dependent STAT-3 activation, and the focal adhesion signaling pathways including FAK and p130Cas.

Clinical evaluation of CX-4945 in humans as a single agent to treat solid tumors and multiple myeloma has established its promising pharmacokinetic, pharmacodynamic, and safety profiles [31]. Targeting this pleiotropic kinase that influences multiple signaling cascades involved in renal carcinoma progression may prove to be beneficial for the design of future clinical studies.

Acknowledgments This research was supported by the INSERM, CEA, Ligue Nationale contre le Cancer (équipe labellisée 2010–2012), UJF, Ligue Régionale contre le Cancer (Isère), Fondation de France, ARC, and Projet CIBLE. We thank Marlène Debiossat for technical assistance.

References

1. Beroukhim R, Brunet JP, Di Napoli A, Mertz KD, Seeley A, Pires MM et al (2009) Patterns of gene expression and copy-number alterations in von-hippel lindau disease-associated and sporadic clear cell carcinoma of the kidney. Cancer Res 69(11):4674–4681, PubMed PMID: 19470766. Pubmed Central PMCID: 2745239
2. Furge KA, MacKeigan JP, Teh BT (2010) Kinase targets in renal-cell carcinomas: reassessing the old and discovering the new. Lancet Oncol 11(6):571–578, PubMed PMID: 20381423
3. Duncan JS, Litchfield DW (2008) Too much of a good thing: the role of protein kinase CK2 in tumorigenesis and prospects for therapeutic inhibition of CK2. Biochim Biophys Acta 1784(1):33–47, PubMed PMID: 17931986
4. Ruzzene M, Pinna LA (2010) Addiction to protein kinase CK2: a common denominator of diverse cancer cells? Biochim Biophys Acta 1804(3):499–504, PubMed PMID: 19665589
5. Siddiqui-Jain A, Drygin D, Streiner N, Chua P, Pierre F, O'Brien SE et al (2010) CX-4945, an orally bioavailable selective inhibitor of protein kinase CK2, inhibits prosurvival and angiogenic signaling and exhibits antitumor efficacy. Cancer Res 70(24):10288–10298, PubMed PMID: 21159648
6. Pierre F, Chua PC, O'Brien SE, Siddiqui-Jain A, Bourbon P, Haddach M et al (2011) Discovery and SAR of 5-(3-chlorophenylamino)benzo[c][2,6]naphthyridine-8-carboxylic acid (CX-4945), the first clinical stage inhibitor of protein kinase CK2 for the treatment of cancer. J Med Chem 54(2):635–654, PubMed PMID: 21174434
7. Deshiere A, Duchemin-Pelletier E, Spreux E, Ciais D, Combes F, Vandenbrouck Y et al (2013) Unbalanced expression of CK2 kinase subunits is sufficient to drive epithelial-to-mesenchymal transition by Snail1 induction. Oncogene 32(11):1373–1383, PubMed PMID: 22562247
8. Livak KJ, Schmittgen TD (2001) Analysis of relative gene expression data using real-time quantitative PCR and the 2(-Delta Delta C(T)) method. Methods 25(4):402–408, PubMed PMID: 11846609

9. Filhol O, Cochet C, Wedegaertner P, Gill GN, Chambaz EM (1991) Coexpression of both alpha and beta subunits is required for assembly of regulated casein kinase II. Biochemistry 30(46):11133–11140, PubMed PMID: 1932033
10. Aparicio-Siegmund S, Sommer J, Monhasery N, Schwanbeck R, Keil E, Finkenstadt D et al (2014) Inhibition of protein kinase II (CK2) prevents induced signal transducer and activator of transcription (STAT) 1/3 and constitutive STAT3 activation. Oncotarget 5(8):2131–2148, PubMed PMID: 24742922
11. Liu S, Qi L, Yu Q, Song Y, Han W, Zu X et al (2014) Survivin and HLA-I expression predicts survival of patients with clear cell renal cell carcinoma. Tumour Biol 35:8281–8288, PubMed PMID: 24852427
12. Ponce DP, Yefi R, Cabello P, Maturana JL, Niechi I, Silva E et al (2011) CK2 functionally interacts with AKT/PKB to promote the beta-catenin-dependent expression of survivin and enhance cell survival. Mol Cell Biochem 356(1–2):127–132, PubMed PMID: 21735093
13. Torres J, Pulido R (2001) The tumor suppressor PTEN is phosphorylated by the protein kinase CK2 at its C terminus. Implications for PTEN stability to proteasome-mediated degradation. J Biol Chem 276(2):993–998, PubMed PMID: 11035045
14. Di Maira G, Salvi M, Arrigoni G, Marin O, Sarno S, Brustolon F et al (2005) Protein kinase CK2 phosphorylates and upregulates Akt/PKB. Cell Death Differ 12(6):668–677, PubMed PMID: 15818404
15. Li Y, Dowbenko D, Lasky LA (2002) AKT/PKB phosphorylation of p21Cip/WAF1 enhances protein stability of p21Cip/WAF1 and promotes cell survival. J Biol Chem 277(13):11352–11361, PubMed PMID: 11756412
16. Ji H, Wang J, Nika H, Hawke D, Keezer S, Ge Q et al (2009) EGF-induced ERK activation promotes CK2-mediated disassociation of alpha-catenin from beta-catenin and transactivation of beta-catenin. Mol Cell 36(4):547–559, PubMed PMID: 19941816. Pubmed Central PMCID: 2784926
17. Sayed M, Kim SO, Salh BS, Issinger OG, Pelech SL (2000) Stress-induced activation of protein kinase CK2 by direct interaction with p38 mitogen-activated protein kinase. J Biol Chem 275(22):16569–16573, PubMed PMID: 10747897
18. Kramerov AA, Golub AG, Bdzhola VG, Yarmoluk SM, Ahmed K, Bretner M et al (2011) Treatment of cultured human astrocytes and vascular endothelial cells with protein kinase CK2 inhibitors induces early changes in cell shape and cytoskeleton. Mol Cell Biochem 349 (1–2):125–137, PubMed PMID: 21125314. Pubmed Central PMCID: 3426611
19. Kim HD, Yu SJ, Kim HS, Kim YJ, Choe JM, Park YG et al (2013) Interleukin-4 induces senescence in human renal carcinoma cell lines through STAT6 and p38 MAPK. J Biol Chem 288(40):28743–28754, PubMed PMID: 23935100. Pubmed Central PMCID: 3789971
20. Hadjal Y, Hadadeh O, Yazidi CE, Barruet E, Binetruy B (2013) A p38MAPK-p53 cascade regulates mesodermal differentiation and neurogenesis of embryonic stem cells. Cell Death Dis 4:e737, PubMed PMID: 23887628. Pubmed Central PMCID: 3730419
21. De Amicis F, Giordano F, Vivacqua A, Pellegrino M, Panno ML, Tramontano D et al (2011) Resveratrol, through NF-Y/p53/Sin3/HDAC1 complex phosphorylation, inhibits estrogen receptor alpha gene expression via p38MAPK/CK2 signaling in human breast cancer cells. FASEB J 25(10):3695–3707, PubMed PMID: 21737614. Pubmed Central PMCID: 3177579
22. Lahlou H, Saint-Laurent N, Esteve JP, Eychene A, Pradayrol L, Pyronnet S et al (2003) sst2 somatostatin receptor inhibits cell proliferation through Ras-, Rap1-, and B-Raf-dependent ERK2 activation. J Biol Chem 278(41):39356–39371, PubMed PMID: 12878607
23. Olsen BB, Svenstrup TH, Guerra B (2012) Downregulation of protein kinase CK2 induces autophagic cell death through modulation of the mTOR and MAPK signaling pathways in human glioblastoma cells. Int J Oncol 41(6):1967–1976, PubMed PMID: 23007634. Pubmed Central PMCID: 3583692
24. Stofas A, Levidou G, Piperi C, Adamopoulos C, Dalagiorgou G, Bamias A et al (2014) The role of CXC-chemokine receptor CXCR2 and suppressor of cytokine signaling-3 (SOCS-3) in renal cell carcinoma. BMC Cancer 14:149, PubMed PMID: 24593195. Pubmed Central PMCID: 4015755

25. Cuadros T, Trilla E, Sarro E, Vila MR, Vilardell J, de Torres I et al (2014) HAVCR/KIM-1 activates the IL-6/STAT-3 pathway in clear cell renal cell carcinoma and determines tumor progression and patient outcome. Cancer Res 74(5):1416–1428, PubMed PMID: 24390735
26. Zheng Y, McFarland BC, Drygin D, Yu H, Bellis SL, Kim H et al (2013) Targeting protein kinase CK2 suppresses prosurvival signaling pathways and growth of glioblastoma. Clin Cancer Res 19(23):6484–6494, PubMed PMID: 24036851. Pubmed Central PMCID: 3932633
27. Zheng Y, Qin H, Frank SJ, Deng L, Litchfield DW, Tefferi A et al (2011) A CK2-dependent mechanism for activation of the JAK-STAT signaling pathway. Blood 118(1):156–166, PubMed PMID: 21527517. Pubmed Central PMCID: 3139382
28. Li P, Yuan S, Galipeau J (2013) A fusion cytokine coupling GMCSF to IL9 induces heterologous receptor clustering and STAT1 hyperactivation through JAK2 promiscuity. PLoS One 8(7):e69405, PubMed PMID: 23840913. Pubmed Central PMCID: 3698169
29. Feliciano A, Castellvi J, Artero-Castro A, Leal JA, Romagosa C, Hernandez-Losa J et al (2013) miR-125b acts as a tumor suppressor in breast tumorigenesis via its novel direct targets ENPEP, CK2-alpha, CCNJ, and MEGF9. PloS One 8(10):e76247, PubMed PMID: 24098452. Pubmed Central PMCID: 3789742
30. Rivera MN, Haber DA (2005) Wilms' tumour: connecting tumorigenesis and organ development in the kidney. Nat Rev Cancer 5(9):699–712, PubMed PMID: 16110318
31. Kim J, Kim SH (2012) Druggability of the CK2 inhibitor CX-4945 as an anticancer drug and beyond. Arch Pharm Res 35(8):1293–1296, PubMed PMID: 22941473

Protein Kinase CK2 and Dysregulated Oncogenic Inflammatory Signaling Pathways

Etty N. Benveniste, G. Kenneth Gray, and Braden C. McFarland

Abstract Near simultaneously with the explosive expansion in the understanding of CK2's prodigious promiscuity, the kinase was and has been implicated in a large number of cancers through a variety of mechanisms. Over the past few years, tremendous progress has been made in describing the diverse ways in which CK2 signaling promotes tumorigenesis, tumor maintenance, and progression. In this chapter, we address CK2's role in cancer generally and then provide a detailed overview of CK2's ability to regulate two oncogenic signaling pathways, NF-κB and JAK/STAT, in myriad contexts.

Keywords JAK/STAT • NF-κB • Inflammation • Cancer • Signaling • Cytokines • Gene transcription • GBM • Breast cancer • TCGA

1 Protein Kinase CK2

CK2 is a ubiquitous, constitutively active serine/threonine kinase most often found as a tetramer composed of two catalytic subunits (α and/or α′) and two β regulatory subunits [1]. CK2 is highly pleiotropic [2]: it has nearly 450 known substrates involved in gene transcription, signal transduction, cell growth, and apoptosis [1–4]. Indeed, it has been estimated that up to 20 % of the human phosphoproteome is generated by CK2 alone, based on recent bioinformatics analyses [4–6]. Every cellular process is affected to some degree by CK2; notable examples include cell cycle progression [7], maintenance of the cytoskeleton and extracellular matrix [8],

E.N. Benveniste, Ph.D. (✉) • B.C. McFarland
Department of Cell, Developmental and Integrative Biology, University of Alabama at Birmingham, Birmingham, AL 35294, USA
e-mail: tika@uab.edu

G.K. Gray
PhD Program in Biological and Biomedical Sciences, Harvard University, Boston, MA 02115, USA

K. Ahmed et al. (eds.), *Protein Kinase CK2 Cellular Function in Normal and Disease States*, Advances in Biochemistry in Health and Disease 12, DOI 10.1007/978-3-319-14544-0_15

transcription and translation [1], chaperone-mediated protein folding [9], signal transduction [3], angiogenesis [10], and general cell survival [7]. Recent studies have shown that CK2 also possesses tyrosine kinase activity [11], with specific substrates such as JAK2 [12] and histone H2A [13]. CK2 is also found in nearly every subcellular compartment, with the thought that nuclear localization may portend poor survival in a variety of cancers. The fundamental importance of CK2 for cell viability is reflected in murine knockout studies that show that deletion of either CK2α or CK2β is embryonic lethal [14, 15]. Importantly, multiple studies demonstrate that the individual subunits of CK2 (α and β) can function independently [16]. CK2α subunits are catalytically active by themselves, and CK2β can bind proteins independent of the tetramer complex [17]. Loss of CK2β permits increased phosphorylation of caspase 3 by CK2α′, which prevents caspase 3 cleavage and thus inhibits apoptosis [18–20]. As such, the balance of CK2 subunit expression is extremely important in maintaining cell homeostasis and has implications for tumor development (described below).

2 CK2 and Cancer

Given the enormous importance of CK2 in normal cells, its importance in neoplasia may hardly be surprising [2, 3, 21–24]. Although mutations in CK2 have not been reported, elevated CK2 expression and kinase activity have been demonstrated in numerous liquid and solid tumors [2, 5, 22]. This provides a growth advantage as CK2 activity counteracts apoptosis and sustains the cell cycle. Indeed, downregulation of CK2 in cancer cells or in preclinical cancer models by antisense RNA, siRNA, overexpression of kinase-dead CK2 mutants, or pharmacological inhibitors results in the induction of apoptosis [2, 25] (to be discussed in more detail below). In mouse models, CK2 cooperatively promotes oncogenesis and tumor progression with overexpression of oncogenes such as c-Myc [26] or with loss of tumor suppressors such as p53 [27]. Prosurvival genes such as β-catenin; oncogenes such as c-Myc, c-Myb, and c-Jun; and tumor suppressors such as promyelocytic leukemia protein (PML), PTEN, and p53 have been shown to be targets and/or interactors of CK2 ([3, 28]; Fig. 1).

With the vast amount of molecular information available from The Cancer Genome Atlas (TCGA), we now have more specific information available about CK2 subunit gene copy number variation (CNV) and associations with disease progression. Glioma tumors are histologically graded according to the World Health Organization System into grades I–IV, with the highest being grade IV or glioblastoma (GBM) being the most common malignant tumor of the brain [29]. Transcriptional profiling of GBM has led to molecular classification into four distinct subtypes: classical, proneural, neural, and mesenchymal [30]. Despite the combination of surgery followed by radiotherapy and chemotherapy, the median survival of GBM patients is only 12–15 months, a very grim prognosis [31]. CNV analysis in 537 GBM tumors from the TCGA database indicates that *CSNK2A1*,

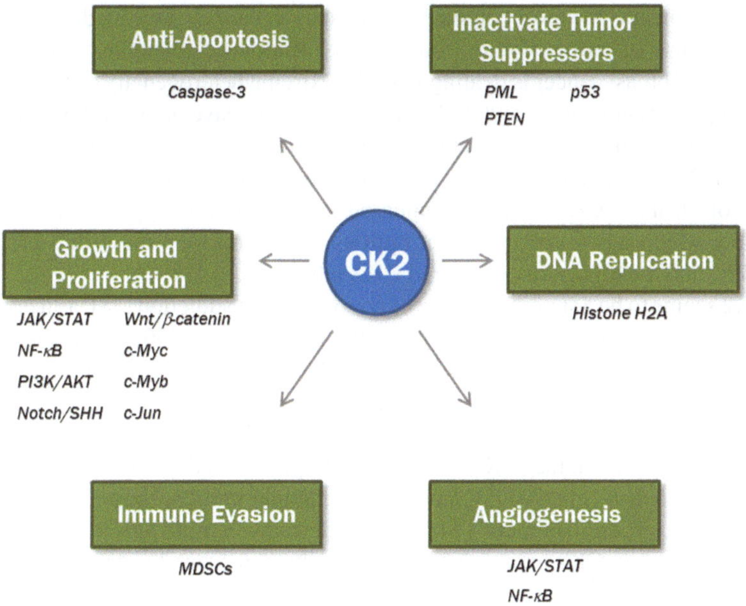

Fig. 1 The impact of CK2 in cancer. CK2 activity enhances many of the pathways described as hallmarks of cancer. CK2 promotes anti-apoptosis through interaction with caspase-3 and inactivates tumor suppressors including PML, PTEN, and p53. DNA replication is enhanced through CK2 interaction with histone H2A, angiogenesis via JAK/STAT and NF-κB activation, and immune evasion through the promotion of MDSCs. CK2 promotes growth and proliferation through multiple signaling pathways involved in oncogenesis, including JAK/STAT, NF-κB, PI3K/AKT, Notch/Sonic Hedgehog (SHH), Wnt/β-catenin, c-Myc, c-Myb, and c-Jun

the gene encoding CK2α and mapping to chromosome 20p13||C, shows low-level amplifications (i.e., gene dosage gains) in 33.7 % of all GBM tumors, and significantly higher *CSNK2A1* mRNA levels were detected in GBMs with *CSNK2A1* gene dosage gains [32]. Among 490 GBM samples with molecular subtype information, *CSNK2A1* gene dosage gain is more common (50.7 %) in classical GBM than in nonclassical GBM (21.3 %). Only minimal changes were noted in *CSNK2A2*, the gene encoding the CK2α′ catalytic subunit. There is relatively little known about the contribution of CK2β to cancer, although recent studies demonstrated that loss of CK2β promotes EMT in breast cancer cells in a SNAIL1-dependent fashion while increasing the expression of TGF-β2 [20, 33, 34]. Thus, CK2β, under certain contexts, appears to function as a tumor suppressor. Interestingly, we observed that *CSNK2B*, the gene encoding CK2β and mapping to chromosome 6p21.3IIc, is deleted in 14.1 % of all GBM tumors analyzed. Importantly, this deletion portends a comparatively poor prognosis (overall survival) in a two-class model (log-rank $p=0.009$) in 553 GBM. The *CSNK2B* deletion is not particularly enriched in any of the four GBM subtypes. As such, loss of CK2β may lead to upregulated CK2α activity and subsequent GBM growth.

Breast cancer is the most common malignancy in women worldwide and remains a major cause of mortality due to metastasis. Another challenging aspect for the treatment of breast cancer is dealing with the extreme heterogeneity of these tumors [35]. Approximately 30 % and 20 % of breast tumors have gains on *CSNK2A1* and *CSNK2B*, respectively, while a large number of tumors have heterozygous deletions of *CSNK2A2* [36]. The basal breast cancer molecular subtype possesses higher levels of all three CK2 subunits compared to non-basal subtypes [36]. The basal subtype, which has fewer treatment options and therefore a more dire prognosis [35, 37], may be highly susceptible to CK2 inhibition.

CK2 seems to be a superb example of non-oncogene addiction [38], as it enhances every one of Hanahan and Weinberg's hallmarks of cancer [39, 40] and underpins multidrug resistance ([41–43]; Fig. 1). CK2 is a "lateral" (nonhierarchical and therefore atypical) kinase [1] that affects numerous pathways vital in cancer, including JAK/STAT, NF-κB, Wnt/β-catenin, PI3K/AKT, and Notch and Hedgehog pathways. It was recently shown that CK2 kinase activity promotes development of myeloid-derived suppressor cells (MDSC), which exert potent immunosuppressive effects in the tumor microenvironment [44]. Thus, there are a multitude of mechanisms by which aberrant CK2 expression and/or activity can promote cancer progression. In this chapter, we review the literature on how CK2 regulates two intertwined tumor-promoting pathways: NF-κB and JAK/STAT.

3 The NF-κB Signaling Pathway

The NF-κB proteins are a family of transcription factors that mediate immune and inflammatory responses [45]. The family contains five structurally similar members that are classified into two groups. The first group, consisting of p65 (RelA), c-Rel, and RelB, is synthesized in their mature forms and contains an N-terminal Rel homology domain (RHD) and a C-terminal transactivation domain (TAD). The second group, consisting of NF-κB1 (p105/p50) and NF-κB2 (p100/p52), is first synthesized as large precursors (p105 and p100) that are later processed into their mature forms (p50 and p52). Both p50 and p52 contain an RHD but neither possesses a TAD. Although these family members can dimerize in numerous combinations, only the NF-κB dimers that contain p65, RelB, or c-Rel are competent transcription factors [46]. The p65/p50 heterodimer predominates, and it is this form that is referred to as NF-κB herein. Typically, all NF-κB molecules are inactive and cytoplasmic. However, activation of these molecules is achieved in response to various stimuli using either the canonical or noncanonical pathway. Although noncanonical NF-κB appears to be involved in certain cancers, its role is currently poorly defined, and canonical signaling is currently considered the major contributor to cancers broadly. In the canonical pathway (Fig. 2a), NF-κB molecules containing p65 and p50 are sequestered through interactions with the inhibitor of NF-κB (IκB) proteins. The IκB proteins contain multiple ankyrin repeat domains that enable them to bind to the RHD and inhibit NF-κB's ability to bind DNA. The pathway is triggered by

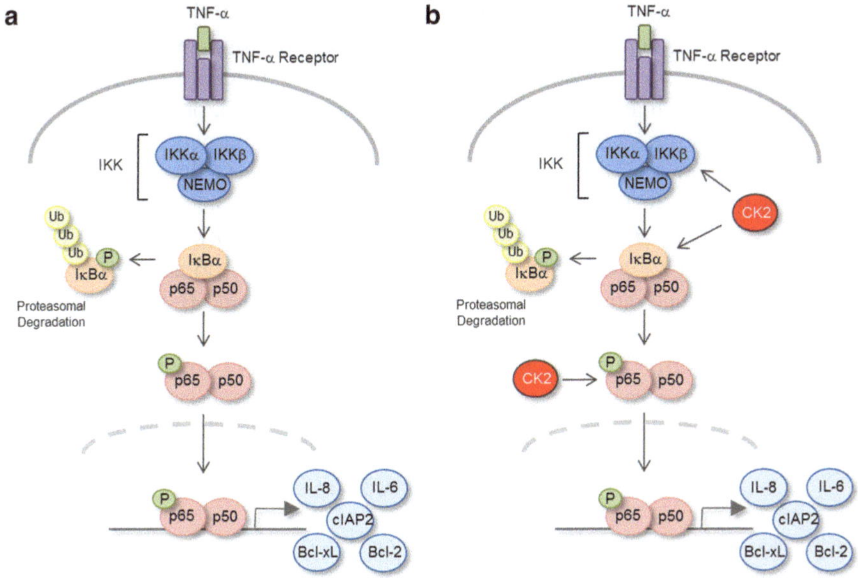

Fig. 2 NF-κB signaling and the contribution of CK2. (**a**) Canonical NF-κB signaling is initiated by inflammatory cytokines, such as TNF-α. Binding of TNF-α to the receptor activates the intracellular signaling complex IKK, which consists of IKKα, IKKβ, and NEMO. IKK phosphorylates IκBα, which results in ubiquitination and degradation of IκBα. The NF-κB dimer, p65/p50, are then phosphorylated and translocate to the nucleus to induce gene expression. (**b**) CK2 interacts with NF-κB signaling in many ways. CK2 phosphorylates IKKβ, which activates IKK and leads to phosphorylation of IκBα and downstream NF-κB activation. IκBα is also directly phosphorylated by CK2 to promote NF-κB activation. Lastly, CK2 phosphorylates p65 on S529, leading to enhanced activation of NF-κB gene transcription

various inflammatory stimuli that activate the inhibitor of NF-κB kinase (IKK) complex, which contains IKKα, IKKβ, and IKKγ/NEMO. In NF-κB signaling, IKKβ phosphorylates the IκB proteins, which targets them for rapid degradation by the proteasome. Then, p65 is phosphorylated at serine 536 and activated; this effectively liberates the NF-κB molecules, which then translocate into the nucleus, bind to their cognate DNA response elements, and induce expression of genes such as *IL-6* and *IL-8*, as well as antiapoptotic genes including *cIAP2*, *Bcl-2*, and *Bcl-xL* (Fig. 2a). A large number of NF-κB target genes have been implicated in cancer pathogenesis, progression, and metastasis.

4 CK2, NF-κB Signaling, and Cancer

The NF-κB signaling pathway is constitutively activated in a large variety of cancers, including breast, pancreatic, prostate, head and neck, multiple myeloma, and GBM, which portends poor patient survival [47–52]. Indeed, NF-κB has been demonstrated to promote pro-cancerous inflammation, prosurvival signals, angiogenesis, metastasis, and radiation resistance [46, 50]. In malignant cells, NF-κB can be activated by the overexpression of proinflammatory mediators such as TNF-α and oncogenes such as Ras and by amplifications, mutations, or deletions in gene loci encoding *c-el*, *NFKBIA* (*IκBα*), and *A20* [53–57]. CK2 targets the NF-κB pathway at multiple points, including IKKs, IκBs, and p65 itself (Fig. 2b). CK2 mediates the activation of IKKβ by phosphorylation, which then leads to the ability of IKKβ to phosphorylate IκBα, promoting NF-κB activation [58]. Additionally, CK2 can directly phosphorylate IκB in the carboxy-terminal PEST domain, leading to its degradation and activation of NF-κB activity [59]. TNF-α and IL-1β induce phosphorylation of p65 Ser529 by CK2, which is associated with full transcriptional activity of p65 [60]. Thus, the NF-κB pathway is positively regulated by CK2 in a manner that promotes tumor cell survival and inhibits apoptosis [3]. There are numerous examples of connections of CK2 and aberrant NF-κB signaling in a variety of cancers.

5 Breast Cancer

Primary human breast cancer samples that display aberrant activation of the NF-κB pathway also exhibit increased CK2 kinase activity [52]. Breast cancer cell lines display higher levels of IKKα and IKKβ, which correlates with increased CK2 and NF-κB activity, compared to untransformed mammary epithelial cells [36, 52]. The inhibition of CK2 activity by selective CK2 inhibitors decreased NF-κB activity in these cell lines [36, 52]. Treatment of triple negative MDA-MB-231 breast cancer cells with the CK2 inhibitor emodin suppressed NF-κB activation and expression of *MMP-9*, an NF-κB target gene [61]. Interestingly, we observed that treatment of MDA-MB-231 cells with the CK2 inhibitor CX-4945 inhibited constitutive phosphorylation of serine 529 in p65, but did not affect TNF-α-induced serine 536 phosphorylation, which is the canonical marker of NF-κB activation and transcriptional capacity [36]. Nonetheless, inhibition of CK2 activity by CX-4945 significantly decreased expression of several NF-κB target genes, including *IL-6* and *IL-8* [36], suggesting that decreasing p65 serine 529 phosphorylation is sufficient to suppress gene transcription. Phase I clinical studies have shown that patients with advanced solid tumors have high circulating levels of IL-6 and IL-8, and treatment with CX-4945 decreases the circulating levels of these cytokines [62]. This may be extremely important since both IL-6 and IL-8 have been connected to stemness and metastasis, as well as the tumor self-seeding phenomenon [63, 64]. CX-4945 displayed antitumor efficacy in preclinical studies of BT-474 breast cancer growth in vivo, significantly inhibiting tumor volume, with no effect on body weight [65]. The Her-2/neu oncogene is overexpressed in approximately 30 % of breast cancers and is associated with poor overall survival. Her-2/neu is able to induce activation of NF-κB in breast cancer cell lines, which is suppressed by inhibition of CK2 [51]. Thus, CK2 plays an important role in Her-2/neu signaling, promoting degradation of IκBα, leading to subsequent NF-κB activation.

6 Pancreatic Cancer

Pancreatic cancer is one of the most common causes of cancer death worldwide, and the 5-year survival rate remains dismal. In the context of pancreatic cancer, the CK2 inhibitor emodin has beneficial effects both in vitro and in vivo. Treatment of SW1990 pancreatic cancer cells with emodin suppressed migration and invasion and also downregulated NF-κB activation and MMP-9 expression [66]. Interestingly, the presence of cleaved caspase 3 was increased, which promoted apoptotic events. In vivo, emodin treatment significantly decreased tumor weight and metastasis, and NF-κB activation and MMP-9 expression were suppressed in the tumor tissue [66]. In another in vivo pancreatic model (BxPC-3 cells), CX-4945 treatment inhibited tumor volume by 93 % [65]. The influence of CK2 inhibitors on repressing metastasis is exceptionally important, as metastatic disease is responsible for the grim 5-year survival rate.

7 Multiple Myeloma (MM)

MM is one of the most frequent hematologic malignancies and is characterized by resistance to drug-induced apoptosis, which underlies poor outcome for MM patients. CK2 is aberrantly active in MM cells and controls their survival [5, 67]. Inhibition of CK2 kinase activity with a variety of compounds (K27, TBB, and IQA) or inhibition of CK2 expression by siRNA promoted apoptosis of MM cells that was associated with activation of caspase cascades and decreased IκB degradation and NF-κB-driven gene transcription. Bortezomib is a first-in-class proteasome inhibitor with efficacy against MM. The CK2 inhibitors CX-4945 and K27 were shown to function in a synergistic manner with bortezomib with respect to inducing apoptosis [68]. This was associated with a decrease in serine 529 phosphorylation of p65 and a reduction in NF-κB target genes such as *COX-2, IL-6, Bcl-2,* and *NOS-2* [68].

8 High-Grade Head and Neck Squamous Cell Carcinomas (HNSCC)

These cancers usually begin in the squamous cells that line the mucosal surfaces inside the mouth, nose, and throat. A subset of HNSCC carries an elevated risk of local occurrence, nodal or distant metastasis (usually to the lungs); 5-year survival is estimated at 25–45 %. Increased nuclear presence of phospho-p65 has been demonstrated in the majority of HNSCC, which correlated with decreased survival. This is also associated with increased CK2 activity [58]. Specifically in HNSCC, CK2 activates IKKβ, leading to NF-κB activation, which can be inhibited by downregulation of the CK2β subunit. In addition, apigenin, a CK2 inhibitor, suppressed the ability of CK2 to phosphorylate IKKβ [58]. The authors suggest that both CK2 and IKKβ may serve as key targets for inhibition of the NF-κB pathway in HNSCC.

9 GBM

NF-κB is constitutively activated in GBM and correlates with increasing grade in astrocytic tumors [69]. The mesenchymal subtype of GBM is characterized by elevated levels of NF-κB signaling components (*TRADD*, *RELB*, *TNFRSF1A*), enhanced chemo- and radiation resistance, and an overall poorer prognosis than patients with other types of GBM [30]. There are numerous proteins and pathways dysregulated in GBM that may cause NF-κB activation. NF-κB is activated by epidermal growth factor (EGF), and/or its receptor, (EGFR), the latter of which is frequently mutated and constitutively activated in GBM [70, 71]. We have shown that ING4, a negative regulator of NF-κB, is expressed at very low levels or is mutated in GBM and that the lack of ING4 activity enables NF-κB to remain constitutively active [72]. Conversely, Pin1, a positive regulator of NF-κB, is overexpressed in GBM and also contributes to constitutive NF-κB activation [73].

In vitro, in a variety of human GBM cell lines, CK2 inhibitors abrogate TNF-α-induced NF-κB activity and sensitize these cells to TNF-α-induced apoptosis [74]. TNFα-induced p65 serine 529 phosphorylation was partially inhibited by CX-4945 in human GBM xenografts, and the NF-κB target genes *IκBα* and *IL-8* were also inhibited by CX-4945 treatment [32]. Treatment with CX-4945 suppressed xenograft cell growth as did knockdown of CK2α and CK2α′ [32]. CX-4945 induced GBM cell apoptosis and caused senescence in GBM cells [32]. Importantly, CX-4945 treatment of normal astrocytes did not affect cell survival or promote apoptosis [32], indicating that transformed cells have enhanced sensitivity to CK2 inhibition. CX-4945 inhibited intracranial GBM tumor growth in vivo. Xenograft X1046 (classical subtype) has gene dosage gain of *CSNK2A1*, and CX-4945 treatment significantly increased survival of X1046 tumor-bearing mice. Constitutively active NF-κB p65 was detected in intracranial tumors of the vehicle-treated group, while activation of NF-κB p65 was diminished in mice treated with CX-4945 [32]. Importantly, oral administration of CX-4945 inhibited serine 529 phosphorylation of p65 in intracranial tumors. Achieving inhibition of this CK2-specific phosphorylation site is indicative of biological activity of CX-4945 within the CNS. As mentioned previously, gene dosage gain of *CSNK2A1* in GBM was significantly associated with the classical subtype [32]. Heterozygous deletion of the *NFKBIA* gene, which encodes IκBα, is associated with the nonclassical subtypes of GBM and correlates with elevated NF-κB levels and poorer patient prognosis [55]. Considering that both alterations can lead to constitutive activation of NF-κB, this suggests that classical and nonclassical subtypes of tumors have distinct genetic aberrations that may result in the same outcome, namely, activation of the NF-κB signaling pathway in GBM.

10 The JAK/STAT Signaling Pathway

The JAK/STAT signaling pathway is the predominant signal transduction cascade utilized by numerous cytokines and is critical for initiating innate immunity, orchestrating adaptive immune systems, and ultimately constraining inflammatory

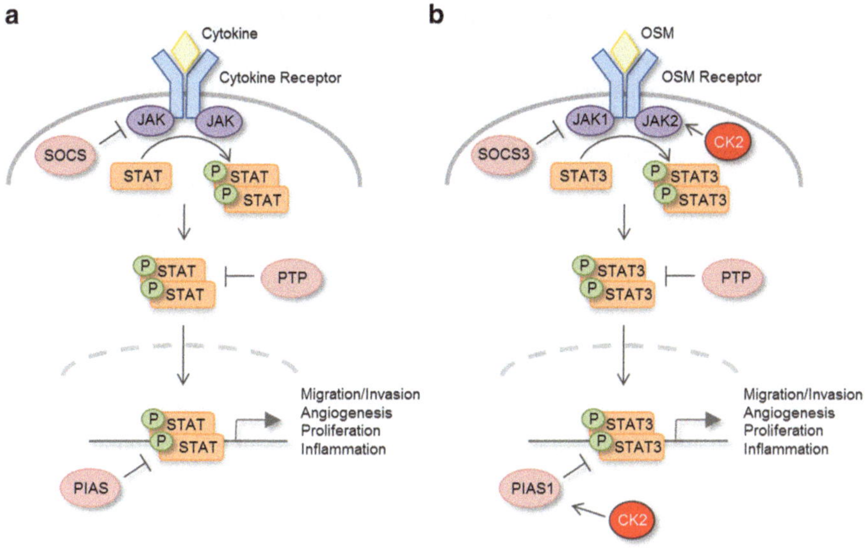

Fig. 3 JAK/STAT signaling and the contribution of CK2. (**a**) Various cytokines bind to their corresponding receptor and activate receptor-associated JAKs, which phosphorylate both themselves and the receptor cytoplasmic domain on tyrosine residues, leading to recruitment of STATs. The JAKs then tyrosine phosphorylate STATs, promoting their activation. Once activated, STATs dimerize, translocate to the nucleus, and bind to regulatory elements to induce transcription of target genes involved in numerous pathways including migration/invasion, angiogenesis, proliferation, and inflammation. There are several negative regulators of JAK/STAT signaling, including SOCS proteins, PIAS proteins, and numerous protein tyrosine phosphatases (PTP). (**b**) CK2 enhances JAK/STAT signaling in many ways. Upon OSM stimulation, CK2 interacts with and phosphorylates JAK2, promoting downstream STAT3 activation. Additionally, CK2 interacts with the negative regulator PIAS1, also promoting enhanced STAT activation

and immune responses [69, 75]. Cytokines activate receptor-associated JAKs, which phosphorylate both themselves and the receptor cytoplasmic domain on tyrosine residues, leading to recruitment of STATs. The JAKs then tyrosine phosphorylate STATs, promoting their activation. Once activated, STATs dimerize, translocate to the nucleus, and bind to regulatory elements to induce transcription of target genes (Fig. 3a). Over 70 cytokines, interferons and growth factors use the JAK/STAT pathway. There are four JAKs (JAK1, JAK2, JAK3, and TYK2) and a total of seven STATs (STAT 1, 2, 3, 4, 5a, 5b, and 6). Various combinations of JAK/STAT usage result in differential gene expression, particularly depending on the STAT transcription factor(s) that is activated. Endogenous negative regulators of the JAK/STAT pathway include suppressors of cytokine signaling (SOCS) proteins, protein inhibitors of activated STATs (PIAS) proteins, and various tyrosine phosphatases [76–78]; Fig. 3a.

11 CK2, JAK/STAT Signaling, and Cancer

JAKs and STATs are essential mediators of almost all biological signaling events initiated by cytokines. As such, unrestrained activation of the JAK/STAT pathway is detrimental and has been associated with many cancers. Of all the STATs, STAT3 is by far the most prominent in cancer with respect to aberrant activation [79–81]. Activating mutations in STAT3 are rare; thus STAT3 hyperactivation is usually due to an overabundance of cytokines and/or dysregulation of negative regulators. STAT3 signaling is commonly activated by the IL-6 family of cytokines, including IL-6, oncostatin M (OSM), and leukemia inhibitory factor, which induces expression of a broad repertoire of genes that regulate antiapoptotic behavior, angiogenesis, migration, invasion, inflammation, and cell proliferation [82] (Fig. 3a). It should be noted that many of the genes induced by the JAK/STAT pathway overlap those induced by the NF-κB pathway [83]. Additionally, IL-6 is an NF-κB-inducible gene and the most potent activator of STAT3, thus highlighting cross talk between these two signal transduction cascades. Interesting, IL-6 is the most dysregulated cytokine in many cancers, being expressed at aberrantly high levels. Activated STAT3 is recognized by phosphorylation of tyrosine 705, which is the canonical marker of STAT3 activation. In addition to tyrosine 705 phosphorylation, STAT3 serine 727 phosphorylation is another important posttranslational modification that affects the transcriptional activity of STAT3. Numerous pathways are capable of phosphorylating serine 727, including the MAPK and PI3K/mTOR cascades, and phosphorylation of serine 727 is necessary for full transcriptional activity of STAT3 [84]. In general, activated STAT3 has a striking ability to promote tumor survival, invasion, and metastasis while suppressing antitumor immunity [80]. We have recently demonstrated that CK2 has the ability to intensify JAK/STAT signaling [12]. OSM-, IFN-γ-, and growth hormone (GH)-induced JAK/STAT activations are dependent on the presence or kinase activity of CK2. Inhibition of CK2 activity by TBB or emodin or knockdown of CK2α or CK2β inhibits OSM-induced JAK1 and JAK2 tyrosine phosphorylation, suggesting that CK2 may function to regulate JAK activation, thus controlling the activation of STATs and downstream gene expression. JAK2 forms a complex with CK2α and CK2β, which is enhanced upon cytokine stimulation. In addition to JAK2, CK2α and CK2β associate with JAK1. This association between JAKs and CK2 suggests that JAKs may be a substrate of CK2 and, indeed, JAK2 can be phosphorylated by CK2 in vitro [12] (Fig. 3b). Preliminary mass spec analysis indicates that CK2 phosphorylates JAK2 on several tyrosine and serine residues. PIAS1, a negative regulator of the JAK/STAT pathway, functions by binding to activated STATs in the nucleus, inhibiting their transcriptional activity [77]. PIAS1 is also a CK2 substrate, and its regulation of transcription factor function is affected by CK2 phosphorylation [85] (Fig. 3b). Additionally, CK2 has been shown to directly phosphorylate STAT1 on serine 727, an oncogenic modification in Wilm's tumor [86]. These data indicate that the JAK/STAT pathway is affected by CK2 in a manner that enhances JAK and STAT activation, which has implications for promoting tumor survival [12, 87, 88].

12 Breast Cancer

The JAK/STAT pathway is involved in both normal mammary development and mammary tumor formation [89]. JAK/STAT signaling has been shown to be important in the progression and metastatic and invasive potential of breast cancer [63, 90–93] and is involved in breast cancer stem cell formation [63]. In particular, constitutive tyrosine phosphorylation of STAT3 is an indicator of poor prognosis and has been associated with late-stage metastatic breast cancer [94, 95]. STAT3 activation is also associated more with advanced breast cancer, as opposed to STAT5, which is more prominently involved in tumor initiation [89]. As mentioned previously, STAT3 can be phosphorylated on serine 727. It has recently been shown that STAT3 functions in the mitochondria to control respiration and Ras transformation, which requires serine 727 phosphorylation, but not that of tyrosine 705 [96, 97]. Serine 727 phosphorylated STAT3 in the mitochondria controls the accumulation of reactive oxygen species and growth of breast cancer in preclinical models [98]. High serum levels of IL-6 predict poor prognosis in breast cancer patients [99], which may be due to its ability to activate STAT3. Inflammatory breast cancer (IBC) is a highly aggressive and proliferative form of breast cancer that disproportionally affects younger women and has a 5-year survival rate of less than 5 %. In an in vivo model of IBC, treatment with CX-4945 inhibited expression of IL-6 [100]. Furthermore, in a phase I clinical trial of CX-4945, an IBC patient exhibited a significant decrease in circulating IL-6 plasma levels [100]. Thus, CK2 activity is able to regulate the expression of IL-6 at the transcriptional level, although the precise mechanisms are not known. We have shown that CK2 inhibition by TBB, emodin, and CX-4945 prevents both constitutive and cytokine-enhanced STAT3 activation and downstream gene expression in the triple negative breast cancer cell line MDA-MB-231 [12, 36]. Further, treatment of MDA-MB-231 cells with CX-4945 caused cell cycle arrest and loss of viability and inhibited the migratory capacity of these cells [36].

13 GBM

STAT3 upregulation, hyperactivation, and nuclear accumulation are well-known features of GBM, and this aberrant STAT3 activation is associated with poor prognosis [69]. We and others have demonstrated that levels of phosphorylated JAK1, JAK2, and STAT3 are elevated in GBM tissues as are STAT3-driven genes, which correlates with increasing tumor grade [69, 82, 101–103]. Loss of endogenous inhibitors of STAT3 is another means by which STAT3 may become hyperactivated. Low expression of the negative regulator PIAS3 has been previously described in GBM by our laboratory, which correlates with increased STAT3 activation and gene expression [101]. PTPRD has been identified as a novel STAT3 tyrosine phosphatase whose activity is lost in approximately 50 % of GBM by means of deletion, epigenetic silencing, and inactivating mutation [104]. Subsequent work demonstrated that in a murine model of glioma, PTPRD heterozygous, but not homozygous, loss could promote gliomagenesis and lead to worse survival. This loss of PTPRD led to increased STAT3 activation and protumorigenic macrophage infiltration [105]. Furthermore, IL-6 and erythropoietin-mediated activation of the STAT3 pathway is crucial for the proliferation and survival of glioma stem cells [106–109]. As well, CK2α has been implicated in GBM tumorigenesis by maintenance of glioma stem cells through regulation of β-catenin [110, 111].

STAT3's involvement in GBM has been highlighted due to its close association with the mesenchymal subtype of GBM. It was shown that STAT3 mRNA levels increase in proneural tumors which recur and thereafter acquire a more mesenchymal-like phenotype [112], and it has been subsequently shown that STAT3 activation and target gene transcription increase in proneural tumors undergoing a mesenchymal shift after radiation treatment [113]. STAT3 and CEBPβ are the two master regulators of the mesenchymal subtype, as identified in a bioinformatic study [114].

Inhibition of JAK/STAT3 signaling has been of considerable interest in preclinical GBM studies. AZD1480, a JAK1/2 inhibitor, potently inhibits activation of STAT3 and JAK2 in human and murine glioma cell lines, blocks glioma cell proliferation, and induces cell death in glioma cells. AZD1480 treatment slowed the growth of tumors in a subcutaneous murine model using human xenografts, and in an intracranial model, AZD1480 significantly extended survival, which was associated with decreased STAT3 activation [103]. G5-7 is a recently described allosteric JAK2 inhibitor which, in GBM, blocks not only JAK2/STAT3 activation but also JAK2 interaction with EGFR, a necessary association for the activation of EGFR and downstream signaling [115].

We have recently shown that CK2 is required for JAK/STAT3 activation in primary human GBM xenografts. siRNA knockdown of CK2α and CK2α′ and pharmacological inhibition of CK2α/α′ kinase activity by CX-4945 reduced IL-6-induced STAT3 activation [32]. IL-6-induced JAK2 activation was also inhibited by CX-4945. Interestingly, total JAK2 levels were reduced after CX-4945 treatment, suggesting CK2 may affect JAK2 stability. Expression of STAT3 target genes such as *SOCS3*, *IL-6, Mcl-1*, and *Pim-1* was inhibited by CX-4945 [32]. A bioinformatics analysis in GBM suggested that CK2 lies downstream of EGFR and EGFRvIII [116], the mutated oncogenic form of EGFR commonly expressed in GBM [117]. EGF-induced STAT5 and STAT3 activation was inhibited by CX-4945, with STAT3 being more effectively inhibited, and knockdown of CK2α and CK2α′ produced the same result [32]. In addition, CX-4945 inhibited expression of EGF-induced c-Myc, a STAT3 target gene. CX-4945 treatment significantly increased survival of GBM tumor-bearing mice. Constitutively active tyrosine phosphorylated STAT3 was detected in intracranial tumors of the vehicle-treated group, while activation was diminished in mice treated with CX-4945 [32]. These findings indicate that CK2 inhibition reduces that transcriptional activity of STAT3, which diminishes brain tumorigenesis [118], suggesting CK2 as a therapeutic target for GBM patients.

Somewhat paradoxically, Mandal et al. [119] report that CK2 negatively regulates phosphorylation of STAT3 at serine 727 in GBM and that this enhances the oncogenic potential of STAT3. They demonstrate that the phosphatase PP2A is activated by CK2 and functions to dephosphorylate serine 727. It is not clear how a reduction in STAT3 serine 727 by CK2/PP2A promotes enhanced STAT3 activity, leading to more aggressive GBM tumor formation; this needs to be determined in future studies.

14 Myeloproliferative Disorders (MPDs)

MPDs are a group of clonal hematopoietic disorders including polycythemia vera (PV), essential thrombocytopenia (ET), and primary myelofibrosis (PMF) [120, 121]. Recent studies have demonstrated that JAK2V617F, a somatic activating point mutation in JAK2, occurs in 95 % of PV patients and in greater than 50 % of ET and PMF patients and is involved in the pathogenesis of MPD [120]. We have shown that CK2α associates with the JAK2 mutant JAK2V617F and that CK2 inhibitors significantly depress the amount of phosphorylated JAK2, STAT3, and STAT5 in primary cells from PV patients [12]. More importantly, CK2 inhibitors promote apoptosis, suppress expression of prosurvival proteins such as Bcl-XL, and decrease levels of pro-caspases 3 and 8 [5, 12, 87]. Furthermore, it was recently shown that JAK2V617F is associated with the GM-CSF receptor at clathrin-coated structures, and this induces a conformational change of JAK2V617F, allowing for activation by CK2 [122]. Patients with MPD are treated with the FDA-approved JAK inhibitor ruxolitinib [75], and these findings suggest that CK2 inhibitors may provide a second, complementary therapeutic for this group of diseases.

15 MM

As mentioned previously, CK2 is involved in MM pathogenesis, in part, via excessive activation of the NF-κB pathway. In addition, CK2 enhances activation of STAT3 in this disease [67]. Analysis of MM patient specimens revealed prominent nuclear phosphoserine 727 STAT3 in association with elevated CK2α expression [68]. Utilizing MM cell lines, Piazza et al. [67] demonstrated that CK2 inhibition by K27 or TBB inhibited IL-6-induced STAT3 activation, suppressing both tyrosine 705 and serine 727 phosphorylation, but with a more striking effect on serine 727. IL-6 is one of the most potent growth-promoting cytokines in MM; as such, CK2 may indeed be a therapeutic target. The drug bortezomib functions in synergy with CK2 inhibitors and CK2 knockdown to suppress STAT3 serine 727 phosphorylation and target genes such as *Bcl2* and *cyclin D1*, which promotes apoptosis of MM cells [68]. It has been proposed that regulation of CK2 may be instrumental for the modulation of MM cell sensitivity to other therapeutic agents [5].

16 Nonsmall Cell Lung Cancer (NSCLC)

NSCLC has traditionally been associated with risk factors such as smoking, but female nonsmokers have an increased incidence of NSCLC. Five-year survival rates based on staging of disease are approximately 30 % for stage II, 10 % for stage III, and 1 % for stage IV. STAT3 is constitutively activated in 50 % of NSCLC primary tumors [123]. Mutations in EGFR mediate STAT3 activation via IL-6 production in human lung adenocarcinoma [124, 125]. CK2 inhibitors block EGF-induced STAT3 tyrosine 705 phosphorylation, which leads to enhanced radiosensitivity of NSCLC cells [126]. Thus, CK2 inhibition may be useful for overcoming radioresistance in lung cancer.

17 T-Cell Large Granular Lymphocytic Leukemia (LGLL)

LGLL is a rare lymphoproliferative disease characterized by the expansion of clonal CD3+CD8+ cytotoxic T-cells. Mutations in STAT3 were found in 40 % of LGLL patients, and all mutations were located in exon 21, encoding the SH2 domain, which is responsible for dimerization and activation of STAT3 [127]. These mutations increase the transcriptional activity of STAT3, suggesting aberrant STAT3 signaling underlies the pathogenesis of this disease. Phosphorylation of the STAT3 variant Y640F was recently shown to be dependent on CK2 activity, as treatment with CK2 inhibitors decreased phosphorylation of STAT3Y640F to basal levels [88]. The authors speculate that STAT3Y640F may be a direct substrate of CK2, although this was not formally shown.

18 Crosstalk Between NF-κB and JAK/STAT3 Signaling in Cancer: CK2 As the Connector?

NF-κB and STAT3 signaling have long been known to be intricately interwoven such that they cooperate in many pathological processes, cancer prominent among these [49, 128]. The astounding promiscuity of protein kinase CK2 concatenates the NF-κB and STAT3 cascades in cancers such as GBM, breast cancer, and MM. CK2 is able to positively regulate both these pathways and thus act as a lateral enhancer of each. CK2 has long been known to enhance NF-κB signaling at many points in the pathway [24]. We now have an appreciation that CK2 does the same for the JAK/STAT pathway [12, 32, 36, 88, 122]. Both NF-κB and STAT3 are activated in the context of GBM, breast cancer, and MM, and pharmacological inhibition of CK2 activity attenuates these pathways simultaneously in in vitro and in vivo models. Thus, CK2 inhibition may provide a novel route by which to inhibit both these oncogenic pathways using a single agent.

19 Conclusions

CK2 is able to regulate numerous cell signaling pathways in the context of cancer in order to promote tumorigenesis, treatment resistance, angiogenesis, prosurvival machinery, and other oncogenic processes. Improving understanding of the functions of individual CK2 subunits is also contributing to a more precise understanding of this kinase's function in disease. In recent years, CK2's ability to positively regulate NF-κB and JAK/STAT signaling in several different cancers has been described, and thus CK2 signaling represents an increasingly attractive therapeutic target for a variety of neoplasms.

Acknowledgments This work was supported in part by NIH grant CA158534 (ENB), NS057563 (ENB), NS050665 (ENB), ABTA Basic Research Fellowship in Honor of Paul Fabbri (BCM), and the William E. Cash Jr. Memorial Fund in Neuro-Oncology Research (BCM).

References

1. Meggio F, Pinna LA (2003) One-thousand-and-one substrates of protein kinase CK2? FASEB J 17(3):349–368
2. Ruzzene M, Pinna LA (2010) Addiction to protein kinase CK2: A common denominator of diverse cancer cells? Biochim Biophys Acta 1804(3):499–504. doi:10.1016/j.bbapap.2009.07.018, Epub 2009 Aug 6
3. Duncan JS, Litchfield DW (2008) Too much of a good thing: The role of protein kinase CK2 in tumorigenesis and prospects for therapeutic inhibition of CK2. Biochim Biophys Acta 1784(1):33–47
4. Bian Y, Ye M, Wang C, Cheng K, Song C, Dong M, Pan Y, Qin H, Zou H (2013) Global screening of CK2 kinase substrates by an integrated phosphoproteomics workflow. Sci Rep 3:3460. doi:10.1038/srep03460
5. Piazza F, Manni S, Ruzzene M, Pinna LA, Gurrieri C, Semenzato G (2012) Protein kinase CK2 in hematologic malignancies: reliance on a pivotal cell survival regulator by oncogenic signaling pathways. Leukemia 26(6):1174–1179. doi:10.1038/leu.2011.385
6. Salvi M, Sarno S, Cesaro L, Nakamura H, Pinna LA (2009) Extraordinary pleiotropy of protein kinase CK2 revealed by weblogo phosphoproteome analysis. Biochim Biophys Acta 1793(5):847–859. doi:10.1016/j.bbamcr.2009.01.013, S0167-4889(09)00029-9 [pii]
7. St-Denis NA, Litchfield DW (2009) Protein kinase CK2 in health and disease: From birth to death: the role of protein kinase CK2 in the regulation of cell proliferation and survival. Cell Mol Life Sci 66(11–12):1817–1829
8. Canton DA, Litchfield DW (2006) The shape of things to come: an emerging role for protein kinase CK2 in the regulation of cell morphology and the cytoskeleton. Cell Signal 18(3):267–275. doi:10.1016/j.cellsig.2005.07.008
9. Miyata Y, Nishida E (2004) CK2 controls multiple protein kinases by phosphorylating a kinase-targeting molecular chaperone, Cdc37. Mol Cell Biol 24(9):4065–4074
10. Montenarh M (2014) Protein kinase CK2 and angiogenesis. Adv Clin Exp Med 23(2):153–158
11. Vilk G, Weber JE, Turowec JP, Duncan JS, Wu C, Derksen DR, Zien P, Sarno S, Donella-Deana A, Lajoie G, Pinna LA, Li SS, Litchfield DW (2008) Protein kinase CK2 catalyzes tyrosine phosphorylation in mammalian cells. Cell Signal 20(11):1942–1951

12. Zheng Y, Qin H, Stuart F, Deng L, Litchfield DW, Terfferi A, Pardanani A, Lin F-T, Li J, Sha B, Benveniste EN (2011) A CK2-dependent mechanism for activation of the JAK-STAT signaling pathway. Blood 118(1):156–166
13. Basnet H, Su XB, Tan Y, Meisenhelder J, Merkurjev D, Ohgi KA, Hunter T, Pillus L, Rosenfeld MG (2014) Tyrosine phosphorylation of histone H2A by CK2 regulates transcriptional elongation. Nature doi:10.1038/nature13736 1-13. doi:10.1038/nature13736
14. Lou DY, Dominguez I, Toselli P, Landesman-Bollag E, O'Brien C, Seldin DC (2008) The alpha catalytic subunit of protein kinase CK2 is required for mouse embryonic development. Mol Cell Biol 28(1):131–139
15. Buchou T, Vernet M, Blond O, Jensen HH, Pointu H, Olsen BB, Cochet C, Issinger OG, Boldyreff B (2003) Disruption of the regulatory β subunit of protein kinase CK2 in mice leads to a cell-autonomous defect and early embryonic lethality. Mol Cell Biol 23(3):908–915
16. Cozza G, Pinna LA, Moro S (2012) Kinase Ck2 inhibition: An update. Curr Med Chem 20(5):671–693. doi:10.2174/092986713804999312
17. Bischoff N, Olsen B, Raaf J, Bretner M, Issinger OG, Niefind K (2011) Structure of the human protein kinase CK2 catalytic subunit CK2alpha' and interaction thermodynamics with the regulatory subunit CK2beta. J Mol Biol 407(1):1–12. doi:10.1016/j.jmb.2011.01.020
18. Duncan JS, Turowec JP, Duncan KE, Vilk G, Wu C, Luscher B, Li SSC, Gloor GB, Litchfield DW (2011) A peptide-based target screen implicates the protein kinase CK2 in the global regulation of caspase signaling. Sci Signal 4(172):30. doi:10.1126/scisignal.2001682
19. Turowec JP, Vilk G, Gabriel M, Litchfield DW (2013) Characterizing the convergence of protein kinase CK2 and caspase-3 reveals isoform-specific phosphorylation of caspase-3 by CK2alpha': implications for pathological roles of CK2 in promoting cancer cell survival. Oncotarget 4(4):560–571
20. Filhol O, Cochet C (2011) Protein kinases curb cell death. Sci Signal 4(172):pe26, 10.1126/scisignal.2001921
21. Hanif IM, Shazib MA, Ahmad KA, Pervaiz S (2011) Casein Kinase II: an attractive target for anti-cancer drug design. Int J Biochem Cell Biol 42(10):1602–1605. doi:10.1016/j.biocel.2010.06.010, S1357-2725(10)00207-4 [pii]
22. Trembley JH, Chen Z, Unger G, Slaton J, Kren BT, Van Waes C, Ahmed K (2010) Emergence of protein kinase CK2 as a key target in cancer therapy. Biofactors 36(3):187–195. doi:10.1002/biof.96
23. Trembley JH, Wang G, Unger G, Slaton J, Ahmed K (2009) Protein kinase CK2 in health and disease: CK2: a key player in cancer biology. Cell Mol Life Sci 66(11–12):1858–1867. doi:10.1007/s00018-009-9154-y
24. Dominguez I, Sonenshein GE, Seldin DC (2009) Protein kinase CK2 in health and disease: CK2 and its role in Wnt and NF-kappaB signaling: linking development and cancer. Cell Mol Life Sci 66(11–12):1850–1857
25. Trembley JH, Unger GM, Korman VL, Tobolt DK, Kazimierczuk Z, Pinna LA, Kren BT, Ahmed K (2012) Nanoencapsulated anti-CK2 small molecule drug or siRNA specifically targets malignant cancer but not benign cells. Cancer Lett 315(1):48–58. doi:10.1016/j.canlet.2011.10.007
26. Channavajhala P, Seldin DC (2002) Functional interaction of protein kinase CK2 and c-Myc in lymphomagenesis. Oncogene 21(34):5280–5288
27. Landesman-Bollag E, Channavajhala PL, Cardiff RD, Seldin DC (1998) p53 deficiency and misexpression of protein kinase CK2α collaborate in the development of thymic lymphomas in mice. Oncogene 16(23):2965–2974
28. Scaglioni PP, Yung TM, Cai LF, Erdjument-Bromage H, Kaufman AJ, Singh B, Teruya-Feldstein J, Tempst P, Pandolfi PP (2006) A CK2-dependent mechanism for degradation of the PML tumor suppressor. Cell 126(2):269–283
29. Omuro A, DeAngelis LM (2013) Glioblastoma and other malignant gliomas: a clinical review. JAMA 310(17):1842–1850. doi:10.1001/jama.2013.280319

30. Verhaak RGW, Hoadley KA, Purdom E, Wang V, Qi Y, Wilkerson MD, Miller CR, Ding L, Golub T, Mesirov JP, Alexe G, Lawrence M, O'Kelly M, Tamayo P, Weir BA, Gabriel S, Winckler W, Gupta S, Jakkula L, Feiler HS, Hodgson JG, James CD, Sarkaria JN, Brennan C, Kahn A, Spellman PT, Wilson RK, Speed TP, Gray JW, Meyerson M, Getz G, Perou CM, Hayes DN (2010) Integrated genomic analysis identifies clinically relevant subtypes of Glioblastoma characterized by abnormalities in PDGFRA, IDH1, EGFR, and NF1. Cancer Cell 17(1):98–110. doi:10.1016/j.ccr.2009.12.020
31. Stupp R, Mason WP, van den Bent MJ, Weller M, Fisher B, Taphoorn MJ, Belanger K, Brandes AA, Marosi C, Bogdahn U, Curschmann J, Janzer RC, Ludwin SK, Gorlia T, Allgeier A, Lacombe D, Cairncross JG, Eisenhauer E, Mirimanoff RO (2005) Radiotherapy plus concomitant and adjuvant temozolomide for glioblastoma. N Engl J Med 352:987–996
32. Zheng Y, McFarland BC, Drygin D, Yu H, Bellis SL, Kim H, Bredel M, Benveniste EN (2013) Targeting protein kinase CK2 suppresses prosurvival signaling pathways and growth of glioblastoma. Clin Cancer Res 19(23):1–11. doi:10.1158/1078-0432.ccr-13-0265
33. Deshiere A, Duchemin-Pelletier E, Spreux E, Ciais D, Forcet C, Cochet C, Filhol O (2011) Regulation of epithelial to mesenchymal transition: CK2beta on stage. Mol Cell Biochem 356(1–2):11–20. doi:10.1007/s11010-011-0942-y
34. Deshiere A, Duchemin-Pelletier E, Spreux E, Ciais D, Combes F, Vandenbrouck Y, Coute Y, Mikaelian I, Giusiano S, Charpin C, Cochet C, Filhol O (2012) Unbalanced expression of CK2 kinase subunits is sufficient to drive epithelial-to-mesenchymal transition by Snail1 induction. Oncogene 32:1373–1383. doi:10.1038/onc.2012.165
35. Curtis C, Shah SP, Chin SF, Turashvili G, Rueda OM, Dunning MJ, Speed D, Lynch AG, Samarajiwa S, Yuan Y, Graf S, Ha G, Haffari G, Bashashati A, Russell R, McKinney S, Group M, Langerod A, Green A, Provenzano E, Wishart G, Pinder S, Watson P, Markowetz F, Murphy L, Ellis I, Purushotham A, Borresen-Dale AL, Brenton JD, Tavare S, Caldas C, Aparicio S (2012) The genomic and transcriptomic architecture of 2,000 breast tumours reveals novel subgroups. Nature 486(7403):346–352. doi:10.1038/nature10983
36. Gray GK, McFarland BC, Rowse AL, Gibson SA, Benveniste EN (2014) Therapeutic CK2 inhibition attenuates diverse prosurvival signaling cascades and decreases cell viability in human breast cancer cells. Oncotarget 5(15):6484–6496
37. Gray J, Druker B (2012) Genomics: the breast cancer landscape. Nature 486(7403):328–329. doi:10.1038/486328a
38. Luo J, Solimini NL, Elledge SJ (2009) Principles of cancer therapy: oncogene and non-oncogene addiction. Cell 136(5):823–837
39. Hanahan D, Weinberg RA (2011) Hallmarks of cancer: the next generation. Cell 144(5): 646–674. doi:10.1016/j.cell.2011.02.013
40. Hanahan D, Weinberg RA (2000) The hallmarks of cancer. Cell 100(1):57–70
41. Di Maira G, Brustolon F, Bertacchini J, Tosoni K, Marmiroli S, Pinna LA, Ruzzene M (2007) Pharmacological inhibition of protein kinase CK2 reverts the multidrug resistance phenotype of a CEM cell line characterized by high CK2 level. Oncogene 26(48):6915–6926. doi:10.1038/sj.onc.1210495, 1210495 [pii]
42. Stolarczyk EI, Reiling CJ, Pickin KA, Coppage R, Knecht MR, Paumi CM (2012) Casein kinase 2alpha (CK2alpha) regulates multidrug resistance associated protein (MRP1) function via phosphorylation of Thr249. Mol Pharmacol 82(3):488–499. doi:10.1124/mol.112.078295
43. Zanin S, Borgo C, Girardi C, O'Brien SE, Miyata Y, Pinna LA, Donella-Deana A, Ruzzene M (2012) Effects of the CK2 inhibitors CX-4945 and CX-5011 on drug-resistant cells. PLoS One 7(11):e49193. doi:10.1371/journal.pone.0049193
44. Cheng P, Kumar V, Liu H, Youn JI, Fishman M, Sherman S, Gabrilovich D (2013) Effects of Notch signaling on regulation of myeloid cell differentiation in cancer. Cancer Res 74(1):141–152. doi:10.1158/0008-5472.CAN-13-1686
45. Hoffmann A, Natoli G, Ghosh G (2006) Transcriptional regulation via the NF-κB signaling module. Oncogene 25(51):6706–6716
46. Perkins ND (2012) The diverse and complex roles of NF-kappaB subunits in cancer. Nat Rev Cancer 12(2):121–132. doi:10.1038/nrc3204

47. Fradet V, Lessard L, Begin LR, Karakiewicz P, Masson AM, Saad F (2004) Nuclear factor-kappaB nuclear localization is predictive of biochemical recurrence in patients with positive margin prostate cancer. Clin Cancer Res 10(24):8460–8464. doi:10.1158/1078-0432.CCR-04-0764
48. Ben-Neriah Y, Karin M (2011) Inflammation meets cancer, with NF-kappaB as the matchmaker. Nat Immunol 12(8):715–723. doi:10.1038/ni.2060
49. Grivennikov SI, Karin M (2010) Dangerous liaisons: STAT3 and NF-kappaB collaboration and crosstalk in cancer. Cytokine Growth Factor Rev 21(1):11–19. doi:10.1016/j.cytogfr.2009.11.005, S1359-6101(09)00112-9 [pii]
50. Karin M (2009) NF-kappaB as a critical link between inflammation and cancer. Cold Spring Harb Perspect Biol 1(5):1–14. doi:10.1101/cshperspect.a000141
51. Romieu-Mourez R, Landesman-Bollag E, Seldin DC, Sonenshein GE (2002) Protein kinase CK2 promotes aberrant activation of nuclear factor-κB, transformed phenotype, and survival of breast cancer cells. Cancer Res 62(22):6770–6778
52. Romieu-Mourez R, Landesman-Bollag E, Seldin DC, Traish AM, Mercurio F, Sonenshein GE (2001) Roles of IKK kinases and protein kinase CK2 in activation of nuclear factor-κB in breast cancer. Cancer Res 61:3810–3818
53. Raychaudhuri B, Han Y, Lu T, Vogelbaum MA (2007) Aberrant constitutive activation of nuclear factor κB in glioblastoma multiforme drives invasive phenotype. J Neuro Oncol 85(1):39–47
54. Zhao X, Laver T, Hong S, Twitty G, DeVos A, DeVos M, Benveniste E, Nozell S (2011) An NF-κB p65-cIAP2 link is necessary for mediating resistance to TNF-α induced cell death in gliomas. J Neuro Oncol 102(3):367–381. doi:10.1007/s11060-010-0346-y
55. Bredel M, Scholtens DM, Yadav AK, Alvarez AA, Renfrow JJ, Chandler JP, Yu IL, Carro MS, Dai F, Tagge MJ, Ferrarese R, Bredel C, Phillips HS, Lukac PJ, Robe PA, Weyerbrock A, Vogel H, Dubner S, Mobley B, He X, Scheck AC, Sikic BI, Aldape KD, Chakravarti A, Harsh GR (2011) NFKBIA deletion in Glioblastomas. N Engl J Med 364(7):627–637. doi:10.1056/NEJMoa1006312
56. Bredel M, Bredel C, Juric D, Duran GE, Yu RX, Harsh GR, Vogel H, Recht LD, Scheck AC, Sikic BI (2006) Tumor necrosis factor-alpha-induced protein 3 as a putative regulator of nuclear factor-kappaB-mediated resistance to O6-alkylating agents in human glioblastomas. J Clin Oncol 24(2):274–287
57. Mayo MW, Wang C-Y, Cogswell PC, Rogers-Graham KS, Lowe SW, Der CJ, Baldwin AS Jr (1997) Requirements of NF-κB activation to suppress p53-independent apoptosis induced by oncogenic Ras. Science 278:1812–1815
58. Yu M, Yeh J, Van Waes C (2006) Protein kinase casein kinase 2 mediates inhibitor-κB kinase and aberrant nuclear factor-κB activation by serum factor(s) in head and neck squamous carcinoma cells. Cancer Res 66(13):6722–6731
59. Kato T Jr, Delhase M, Hoffmann A, Karin M (2003) CK2 is a C-terminal IκB kinase responsible for NF-κB activation during the UV response. Mol Cell 12(4):829–839
60. Wang D, Westerheide SD, Hanson JL, Baldwin AS Jr (2000) Tumor necrosis factor α-induced phosphorylation of RelA/p65 on Ser529 is controlled by casein kinase II. J Biol Chem 275(42):32592–32597
61. Huang Q, Shen HM, Ong CN (2004) Inhibitory effect of emodin on tumor invasion through suppression of activator protein-1 and nuclear factor-kappaB. Biochem Pharmacol 68(2):361–371. doi:10.1016/j.bcp.2004.03.032
62. Marschke RF, Borad MJ, McFarland RW, Alvarez RH, Lim JK, Padgett CS, Von Hoff DD, O'Brien SE, Northfelt DW (2011) Findings from the phase I clinical trials of CX-4945, an orally available inhibitor of CK2. ASCO Meeting Abstracts 29(15 suppl):3087
63. Sansone P, Storci G, Tavolari S, Guarnieri T, Giovannini C, Taffurelli M, Ceccarelli C, Santini D, Paterini P, Marcu KB, Chieco P, Bonafe M (2007) IL-6 triggers malignant features in mammospheres from human ductal breast carcinoma and normal mammary gland. J Clin Invest 117(12):3988–4002
64. Kim MY, Oskarsson T, Acharyya S, Nguyen DX, Zhang XH, Norton L, Massague J (2009) Tumor self-seeding by circulating cancer cells. Cell 139(7):1315–1326. doi:10.1016/j.cell.2009.11.025, S0092-8674(09)01437-8 [pii]

65. Siddiqui-Jain A, Drygin D, Streiner N, Chua P, Pierre F, O'Brien SE, Bliesath J, Omori M, Huser N, Ho C, Proffitt C, Schwaebe MK, Ryckman DM, Rice WG, Anderes K (2011) CX-4945, an orally bioavailable selective inhibitor of protein kinase CK2, inhibits prosurvival and angiogenic signaling and exhibits antitumor efficacy. Cancer Res 70(24):10288–10298. doi:10.1158/0008-5472.CAN-10-1893, 70/24/10288 [pii]
66. Liu A, Chen H, Wei W, Ye S, Liao W, Gong J, Jiang Z, Wang L, Lin S (2011) Antiproliferative and antimetastatic effects of emodin on human pancreatic cancer. Oncol Rep 26(1):81–89. doi:10.3892/or.2011.1257
67. Piazza FA, Ruzzene M, Gurrieri C, Montini B, Bonanni L, Chioetto G, Di Maira G, Barbon F, Cabrelle A, Zambello R, Adami F, Trentin L, Pinna LA, Semenzato G (2006) Multiple myeloma cell survival relies on high activity of protein kinase CK2. Blood 108(5):1698–1707
68. Manni S, Brancalion A, Mandato E, Tubi LQ, Colpo A, Pizzi M, Cappellesso R, Zaffino F, Di Maggio SA, Cabrelle A, Marino F, Zambello R, Trentin L, Adami F, Gurrieri C, Semenzato G, Piazza F (2013) Protein kinase CK2 inhibition down modulates the NF-kappaB and STAT3 survival pathways, enhances the cellular proteotoxic stress and synergistically boosts the cytotoxic effect of bortezomib on multiple myeloma and mantle cell lymphoma cells. PLoS One 8(9):1–16. doi:10.1371/journal.pone.0075280
69. Gray GK, McFarland BC, Nozell SE, Benveniste EN (2014) NF-kappaB and STAT3 in glioblastoma: therapeutic targets coming of age. Expert Rev Neurother doi; 10.1586/14737175.2014.964211 (Early online):1-14. doi:10.1586/14737175.2014.964211
70. Nogueira L, Ruiz-Ontanon P, Vazquez-Barquero A, Lafarga M, Berciano MT, Aldaz B, Grande L, Casafont I, Segura V, Robles EF, Suarez D, Garcia LF, Martinez-Climent JA, Fernandez-Luna JL (2011) Blockade of the NFkappaB pathway drives differentiating glioblastoma-initiating cells into senescence both in vitro and in vivo. Oncogene 30(32):3537–3548. doi:10.1038/onc.2011.74
71. Bonavia R, Inda MM, Vandenberg S, Cheng SY, Nagane M, Hadwiger P, Tan P, Sah DW, Cavenee WK, Furnari FB (2011) EGFRvIII promotes glioma angiogenesis and growth through the NF-kappaB, interleukin-8 pathway. Oncogene 31:4054–4066. doi:10.1038/onc.2011.563
72. Nozell S, Laver T, Moseley D, Nowoslawski L, DeVos M, Atkinson GP, Harrison K, Nabors LB, Benveniste EN (2008) The ING4 tumor suppressor attenuates NF-κB activity at the promoter of target genes. Mol Cell Biol 28(21):6635–6642
73. Atkinson GP, Nozell SE, Harrison DK, Stonecypher MS, Chen D, Benveniste EN (2009) The prolyl isomerase Pin1 regulates the NF-kappaB signaling pathway and interleukin-8 expression in glioblastoma. Oncogene 28:3735–3745. doi:10.1038/onc.2009.232
74. Dixit D, Sharma V, Ghosh S, Mehta VS, Sen E (2012) Inhibition of casein kinase-2 induces p53-dependent cell cycle arrest and sensitizes glioblastoma cells to tumor necrosis factor (TNFalpha)-induced apoptosis through SIRT1 inhibition. Cell Death Dis 3:e271. doi:10.1038/cddis.2012.10
75. O'Shea JJ, Plenge R (2012) JAK and STAT signaling molecules in immunoregulation and immune-mediated disease. Immunity 36(4):542–550. doi:10.1016/j.immuni.2012.03.014
76. Baker BJ, Akhtar LN, Benveniste EN (2009) SOCS1 and SOCS3 in the control of CNS immunity. Trends Immunol 30(8):392–400
77. Shuai K (2006) Regulation of cytokine signaling pathways by PIAS proteins. Cell Res 16(2):196–202
78. Yoshimura A, Suzuki M, Sakaguchi R, Hanada T, Yasukawa H (2012) SOCS, Inflammation, and Autoimmunity. Front Immunol 3(20):1–9. doi:10.3389/fimmu.2012.00020
79. Stark GR, Darnell JE Jr (2012) The JAK-STAT pathway at twenty. Immunity 36(4):503–514. doi:10.1016/j.immuni.2012.03.013
80. Yu H, Pardoll D, Jove R (2009) STATs in cancer inflammation and immunity: a leading role for STAT3. Nat Rev Cancer 9(11):798–809
81. Frank DA (2013) Transcription factor STAT3 as a prognostic marker and therapeutic target in cancer. J Clin Oncol 31(36):4560–4561. doi:10.1200/JCO.2013.52.8414
82. Brantley E, Benveniste EN (2008) Signal transducer and activator of transcription-3: A molecular hub for signaling pathways in gliomas. Mol Cancer Res 6(5):675–684

83. McFarland BC, Hong SW, Rajbhandari R, Twitty GB, Gray GK, Yu H, Benveniste EN, Nozell SE (2013) NF-κB induced IL-6 ensures STAT3 activation and tumor aggressiveness in glioblastoma. PLoS One 8(11):e78728. doi:10.1371/journal.pone.0078728
84. Decker T, Kovarik P (2000) Serine phosphorylation of STATs. Oncogene 19:2628–2637
85. Stehmeier P, Muller S (2009) Phospho-regulated SUMO interaction modules connect the SUMO system to CK2 signaling. Mol Cell 33(3):400–409
86. Timofeeva OA, Plisov S, Evseev AA, Peng S, Jose-Kampfner M, Lovvorn HN, Dome JS, Perantoni AO (2006) Serine-phosphorylated STAT1 is a prosurvival factor in Wilms' tumor pathogenesis. Oncogene 25(58):7555–7564
87. Carter-Su C, Argetsinger LS (2011) JAKs, Stats, and CK2? Blood 118(1):5–6. doi:10.1182/blood-2011-05-352542, 118/1/5[pii]
88. Aparicio-Siegmund S, Sommer J, Monhasery N, Schwanbeck R, Keil E, Finkenstadt D, Pfeffer K, Rose-John S, Scheller J, Garbers C (2014) Inhibition of protein kinase II (CK2) prevents induced signal transducer and activator of transcription (STAT) 1/3 and constitutive STAT3 activation. Oncotarget 5(8):2131–2148
89. Wagner KU, Schmidt JW (2011) The two faces of Janus kinases and their respective STATs in mammary gland development and cancer. J Carcinog 10:32. doi:10.4103/1477-3163.90677
90. Berishaj M, Gao SP, Ahmed S, Leslie K, Al-Ahmadie H, Gerald WL, Bornmann W, Bromberg JF (2007) Stat3 is tyrosine-phosphorylated through the interleukin-6/glycoprotein 130/Janus kinase pathway in breast cancer. Breast Cancer Res 9(3):32. doi:10.1186/bcr1680, bcr1680 [pii]
91. Pedranzini L, Dechow T, Berishaj M, Comenzo R, Zhou P, Azare J, Bornmann W, Bromberg J (2006) Pyridone 6, a pan-janus-activated kinase inhibitor, induces growth inhibition of multiple myeloma cells. Cancer Res 66(19):9714–9721
92. Sansone P, Bromberg J (2012) Targeting the interleukin-6/Jak/Stat pathway in human malignancies. J Clin Oncol 30(9):1005–1014. doi:10.1200/JCO.2010.31.8907
93. Sansone P, Bromberg J (2011) Environment, inflammation, and cancer. Curr Opin Genet Dev 21(1):80–85. doi:10.1016/j.gde.2010.11.001, S0959-437X(10)00195-4 [pii]
94. Chen Y, Wang J, Wang X, Liu X, Li H, Lv Q, Zhu J, Wei B, Tang Y (2013) STAT3, a poor survival predicator, is associated with lymph node metastasis from breast cancer. J Breast Cancer 16(1):40–49. doi:10.4048/jbc.2013.16.1.40
95. Chang Q, Bournazou E, Sansone P, Berishaj M, Gao SP, Daly L, Wels J, Theilen T, Granitto S, Zhang X, Cotari J, Alpaugh ML, de Stanchina E, Manova K, Li M, Bonafe M, Ceccarelli C, Taffurelli M, Santini D, Altan-Bonnet G, Kaplan R, Norton L, Nishimoto N, Huszar D, Lyden D, Bromberg J (2013) The IL-6/JAK/Stat3 feed-forward loop drives tumorigenesis and metastasis. Neoplasia 15(7):848–862
96. Wegrzyn J, Potla R, Chwae YJ, Sepuri NB, Zhang Q, Koeck T, Derecka M, Szczepanek K, Szelag M, Gornicka A, Moh A, Moghaddas S, Chen Q, Bobbili S, Cichy J, Dulak J, Baker DP, Wolfman A, Stuehr D, Hassan MO, Fu XY, Avadhani N, Drake JI, Fawcett P, Lesnefsky EJ, Larner AC (2009) Function of mitochondrial Stat3 in cellular respiration. Science 323(5915):793–797
97. Gough DJ, Corlett A, Schlessinger K, Wegrzyn J, Larner AC, Levy DE (2009) Mitochondrial STAT3 supports Ras-dependent oncogenic transformation. Science 324(5935):1713–1716
98. Zhang Q, Raje V, Yakovlev VA, Yacoub A, Szczepanek K, Meier J, Derecka M, Chen Q, Hu Y, Sisler J, Hamed H, Lesnefsky EJ, Valerie K, Dent P, Larner AC (2013) Mitochondrial localized Stat3 promotes breast cancer growth via phosphorylation of serine 727. J Biol Chem 288(43):31280–31288. doi:10.1074/jbc.M113.505057
99. Knupfer H, Preiss R (2007) Significance of interleukin-6 (IL-6) in breast cancer (review). Breast Cancer Res Treat 102(2):129–135. doi:10.1007/s10549-006-9328-3
100. Drygin D, Ho CB, Omori M, Bliesath J, Proffitt C, Rice R, Siddiqui-Jain A, O'Brien S, Padgett C, Lim JK, Anderes K, Rice WG, Ryckman D (2011) Protein kinase CK2 modulates IL-6 expression in inflammatory breast cancer. Biochem Biophys Res Commun 415(1):163–167. doi:10.1016/j.bbrc.2011.10.046

101. Brantley EC, Nabors LB, Gillespie GY, Choi Y-H, Palmer C, Harrison K, Roraty K, Benveniste EN (2008) Loss of PIAS3 expression in glioblastoma multiforme tumors: Implications for STAT-3 activation and gene expression. Clin Cancer Res 14(15):4694–4704
102. Lo HW, Cao X, Zhu H, Ali-Osman F (2008) Constitutively activated STAT3 frequently coexpresses with epidermal growth factor receptor in high-grade gliomas and targeting STAT3 sensitizes them to Iressa and alkylators. Clin Cancer Res 14(19):6042–6054
103. McFarland BC, Ma JY, Langford CP, Gillespie GY, Yu H, Zheng Y, Nozell SE, Huszar D, Benveniste EN (2011) Therapeutic potential of AZD1480 for the treatment of human glioblastoma. Mol Cancer Ther 10(12):2384–2393. doi:10.1158/1535-7163.MCT-11-0480
104. Veeriah S, Brennan C, Meng S, Singh B, Fagin JA, Solit DB, Paty PB, Rohle D, Vivanco I, Chmielecki J, Pao W, Ladanyi M, Gerald WL, Liau L, Cloughesy TC, Mischel PS, Sander C, Taylor B, Schultz N, Major J, Heguy A, Fang F, Mellinghoff IK, Chan TA (2009) The tyrosine phosphatase PTPRD is a tumor suppressor that is frequently inactivated and mutated in glioblastoma and other human cancers. Proc Natl Acad Sci U S A 106(23):9435–9440
105. Ortiz B, Fabius AW, Wu WH, Pedraza A, Brennan CW, Schultz N, Pitter KL, Bromberg JF, Huse JT, Holland EC, Chan TA (2014) Loss of the tyrosine phosphatase PTPRD leads to aberrant STAT3 activation and promotes gliomagenesis. Proc Natl Acad Sci U S A 111(22):8149–8154. doi:10.1073/pnas.1401952111
106. Sherry MM, Reeves A, Wu JK, Cochran BH (2009) STAT3 is required for proliferation and maintenance of multipotency in glioblastoma stem cells. Stem Cells 27(10):2383–2392
107. Wang H, Lathia JD, Wu Q, Wang J, Li Z, Heddleston JM, Eyler CE, Elderbroom J, Gallagher J, Schuschu J, Macswords J, Cao Y, McLendon RE, Wang XF, Hjelmeland AB, Rich JN (2009) Targeting interleukin 6 signaling suppresses glioma stem cell survival and tumor growth. Stem Cells 27(10):2393–2404
108. Cao Y, Lathia JD, Eyler CE, Wu Q, Li Z, Wang H, McLendon RE, Hjelmeland AB, Rich JN (2010) Erythropoietin receptor signaling through STAT3 is required for glioma stem cell maintenance. Genes Cancer 1(1):50–61. doi:10.1177/1947601909356352
109. Guryanova OA, Wu Q, Cheng L, Lathia JD, Huang Z, Yang J, Macswords J, Eyler CE, McLendon RE, Heddleston JM, Shou W, Hambardzumyan D, Lee J, Hjelmeland AB, Sloan AE, Bredel M, Stark GR, Rich JN, Bao S (2011) Nonreceptor tyrosine kinase BMX maintains self-renewal and tumorigenic potential of glioblastoma stem cells by activating STAT3. Cancer Cell 19(4):498–511. doi:10.1016/j.ccr.2011.03.004, S1535-6108(11)00093-6[pii]
110. Nitta RT, Gholamin S, Feroze AH, Agarwal M, Cheshier SH, Mitra SS, Li G (2014) Casein kinase 2[alpha] regulates glioblastoma brain tumor-initiating cell growth through the [beta]-catenin pathway. Oncogene. doi:10.1038/onc.2014.299:1-12
111. Agarwal M, Nitta R, Li G (2014) Casein kinase 2: A novel player in glioblastoma therapy and cancer stem cells. J Mol Genet Med 8(1):1–6. doi:10.4172/1747-0862.1000094
112. Phillips HS, Kharbanda S, Chen R, Forrest WF, Soriano RH, Wu TD, Misra A, Nigro JM, Colman H, Soroceanu L, Williams PM, Modrusan Z, Feuerstein BG, Aldape K (2006) Molecular subclasses of high-grade glioma predict prognosis, delineate a pattern of disease progression, and resemble stages in neurogenesis. Cancer Cell 9:157–173
113. Halliday J, Helmy K, Pattwell SS, Pitter KL, LaPlant Q, Ozawa T, Holland EC (2014) In vivo radiation response of proneural glioma characterized by protective p53 transcriptional program and proneural-mesenchymal shift. Proc Natl Acad Sci U S A 111(14):5248–5253. doi:10.1073/pnas.1321014111
114. Carro MS, Lim WK, Alvarez MJ, Bollo RJ, Zhao X, Snyder EY, Sulman EP, Anne SL, Doetsch F, Colman H, Lasorella A, Aldape K, Califano A, Iavarone A (2010) The transcriptional network for mesenchymal transformation of brain tumours. Nature 463(7279): 318–325
115. He K, Qi Q, Chan CB, Xiao G, Liu X, Tucker-Burden C, Wang L, Mao H, Lu X, McDonald FE, Luo H, Fan QW, Weiss WA, Sun SY, Brat DJ, Ye K (2013) Blockade of glioma proliferation through allosteric inhibition of JAK2. Sci Signal 6(283):ra55, 10.1126/scisignal.2003900

116. Joughin BA, Naegle KM, Huang PH, Yaffe MB, Lauffenburger DA, White FM (2009) An integrated comparative phosphoproteomic and bioinformatic approach reveals a novel class of MPM-2 motifs upregulated in EGFRvIII-expressing glioblastoma cells. Mol Biosyst 5(1):59–67
117. Gadji M, Crous AM, Fortin D, Krcek J, Torchia M, Mai S, Drouin R, Klonisch T (2009) EGF receptor inhibitors in the treatment of glioblastoma multiform: old clinical allies and newly emerging therapeutic concepts. Eur J Pharmacol 625(1–3):23–30. doi:10.1016/j.ejphar.2009.10.010
118. Ji H, Lu Z (2013) The role of protein kinase CK2 in glioblastoma development. Clin Cancer Res 6335–6337, 10.1158/1078-0432.CCR-13-2478
119. Mandal T, Bhowmik A, Chatterjee A, Chatterjee U, Chatterjee S, Ghosh MK (2014) Reduced phosphorylation of Stat3 at Ser-727 mediated by casein kinase 2—Protein phosphatase 2A enhances Stat3 Tyr-705 induced tumorigenic potential of glioma cells. Cell Signal 26(8):1725–1734. doi:10.1016/j.cellsig.2014.04.003
120. Levine RL, Pardanani A, Tefferi A, Gilliland DG (2007) Role of JAK2 in the pathogenesis and therapy of myeloproliferative disorders. Nat Rev Cancer 7(9):673–683. doi:10.1038/nrc2210, nrc2210 [pii]
121. Geron I, Abrahamsson AE, Barroga CF, Kavalerchik E, Gotlib J, Hood JD, Durocher J, Mak CC, Noronha G, Soll RM, Tefferi A, Kaushansky K, Jamieson CH (2008) Selective inhibition of JAK2-driven erythroid differentiation of polycythemia vera progenitors. Cancer Cell 13(4):321–330
122. Chen PH, Chien FC, Lee SP, Chan WE, Lin IH, Liu CS, Lee FJ, Lai JS, Chen P, Yang-Yen HF, Yen JJ (2012) Identification of a novel function of the clathrin-coated structure at the plasma membrane in facilitating GM-CSF receptor-mediated activation of JAK2. Cell Cycle 11(19):3611–3626. doi:10.4161/cc.21920
123. Haura EB, Zheng Z, Song L, Cantor A, Bepler G (2005) Activated epidermal growth factor receptor-Stat-3 signaling promotes tumor survival in vivo in non-small cell lung cancer. Clin Cancer Res 11(23):8288–8294
124. Alvarez JV, Greulich H, Sellers WR, Meyerson M, Frank DA (2006) Signal transducer and activator of transcription 3 is required for the oncogenic effects of non-small-cell lung cancer-associated mutations of the epidermal growth factor receptor. Cancer Res 66(6):3162–3168
125. Gao SP, Mark KG, Leslie K, Pao W, Motoi N, Gerald WL, Travis WD, Bornmann W, Veach D, Clarkson B, Bromberg JF (2007) Mutations in the EGFR kinase domain mediate STAT3 activation via IL-6 production in human lung adenocarcinomas. J Clin Invest 117(12):3846–3856
126. Lin YC, Hung MS, Lin CK, Li JM, Lee KD, Li YC, Chen MF, Chen JK, Yang CT (2011) CK2 inhibitors enhance the radiosensitivity of human non-small cell lung cancer cells through inhibition of stat3 activation. Cancer Biother Radiopharm 26(3):381–388. doi:10.1089/cbr.2010.0917
127. Koskela HL, Eldfors S, Ellonen P, van Adrichem AJ, Kuusanmaki H, Andersson EI, Lagstrom S, Clemente MJ, Olson T, Jalkanen SE, Majumder MM, Almusa H, Edgren H, Lepisto M, Mattila P, Guinta K, Koistinen P, Kuittinen T, Penttinen K, Parsons A, Knowles J, Saarela J, Wennerberg K, Kallioniemi O, Porkka K, Loughran TP Jr, Heckman CA, Maciejewski JP, Mustjoki S (2012) Somatic STAT3 mutations in large granular lymphocytic leukemia. N Engl J Med 366(20):1905–1913. doi:10.1056/NEJMoa1114885
128. Lee H, Herrmann A, Deng JH, Kujawski M, Niu G, Li Z, Forman S, Jove R, Pardoll DM, Yu H (2009) Persistently activated Stat3 maintains constitutive NF-kappaB activity in tumors. Cancer Cell 15(4):283–293. doi:10.1016/j.ccr.2009.02.015

CIGB-300: A Promising Anti-Casein Kinase 2 (CK2) Peptide for Cancer Targeted Therapy

Silvio E. Perea, Yasser Perera, Idania Baladrón, Lidia González, Fernando Benavent, Hernán G. Fariña, Idrián García, Arielis Rodríguez, Vilcy Reyes, Yanelda García, Roberto Gómez, Daniel F. Alonso, and Carmen Valenzuela

Abstract Over the past few years, the development of CK2 inhibitors using small molecules has emerged as a paradigmatic approach for blocking the enzymatic activity. However, despite successful experimental validation, so far only one of such chemical compounds has entered into clinical trials. Using a different rationale to inhibit CK2, we have developed CIGB-300 as a novel hypothesis-driven peptide targeting the CK2 phosphoacceptor domain instead of the ATP-binding site. Data from in vitro studies have revealed that at least in human cell lines from solid tumors, CIGB-300 binds mainly to and inhibits CK2-mediated phosphorylation of B23/npm. Studies of the molecular and cellular events downstream this interaction have demonstrated that CIGB-300 induces apoptosis in vitro and in vivo, modulating a wide array of proteins involved in cell proliferation, apoptosis, ribosome biogenesis, drug resistance, cell motility, and adhesion among other processes. Accordingly, CIGB-300 has shown synergistic interaction with anticancer drugs, suppressing angiogenesis and exhibiting antimetastatic properties. The pharmacology of this peptide-based drug has already been investigated in cancer patients. Different Phase 1 clinical trials have shown CIGB-300 to be safe and well tolerated and have studied its pharmacokinetics after either local or systemic administration.

S.E. Perea (✉) • Y. Perera • V. Reyes
Molecular Oncology Laboratory, Biomedical Research Area,
Center for Genetic Engineering and Biotechnology, Havana 10600, Cuba
e-mail: silvio.perea@cigb.edu.cu

I. Baladrón • L. González • I. García • Y. García • C. Valenzuela
Clinical Research Area, Center for Genetic Engineering and Biotechnology,
Havana 10600, Cuba

A. Rodríguez
Proteomic Department, Biomedical Research Area, Center for Genetic Engineering and Biotechnology, Havana 10600, Cuba

F. Benavent • H.G. Fariña • D.F. Alonso
Molecular Oncology Laboratory, National University of Quilmes, Buenos Aires, Argentina

R. Gómez
ELEA laboratories, Buenos Aires, Argentina

K. Ahmed et al. (eds.), *Protein Kinase CK2 Cellular Function in Normal and Disease States*, Advances in Biochemistry in Health and Disease 12, DOI 10.1007/978-3-319-14544-0_16

Remarkably, during a dose-finding Phase 2 trial in women with cervical cancer, cohorts receiving CIGB-300 and chemoradiotherapy concomitantly had a higher frequency of complete response than those receiving chemoradiotherapy alone. Taken together, the data presented here summarize all relevant preclinical and clinical findings that make CIGB-300 a promising peptide-based drug for the treatment of cancer patients.

Keywords CIGB-300 • Casein kinase 2 • CK2 inhibitors • Targeted therapy • B23 • Nucleophosmin • Apoptosis • Anticancer peptides

1 Rationale and Concept Behind CIGB-300

Over the past decade, different approaches have been followed to discover and develop inhibitors of the enzymatic activity of CK2. These have resulted in a number of successful proof of concepts exhibiting antineoplastic effect both in vitro and in vivo. The most successful strategies have used small molecules, peptides, or oligonucleotides that inhibit CK2 through different mechanisms. For instance, relatively small organic compounds such as TBB, DRB, DMAT, IQA, emodin, CX-4945, and the flavonoids apigenin, luteolin, and quercetin have been shown to target the ATP-binding site of the CK2 catalytic subunit, leading to apoptosis in tumor cells [1–8]. Other examples are Pc, a cyclic peptide that antagonizes the interaction between CK2 subunits [9], and antisense oligonucleotides that target transcription of the CK2 alpha gene, thereby inducing cell death and antitumoral activity [10].

CIGB-300 is a new anti-CK2 peptide that explores a novel concept to impair CK2 activity: targeting its substrates, rather than the enzyme per se. Taking into account the presence of strong sequence homologies across acidic phosphoacceptor domains from over 300 CK2 substrates, we speculated that a cyclic peptide targeting such a domain could inhibit its phosphorylation by outcompeting the enzyme, thus exhibiting antineoplastic properties. Under this hypothesis, a random 9-mer cyclic peptide phage display library was screened using the human papillomavirus type 16 E7 (HPV-16 E7) phosphoacceptor domain as target [11]. Cyclic peptides from those phage clones with the strongest binding and inhibitory effect on CK2-mediated phosphorylation were synthesized chemically for independent confirmation. Afterward, the peptide from phage clone 15 (P15) was selected for a subsequent proof *of* concept, where it was synthesized as a fusion to the cell-penetrating peptide Tat (48–68). Data from those first in vitro experiments showed that the P15–Tat chimera (termed CIGB-300), but not Tat (48–68), significantly inhibited CK2-mediated phosphorylation of the HPV-16 E7 oncoprotein [11]. Importantly, subsequent in vivo studies demonstrated that CIGB-300 induced apoptosis, as determined by caspase activation, and halted tumor growth in an HPV-16 syngeneic murine tumor animal model [11]. These preliminary results indicated that the concept behind CIGB-300 was proved and led to further investigations on its mechanism of action and potential application to a clinical setting.

2 Molecular and Functional Characterization of CIGB-300

2.1 Proteins Interacting with CIGB-300

As CIGB-300 was initially selected for its ability to impair CK2-mediated phosphorylation in vitro of only one specific acidic phosphoacceptor domain, it was first necessary to investigate whether CIGB-300 targeted other CK2 substrates in the cellular context. This issue was addressed primarily through both in vitro and in vivo pulldown experiments performed in NCI-H82 (SCLC) cells using biotin-tagged CIGB-300, followed by mass spectrometry analysis and/or Western blotting of the protein species retained by the streptavidin–Sepharose resin. According to the data from the in vitro studies, CIGB-300 bound 21 different CK2 substrates, mostly linked to cell signaling, translation, ribosome biogenesis, cytoskeletal dynamics, and DNA/RNA processing, among others [12]. However, in vivo, CIGB-300 bound primarily only two major nucleolar CK2 substrates: B23/nucleophosmin and C23/nucleolin [13]. Interestingly, 13 structural ribosomal proteins from the small subunit and one from the large subunit were also identified with this experimental approach, probably owing to their interaction with B23/npm. Considering that B23 and C23 are usually found associated with ribosomal RNA or small nucleolar RNA complexes that form part of the ribosomal biogenesis pathway [14], the in vivo pulldown fractions were further treated with excess RNase. This demonstrated that the CIGB-300–C23 interaction (but not that of CIGB-300–B23) is mediated by RNA, suggesting B23, rather than C23, as a true target of this peptide in NCI-H82 cells [13]. The physical interaction between CIGB-300 and B23/npm was further confirmed by confocal fluorescence microscopy, which revealed that both co-localized at the nucleolar compartment in this cell line [13]. These findings prompted us to investigate whether CIGB-300 was able to bind B23/npm in other human tumor cell lines of different origins. The data yielded by these additional in vivo pulldown assays (plus Western blotting analysis) confirmed that both molecules interact not only in NCI-82 but also in the NCI-H125 (non-small cell lung cancer), Hep-2C (uterine cervix), SiHa (uterine cervix), PC-3 (prostate), and SW948 (colon) cell lines [15] after incubation periods as short as 10 min. Therefore, targeting of B23/npm by CIGB-300 seems to take place irrespectively of tumor cell origin, and this molecular event may constitute a relevant step for the proapoptotic effect of this molecule.

2.2 Inhibition of CK2-Mediated Phosphorylation In Vitro and In Vivo

Once it was confirmed that CIGB-300 directly binds B23/npm in NCI-H82 and other tumor cell lines, in vivo metabolic labeling plus immunoprecipitation was used to determine whether this interaction affects CK2-mediated phosphorylation of B23. These experiments revealed that equipotent doses of CIGB-300, but not of negative control peptides, impaired the phosphorylation of B23/npm to varying degrees after a 30-min incubation in all the assayed cell lines [13, 15]. We also took advantage of the fact that residue Ser125 has been identified as a CK phosphorylation target on B23/npm [16] to verify the in vivo inhibitory effect of CIGB-300 by isotopic labeling coupled to mass spectrometry (unpublished data). Likewise, CIGB-300 was shown to inhibit CK2-mediated phosphorylation of recombinant B23/npm–GST in a dose-dependent manner [13].

Taking into account that C23/nucleolin was the other nucleolar CK2 substrate identified by the in vivo pulldown experiments in CIGB-300-treated NCI-H82 cells, the effect of CIGB-300 on CK2-mediated phosphorylation of this protein has been also investigated in vivo using the same experimental approach. In contrast to the behavior observed for B23/npm, CIGB-300 inhibited CK2-mediated phosphorylation of C23/nucleolin only marginally (25 %) even at the highest dose (200 μM) (unpublished data), a finding that could be explained by the evidence above suggesting that the interaction between CIGB-300 and C23/nucleolin is indirect [13].

Interestingly, other experiments using blood cancer cells have shown that CIGB-300 can impair CK2-mediated phosphorylation of specific substrates other than B23/npm. For instance, in the MO1043 (chronic lymphocytic leukemia) cell line, which is highly sensitive for the antiproliferative effect of CIGB-300, it prevented the phosphorylation of Ser129 on Akt/PKB and Ser380 on PTEN [17]. Accordingly, phosphorylation of PI3K downstream targets Akt/PKB and GSK-3β decreased after treatment with CIGB-300. Interestingly, a similar inhibitory effect was also observed when primary CLL cells from cancer patients were used in this investigation [17].

Altogether, it is clear that CIGB-300 impairs CK2-mediated phosphorylation of different proteins connected to processes and pathways that are critical for cell survival. It should be addressed, though, that the abovementioned CK2 substrates may not be its only in vivo targets, an issue that can only be addressed through experimental strategies such as comprehensive analyses of the phosphoproteome (which will not only expand the list of CK2 substrates potentially inhibited by CIGB-300, but enable a thorough exploration of potential off-target effects).

2.3 *Effect of CIGB-300 in Cell Proliferation and Survival*

The sensitivity of tumor cells toward CIGB-300 has also been extensively examined [11, 12]. One study evaluated the antiproliferative effect of CIGB-300 in a panel of human cancer cell lines derived from four solid tumor types, compared to a panel of nontumorigenic cell lines [15]. In this case, CIGB-300 exerted an antiproliferative effect that varied across the micromolar range (IC_{50} 20–300 μM) in cells derived from lung, cervix, prostate, and human colon cancers, although the effect on lung cancer cells was much more uniform (mean IC_{50}=60 μM). Importantly, its potency in nontumorigenic cell lines was nearly twofold lower (mean IC_{50}= 190 μM compared to mean IC_{50}= 100 μM for tumor cell lines). CIGB-300 has also been shown to exert an antiproliferative effect in blood cancer cells such as CLL [17]. In that report, the mean IC_{50} value for CIGB-300 across all the assayed cell lines was 29 ± 7 μM, an important finding in light of the fact that CIGB-300 at concentrations as low as 12.5 μM decreased dramatically the viability of all primary CLL patient samples analyzed, including those from poor prognosis cases [17]. Overall, CIGB-300 exhibits nearly fourfold higher potency in blood cancer cells than in solid tumor cell lines in terms of its antiproliferative effect. A plausible explanation for such a differential effect that fits the available data has not been found so far. As will be discussed later in this chapter, the mechanism through which CIGB-300 is internalized and transported to the nucleolar compartment seems to influence the sensitivity of solid tumor cell lines toward this compound. Whether this factor also modulates the sensitivity of blood cancer cells toward CIGB-300 remains to be elucidated.

CK2-mediated phosphorylation is one of the mechanisms used at the cellular level to avoid apoptosis and promote cell survival. Therefore, one of the first lines of inquiry pursued once the first proof-*of*-concept experiments on CIGB-300 concluded successfully was to determine the potential proapoptotic effect of this molecule. Using a variety of experimental approaches, it was demonstrated that the in vitro treatment of cultured cells with CIGB-300 leads to a rapid (within a few minutes of incubation) onset of apoptosis, a finding that was replicated in vivo using tumor-bearing mice [11, 18, 12, 15]. A detailed study of the molecular and cellular events leading to apoptosis in CIGB-300-treated NCI-H82 cells suggested that the inhibition of CK2-mediated phosphorylation of B23/npm and the ensuing nucleolar disassembly constituted the crucial intracellular events triggering CIGB-300-induced cell death [13]. In fact, data from other authors, obtained by using the DRB inhibitor, suggest that CK2 plays a relevant role in nucleolar structure and ribosomal biogenesis [19], and it has been shown that the mutation of Ser125 in B23/npm leads to nucleolar breakdown [20].

Regardless, it is plausible that CIGB-300 induces apoptosis—at least in CLL cells—by other mechanisms, including the activation of the tumor suppressor gene PTEN and the abrogation of PI3K-mediated downstream signaling. Besides, signs of mitochondrial damage have been observed in preliminary experiments where CIGB-300-treated cells are incubated with the fluorescent reagent JC-1, suggesting that CIGB-300 may also induce cell death by activating the intrinsic apoptotic pathways (unpublished data). In any case, it should be underlined that the proapoptotic nature of CIGB-300 has been confirmed by analyzing the set of proteins regulated downstream the regulatory cascade triggered by the presence of this peptide, as will be described later in this chapter.

As mentioned above, CIGB-300 inhibits cell proliferation and induces apoptosis both in vitro and in vivo to different extents. However, its impact on the cell cycle seems to be more subdued and less uniform. For instance, in time course experiments, equipotent doses of CIGB-300 induced subtle accumulation at G2/M, Go/G1, S, and G2/M in the cell lines NC-H125, SiHa, Hep-2C, and PC-3, respectively [15]. However, no significant changes in cell cycle distribution were detected in the SW948 cell line. A differential effect on cell cycle arrest has been also found for other inhibitors of CK2, such as the orally bioavailable compound CX-4945 [8].

Thus, while CK2 has been identified as a cell cycle regulator [21–23], the effect of CK2 inhibitors such as CIGB-300 and also CX-4945 is cell line dependent, probably determined by specific genetic and/or epigenetic traits of the assayed cell host.

2.4 CIGB-300 and p53

The relationships at the molecular level between CK2 and p53 are not fully understood yet. However, an antiapoptotic role for CK2-mediated phosphorylation of p53 has been documented by dampening CK2 activity with either pharmacological or genetic approaches in different tumor cell lines [24–26].

As of this moment, no study exists fully addressing the role of p53 in the proapoptotic effect of CIGB-300. However, an inference can be made that CIGB-300 acts irrespectively of p53 baseline levels in tumor cells, based solely on available data. For instance, the cell lines NCI-H125, containing a single p53 mutation, and HL-60, containing a homozygotic p53 deletion, are both highly sensitive toward the antiproliferative effect of CIGB-300 (IC_{50}, 48 and 59 μM, respectively). This behavior stands in contrast to that observed for other inhibitors such as CX-4945, which exhibit an evident p53-dependent effect in acute myeloid leukemia (AML) cell lines [27].

Interestingly, CIGB-300 treatment increased the transcriptional activity of p53 in two cell lines with wild-type p53 genes (Hep-2C and SiHa) as determined by a specific ELISA system (Panomics) (Fig. 1). This finding is consistent with previous

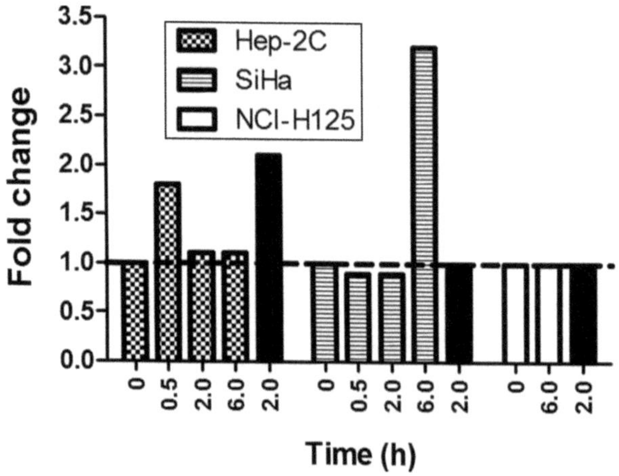

Fig. 1 CIGB-300 increases p53 DNA-binding activity: Hep-2C, SiHa, and NCI-H125 cells were treated with equipotent doses of CIGB-300 (IC50) and nuclear extracts were prepared at the indicated time points. Subsequently, 5 μg of total protein from each extract was added to the wells of an ELISA plate to quantify the specific binding activity of p53 using the transcription factor ELISA kit (Panomics). Absorbance ratios (AR, 450 nm) (positive control/blank wells) >2.5 and CV <15 % were obtained for all assays

reports demonstrating that the treatment of glioblastoma cell lines with CK2 inhibitors such as DRB and Apigenin increases p53 DNA-binding activity and its transcriptional activation [24]. In contrast, no p53 DNA-binding activity was detected after treating NCI-H125 (a cell line where p53 is mutated) with CIGB-300.

In summary, CIGB-300, as other CK2 inhibitors, increases p53 DNA-binding activity, although the baseline status of p53 does not appear to hamper the anticellular effect of this anti-CK2 peptide. Full confirmation of the latter inference, however, still awaits further research.

2.5 *The CIGB-300-Regulated Proteome*

Despite the availability of a huge body of research on the biochemistry and biology of CK2 inhibitors, the proteomics of the downstream cascade triggered by the inhibition of CK2-mediated phosphorylation still remains a relatively unexplored field. Lack of relevance is not the cause of this state of affairs, as knowledge of the exact nature of these changes would no doubt change our understanding of both the on-target and off-target effects of these inhibitors. In other words, the very proteins that are regulated downstream from CK2 inhibition might actually constitute the mediators ultimately leading to an antitumor effect and/or to any potential toxicity that might arise after this event.

To better understand the mechanisms involved in the anticancer activity of CIGB-300, the nuclear proteome of CIGB-300-treated NCI-H125 cells was investigated after a 45-min incubation period. A total of 137 proteins exhibited abundance changes larger than twofold in the presence of CIGB-300, with those linked to ribosomal biogenesis, metastasis/angiogenesis, cell survival, proliferation, apoptosis, and drug resistance being significantly overrepresented [28]. Interestingly, CIGB-300-mediated downregulation of the B23/npm nucleolar chaperone seemed to take place at the level of protein stability, as evidenced by the presence of degraded species upon Western blotting and mass spectrometry. This molecular event could be interpreted as a consequence of the inhibition of CK2-mediated phosphorylation of B23/npm, as suggested by others [29]. The downregulation of this and other B23/npm-related proteins in this experimental setting further supports the hypothesis of a proapoptotic/antiproliferative effect for CIGB-300 and in addition suggests that it might be involved in the impairment of ribosomal biogenesis.

One of the advantages of comparative proteomics is the ease with which it unearths putative associations to otherwise difficult to detect biological effects or cellular processes, simply by looking at changes in protein abundance triggered by the assayed experimental treatment. In case of CIGB-300, there were also changes in protein species associated with processes other than apoptosis and cell proliferation, such as angiogenesis. The putative antiangiogenic effect of CIGB-300 was subjected to experimental validation, which confirmed that at least in vitro, CIGB-300 affected the adhesion, migration, and tubular network formation of human umbilical vein endothelial cells (HUVEC) [30]. Also, the systemic administration of CIGB-300 impaired the angiogenesis of tumors growing in matrigel plugs in mice (unpublished data). These findings, along with those demonstrating the antiangiogenic effect of CK2 inhibitor CX-4945 [8], confirm the role that CK2-mediated phosphorylation plays in tumor angiogenesis.

Further head-*to*-head studies of the CIGB-300-regulated proteome and phosphoproteome will be required to dissect the connection between CIGB-300-inhibited substrates, its downstream-regulated proteins, and the cellular processes affected by this compound.

3 Subcellular Localization and Internalization Mechanism of CIGB-300

As mentioned above, CIGB-300 is a synthetic chimera comprised of the P15 cyclic peptide linked to the Tat cell-penetrating peptide (CPP) to facilitate its intracellular delivery. The advantages and limitations of CPPs as delivery vehicles have been widely discussed elsewhere [31]. CPPs can efficiently deliver biological cargoes by mechanisms that include different types of endocytosis or even direct translocation through the cell membrane [31]. The relative contribution of each internalization pathway to global cellular uptake of a single CPP may depend even on cell type or CPP concentration used; these parameters are, therefore, important for intracellular stability of the cargo and/or access to the target [32, 33].

Due to the major role played by CPPs, studies on CIGB-300 have not been circumscribed to its intracellular targets, but have included its internalization pathway and subcellular distribution. CIGB-300 was synthesized for this purpose with an N-terminal carboxyfluorescein tag, analyzing its cellular uptake by both flow cytometry and fluorescence microscopy. Interestingly, the number of transduced cells reached a plateau by 3 min after the addition of the fluorescent peptide and remained unaltered for the following 60 min of incubation [15]. However, peptide uptake, as measured by the geometric mean of fluorescence emission, differed among the five tumor cell lines studied in these experiments. According to the subcellular distribution data, CIGB-300 mostly locates to the nucleus, with a preferentially nucleolar residency [13] (Fig. 2). This distribution has been confirmed by confocal microscopy, which showed that CIGB-300 co-localizes with B23/npm at the nucleolar compartment [13]. An important finding was that these experiments also demonstrated that the antiproliferative activity of CIGB-300 did correlate with its nucleolar localization in solid tumor cell lines [15].

To better understand whether internalization mechanisms accounted for the varying sensitivities exhibited by solid tumor cell lines toward the antiproliferative effect of CIGB-300, the cellular uptake, intracellular transportation, and degradation events of this peptide were analyzed in this scenario. The data suggested, first, that CIGB-300 uses cell membrane heparan sulfate proteoglycans as the main receptor for extracellular uptake. Second, cellular sensitivity toward CIGB-300 was found to correlate positively with intracellular incorporation. Interestingly, the internalization

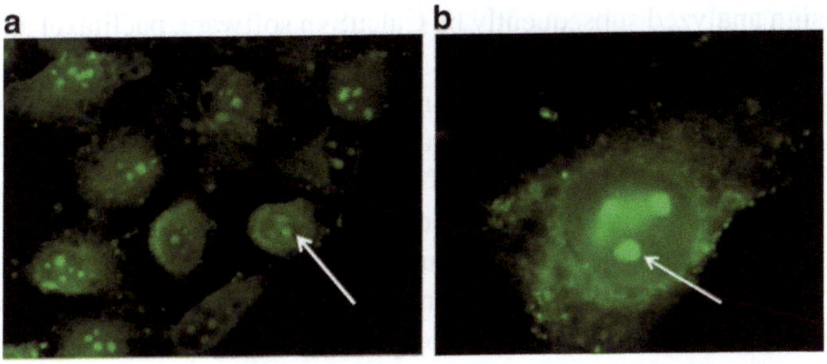

Fig. 2 Nucleolar deposit of CIGB-300: HeLa cells were incubated with CIGB-300–biotin (100 μM) at varying intervals. Subsequently, the cells were fixed with paraformaldehyde 4 % (Sigma, USA) and incubated with an FITC–streptavidin conjugate for 1 h at 4 °C. After washing with PBS, the slides were mounted in 40 % glycerol (Plusone, Uppsala, Sweden). The fixed cells were immediately analyzed by fluorescence microscopy at 40× (**a**) and 100× (**b**) magnification. The *arrows point* at nucleolar structures

of CIGB-300 was shown to take place mainly by direct membrane translocation, although energy-dependent endocytic pathway mechanisms had a minor contribution. In the latter case, caveolae-mediated endocytosis seems to be the preferred mechanism [34].

Collectively, these findings provide novel insights about CIGB-300 uptake that could, in part, explain the differential sensitivity of solid tumor cell lines toward the antiproliferative effect of CIGB-300. However, as blood cancer cell lines have been shown to be some fourfold more sensitive to CIGB-300, there remains the intriguing question of whether these peptide uptake mechanisms operate in the same manner in that case. Further experimentation will be required to address this issue.

4 Combination of CIGB-300 with Anticancer Compounds

As a single agent, CIGB-300 induces apoptosis in vitro and in vivo in different experimental settings. However, cumulative experience in clinical oncology, together with current knowledge on tumor biology, suggests that drug combinations, rather than monotherapy, constitute the best fit for the high complexity of tumoral disorders. In addition, CIGB-300 may represent an excellent candidate for combination therapies, since previous proteomic studies indicate that some of the proteins that have been linked to the appearance of resistance to anticancer drugs, such as T-plastin, glutathione S-transferase P, lactoylglutathione lyase, and the nuclease-sensitive element-binding protein 1 (YB-1), are downregulated by CIGB-300 [28]. Taking the above into account, we have examined the antiproliferative effect of CIGB-300 in combination with standard anticancer drugs, such as cisplatin (alkylating), paclitaxel (antimitotic), doxorubicin (anti-topoisomerase II), or 5-fluorouracil (DNA/RNA antimetabolite), in solid tumor cell lines. Using a Latin square design analyzed subsequently by CalcuSyn software, paclitaxel and cisplatin exhibited the best synergistic/additive profile when combined with CIGB-300, according to the combination and dose reduction indexes [35]. Importantly, during in vivo dose-finding schedules using human cervical tumors xenografted in nude mice, the concomitant administration of CIGB-300 and cisplatin increased survival to a higher extent than either agent alone [35].

Although CIGB-300-mediated downregulation of the proteins mentioned above could explain by itself the synergistic and nearly additive interactions between this peptide and cisplatin/paclitaxel, more comprehensive experimentation is required to dissect the molecular basis of this phenomenon. It should also be noted, in addition, that the preclinical findings described above have been successfully replicated in clinical settings, as mentioned later in this chapter.

The exacerbation of the PI3K signaling pathway has been shown to play an important role in tumor cell survival, angiogenesis, metastasis, and other cancer-related processes [36–38]. Some molecular target therapies, such as erlotinib, dim the PI3K signaling pathway by interfering with upstream EGFR autophosphorylation and, ultimately, inducing apoptosis in tumor cells [39]. Considering that CIGB-300 inhibits the CK2-mediated phosphorylation of Akt/PKB, which should subsequently affect the PI3K signaling pathway, we hypothesized that this peptide would synergize with erlotinib. To test this hypothesis, the H460 and A549 lung cancer cell lines were incubated with different erlotinib doses together with CIGB-300 (12.5 or 25 μM) for 48 h. According to the preliminary results, there was significant synergism between CIGB-300 and erlotinib, based on the combination indexes (Fig. 3). Although these data need to be confirmed in different experimental settings, they do provide a potential route forward for developing a novel therapeutic strategy in clinical oncology.

5 Antitumor Effect of CIGB-300 in Animal Models of Cancer

The antitumor activity of CIGB-300 has been tested in a variety of animal models of cancer, using different administration routes. For instance, it has been shown that daily intratumoral injections of CIGB-300 (200 μg) for 5 consecutive days can halt tumor growth in both syngeneic murine tumors implanted in syngeneic mice and human tumors xenografted in nude mice [11, 12]. In order to determine the minimum number of intratumoral injections required to elicit significant antitumor activity, NCI-H82 tumor-bearing nude mice were injected one, three, or five times with 200 μg of CIGB-300. Three and five consecutive intratumoral injections reduced tumor growth to a similar extent; however, single CIGB-300 administrations did not exhibit a detectable effect (Fig. 4). These findings encouraged subsequent experimentation to investigate whether multiple cycles of three consecutive intratumoral injections exhibited higher tumor growth delay than single cycles. SiHa tumor-bearing nude mice were treated with 200 μg of CIGB-300 in 1, 3, or 5 cycles of three consecutive intratumoral injections. Although the single-cycle

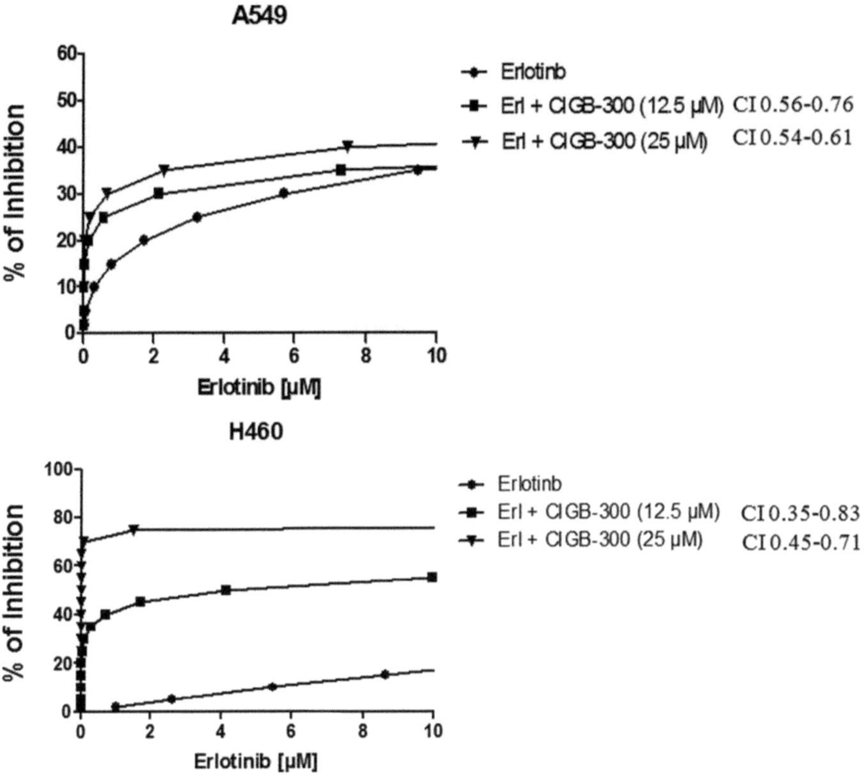

Fig. 3 CIGB-300 synergizes with erlotinib in vitro: A six-point curve of serial dilutions (1:2) was prepared for each drug, mixed following a Latin square design, and added to cell plates seeded the day before at 6×10^3 cells/well. After 48 h incubation, absorbance was read and the data were analyzed with the CalcuSyn software package to classify the interaction as synergistic, additive, or antagonistic according to the combination index (CI) value. The data shown correspond only to two fixed CIGB-300 concentrations (12.5 and 25 μM) (nonconstant combination analysis)

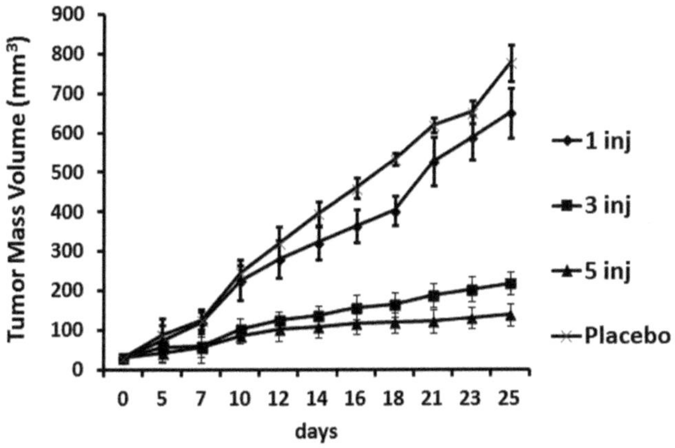

Fig. 4 Antitumor effect of CIGB-300 administered in multiple cycles: Four million NCI-82 (SCLCC) cells were subcutaneously inoculated into the dorsal region of 8-week-old female nude mice. Once tumors became detectable, the mice received 1, 3, or 5 daily intratumoral injections of CIGB-300 (200 μg). Tumors were measured with a caliper to calculate their volume

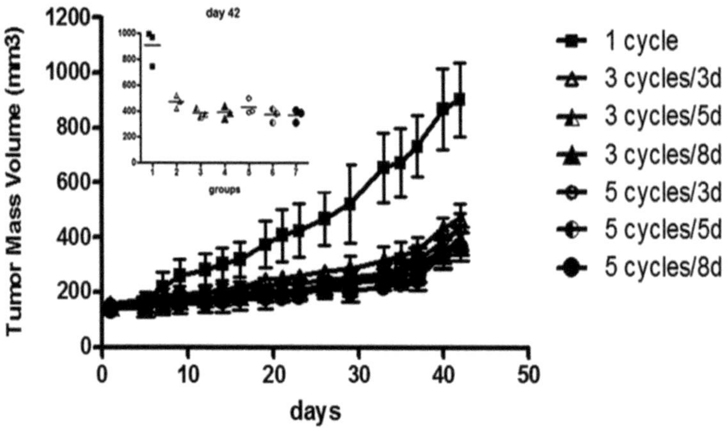

Fig. 5 Optimization of the local administration of CIGB-300 using a factorial design: CIGB-300 (200 μg) was directly injected into the tumor mass (150 mm^3) of SiHa tumor-bearing mice following a schedule of one, three, or five cycles of three consecutive injections, each cycle separated by 3, 5, or 8 days as indicated in the figure. The inset shows individual and mean tumor volumes estimated at day 42 for each group (before the first death at any of the groups). The analyses were done using GraphPad Prism version 4.00

treatment was less effective than multiple cycles in terms of antitumor effect, no significant differences were observed between three and five cycles (Fig. 5).

Stimulating as the results above might be, it was decided to confirm them using systemic rather than intratumoral administration, taking into account the poor anatomical accessibility of most solid tumors. The data from these assays showed that tumor growth was delayed by CIGB-300 administered at 2 mg/kg ($p<0.05$), 10 mg/kg ($p<0.01$), or 40 mg/kg ($p<0.001$) to mice bearing syngeneic murine tumors or nude mice xenografted with human tumors. These tumors exhibited detectable apoptosis, as evidenced by in situ DNA fragmentation, and experiments using 99mTc-labeled CIGB-300 showed that they accumulated the peptide [18]. A factorial design implemented to further optimize the delivery schedule indicated that three consecutive injections were less effective than 5, 10, or 15 consecutive injections of CIGB-300 when administered to NCI-H125 tumor-bearing nude mice (Fig. 6). Interestingly, the antitumor activity of CIGB-300 plateaued among the groups receiving the largest number of systemic injections, and CIGB-300 doses of 2 and 10 mg/kg of CIGB-300 exhibited very similar effects, as previously observed for intratumoral injections [18].

Summarizing, in animal models CIGB-300 reduces tumor growth when administered either by the local or systemic route, which underscores the potential applicability of therapies using this peptide to a wide variety of tumor types, especially those in hard-*to*-access locations. Furthermore, these findings indicate that three consecutive intratumoral injections of CIGB-300 (200 μg) are sufficient to elicit a strong antitumor activity and that a 3-cycle approach (each consisting of three consecutive intratumoral injections) is best to halt tumor growth significantly.

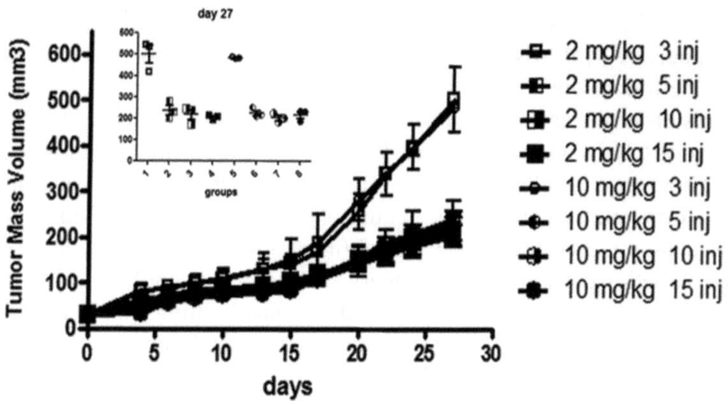

Fig. 6 Optimization of the systemic administration of CIGB-300 using a factorial design: CIGB-300 was intraperitoneally administered to NCI-H125 tumor-bearing mice (50 mm^3) at two dose levels (2 and 10 mg/kg), following a schedule of 3, 5, 10, or 15 consecutive injections as indicated in the figure. The inset shows individual and mean tumor volumes estimated at day 27 for each group (before the first death at any of the groups). The analyses were done using GraphPad Prism version 4.00

When administered systemically, five consecutive injections of CIGB-300 represented the simplest delivery regime still reducing tumor growth significantly, with no major differences between dosages of 2 and 10 mg/kg. These preclinical findings also provide useful guidelines for the risk–benefit and cost–benefit analyses required for future clinical trials of CIGB-300 in cancer patients.

6 Clinical Research on CIGB-300

CIGB-300 and the oral ATP competitor CX-4945 are the only CK2 inhibitors ever to have reached clinical trials. The clinical development of CIGB-300 has been guided in part by the data collected from preclinical testing; therefore, both local and intravenous administration routes were selected for Phase 1 trials intended to evaluate the safety and tolerability of CIGB-300.

Local administration was examined in a first-*in*-human clinical trial that recruited 31 women with high-grade squamous intraepithelial lesions. This trial indicated that the local administration of five consecutive doses of CIGB-300, 1 per day, was safe and well tolerated at doses ranging from 14 to 490 mg [40]. Neither a maximum tolerated dose nor dose-limiting toxicity was identified, and most of the local and systemic adverse events were grades 1–2. Although Phase 1 clinical trials are largely circumscribed to a preliminary exploration of the human toxicology of the test compound, this study did provide some indications of the potential efficacy of CIGB-300. Noticeably, in 75 % of the patients, the lesion was reduced significantly by colposcopy, 19 % exhibited full histological regression, and 48 % of all patients previously positive for human papillomavirus (HPV) DNA became negative [40]. Also, no recurrences or adverse events were reported long term during follow-up.

The results of the abovementioned trial prompted us to further explore the pharmacology of CIGB-300 administered locally to women with cervical stage IB2/II cancer. Three separated Phase 1 trials were carried for this purpose, each answering different clinical, scientific, and regulatory questions. One of them was the dose-escalation CERVIFARM-I-300 trial ($n = 12$) where safety, tumor uptake, and pharmacokinetics constituted the primary end points. According to the data, the maximum tolerated dose of CIGB-300 in this clinical setting was 70 mg [41]. This trial also identified an allergic-like syndrome as the dose-limiting toxicity and mean tumor uptake of CIGB-300 were 14.9 mg and 10.4 mg after the administration of 35 and 70 mg of CIGB-300, respectively. Finally, the treatment with CIGB-300 reduced B23/npm protein levels in tumor specimens, as determined by immunohistochemistry. These findings are in close agreement with those from the NCI-H125 experiments mentioned above, where CIGB-300 downregulated B23/npm according to the Western blotting and mass spectrometry data [28], and can be construed as the logical consequence of inhibiting CK2-mediated phosphorylation, as mentioned by other authors [29].

As pointed out earlier, the local administration of CIGB-300 in 3 cycles of 3 consecutive intratumoral injections was enough to achieve significant antitumor activity in human cervical SiHa tumors implanted in nude mice. The second of the three Phase 1 trials mentioned earlier, denominated CERVIFARM-II-300, was based precisely on this observation. This trial recruited 14 female subjects with stage IB2/II cervical cancer, using the same treatment schedule of three consecutive cycles of three administrations each at two CIGB-300 dose levels, 35 and 70 mg. The assayed therapeutic regime was safe, resulting in significant uptake of CIGB-300 by the tumors along with a pronounced reduction of B23/npm protein levels in tumor biopsies (manuscript in preparation). Thus, this treatment schedule, originally devised during preclinical testing, appears to be suitable for cervical cancer patients and merits further study in Phase 2 trials.

The third Phase 1 study was another dose-escalation trial (CERVISEG-300) using 70, 105, and 175 mg of CIGB-300 with prior antihistamine administration, designed to answer the issues raised by the allergic-like syndrome identified as the dose-limiting toxicity of CIGB-300 during clinical research. This trial, unlike CERVIFARM-I-300, never reached the maximum tolerated dose. Importantly, the prophylactic administration of antihistaminic medication enabled increasing the dosage of CIGB-300 and enlarged the existing safety margins for the administration of this compound.

By design, the CERVIFARM-I-300, CERVIFARM-II-300, and CERVISEG-300 Phase 1 trials addressed the safety and pharmacology of CIGB-300 exclusively in a neoadjuvant setting, that is, as a monotherapy prior to chemoradiotherapy in women with cervical cancer stage 1B2/II. However, as CIGB-300 has shown synergistic and/or additive interactions in tumor cell lines when combined with standard anticancer drugs, additional clinical trials were designed where intratumoral injections of CIGB-300 were administered concomitantly with chemoradiotherapy. According to data from a dose-finding Phase 2 trial using this approach (CERVICON-300), the combination of 15, 35, or 70 mg of CIGB-300 with chemoradiotherapy was safe and well tolerated. Additionally, the patients enrolled in the cohort receiving 15 mg of CIGB-300 and chemoradiotherapy had a higher frequency of complete response than those of the cohort receiving chemoradiotherapy alone, as determined by magnetic resonance imaging (unpublished data). Likewise, a separate trial where antitumor responses were determined by positron emission tomography (PET) also found that the concomitant administration of CIGB-300 and chemoradiotherapy was safe and well tolerated (unpublished data).

Collectively, the existing data from human pharmacology and therapeutic exploration have enabled determining the safety profile as well as optimal treatment schedules for the local administration of CIGB-300 to patients with stage IB2/II cervical cancer. These findings will guide the design of future Phase 3 clinical trial in the same patient population, where efficacy will be the major end point.

Taking into account the poor anatomical accessibility of most tumors, and in light of preclinical findings demonstrating that the systemic delivery of CIGB-300 successfully hampers tumor growth in mice, early stage clinical trials of CIGB-300 have also addressed the possibility of intravenous administration. The first Phase 1 trial that investigated this delivery route used 16 patients with advanced cancer, delivering CIGB-300 in a dose-escalation scheme that went from 0.2 to 1.6 mg/kg. The drug was administered in 3 cycles, each consisting of a daily 15-min intravenous infusion for 5 consecutive days of the test compound. This therapeutic regime failed to reach the maximum tolerated dose and yielded only grade 1/2 adverse events. Although CIGB-300, like other peptide-based drugs, had a relative quick clearance and a short half-life in blood after intravenous delivery, it did exhibit a large volume of distribution. Most importantly, survival was increased in 12 of 16 patients beyond the life expectancy of their advanced cancers (manuscript in preparation). A subsequent trial was also implemented in patients with blood cancer, using an intra-patient dose-escalation scheme from 1.6 to 12.8 mg/kg that administered CIGB-300 in 15-min intravenous infusions. Each CIGB-300 dose level was delivered in cycles of 3 consecutive days during 5 weeks. Satisfyingly, it was possible to explore the safety and tolerability of CIGB-300 across this wide dose range, and signs of therapeutic efficacy were also observed (unpublished data).

CIGB-300 has also been used to treat isolated patients under compassionate use clauses. Remarkably, a successful antitumor response was observed in a chemoradio refractory patient diagnosed with germinoma who experienced significant tumor shrinking with a sustained improvement of his quality of life after CIGB-300 treatment [12].

In summary, clinical research on CIGB-300 is still undergoing, and data from further trials exploring its efficacy will undoubtedly contribute to turn CK2-mediated phosphorylation into a validated clinical target.

7 Concluding Remarks

CIGB-300 is an anti-CK2 peptide that induces apoptosis in vitro and in vivo and is under clinical research in cancer patients. So far, CIGB-300 has been shown to inhibit CK2-mediated phosphorylation of B23/npm, AKT, and PTEN, although other CK2 substrates could be also affected. Of note, downstream modulation by CIGB-300 includes a diverse array of proteins involved in processes that go from apoptosis and cell proliferation to ribosomal biogenesis, drug resistance, and angiogenesis, among others. Accordingly, CIGB-300 not only affects cell viability and proliferation but also exhibits an antiangiogenic effect and synergizes with standard anticancer drugs, such as paclitaxel, cisplatin, and erlotinib. Remarkably, CIGB-300 can be delivered either by the systemic route or by local intratumoral administration, highlighting promising features of this investigational drug for cancer therapy. The available human pharmacology data for CIGB-300 not only demonstrate its safety and tolerability but also provide important clues for designing future clinical trials focusing on its therapeutic efficacy. Overall, CIGB-300 meets several important requirements as an anticancer targeted agent and may serve as a tool to validate CK2 as a promising clinical target for cancer therapy.Acknowledgements We thank to C.I.G.B. and Biorec for supporting all of the work compiled in this review article. We also thank Alejandro Martin for reviewing the manuscript.

References

1. Battistutta R, Sarno S, De Moliner E et al (2000) The replacement of ATP by the competitive inhibitor emodin induces conformational modifications in the catalytic site of protein kinase CK2. J Biol Chem 275:29618–29622
2. Li C, Liu X, Lin X, Chen X (2009) Structure-activity relationship of 7 flavonoids on recombinant human protein kinase CK2 holoenzyme. Zhong Nan Da Xue Xue Bao Yi Xue Ban 34:20–26
3. Pagano MA, Meggio F, Ruzzene M et al (2004) 2-Dimethylamino-4,5,6,7-tetrabromo-1H-benzimidazole: a novel powerful and selective inhibitor of protein kinase CK2. Biochem Biophys Res Commun 321:1040–1044
4. Sarno S, de Moliner E, Ruzzene M et al (2003) Biochemical and three-dimensional-structural study of the specific inhibition of protein kinase CK2 by [5-oxo-5,6-dihydroindolo-(1,2-a) quinazolin-7-yl]acetic acid (IQA). Biochem J 374:639–646
5. Sarno S, Reddy H, Meggio F et al (2001) Selectivity of 4,5,6,7-tetrabromobenzotriazole, an ATP site-directed inhibitor of protein kinase CK2 ('casein kinase-2'). FEBS Lett 496:44–48
6. Pierre F, Chua PC, O'Brien SE et al (2011) Discovery and SAR of 5-(3-chlorophenylamino) benzo[c][2,6]naphthyridine-8-carboxylic acid (CX-4945), the first clinical stage inhibitor of protein kinase CK2 for the treatment of cancer. J Med Chem 54:635–654
7. Pierre F, Chua PC, O'Brien SE et al (2011) Pre-clinical characterization of CX-4945, a potent and selective small molecule inhibitor of CK2 for the treatment of cancer. Mol Cell Biochem 356:37–43
8. Siddiqui-Jain A, Drygin D, Streiner N et al (2010) CX-4945, an orally bioavailable selective inhibitor of protein kinase CK2, inhibits pro- survival and angiogenic signaling and exhibits antitumor efficacy. Cancer Res 70:10288–10298

9. Laudet B, Barette C, Dulery V et al (2007) Structure-based design of small peptide inhibitors of protein kinase CK2 subunit interaction. Biochem J 408:363–373
10. Slaton JW, Unger GM, Sloper DT et al (2004) Induction of apoptosis by antisense CK2 in human prostate cancer xenograft model. Mol Cancer Res 2:712–721
11. Perea SE, Reyes O, Puchades Y et al (2004) Antitumor effect of a novel proapoptotic peptide that impairs the phosphorylation by the Casein Kinase 2 (CK2). Cancer Res 64:7127–7129
12. Perea SE, Baladron I, Garcia Y et al (2011) CIGB-300, a synthetic peptide-based drug that targets the CK2 phosphoaceptor domain. Translational and clinical Research. Mol Cell Biochem 356:45–50
13. Perera Y, Farina HG, Gil J et al (2009) Anticancer peptide CIGB-300 binds to nucleophosmin/B23, impairs its CK2-mediated phosphorylation, and leads to apoptosis through its nucleolar disassembly activity. Mol Cancer Ther 8:1189–1196
14. Piñol-Roma S (1999) Association of nonribosomal nucleolar proteins in ribonucleoprotein complexes during interphase and mitosis. Mol Biol Cell 10:77–90
15. Perera Y, Costales HC, Díaz Y et al (2012) Sensitivity of tumor cells towards CIGB-300 anti-cancer peptide relies on its nucleolar localization. J Pept Sci 18:215–223
16. Meggio F, Pinna L (2003) One-thousand-and-one substrates of protein kinase CK2? FASEB J 17:349–368
17. Martins LR, Lúcio P, Silva MG et al (2014) Targeting chronic lymphocytic leukemia using CIGB-300, a clinical-stage CK2-specific cell-permeable peptide inhibitor. Oncotarget 5:258–263
18. Perera Y, Farina HG, Hernández I et al (2008) Systemic administration of a peptide that impairs the Protein Kinase (CK2) phosphorylation reduces solid tumor growth in mice. Int J Cancer 122:57–62
19. Louvet E, Junera HR, Le Panse S et al (2005) Dynamics and compartmentation of the nucleolar processing machinery. Exp Cell Res 304:457–470
20. Louvet E, Junéra HR, Berthuy I et al (2006) Compartmentation of the nucleolar processing proteins in the granular component is a CK2-driven process. Mol Biol Cell 17:2537–2546
21. Homma MK, Homma Y (2008) Cell cycle and activation of CK2. Mol Cell Biochem 316:49–55
22. Homma MK, Homma Y (2005) Regulatory role of CK2 during the progression of cell cycle. Mol Cell Biochem 247:47–52
23. Yamane K, Kisella TJ (2005) Casein kinase 2 regulates both apoptosis and the cell cycle following DNA damage induced by 6-thioguanine. Clin Cancer Res 11:2355–2363
24. Dixit D, Sharma V, Ghosh S et al (2012) Inhibition of Casein kinase-2 induces p53-dependent cell cycle arrest and sensitizes glioblastoma cells to tumor necrosis factor (TNFa)-induced apoptosis through SIRT1 inhibition. Cell Death Dis 3:e271. doi:10.1038/cddis.2012.10
25. Keller DM, Lu H (2002) p53 serine 392 phosphorylation increases after UV through induction of the assembly of the CK2·hSPT16·SSRP1 complex. J Biol Chem 277:50206–50213
26. McKendrick L, Milne D, Meek D (1999) Protein kinase CK2-dependent regulation of p53 function: evidence that the phosphorylation status of the serine 386 (CK2) site of p53 is constitutive and stable. Mol Cell Biochem 191:187–199
27. Quotti L, Gurrieri C, Brancalion A et al (2013) Inhibition of protein kinase CK2 with the clinical-grade small ATP-competitive compound CX-4945 or by RNA interference unveils its role in acute myeloid leukemia cell survival, p53-dependent apoptosis and daunorubicin-induced cytotoxicity. J Hematol Oncol 6:78, http://www.jhoonline.org/content/6/1/78
28. Rodriguez-Ulloa A, Ramos Y, Gil J et al (2010) Proteomic profile regulated by the anticancer peptide CIGB-300 in non-small cell lung cancer (NSCLC) cells. J Proteome Res 9: 5473–5483
29. Tawfic S, Olson MO, Ahmed KJ (1995) Role of protein phosphorylation in post-translational regulation of protein B23 during programmed cell death in the prostate gland. J Biol Chem 270:21009–21015
30. Farina HG, Benavent F, Perera Y et al (2011) CIGB-300, a proapoptotic peptide, inhibits angiogenesis in vitro and in vivo. Exp Cell Res 317:1677–1688

31. Langel U (2006) Handbook of cell-penetrating peptides. Taylor & Francis Group, Stockholm, Sweden, pp 29–43
32. Patel LN, Zaro JL, Shen WC (2007) Cell penetrating peptides: intracellular pathways and pharmaceutical perspectives. Pharm Res 24:1977–1992
33. Tünnemann G, Martin RM, Haupt S et al (2006) Cargo-dependent mode of uptake and bioavailability of TAT-containing proteins and peptides in living cells. FASEB J 20:1775–1784
34. Benavent F, Perera Y, Alonso DF et al (2014) Mechanisms of cellular uptake, intracellular transportation, and degradation of CIGB-300, a Tat-conjugated peptide, in tumor cell lines. Mol Pharm 11:1798–1807
35. Perera Y, Del Toro N, Gorovaya L (2014) Synergistic interactions of the anti-casein kinase 2 CIGB-300 peptide and chemotherapeutic agents in lung and cervical preclinical cancer models. Mol Clin Oncol 2(6):935–944
36. Datta SR, Brunet A, Greenberg ME (1999) Cellular survival: a play in three Akts. Genes Dev 13:2905–2927
37. Jiang BH, Liu LZ (2009) PI3K/PTEN signaling in angiogenesis and tumorigenesis. Adv Cancer Res 102:19–65
38. Chen JC, Wang Q, Huang XH et al (2009) Involvement of PI3K/PTEN/AKT/mTOR pathway in invasion and metastasis in hepatocellular carcinoma: Association with MMP-9. Hepatol Res 39:177–186
39. Huether A, Hopfner M, Sutter AP et al (2005) Erlotinib induces cell cycle arrest and apoptosis in hepatocellular cancer cells and enhances chemosensitivity towards cytostatics. J Hepatol 43:661–669
40. Solares AM, Santana A, Baladrón I et al (2009) Safety and preliminary efficacy data of a novel casein kinase 2 (CK2) peptide inhibitor administered intralesionally at four dose levels in patients with cervical malignancies. BMC Cancer 9(1):146
41. Soriano-Garcia JL, Lopez-Diaz A, Solares-Asteasuainzarra M et al (2013) Pharmacological and safety evaluation of CIGB-300, a casein kinase 2 inhibitor peptide, administered intralesionally to patients with cervical cancer stage IB2/II. J Cancer Res Ther 1:163–173

Targeting CK2 for Cancer Therapy Using a Nanomedicine Approach

Khalil Ahmed, Gretchen Unger, Betsy T. Kren, and Janeen H. Trembley

Abstract CK2 is a signal-responsive serine/threonine protein kinase which promotes cell proliferation, suppresses apoptosis, and demonstrates increased expression in numerous cancers. Here, we present information on investigations into CK2-focused cancer therapy in general and discuss in detail a nanomedicine approach to targeting CK2 in a cancer cell-specific manner. Specifically, we summarize data on biodistribution and therapeutic efficacy of a tenfibgen (TBG) nanoencapsulation technology for the delivery of anti-CK2 cargos to malignant cells. The TBG nanocapsule cargos discussed include siRNA (siCK2), single-stranded DNA/RNA chimeric oligonucleotides (RNAi-CK2), and a small-molecule CK2 inhibitor (DMAT). Systemic administration of TBG-RNAi-CK2 resulted in xenograft tumor reduction using low doses with concomitant reduction in CK2 protein expression. Systemic TBG-DMAT treatment decreased xenograft tumor proliferation. No toxicity or early inflammation response was observed after using any of the TBG encapsulated anti-CK2 cargos. The utility of this therapy approach for targeting metastatic cancer sites and on overall survival is also discussed. Protected and malignant cell-specific delivery of a therapeutic is a promising target-specific and

---

K. Ahmed
Department of Laboratory Medicine and Pathology, University of Minnesota Medical School and Minneapolis VA Health Care System, Minneapolis, MN, USA

G. Unger
GeneSegues, Inc., Chaska, MN 55318, USA

B.T. Kren • J.H. Trembley (✉)
Cellular and Molecular Biochemistry Research Laboratory (151), Minneapolis Veterans Affairs Health Care System , One Veterans Drive, Minneapolis, MN 55417, USA

Department of Laboratory Medicine and Pathology, University of Minnesota, Minneapolis, MN 55455, USA

Masonic Cancer Center, University of Minnesota, Minneapolis, MN 55455, USA

Cellular and Molecular Biochemistry Research Laboratory (151), Veterans Affairs Medical Center, Minneapolis, MN, USA
e-mail: trem0005@umn.edu

K. Ahmed et al. (eds.), *Protein Kinase CK2 Cellular Function in Normal and Disease States*, Advances in Biochemistry in Health and Disease 12, DOI 10.1007/978-3-319-14544-0_17

versatile approach for cancer therapy, and both the TBG encapsulation technology and the anti-CK2 oligonucleotide approach demonstrate substantial potential for the treatment of malignancy.

Keywords CK2 • Antisense • RNAi • siRNA • DMAT • Nanocapsule • Nanoparticle • Nanomedicine • Xenograft • Prostate • HNSCC • Biodistribution • Metastasis

1 Introduction

Our laboratories are focused on disabling the expression and function of the CK2 protein kinase as an approach to cancer therapy. Early studies in this laboratory on prostate and in head and neck squamous cell cancers led to the conclusion that not only was CK2 expression and activity associated with the malignant phenotype, but that moderate loss of CK2 activity could tip the balance of cancer cell survival toward cell death [1–6]. As is detailed in another chapter (see "CK2: A Global Regulator of Cell Death"), coincident experimental work demonstrated that the central role for higher CK2 expression and activity in cancer cells is the suppression of cell death [7–11]. These studies culminated in our original proposal to target CK2 for cancer therapy [12, 13], a concept that has gained considerable attention in recent years. The discussion in this chapter is focused on the rationale and methodology behind taking a nanomedicine approach for suppressing CK2 function as well as a discussion of results from targeting CK2 in xenograft models of cancer in mice.

2 Why Is CK2 a Meaningful Target?

CK2 (formerly casein kinase II or CKII) is predominantly a protein Ser/Thr kinase with a consensus amino acid target sequence favoring a phosphorylation site surrounded by negatively charged amino acids, such as X-S/T-X-X-D/E/Sp/Yp [14, 15]. The CK2 heterotetrameric enzyme typically contains two catalytic subunits (CK2α and/or CK2α') and two regulatory subunits (CK2β). This kinase is ubiquitously expressed and is associated with various cellular compartments and organelles, including chromatin, nuclear matrix, nucleoli, endoplasmic reticulum, Golgi, mitochondria, and cytoskeleton as well as cell membrane [16–18]. CK2 phosphorylates a large number of substrates with functions relating to the majority of cellular activities. Additionally, although aspects of its regulation which modulate CK2 activity continue to emerge, overall CK2 exhibits constitutive activity in cells [19–23]. Thus, specific activating events such as phosphorylation of the activation segment observed in cyclin-dependent kinase 2 (CDK2) or protein kinase A (PKA) are not required for CK2 enzymatic function [24].

Studies from our laboratory and others have resulted in the labeling of CK2 as a "master regulator" of cell survival, especially in the context of cancer. In non-transformed cells, expression of CK2α and CK2β is essential for survival [19, 25–28]. At the same time, in cells under normal growth or maintenance conditions, CK2 steady-state expression levels are generally moderate and stable. When considering transformed cells, CK2 expression and activity are consistently elevated [6, 29–33]; moreover, the increase in CK2 expression in malignant cells is often associated with a more prominent nuclear localization [3, 34, 35]. These expression characteristics correlate with disease severity and prognosis [1–3, 35–40]. Interestingly, in contrast to many oncogenes or cancer-associated genes, mutational changes in CK2 genes are not generally observed in cancers. Moreover, recent publications demonstrated through database analyses that copy number variations for the CK2 genes are observed in glioma and breast cancer [41, 42]. Thus, response to anti-CK2 therapy is not dependent on the presence of particular mutational forms of the enzyme. Overall, CK2 is undoubtedly an important contributor to the cancer phenotype, making it an attractive target for anticancer therapeutic development based on the increased dependence of cancer cells on CK2 expression.

3 Targeting CK2 Protein Expression and Kinase Activity in Tumors

Both chemical and molecular biological methods have been developed for targeting CK2. Protein expression can be blocked using siRNA and antisense oligonucleotides to promote cleavage and degradation or hinder translation of the mRNA transcript. The majority of nucleic acid-based CK2-targeting research has been performed in cultured cancer cells. Antisense strategies targeting CK2 have been applied in breast and colon cancer cells, head and neck squamous cell carcinoma (HNSCC), leukemia, neuroblastoma, and prostate cancer cells [4, 5, 43–46]. Transfection of anti-CK2 siRNAs has also been performed in many and varied cancer cell lines [38, 40, 47–53]. Overall, the results from these studies have demonstrated the loss of activity in key signaling pathways such as Notch1, Hedgehog/Gli1, Akt, and NF-κB. Further, loss of CK2 expression caused growth inhibition, increased sensitivity to anoikis and chemotherapeutic drugs, induction of cell death, and loss of stem-like cancer cell populations. Some studies have shown that CK2 downregulation caused differing effects with respect to cellular p53 status. In colon cancer HCT-116 cells which express wild-type p53, loss of CK2 caused cellular senescence [54]. In contrast, loss of CK2 in acute myeloid leukemia cells caused p53-dependent apoptosis [47]. However, in prostate cancer cells it has been shown that p53 is not required for the induction of apoptosis [55]. Results from studies in which a nucleic acid therapy approach was utilized in mouse xenograft tumor studies will be discussed later in this chapter.

As an alternate to a nucleic acid-based approach, protein catalytic activity can be suppressed using small-molecule chemical inhibitors. Numerous CK2 small-molecule inhibitors exist and continue to be developed [56]. CK2 inhibitors are typically type I ATP-competitive molecules. Similar to the effects of anti-CK2 antisense and siRNA treatment of cells, the addition of CK2 inhibitors to cultured patient cells and established cell lines induced loss of cell viability, enhanced radio- and chemosensitivity, overcame drug resistance, caused loss of key survival signaling activity, and induced cell death [47, 51, 57–74]. CK2 small-molecule inhibitors have been employed to assess their effects in mouse xenograft models of human cancer [41, 67, 75–80]. These studies have included single agent testing of the CK2 inhibitors DMAT, TBB, and CX-4945 as well as combinatorial testing of CK2 inhibitors with other agents such as the EGFR inhibitor erlotinib or standard chemotherapy drugs. CX-4945 is an orally available ATP-competitive anti-CK2 drug, and much of the published mouse xenograft tumor response data is from the use of this drug delivered orally twice per day at 75 mg/kg. Used as a single agent, CX-4945 has shown tumor growth inhibition effects of 76–97 % in breast, prostate, pancreatic, and glioblastoma tumors [41, 67, 76]. The use of CX-4945 in combinatorial testing has shown improved results with cisplatin and gemcitabine in ovarian cancer and with erlotinib in squamous cell and non-small cell lung cancers [79, 80]. CX-4945 has also undergone testing in solid tumors in Phase 1 clinical trials. No complete or partial RECIST responses were observed, but stable disease for at least 6 months was evident in 15 % of treated patients [81]. Recently, Senhwa Biosciences, Inc., announced a new randomized Phase 1b/2 trial of CX-4945 in combination with gemcitabine and cisplatin for the treatment of bile duct cancers (ClinicalTrials.gov NCT02128282).

Another treatment approach for modifying CK2 activity in cancer cells has been taken by the development of a peptide-based drug called CIGB-300 (formerly P15-Tat) (discussed in detail in chapter "CIGB-300: A Promising Anti-Casein Kinase 2 (CK2) Peptide for Cancer Targeted Therapy" by Perea et al.). CIGB-300 targets the CK2 phospho-acceptor site, predominantly blocking phosphorylation of B23/nucleophosmin, thereby causing nucleolar disassembly and apoptosis [82]. This drug was tested by direct injection into tumors as well as systemic administration in mice [83–85]. Importantly, CIGB-300 has also undergone human testing by intralesional injection for safety and efficacy in patients with cervical cancer [86, 87]. In these Phase 1 trials, patients experienced significant lesion reduction and reduced nuclear B23 protein levels.

4 Delivery of an Anti-CK2 Drug to Malignant Cells by Nanoencapsulation

An advantage to targeting CK2 as an anticancer strategy is the universally increased dependence of malignant cells compared to normal cells on CK2 expression. However, considering that normal cells still require its expression and activity for survival, it should be preferable to deliver a CK2-targeting drug specifically to the malignant cells in the body. Nucleic acid-based cancer therapy is an approach with exquisite gene target specificity. However, when considering an oligonucleotide therapeutic approach for systemic treatment, the delivery vehicle for the nucleic acids becomes critical. Not only do the nucleic acids need to be protected from degradation in the blood stream, but ideally they need to specifically enter cancer and not normal cells and induce the desired response within the targeted population, thus avoiding the issues of host toxicity. We have developed a tenfibgen nanoencapsulation

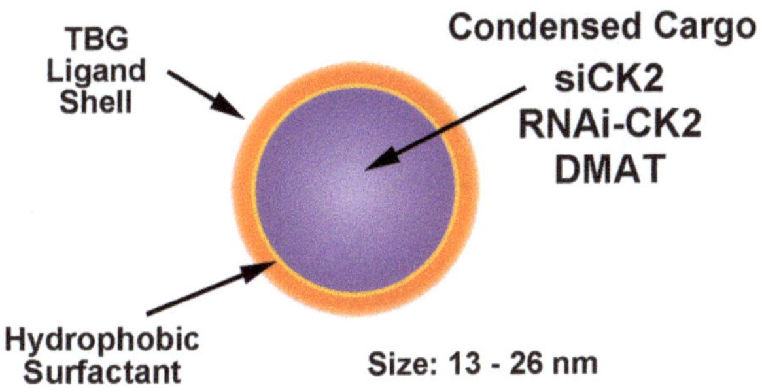

Fig. 1 Tenfibgen nanocapsule design. Cartoon depiction of tenfibgen (TBG) nanocapsule components including the cargo

technology that allows for the delivery of drug cargos into malignant cells in vivo while avoiding accumulation in normal cells [51, 88]. The drug cargo can be nucleic acids or small-molecule drugs.

The sub-50 nm (i.e., less than 50 nm) size nanocapsules used by us for cancer therapy are comprised of the ligand tenfibgen (TBG), which is the carboxy-terminal fibrinogen globe domain of tenascin-C, and a condensed drug cargo. To allow for receptor-mediated targeting via tenascin-C receptors which are elevated in cancer cells [89–95], the shell of the nanocapsule is comprised of the protein ligand TBG. A diagrammatic representation of the TBG nanocapsule is shown in Fig. 1. The nanocapsules are formulated by complexing oligonucleotides or DMAT with polyornithine or spermine, respectively. The complexes are then emulsified using a water-insoluble surfactant dispersed into a water-miscible solvent and then inverted by dilution into a suspension with addition of PBS. The resultant hydrophobic micelles are coated by adsorption of TBG dispersed into the solution and then subjected to spray dispersion atomization into a LiCl salt solution. Following incubation at 4–6 °C with rotation in the salt solution for more than 14 h, the nanocapsules are recovered by centrifugation or diafiltration and finally exchanged into PBS containing 10 % lactitol (w/v). This process results in uniform rounded nanocapsules with a size range of 13–26 nm for oligonucleotide and small-molecule cargos (see Table 1 and [88, 96, 97]).

5 Malignant Cell Specificity and Access by TBG Nanocapsules

TBG nanocapsules are specifically taken up by malignant cells but not by normal cells. This uptake specificity has been demonstrated in cultured cells by measuring cell viability following treatment of prostate epithelial cells (PrEC), benign prostatic hyperplasia (BPH-1), and malignant prostate cells (C4-2, PC3-LN4) with TBG nanoencapsulated anti-CK2 small-molecule inhibitor DMAT, TBG

Table 1 General characteristics of tenfibgen nanocapsules

Shell ligand	Particle size (nm)[a]	Zeta potential (meV)[b]	Morphology[c]	Cargo	Sequence information[d]
Tenfibgen (TBG) 27 kDa	20.7±5.2	−1.7±2.4	Uniform, single capsules	siCK2[e]	5′-auacaacccaaacucca caudTdT-3′ (guide strand)
Tenfibgen (TBG) 27 kDa	17.6±2.5	−6.2±3.1	Uniform, single capsules	RNAi-CK2[f]	5′-ATACAACCCAAACT *ccacau*(propyl)-3′
Tenfibgen (TBG) 27 kDa	14.6±1.2	−8.6±2.2	Uniform, single capsules	DMAT[g]	N/A

[a]Mean±SD of the average elliptical diameter determined from AFM or TEM micrographs
[b]Mean±SE of the average surface charge measured by DLS from two different preparations across a 20 V potential in 1 mM KCl at 2 μg/ml
[c]Morphology of all nanocapsules determined by visual AFM and TEM observation as uniform, single capsules [88, 96, 97]
[d]Upper case letters represent phosphodiester DNA bases; small case letters represent RNA bases; small case italic letters represent 2′ *O*-methyl RNA bases
[e]TBG-siCK2 encapsulation efficiency mean of 86.8 % observed by Burton analysis relative to unencapsulated oligomer
[f]TBG-RNAi-CK2 encapsulation efficiency mean of 79.8 % observed by Burton analysis relative to unencapsulated oligomer
[g]TBG-DMAT encapsulation efficiency mean of 0.24 % determined by neutron activation analysis of the four Br residues in DMAT

nanoencapsulated anti-CK2 siRNA (siCK2), or TBG nanoencapsulated single-stranded DNA/RNA oligonucleotide (RNAi-CK2) [51, 97]. The cancer cells showed reduced viability, whereas the benign cells did not exhibit any loss of viability. In comparison, both the benign and malignant cells showed reduced viability after treatment with naked DMAT or transfection with siCK2 using commercial lipid reagents.

To ascertain biodistribution and malignant cell specificity in vivo, the uptake of naked antisense oligonucleotides and of TBG encapsulated oligonucleotides was examined in tumors and other tissues after systemic administration in mice. Naked FITC-labeled antisense (AS) phosphorothioate oligonucleotides directed against CK2αα′ were injected into nude mice carrying orthotopic PC3-LN4 xenograft prostate tumors. Tumors and other tissues were collected 24 h after the second injection and the FITC signal quantitated in tissue lysates. The highest level of accumulation of naked FITC-AS-CK2 was in kidney, followed by tumor, then spleen, liver, blood, and heart. In order to measure an encapsulated oligonucleotide, we utilized an iodine-derivatized oligonucleotide that we could measure via neutron activation analysis regardless of whether it was released from the nanocapsule. In this study, nude mice carrying HNSCC tumors were injected intravenously (i.v.) with TBG-I^{127}-siRNA or TBG-sugar (background control) nanocapsules. Two hours after administration of the nanocapsules, the mice were sacrificed, and collected

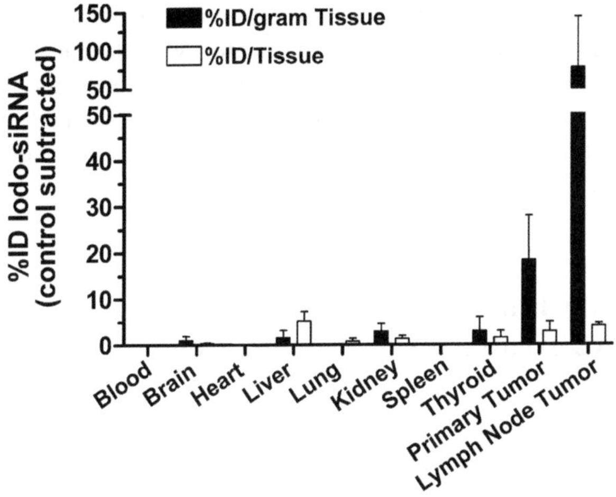

Fig. 2 Biodistribution of TBG nanocapsules in FaDu tumor xenograft mice. Biodistribution was determined in vivo 2 h after i.v. administration of TBG nanocapsules containing I^{127}-siRNA to nude mice bearing FaDu tumors ($n=3$) by neutron activation analysis of the tissues. Levels of endogenous tissue iodine were measured in FaDu tumor mice treated with TBG-sugar ($n=2$), and these values were subtracted from those of the TBG-I^{127}-siRNA-treated mice. The results are expressed as % injected dose (ID) per tissue (*white bars*) and % ID per gram of tissue (*black bars*)

tissues were subjected to neutron bombardment which converted I^{127} to I^{128}. TBG nanoencapsulated oligonucleotide accumulation was observed at highest levels per gram of tissues in lymph node tumor and prostate tumor. Kidney, thyroid, liver, and brain showed minimal accumulation (Fig. 2).

In another study, the CK2 inhibitor DMAT was TBG nanoencapsulated and used to treat PC3-LN4 subcutaneous xenograft tumors in mice. These nanocapsules contained the $F(ab')_2$ fragment of Syrian hamster IgG within the TBG protein coat; this IgG was used as a marker for the identification of the nanocapsule within tissues. By indirect immunofluorescence, the presence of Syrian hamster IgG was detected in tumors of TBG-DMAT-treated mice, but not in the liver and testis from the same mice [96]. As testament to the cross-species conservation of the TBG protein, it was documented that nanocapsules made using the same purified recombinant human TBG were taken up by tumors of mouse origin [19]. Thus, we anticipate that non-tumor uptake of TBG nanocapsules in mouse tissues would be readily detected.

Further evidence of TBG-directed delivery of the nanocapsules to metastatic sites was seen using TBG encapsulated luciferase expression plasmid. Mice carrying metastatic HNSCC xenograft tumors were injected with TBG-luciferase nanocapsules, and lung and abdominal metastatic signals were detected by imaging 5 days later. Upon sacrifice, tumor and tissue lysates were subjected to immunoblot analysis for keratin-14, a HNSCC marker protein. The lysates were also immunoprecipitated for luciferase. Keratin-14 and luciferase signal were co-detected in brain, lung, and spleen [29]. Further, it was shown that TBG nanocapsules localize to prostate tumor located in bone [97]. These data demonstrate co-localization of TBG-delivered cargos to sites of metastatic tumor growth. Finally, other data also suggest that the TBG nanocapsules are taken up by tumor-derived microvessels [13, 29]. Taken together, these studies demonstrate that TBG nanoencapsulated molecules successfully and preferentially enter malignant-derived cells when systemically administered in vivo.

6 Acute Effects of Naked and TBG Nanocapsule Anti-CK2 Treatments in Primary Tumors

In addition to biodistribution studies in mice, we have also examined the effects of TBG nanocapsules containing various anti-CK2 cargos on primary tumor growth and death. Initial therapy studies were conducted in prostate xenograft tumors to examine CK2 as a target for oligonucleotide-based therapy. An antisense oligonucleotide targeted to the 5′ end of the CK2α mRNA was directly injected into PC3-LN4 subcutaneous tumors as a single treatment. A dose-response was carried out from 5 to 20 μg, with the 20 μg injection resulting in complete elimination of the tumor in 7 days [98]. At the lower doses, tumor size was significantly reduced with corresponding loss of CK2α mRNA and nuclear matrix-associated protein.

In the next study, we tested systemic administration of an antisense oligonucleotide. Because we frequently observed increased CK2α′ protein expression after downregulation of CK2α, we redesigned the oligonucleotide so that it perfectly targeted CK2α (20/20 residues) and imperfectly targeted CK2α′ (19/20 residues) within the coding region of both transcripts [46]. The oligonucleotide was used as phosphorothioate chemistry to improve in vivo stability (termed bs-AS-CK2). In a dose-response design study, nude mice bearing orthotopic PC3-LN4 prostate tumors were treated twice by intraperitoneal (i.p.) injections given 24 h apart with cumula-

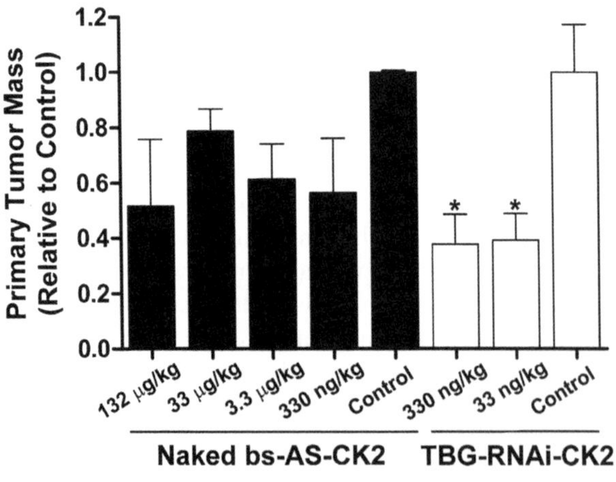

Fig. 3 Comparison of therapeutic efficacy of naked bs-As-CK2 and TBG-RNAi-CK2 in orthotopic prostate cancer xenograft tumors. Primary tumor masses are shown following treatments as indicated on the chart. Means ± SE are presented. *$p < 0.05$

tive doses ranging from 0.33 to 132 μg/kg. Decreased tumor weight (Fig. 3) and CK2αα′ mRNA and protein expression were observed in treated mice relative to saline and GAPDH antisense controls [46]. In a further study in which a cumulative dose of 66 μg/kg was delivered, divided into different delivery schedules over two or four injections, protein level therapeutic responses were seen for CK2α, CK2α′, CK2β, NF-κB, and AKT [46]. To sum, these experiments showed the efficacy of targeting CK2 in prostate cancer using an oligonucleotide-based approach.

Having established that using oligonucleotide-based methods to downregulate CK2 expression is effective in vivo and further that TBG nanoencapsulation functions to deliver cargo preferentially to tumor sites, we undertook in vivo therapeutic studies employing TBG-anti-CK2. In the subsequent examples, the encapsulated oligonucleotide has the identical bispecific sequence targeting CK2αα′ as the systemically delivered naked bs-AS-CK2 oligonucleotide described above. However, the nanoencapsulated cargo, RNAi-CK2, utilized in the next studies differs in chemical composition; the first 14 bases (5′) are standard DNA nucleotides and the last six residues (3′) are 2′ *O*-methyl RNA oligonucleotides, terminated with a 3′ propyl group. This DNA/RNA hybrid RNAi molecule is stable and can potentially use both antisense-like RNase H-mediated and siRNA-like RISC-mediated pathways [88, 97]. Detailed information on the general characteristics of TBG-RNAi-CK2 nanocapsules is presented in Table 1.

Our initial proof-of-concept mouse studies using TBG-RNAi-CK2 nanocapsules were carried out using PC3-LN4 orthotopic prostate tumors in nude mice. These mice received two i.p. injections, given with an interval of 24 h, of either TBG-RNAi-CK2 or vehicle. At 13 days after the initiation of treatment, a decrease of greater than 60 % in tumor mass was observed using doses of 33 and 330 ng/kg (Fig. 3). Quantitative analysis of the CK2 mRNA levels showed a lesser reduction than observed with naked bs-AS-CK2 treatments, whereas reduction in CK2αα′ proteins was very comparable [46, 97]. Overall, the results observed using TBG-RNAi-CK2 at doses from 33 to 330 ng/kg were equal to or better than those we observed using the naked bs-AS-CK2 at doses from 330 ng/kg to 132 μg/kg. In addition, in order to assess the potential for an early inflammatory response, we measured the serum concentration of interferon-γ and the tissue weight ratios for spleen, liver, and thymus in non-tumor-bearing, immunocompetent mice. No differences in the tissue weights for the animals treated with the TBG-RNAi-CK2 or TBG-sugar relative to vehicle control were noted. We also found no evidence of interferon-γ elevation, which typically is observed in particle-mediated early inflammatory responses [99] thus suggesting the relative safety of these nanocapsules.

We have also examined the effects of a TBG nanoencapsulated CK2 inhibitor DMAT on prostate xenograft tumor growth in nude mice. Formulation of DMAT into TBG nanocapsules was much less efficient than the various anti-CK2 oligonucleotides, and thus we were limited in the dose level and number of treatments we could administer. Information on the TBG-DMAT nanocapsules is detailed in Table 1. We studied four treatment groups: TBG-DMAT 100 μg/kg and vehicle control groups received daily injections for 8 days and were sacrificed on day 9, and TBG-DMAT 20 μg/kg and naked DMAT 500 μg/kg groups received daily injections for 6 days and were sacrificed on day 7. Treatment with TBG-DMAT at either dose level caused increased death within the tumor masses and significantly decreased tumor cell proliferation as measured by Ki-67 staining [96]. NF-κB p65 protein expression was also reduced by TBG-DMAT treatment. No tissue damage was observed in liver or testis after TBG-DMAT treatment.

We carried out similar acute response studies to evaluate the therapeutic effects of TBG-RNAi-CK2 treatment in models of HNSCC. In both the UM-SCC 11A and Fadu models of HNSCC, two to three treatment injections of TBG-RNAi-CK2 at 10 μg/kg resulted in significantly decreased tumor volumes (greater than 50 %) relative to control [53]. Reduced CK2αα′ protein expression was observed in the treated tumors by immunohistochemical staining. Markers of apoptosis were detected such as increased TUNEL, cleaved caspase-3, and reduced expression of Bcl-2 and Bcl-xL. A negative impact on proliferative signaling was evidenced by reduced NF-κB p65 expression, reduced NF-κB p65 phosphorylation on S536 and S529, loss of Cyclin D1 expression, and increased p53 and p63 expression [53].

7 Delivery and Efficacy of TBG Anti-CK2 Nanocapsule Treatment in Metastatic Tumors

The majority of deaths related to cancer are caused by metastatic growth of cancer cells in alternate sites both near and far from the primary tumor. We have data indicating the treatment effects of TBG-RNAi-CK2 nanocapsules on metastatic cancer originating from orthotopic prostate and HNSCC tumors in mice. First, in the case of orthotopic PC3-LN4 prostate tumors, there was growth of regional lymph node tumors in the peritoneal cavity as well as the presence of distant metastases. Treatment with 33 and 330 ng/kg TBG-RNAi-CK2 resulted in greater than 70 % reduction in volume of the largest retroperitoneal lymph node tumor in each mouse as well as decreased mean length of all collected retroperitoneal lymph node tumors compared to control. Additionally, there was a significant difference as to the presence or absence of distant metastases when comparing TBG-RNAi-CK2 to control-treated mice. Indirect immunofluorescence analysis of the lymph node tumors demonstrated antiproliferative treatment effects on both the AKT and NF-κB pathways [97].

Second, in the case of HNSCC Fadu tumors initiated intradermally, the mice were treated one time with a high dose of TBG-RNAi-CK2 or TBG-sense-CK2 (nanoencapsulated oligonucleotide control) at 25 mg/kg or with vehicle. Immunoblot analysis of spleens from control-treated mice showed the presence of keratin-14 signal indicating metastases, as well as stronger CK2αα′ signal than spleens from naïve, non-tumor mice. In contrast, the expression of CK2αα′ mRNA and protein and of keratin-14 protein was significantly reduced in the mice treated with TBG-RNAi-CK2 [88]. Moreover, hematological profile and serum chemistry analyses indicated no changes between 25 mg/kg TBG-RNAi-CK2-treated and TBG-sense-CK2 control mice. Together, these data suggest the strong potential for TBG-RNAi-CK2 to effectively treat both primary tumors and metastases.

8 Survival of TBG-RNAi-CK2-Treated Mice

Survival studies were performed in three orthotopic xenograft models of HNSCC. These mice were treated with one to two injections of TBG-RNAi-CK2 ranging in dose level from 0.01 to 50 mg/kg. Control treatments included TBG-sense-CK2, TBG-sugar, and vehicle. Tumor volumes were tracked for 35 days and demonstrated a significant difference in two models at day 35 for TBG-RNAi-CK2-treated mice compared to controls. At 200 days posttreatment, 25 of 46 TBG-RNAi-CK2-treated mice survived without local recurrence, whereas only 12 of 53 control mice survived. Survival for more than 6 months was statistically significant for two of the three models [88].

9 Concluding Remarks

CK2 is a remarkably universal target for cancer drug development research. Multiple approaches are being pursued across the world, focusing on reducing CK2 activity or expression to an extent where malignant cells no longer survive. It is possible that several anti-CK2 therapeutic strategies will be employed in the future, with the strategy chosen dependent on the cancer to be treated and the combination of drugs to be employed. In any case, because CK2 is a ubiquitous and essential cell survival signal, it would be advantageous to target CK2 in a protected and cancer cell-directed manner. To that end, we have summarized here our published studies showing the efficacy of TBG nanoencapsulated anti-CK2 drugs in two different types of cancer—prostate adenocarcinoma and head and neck squamous cell carcinoma. We are currently in the final stages of a study evaluating the safety of a modified TBG-RNAi-CK2 nanocapsule therapy in companion cats with naturally occurring HNSCC. The results are very encouraging and will help us to move this therapeutic concept forward toward human translation.

References

1. Gapany M, Faust RA, Tawfic S, Davis A, Adams GL, Ahmed K (1995) Association of elevated protein kinase CK2 activity with aggressive behavior of squamous cell carcinoma of the head and neck. Mol Med 1(6):659–666
2. Faust RA, Gapany M, Tristani P, Davis A, Adams GL, Ahmed K (1996) Elevated protein kinase CK2 activity in chromatin of head and neck tumors: association with malignant transformation. Cancer Lett 101(1):31–35. doi:10.1016/0304-3835(96)04110-9

3. Faust RA, Niehans G, Gapany M, Hoistad D, Knapp D, Cherwitz D, Davis A, Adams GL, Ahmed K (1999) Subcellular immunolocalization of protein kinase CK2 in normal and carcinoma cells. Int J Biochem Cell Biol 31(9):941–949. doi:10.1016/S1357-2725(99)00050-3
4. Faust RA, Tawfic S, Davis AT, Bubash LA, Ahmed K (2000) Antisense oligonucleotides against protein kinase CK2-α inhibit growth of squamous cell carcinoma of the head and neck in vitro. Head Neck 22(4):341–346. doi:10.1002/1097-0347(200007)22:4<341::AID-HED5>3.0.CO;2-3 [pii]
5. Wang H, Davis A, Yu S, Ahmed K (2001) Response of cancer cells to molecular interruption of the CK2 signal. Mol Cell Biochem 227(1–2):167–174. doi:10.1023/A:1013112908734
6. Tawfic S, Yu S, Wang H, Faust R, Davis A, Ahmed K (2001) Protein kinase CK2 signal in neoplasia. Histol Histopathol 16(2):573–582
7. Guo C, Yu S, Davis AT, Wang H, Green JE, Ahmed K (2001) A potential role of nuclear matrix-associated protein kinase CK2 in protection against drug-induced apoptosis in cancer cells. J Biol Chem 276(8):5992–5999. doi:10.1074/jbc.M004862200
8. Wang G, Ahmad KA, Ahmed K (2006) Role of protein kinase CK2 in the regulation of tumor necrosis factor-related apoptosis inducing ligand-induced apoptosis in prostate cancer cells. Cancer Res 66(4):2242–2249. doi:10.1158/0008-5472.CAN-05-2772, 66/4/2242 [pii]
9. Yu S, Wang H, Davis A, Ahmed K (2001) Consequences of CK2 signaling to the nuclear matrix. Mol Cell Biochem 227(1–2):67–71. doi:10.1023/A:1013156721938
10. Ahmed K, Gerber DA, Cochet C (2002) Joining the cell survival squad: an emerging role for protein kinase CK2. Trends Cell Biol 12(5):226–230. doi:10.1016/S0962-8924(02)02279-1
11. Wang G, Ahmad KA, Ahmed K (2005) Modulation of death receptor-mediated apoptosis by CK2. Mol Cell Biochem 274(1–2):201–205
12. Wang G, Unger G, Ahmad KA, Slaton JW, Ahmed K (2005) Downregulation of CK2 induces apoptosis in cancer cells–a potential approach to cancer therapy. Mol Cell Biochem 274(1–2):77–84
13. Ahmad KA, Wang G, Slaton J, Unger G, Ahmed K (2005) Targeting CK2 for cancer therapy. Anticancer Drugs 16(10):1037–1043. doi:10.1097/00001813-200511000-00001
14. Meggio F, Pinna LA (2003) One-thousand-and-one substrates of protein kinase CK2? FASEB J 17(3):349–368. doi:10.1096/fj.02-0473rev
15. Marchiori F, Meggio F, Marin O, Borin G, Calderan A, Ruzza P, Pinna LA (1988) Synthetic peptide substrates for casein kinase 2. Assessment of minimum structural requirements for phosphorylation. Biochim Biophys Acta 971(3):332–338
16. Faust M, Montenarh M (2000) Subcellular localization of protein kinase CK2. A key to its function? Cell Tissue Res 301(3):329–340
17. Yu S, Davis AT, Guo C, Green JE, Ahmed K (1999) Differential targeting of protein kinase CK2 to the nuclear matrix upon transient overexpression of its subunits. J Cell Biochem 74(1):127–134. doi:10.1002/(SICI)1097-4644(19990701)74:1<127::AID-JCB14>3.0.CO;2-3
18. Ahmed K, Yenice S, Davis A, Goueli SA (1993) Association of casein kinase 2 with nuclear chromatin in relation to androgenic regulation of rat prostate. Proc Natl Acad Sci U S A 90(10):4426–4430
19. Trembley JH, Wang G, Unger G, Slaton J, Ahmed K (2009) Protein kinase CK2 in health and disease: CK2: a key player in cancer biology. Cell Mol Life Sci 66(11–12):1858–1867. doi:10.1007/s00018-009-9154-y
20. Niefind K, Issinger OG (2010) Conformational plasticity of the catalytic subunit of protein kinase CK2 and its consequences for regulation and drug design. Biochim Biophys Acta 1804(3):484–492. doi:10.1016/j.bbapap.2009.09.022, S1570-9639(09)00270-2 [pii]
21. le Nguyen XT, Mitchell BS (2013) Akt activation enhances ribosomal RNA synthesis through casein kinase II and TIF-IA. Proc Natl Acad Sci U S A 110(51):20681–20686. doi:10.1073/pnas.1313097110
22. Tarrant MK, Rho HS, Xie Z, Jiang YL, Gross C, Culhane JC, Yan G, Qian J, Ichikawa Y, Matsuoka T, Zachara N, Etzkorn FA, Hart GW, Jeong JS, Blackshaw S, Zhu H, Cole PA (2012) Regulation of CK2 by phosphorylation and O-GlcNAcylation revealed by semisynthesis. Nat Chem Biol 8(3):262–269. doi:10.1038/nchembio.771

23. Patsoukis N, Li L, Sari D, Petkova V, Boussiotis VA (2013) PD-1 increases PTEN phosphatase activity while decreasing PTEN protein stability by inhibiting casein kinase 2. Mol Cell Biol 33(16):3091–3098. doi:10.1128/MCB. 00319-13
24. Nolen B, Taylor S, Ghosh G (2004) Regulation of protein kinases; controlling activity through activation segment conformation. Mol Cell 15(5):661–675. doi:10.1016/j.molcel.2004.08.024
25. Padmanabha R, Chen-Wu JL, Hanna DE, Glover CV (1990) Isolation, sequencing, and disruption of the yeast CKA2 gene: casein kinase II is essential for viability in *Saccharomyces cerevisiae*. Mol Cell Biol 10(8):4089–4099
26. Buchou T, Vernet M, Blond O, Jensen HH, Pointu H, Olsen BB, Cochet C, Issinger OG, Boldyreff B (2003) Disruption of the regulatory β subunit of protein kinase CK2 in mice leads to a cell-autonomous defect and early embryonic lethality. Mol Cell Biol 23(3):908–915
27. Seldin DC, Lou DY, Toselli P, Landesman-Bollag E, Dominguez I (2008) Gene targeting of CK2 catalytic subunits. Mol Cell Biochem 316(1–2):141–147. doi:10.1007/s11010-008-9811-8
28. Fraser AG, Kamath RS, Zipperlen P, Martinez-Campos M, Sohrmann M, Ahringer J (2000) Functional genomic analysis of *C. elegans* chromosome I by systematic RNA interference. Nature 408(6810):325–330
29. Trembley JH, Chen Z, Unger G, Slaton J, Kren BT, Van Waes C, Ahmed K (2010) Emergence of protein kinase CK2 as a key target in cancer therapy. Biofactors 36(3):187–195. doi:10.1002/biof.96
30. Ruzzene M, Pinna LA (2010) Addiction to protein kinase CK2: a common denominator of diverse cancer cells? Biochim Biophys Acta 1804(3):499–504. doi:10.1016/j.bbapap.2009.07.018
31. Guerra B, Issinger OG (2008) Protein kinase CK2 in human diseases. Curr Med Chem 15(19):1870–1886. doi:10.1007/s00018-009-9148-9
32. Munstermann U, Fritz G, Seitz G, Lu YP, Schneider HR, Issinger OG (1990) Casein kinase II is elevated in solid human tumours and rapidly proliferating non-neoplastic tissue. Eur J Biochem 189(2):251–257
33. Pistorius K, Seitz G, Remberger K, Issinger OG (1991) Differential CKII activities in human colorectal mucosa, adenomas and carcinomas. Onkologie 14(3):256–260
34. Mandal T, Bhowmik A, Chatterjee A, Chatterjee U, Chatterjee S, Ghosh MK (2014) Reduced phosphorylation of Stat3 at Ser-727 mediated by casein kinase 2 - protein phosphatase 2A enhances Stat3 Tyr-705 induced tumorigenic potential of glioma cells. Cell Signal 26(8):1725–1734. doi:10.1016/j.cellsig.2014.04.003
35. Giusiano S, Cochet C, Filhol O, Duchemin-Pelletier E, Secq V, Bonnier P, Carcopino X, Boubli L, Birnbaum D, Garcia S, Iovanna J, Charpin C (2011) Protein kinase CK2alpha subunit over-expression correlates with metastatic risk in breast carcinomas: quantitative immunohistochemistry in tissue microarrays. Eur J Cancer 47(5):792–801. doi:10.1016/j.ejca.2010.11.028
36. O-charoenrat P, Rusch V, Talbot SG, Sarkaria I, Viale A, Socci N, Ngai I, Rao P, Singh B (2004) Casein kinase II alpha subunit and C1-inhibitor are independent predictors of outcome in patients with squamous cell carcinoma of the lung. Clin Cancer Res 10(17):5792–5803. doi:10.1158/1078-0432.CCR-03-0317
37. Laramas M, Pasquier D, Filhol O, Ringeisen F, Descotes JL, Cochet C (2007) Nuclear localization of protein kinase CK2 catalytic subunit (CK2α) is associated with poor prognostic factors in human prostate cancer. Eur J Cancer 43(5):928–934. doi:10.1016/j.ejca.2006.11.021
38. Kim JS, Eom JI, Cheong J-W, Choi AJ, Lee JK, Yang WI, Min YH (2007) Protein kinase CK2α as an unfavorable prognostic marker and novel therapeutic target in acute myeloid leukemia. Clin Cancer Res 13(3):1019–1028. doi:10.1158/1078-0432.ccr-06-1602
39. Lin KY, Fang CL, Chen Y, Li CF, Chen SH, Kuo CY, Tai C, Uen YH (2010) Overexpression of nuclear protein kinase CK2 Beta subunit and prognosis in human gastric carcinoma. Ann Surg Oncol 17(6):1695–1702. doi:10.1245/s10434-010-0911-9
40. Lin KY, Tai C, Hsu JC, Li CF, Fang CL, Lai HC, Hseu YC, Lin YF, Uen YH (2011) Overexpression of nuclear protein kinase CK2 alpha catalytic subunit (CK2alpha) as a poor prognosticator in human colorectal cancer. PLoS One 6(2):e17193. doi:10.1371/journal.pone.0017193

41. Zheng Y, McFarland BC, Drygin D, Yu H, Bellis SL, Kim H, Bredel M, Benveniste EN (2013) Targeting protein kinase CK2 suppresses prosurvival signaling pathways and growth of glioblastoma. Clin Cancer Res 19(23):6484–6494. doi:10.1158/1078-0432.ccr-13-0265
42. Gray GK, McFarland BC, Rowse AL, Gibson SA, Benveniste EN (2014) Therapeutic CK2 inhibition attenuates diverse prosurvival signaling cascades and decreases cell viability in human breast cancer cells. Oncotarget 5(15):6484–6496
43. Ulloa L, Diaz-Nido J, Avila J (1993) Depletion of casein kinase II by antisense oligonucleotide prevents neuritogenesis in neuroblastoma cells. EMBO J 12(4):1633–1640
44. Formby B, Stern R (1998) Phosphorylation stabilizes alternatively spliced CD44 mRNA transcripts in breast cancer cells: inhibition by antisense complementary to casein kinase II mRNA. Mol Cell Biochem 187(1):23–30
45. Seeber S, Issinger OG, Holm T, Kristensen LP, Guerra B (2005) Validation of protein kinase CK2 as oncological target. Apoptosis 10(4):875–885. doi:10.1007/s10495-005-0380-y
46. Trembley JH, Unger GM, Tobolt DK, Korman VL, Wang G, Ahmad KA, Slaton JW, Kren BT, Ahmed K (2011) Systemic administration of antisense oligonucleotides simultaneously targeting CK2alpha and alpha' subunits reduces orthotopic xenograft prostate tumors in mice. Mol Cell Biochem 356:21–35. doi:10.1007/s11010-011-0943-x
47. Quotti Tubi L, Gurrieri C, Brancalion A, Bonaldi L, Bertorelle R, Manni S, Pavan L, Lessi F, Zambello R, Trentin L, Adami F, Ruzzene M, Pinna LA, Semenzato G, Piazza F (2013) Inhibition of protein kinase CK2 with the clinical-grade small ATP-competitive compound CX-4945 or by RNA interference unveils its role in acute myeloid leukemia cell survival, p53-dependent apoptosis and daunorubicin-induced cytotoxicity. J Hematol Oncol 6:78. doi:10.1186/1756-8722-6-78
48. Zhang S, Wang Y, Mao J-H, Hsieh D, Kim I-J, Hu L-M, Xu Z, Long H, Jablons DM, You L (2012) Inhibition of CK2α down-regulates Hedgehog/Gli signaling leading to a reduction of a stem-like side population in human lung cancer cells. PLoS One 7(6):e38996. doi:10.1371/journal.pone.0038996
49. Ko H, Kim S, Jin CH, Lee E, Ham S, Yook JI, Kim K (2012) Protein kinase casein kinase 2-mediated upregulation of N-cadherin confers anoikis resistance on esophageal carcinoma cells. Mol Cancer Res 10(8):1032–1038. doi:10.1158/1541-7786.MCR-12-0261
50. Zhang S, Long H, Yang Y-L, Wang Y, Hsieh D, Li W, Au A, Stoppler HJ, Xu Z, Jablons DM, You L (2013) Inhibition of CK2α down-regulates Notch1 signalling in lung cancer cells. J Cell Mol Med 17(7):854–862. doi:10.1111/jcmm.12068
51. Trembley JH, Unger GM, Korman VL, Tobolt DK, Kazimierczuk Z, Pinna LA, Kren BT, Ahmed K (2012) Nanoencapsulated anti-CK2 small molecule drug or siRNA specifically targets malignant cancer but not benign cells. Cancer Lett 315(1):48–58. doi:10.1016/j.canlet.2011.10.007
52. Kreutzer J, Ruzzene M, Guerra B (2010) Enhancing chemosensitivity to gemcitabine via RNA interference targeting the catalytic subunits of protein kinase CK2 in human pancreatic cancer cells. BMC Cancer 10(1):440
53. Brown MS, Diallo OT, Hu M, Ehsanian R, Yang X, Arun P, Lu H, Korman V, Unger G, Ahmed K, Van Waes C, Chen Z (2010) CK2 modulation of NF-κB, TP53, and the malignant phenotype in head and neck cancer by anti-CK2 oligonucleotides in vitro or in vivo via sub–50-nm nanocapsules. Clin Cancer Res 16(8):2295–2307. doi:10.1158/1078-0432.ccr-09-3200
54. Kang JY, Kim JJ, Jang SY, Bae YS (2009) The p53-p21(Cip1/WAF1) pathway is necessary for cellular senescence induced by the inhibition of protein kinase CKII in human colon cancer cells. Mol Cells 28(5):489–494. doi:10.1007/s10059-009-0141-9
55. Schneider CC, Hessenauer A, Montenarh M, Gotz C (2010) p53 is dispensable for the induction of apoptosis after inhibition of protein kinase CK2. Prostate 70(2):126–134. doi:10.1002/pros.21044
56. Cozza G, Pinna LA, Moro S (2013) Kinase CK2 inhibition: an update. Curr Med Chem 20(5):671–693
57. Zhu D, Hensel J, Hilgraf R, Abbasian M, Pornillos O, Deyanat-Yazdi G, Hua XH, Cox S (2010) Inhibition of protein kinase CK2 expression and activity blocks tumor cell growth. Mol Cell Biochem 333(1–2):159–167. doi:10.1007/s11010-009-0216-0

58. Hamacher R, Saur D, Fritsch R, Reichert M, Schmid RM, Schneider G (2007) Casein kinase II inhibition induces apoptosis in pancreatic cancer cells. Oncol Rep 18(3):695–701
59. Piazza FA, Ruzzene M, Gurrieri C, Montini B, Bonanni L, Chioetto G, Di Maira G, Barbon F, Cabrelle A, Zambello R, Adami F, Trentin L, Pinna LA, Semenzato G (2006) Multiple myeloma cell survival relies on high activity of protein kinase CK2. Blood 108(5):1698–1707. doi:10.1182/blood-2005-11-013672, blood-2005-11-013672 [pii]
60. Yde CW, Frogne T, Lykkesfeldt AE, Fichtner I, Issinger OG, Stenvang J (2007) Induction of cell death in antiestrogen resistant human breast cancer cells by the protein kinase CK2 inhibitor DMAT. Cancer Lett 256(2):229–237. doi:10.1016/j.canlet.2007.06.010, S0304-3835(07)00287-X [pii]
61. Lawnicka H, Kowalewicz-Kulbat M, Sicinska P, Kazimierczuk Z, Grieb P, Stepien H (2010) Anti-neoplastic effect of protein kinase CK2 inhibitor, 2-dimethylamino-4,5,6,7-tetrabromobenzimidazole (DMAT), on growth and hormonal activity of human adrenocortical carcinoma cell line (H295R) in vitro. Cell Tissue Res 340(2):371–379. doi:10.1007/s00441-010-0960-1
62. Ruzzene M, Penzo D, Pinna LA (2002) Protein kinase CK2 inhibitor 4,5,6,7-tetrabromobenzotriazole (TBB) induces apoptosis and caspase-dependent degradation of haematopoietic lineage cell-specific protein 1 (HS1) in Jurkat cells. Biochem J 364(Pt 1):41–47
63. Pagano MA, Meggio F, Ruzzene M, Andrzejewska M, Kazimierczuk Z, Pinna LA (2004) 2-Dimethylamino-4,5,6,7-tetrabromo-1H-benzimidazole: a novel powerful and selective inhibitor of protein kinase CK2. Biochem Biophys Res Commun 321(4):1040–1044. doi:10.1016/j.bbrc.2004.07.067, S0006-291X(04)01549-9 [pii]
64. Prudent R, Moucadel V, Lopez-Ramos M, Aci S, Laudet B, Mouawad L, Barette C, Einhorn J, Einhorn C, Denis JN, Bisson G, Schmidt F, Roy S, Lafanechere L, Florent JC, Cochet C (2008) Expanding the chemical diversity of CK2 inhibitors. Mol Cell Biochem 316(1–2):71–85. doi:10.1007/s11010-008-9828-z
65. Gianoncelli A, Cozza G, Orzeszko A, Meggio F, Kazimierczuk Z, Pinna LA (2009) Tetraiodobenzimidazoles are potent inhibitors of protein kinase CK2. Bioorg Med Chem 17(20):7281–7289. doi:10.1016/j.bmc.2009.08.047
66. Lopez-Ramos M, Prudent R, Moucadel V, Sautel CF, Barette C, Lafanechere L, Mouawad L, Grierson D, Schmidt F, Florent JC, Filippakopoulos P, Bullock AN, Knapp S, Reiser JB, Cochet C (2010) New potent dual inhibitors of CK2 and Pim kinases: discovery and structural insights. FASEB J 24(9):3171–3185. doi:10.1096/fj.09-143743, fj.09-143743 [pii]
67. Siddiqui-Jain A, Drygin D, Streiner N, Chua P, Pierre F, O'Brien SE, Bliesath J, Omori M, Huser N, Ho C, Proffitt C, Schwaebe MK, Ryckman DM, Rice WG, Anderes K (2010) CX-4945, an orally bioavailable selective inhibitor of protein kinase CK2, inhibits prosurvival and angiogenic signaling and exhibits antitumor efficacy. Cancer Res 70(24):10288–10298. doi:10.1158/0008-5472.can-10-1893
68. Manni S, Brancalion A, Tubi LQ, Colpo A, Pavan L, Cabrelle A, Ave E, Zaffino F, Di Maira G, Ruzzene M, Adami F, Zambello R, Pitari MR, Tassone P, Pinna LA, Gurrieri C, Semenzato G, Piazza F (2012) Protein kinase CK2 protects multiple myeloma cells from ER stress-induced apoptosis and from the cytotoxic effect of HSP90 inhibition through regulation of the unfolded protein response. Clin Cancer Res 18(7):1888–1900. doi:10.1158/1078-0432.CCR-11-1789
69. Lin YC, Hung MS, Lin CK, Li JM, Lee KD, Li YC, Chen MF, Chen JK, Yang CT (2011) CK2 inhibitors enhance the radiosensitivity of human non-small cell lung cancer cells through inhibition of stat3 activation. Cancer Biother Radiopharm 26(3):381–388. doi:10.1089/cbr.2010.0917
70. Gotz C, Gratz A, Kucklaender U, Jose J (2012) TF–a novel cell-permeable and selective inhibitor of human protein kinase CK2 induces apoptosis in the prostate cancer cell line LNCaP. Biochim Biophys Acta 1820(7):970–977. doi:10.1016/j.bbagen.2012.02.009
71. Zhao M, Ma J, Zhu HY, Zhang XH, Du ZY, Xu YJ, Yu XD (2011) Apigenin inhibits proliferation and induces apoptosis in human multiple myeloma cells through targeting the trinity of CK2, Cdc37 and Hsp90. Mol Cancer 10:104. doi:10.1186/1476-4598-10-104

72. Yao K, Youn H, Gao X, Huang B, Zhou F, Li B, Han H (2012) Casein kinase 2 inhibition attenuates androgen receptor function and cell proliferation in prostate cancer cells. Prostate 72(13):1423–1430. doi:10.1002/pros.22493
73. Borgo C, Cesaro L, Salizzato V, Ruzzene M, Massimino ML, Pinna LA, Donella-Deana A (2013) Aberrant signalling by protein kinase CK2 in imatinib-resistant chronic myeloid leukaemia cells: biochemical evidence and therapeutic perspectives. Mol Oncol 7(6):1103–1115. doi:10.1016/j.molonc.2013.08.006
74. Di Maira G, Brustolon F, Bertacchini J, Tosoni K, Marmiroli S, Pinna LA, Ruzzene M (2007) Pharmacological inhibition of protein kinase CK2 reverts the multidrug resistance phenotype of a CEM cell line characterized by high CK2 level. Oncogene 26(48):6915–6926. doi:10.1038/sj.onc.1210495, 1210495 [pii]
75. Sass G, Klinger N, Sirma H, Hashemolhosseini S, Hellerbrand C, Neureiter D, Wege H, Ocker M, Tiegs G (2011) Inhibition of experimental HCC growth in mice by use of the kinase inhibitor DMAT. Int J Oncol 39(2):433–442. doi:10.3892/ijo.2011.1037
76. Pierre F, Chua PC, O'Brien SE, Siddiqui-Jain A, Bourbon P, Haddach M, Michaux J, Nagasawa J, Schwaebe MK, Stefan E, Vialettes A, Whitten JP, Chen TK, Darjania L, Stansfield R, Bliesath J, Drygin D, Ho C, Omori M, Proffitt C, Streiner N, Rice WG, Ryckman DM, Anderes K (2011) Pre-clinical characterization of CX-4945, a potent and selective small molecule inhibitor of CK2 for the treatment of cancer. Mol Cell Biochem 356(1–2):37–43. doi:10.1007/s11010-011-0956-5
77. Prudent R, Moucadel V, Nguyen CH, Barette C, Schmidt F, Florent JC, Lafanechere L, Sautel CF, Duchemin-Pelletier E, Spreux E, Filhol O, Reiser JB, Cochet C (2010) Antitumor activity of pyridocarbazole and benzopyridoindole derivatives that inhibit protein kinase CK2. Cancer Res 70(23):9865–9874. doi:10.1158/0008-5472.CAN-10-0917, 0008-5472.CAN-10-0917 [pii]
78. Yoo JY, Lim BJ, Choi HK, Hong SW, Jang HS, Kim C, Chun KH, Choi KC, Yoon HG (2013) CK2-NCoR signaling cascade promotes prostate tumorigenesis. Oncotarget 4(7):972–983
79. Bliesath J, Huser N, Omori M, Bunag D, Proffitt C, Streiner N, Ho C, Siddiqui-Jain A, O'Brien SE, Lim JK, Ryckman DM, Anderes K, Rice WG, Drygin D (2012) Combined inhibition of EGFR and CK2 augments the attenuation of PI3K-Akt-mTOR signaling and the killing of cancer cells. Cancer Lett 322(1):113–118. doi:10.1016/j.canlet.2012.02.032
80. Siddiqui-Jain A, Bliesath J, Macalino D, Omori M, Huser N, Streiner N, Ho CB, Anderes K, Proffitt C, O'Brien SE, Lim JK, Von Hoff DD, Ryckman DM, Rice WG, Drygin D (2012) CK2 inhibitor CX-4945 suppresses DNA repair response triggered by DNA-targeted anticancer drugs and augments efficacy: mechanistic rationale for drug combination therapy. Mol Cancer Ther 11(4):994–1005. doi:10.1158/1535-7163.MCT-11-0613
81. Marschke RF, Borad MJ, McFarland RW, Alvarez RH, Lim JK, Padgett CS, Von Hoff DD, O'Brien SE, Northfelt DW (2011) Findings from the phase I clinical trials of CX-4945, an orally bioavailable inhibitor of CK2. J Clin Oncol 29 (suppl):abstr 3087
82. Perera Y, Farina HG, Gil J, Rodriguez A, Benavent F, Castellanos L, Gomez RE, Acevedo BE, Alonso DF, Perea SE (2009) Anticancer peptide CIGB-300 binds to nucleophosmin/B23, impairs its CK2-mediated phosphorylation, and leads to apoptosis through its nucleolar disassembly activity. Mol Cancer Ther 8(5):1189–1196. doi:10.1158/1535-7163.MCT-08-1056, 1535-7163.MCT-08-1056 [pii]
83. Perea SE, Reyes O, Puchades Y, Mendoza O, Vispo NS, Torrens I, Santos A, Silva R, Acevedo B, Lopez E, Falcon V, Alonso DF (2004) Antitumor effect of a novel proapoptotic peptide that impairs the phosphorylation by the protein kinase 2 (casein kinase 2). Cancer Res 64(19):7127–7129. doi:10.1158/0008-5472.CAN-04-2086, 64/19/7127 [pii]
84. Perera Y, Farina HG, Hernandez I, Mendoza O, Serrano JM, Reyes O, Gomez DE, Gomez RE, Acevedo BE, Alonso DF, Perea SE (2008) Systemic administration of a peptide that impairs the protein kinase (CK2) phosphorylation reduces solid tumor growth in mice. Int J Cancer 122(1):57–62. doi:10.1002/ijc.23013
85. Martins LR, Perera Y, Lucio P, Silva MG, Perea SE, Barata JT (2014) Targeting chronic lymphocytic leukemia using CIGB-300, a clinical-stage CK2-specific cell-permeable peptide inhibitor. Oncotarget 5(1):258–263

86. Perea SE, Baladron I, Garcia Y, Perera Y, Lopez A, Soriano JL, Batista N, Palau A, Hernandez I, Farina H, Garcia I, Gonzalez L, Gil J, Rodriguez A, Solares M, Santana A, Cruz M, Lopez M, Valenzuela C, Reyes O, Lopez-Saura PA, Gonzalez CA, Diaz A, Castellanos L, Sanchez A, Betancourt L, Besada V, Gonzalez LJ, Garay H, Gomez R, Gomez DE, Alonso DF, Perrin P, Renualt JY, Sigman H, Herrera L, Acevedo B (2011) CIGB-300, a synthetic peptide-based drug that targets the CK2 phosphoaceptor domain. Translational and clinical research. Mol Cell Biochem 356(1–2):45–50. doi:10.1007/s11010-011-0950-y
87. Solares AM, Santana A, Baladron I, Valenzuela C, Gonzalez CA, Diaz A, Castillo D, Ramos T, Gomez R, Alonso DF, Herrera L, Sigman H, Perea SE, Acevedo BE, Lopez-Saura P (2009) Safety and preliminary efficacy data of a novel casein kinase 2 (CK2) peptide inhibitor administered intralesionally at four dose levels in patients with cervical malignancies. BMC Cancer 9:146. doi:10.1186/1471-2407-9-146
88. Unger GM, Kren BT, Korman VL, Kimbrough TG, Vogel RI, Ondrey FG, Trembley JH, Ahmed K (2014) Mechanism and efficacy of sub-50 nm tenfibgen nanocapsules for cancer cell-directed delivery of anti-CK2 RNAi to primary and metastatic squamous cell carcinoma. Mol Cancer Ther 13(8):2018–2029
89. Aukhil I, Joshi P, Yan Y, Erickson HP (1993) Cell- and heparin-binding domains of the hexabrachion arm identified by tenascin expression proteins. J Biol Chem 268(4):2542–2553
90. Erickson HP, Bourdon MA (1989) Tenascin: an extracellular matrix protein prominent in specialized embryonic tissues and tumors. Annu Rev Cell Biol 5:71–92. doi:10.1146/annurev.cb.05.110189.000443
91. Yokoyama K, Erickson HP, Ikeda Y, Takada Y (2000) Identification of amino acid sequences in fibrinogen gamma-chain and tenascin C C-terminal domains critical for binding to integrin alpha vbeta 3. J Biol Chem 275(22):16891–16898. doi:10.1074/jbc.M000610200, M000610200 [pii]
92. Desgrosellier JS, Cheresh DA (2010) Integrins in cancer: biological implications and therapeutic opportunities. Nat Rev Cancer 10(1):9–22. doi:10.1038/nrc2748
93. Oskarsson T, Acharyya S, Zhang XH, Vanharanta S, Tavazoie SF, Morris PG, Downey RJ, Manova-Todorova K, Brogi E, Massague J (2011) Breast cancer cells produce tenascin C as a metastatic niche component to colonize the lungs. Nat Med 17(7):867–874. doi:10.1038/nm.2379
94. Van Obberghen-Schilling E, Tucker RP, Saupe F, Gasser I, Cseh B, Orend G (2011) Fibronectin and tenascin-C: accomplices in vascular morphogenesis during development and tumor growth. Int J Dev Biol 55(4–5):511–525. doi:10.1387/ijdb.103243eo
95. Midwood KS, Orend G (2009) The role of tenascin-C in tissue injury and tumorigenesis. J Cell Commun Signal 3(3–4):287–310. doi:10.1007/s12079-009-0075-1
96. Trembley JH, Unger GM, Cespedes Gomez O, Abedin MJ, Korman VL, Vogel RI, Niehans G, Kren BT, Ahmed K (2014) Tenfibgen-DMAT nanocapsule delivers CK2 inhibitor DMAT to prostate cancer xenograft tumors causing inhibition of cell proliferation. Mol Cell Pharmacol 6:15–25
97. Trembley JH, Unger GM, Korman VL, Abedin MJ, Nacusi LP, Vogel RI, Slaton JW, Kren BT, Ahmed K (2014) Tenfibgen ligand nanoencapsulation delivers bi-functional anti-CK2 RNAi oligomer to key sites for prostate cancer targeting. PLoS One. doi:10.1371/journal.pone.0109970
98. Slaton JW, Unger GM, Sloper DT, Davis AT, Ahmed K (2004) Induction of apoptosis by antisense CK2 in human prostate cancer xenograft model. Mol Cancer Res 2(12):712–721
99. Tousignant JD, Gates AL, Ingram LA, Johnson CL, Nietupski JB, Cheng SH, Eastman SJ, Scheule RK (2000) Comprehensive acute toxicities induced by systemic administration of cationic lipid:plasmid DNA complexes in mice. Hum Gene Ther 11:2493–2513

Part IV
Studies Involving CK2 Small Molecule Inhibitors

Screening of DTP Compound Libraries for CK2 Inhibitors with Focus on Natural Products

Tine D. Rasmussen, Barbara Guerra, and Olaf-Georg Issinger

Abstract Various compound libraries of the Drug Therapeutic Program (DTP) of the NIH/NCI were screened against the catalytic subunit and the tetrameric holoenzyme of protein kinase CK2. Different IC50 values were obtained for the two CK2 molecules. In the case of nortangeretin, the IC50 value was 0.34 μM for the catalytic subunit and 15 μM for the holoenzyme. In the case of coumestrol, opposite results were obtained, i.e., high IC50 for the CK2α subunit (2.7 μM) and a lower IC50 value for the holoenzyme (0.19 μM).

From the many compounds identified to inhibit CK2, we have selected 14 different compounds and listed them according to their CK2α/CK2 holoenzyme IC50 ratio.

Four compounds were tested on a panel of seven cell lines revealing considerable differences in the degree of CK2 inhibition inside the cells.

Keywords Protein kinase CK2 • Casein kinase 2 • Eukaryotic protein kinases • CMGC kinases • Signal transduction pathways • Protein kinase inhibitors • Small molecule compounds • DTP compound libraries

T.D. Rasmussen
Clinical Pathology, OUH, J.B. Winsløws Vej 15, 5000 Odense C, Denmark

B. Guerra
BMB, SDU, Campusvej 55, 5230 Odense, Denmark

O.-G. Issinger (✉)
KinaseDetect ApS, Skovvej 22, 6340 Kruså, Denmark
e-mail: ogi@kinasedetect.com

K. Ahmed et al. (eds.), *Protein Kinase CK2 Cellular Function in Normal and Disease States*, Advances in Biochemistry in Health and Disease 12, DOI 10.1007/978-3-319-14544-0_18

1 Importance of Secondary Natural Compounds for Their Host

Cancer is from a scientific perspective caused by external or internal factors such as radiation, pollution, sun, and heritage, among others. In this chapter, we want to shed some light on the possibility of using nature's own "products" for the treatment of cancer, i.e., "a natural approach of curing cancer."

The role of secondary plant compounds has long been unclear. It was assumed that secondary metabolic pathways served to make useless or toxic metabolic by-products or end products of primary metabolites harmless, because it was not possible to assign them a direct role in metabolism. Today, it is believed that secondary plant metabolites have important ecological functions. As a result of an intense interaction between plants and their environment, secondary metabolites have been developed especially against predators. Many secondary plant metabolites are used as an effective chemical defense against herbivores and pathogens. On the other hand, they attract through their coloring and flavoring compounds pollen-spreading insects and seed-spreading fruit eaters.

Plants use secondary metabolites, such as tannins, iridoids, phytosterols, pyrethroids, alkaloids, cannabinoids, and cardenolides to protect themselves against herbivores; they use carotenoids, flavonoids, and anthocyanins for attracting pollinators; they use monoterpenes, suberin, and cutin to prevent evaporation, or lignins as mechanical strengthening.

Natural compounds evolved during millions of years. They have conquered their place in the various organisms' survival programs and are now a permanent part of the individual organisms' metabolism. Given the fact that the abovementioned secondary plant compounds have evolved specifically to serve plant survival and propagation, so how shall we consider their possible benefits for human health that is inherent to some of these compounds? Or in other words, shall we rely/trust/apply these compounds simply because they are natural products?

There is a plethora of anecdotal information documenting the historical use of herbs and spices for their health benefits (reviewed in Johnson and Balick from [1]). Reports on the use of herbs and spices are found in Egyptian, Chinese, and Indian literature dating back thousands of years. One can only speculate how the knowledge was acquired.

2 Why the Focus on Natural Compounds as Potential Drug Leads for Kinase Inhibitors?

We focus on plants and specifically on their secondary compounds/metabolites, which have been proposed to exert beneficial effects in a multitude of disease states, including cancer, cardiovascular disease, and neurodegenerative disorder.

Many of the biological actions of secondary natural compounds are attributed to their antioxidant properties, e.g., interfering with reactive oxygen species (ROS) production [2]. However, recent studies showed that the concentration of secondary natural compounds, e.g., flavonoids in the body, is too low to compete with the small molecule antioxidants such as ascorbate and α-tocopherol. Hence, additional functions for these natural secondary compounds are postulated such as the direct interaction with specific proteins in signaling cascades, e.g., protein kinases. Indeed, bioactive dietary compounds have been shown to modify all of the major signaling pathways deregulated in cancer. Estimates suggest that 30–70 % of all cancer cases might be preventable by diet, depending on the dietary compounds and the specific type of cancer (reviewed in Cindy D. Davis from [1]). Compounds such as apigenin, curcumin, resveratrol, and quercetin inhibit the progression through the cell cycle. These compounds, together with others, also have been shown to influence the two major pathways of programmed cell death, either by binding to the death receptor (extrinsic pathway) or by interfering with AKT, p53, and SMAC/Diablo (intrinsic pathway). There are many overlapping compound activities which do not only involve interactions with protein kinases but also with other unrelated molecules. But even when focusing only on protein kinases, the targets are many and the selectivity and potency usually are low. Indeed, it has been shown for various protein kinases that plant-derived compounds such as quercetin and apigenin inhibit recombinant protein kinases at low micromolar concentrations [3]. An example for the affinity of flavonoids for protein kinases is the binding of quercetin and myricetin to the ATP-binding site of PI3K [4]. The inhibitory activity expressed as IC50 was 3.8 μM and 1.8 μM, respectively. In the case of protein kinase CK2, the values were 0.55 μM and 0.92 μM, respectively [5]. However, these concentrations are still too high to expect a therapeutic effect when fed to whole organisms. Yet, some flavonoids have served as lead compounds and as such made to be 100 times more potent [6].

3 Screening Platforms for Kinases

In a first screening attempt, a luminometric assay platform was used which has been described earlier for testing PI3K [7]. We used the Perkin Elmer, VICTOR luminometer with semiautomated sample handling, which can handle several thousand compounds daily. For target screening the luminometric kinase activity assay, EasyLite Luminescence ATP Detection Assay System, was used. The principle of the reaction is that in a kinase reaction, the γ-phosphate moiety of ATP is transferred to the substrate resulting in a phosphorylated substrate and ADP. During this reaction, the amount of ATP in the kinase reaction solution declines. After the completion of the kinase reaction, the remaining ATP can be measured by the addition of the firefly luciferase and D-luciferin, according to the scheme below.

The emitted light is proportional to the ATP concentration within a certain range of concentrations:

$$ATP + D - Luciferin + O_2 \overset{Luciferase/Mg^{2+}}{\rightarrow} Oxyluciferin + AMP + PPi + CO_2 + Light$$

The more active the protein kinase is, the less ATP is available for the subsequent luciferase-catalyzed reaction; hence, the signal intensity (CPS) will be low. If the kinase reaction is inhibited, lesser or no ATP is used by the kinase, and most ATP is available for the luciferase; hence, the signal intensity (CPS) will be high.

Since this is an indirect method, to measure kinase activity, where the use of ATP is measured, promising compounds have to be retested using a radioactivity-based assay where phosphorylation of a substrate harboring the consensus sequence for the kinase under investigation is monitored.

Before the screening was initialized, we have established the luminometric test with the catalytic CK2α1-335 subunit and the heterotetrameric holoenzyme, (CK2α1-335)2(CK2β)2. Figure 1 shows the results for the CK2α activity. The activity is expressed in percent (y-axis). No addition of substrate peptide or enzyme is indicated by "0" (i.e., only ATP is present); negative control indicates the presence of 25 ng (0.66 pmol) CK2α in the absence of substrate. Usually, an amount of CK2α enzyme, resulting in more than 80 % ATP consumption, was used for the screening. Here, 25 ng of CK2α led to 83 % reduction of ATP in the assay. Although 50 ng CK2α depleted the ATP content by 97 %, we preferred to choose the lower amount of enzyme for several reasons, (1) to hold the use of inhibitor low, (2) to detect compounds which might exert a stimulatory effect, and (3) for economical reasons.

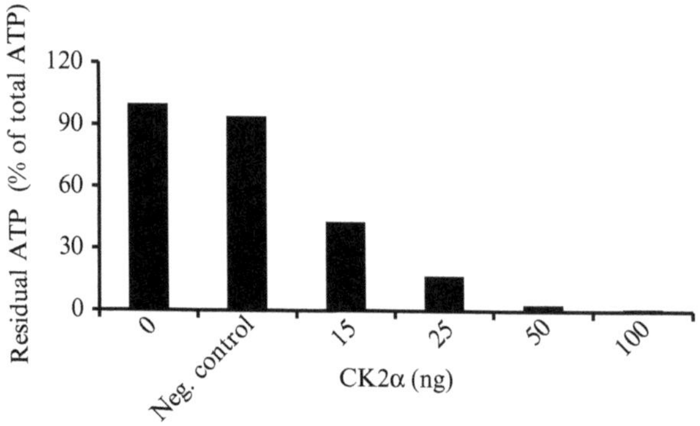

Fig. 1 Luminometric-based assay for the determination of optimal kinetic conditions. Increasing amounts of the catalytic CK2α subunit were tested, and optimal enzyme concentration was determined by quantifying residual ATP available to the luciferase in the presence of D-luciferin. Total amount of ATP in the assay is expressed in percentage relative to the control experiment performed in the absence of CK2α (0). Negative control (Neg. control) refers to an assay in the presence of 25 ng CK2α and absence of substrate

4 Screening of Compound Libraries from the DTP/NIH/NCI

DTP is the drug discovery and therapeutic program (DTP) of the National Cancer Institute (NCI). DTP plans, conducts, and facilitates the development of therapeutic agents for cancer and AIDS treatment and provides compounds for the researcher directly to use. In this study, different compound libraries were used including a natural compound library.

In the latter case, compounds were acquired from terrestrial or marine environments (see: http://dtp.nci.nih.gov/about.html).

5 Screening for Kinases with Focus on CK2

Since protein kinase CK2 is an antiapoptotic kinase which is upregulated in all so far investigated cancers, it is an accepted drug target (reviewed in [8–11]).

Different compound libraries from the NIH/NCI were screened for CK2 inhibitors against both the recombinant human CK2α subunit and the tetrameric CK2α2β2 holoenzyme.

Figure 2 shows an example of a screening of 30 compounds out of 90 from one screening plate. The negative control shows the ATP status in the presence of 25 ng of CK2 holoenzyme with no substrate in the assay. This value is ca. 250,000 CPS. Positive control is 25 ng CK2 holoenzyme in the presence of a substrate.

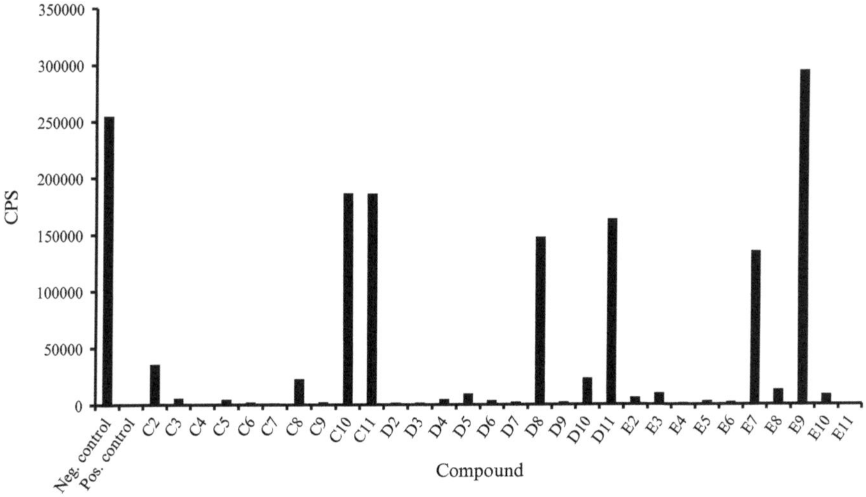

Fig. 2 Inhibitory effects of natural compounds on CK2 holoenzyme kinase activity. Natural compounds were tested at 10 μM final concentration in a luminometric-based assay. Negative control refers to an assay in the presence of CK2 and absence of substrate. Positive control refers to an assay in the presence of CK2 and substrate peptide. Values are expressed in arbitrary units (CPS)

The highest CPS values were found with compound E9 (i.e., up to 300,000 CPS). How can this value, which is higher than the negative control value, be explained?

CK2 holoenzyme shows in the absence of a substrate substantial autophosphorylation of the CK2β subunits at the N-terminal serine residues, i.e., some ATP is used (this is not the case for the catalytic α-subunit [12]).

A potent inhibitor will, therefore, interfere with such an autophosphorylation.

6 Identification of CK2 Inhibitors

Our compound screening of the DTP natural compound library led to the identification of various molecules, e.g., fisetin, pomiferin, coumestrol, stictic acid, β-carotene, hypericin, emodin, and finally resorufin, being the latter in sensu stricto not a natural compound but a derivative. Other natural compounds have been reported to target CK2 [5, 13].

Table 1 shows 14 different compounds which were identified in our screens using the diversity, mechanistic, and natural compound libraries from DTP. The compounds were tested against the catalytic CK2α subunit and the tetrameric CK2 holoenzyme. During the screening, we observed that many of the compounds detected show significant differences in the IC50 values between CK2α and the holoenzyme. Hence, we arranged the compounds not according to their chemical relationship or with respect to their origin. We have arranged them according to the IC50 ratio CK2α/CK2 holoenzyme (Table 1) in two groups, i.e., IC50 ratio >1 and IC50 ratio <1.

Hence, coumestrol with an IC50 of 2.7 μM obtained for the CK2α subunit in comparison to an IC50 value of 0.185 μM for the CK2 holoenzyme has the highest ratio of 14. Pomiferin with an IC50 value of 4 μM for the CK2α subunit and 63 μM for the CK2 holoenzyme resulting in a ratio of 0.06 is found at the lower end of the scale. This classification is arbitrary and might not have a practical value momentarily. Yet, it can be helpful in selecting compounds for specific investigations where differences in the IC50 ratio between the CK2α subunit and the holoenzyme may be desirable and advantageous, e.g., when searching for compounds with the potential to target allosteric sites on a kinase.

The molecular mass of most compounds investigated is in a range between 250 and 400 Da, apart from a few exceptions such as naganol (Table 1).

This is, to our knowledge, the first systematic comparison of IC50 values of the catalytic CK2α subunit with those for the CK2 holoenzyme.

Interactions between subunits of CK2 and their protein substrates toward sensitivity to specific inhibitors were investigated by Janeczko et al. [14]. The experiments were performed using yeast CK2 isoforms and the ATP-competitive inhibitors TBB and TBI. Since no IC50 values were determined, a direct comparison to our results is difficult to assess.

So far, all evidence suggests that the majority of CK2 molecules is present within the cells as a tetramer; however, there is an evidence for unbalanced expression of the subunits [15, 16]. Since most of the screenings for new CK2 inhibiting

Table 1 IC_{50} values for CK2 inhibitory compounds determined by dose-response curves

Compound	Structure	NSC no.	M_w	IC_{50} CK2α (μM)	IC_{50} $CK2\alpha_2\beta_2$ (μM)	Ratio $CK2\alpha/CK2\alpha_2\beta_2$
Coumestrol	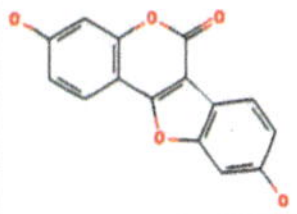	22842	268.22	2.70	0.19	14.2
Naganol	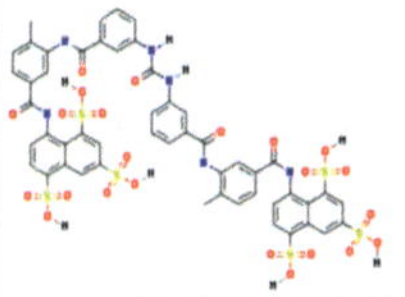	34936	1429.2	0.75	0.25	3.00
Resorufin		12097	213.19	4.10	1.50	2.73
Boldine	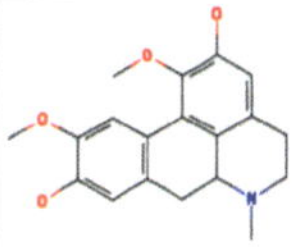	65689	327.37	1.70	0.70	2.43
B8-4722	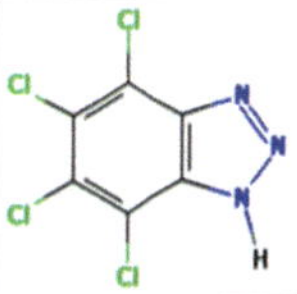	134674	257.0	4.72	2.00	2.36
Galloflavin	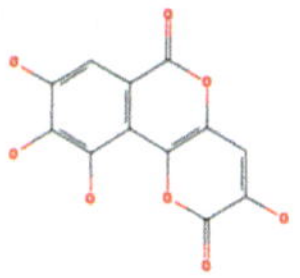	107022	278.0	0.39	0.20	1.70
Stictic acid	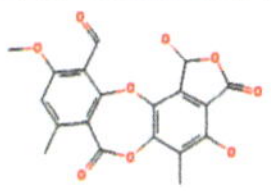	87511	386.31	9.70	10.7	0.91
G11-4726	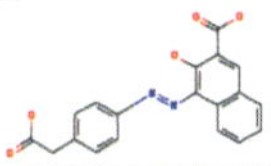	156563-L/2	350	1.30	1.60	0.81
H11-4245	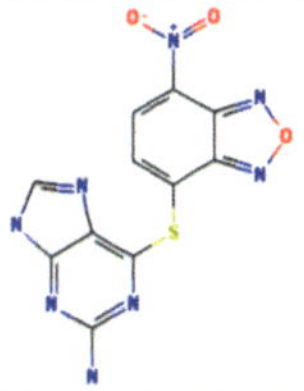	348401	330.28	2.50	7.70	0.32
Fisetin	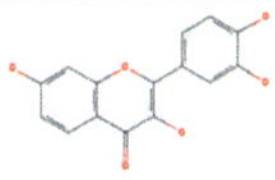	407010	286.24	0.16	1.19	0.13

(continued)

Table 1 (continued)

Compound	Structure	NSC no.	M_w	IC_{50} CK2α (μM)	IC_{50} $CK2\alpha_2\beta_2$ (μM)	Ratio $CK2\alpha/CK2\alpha_2\beta_2$
Hypericin	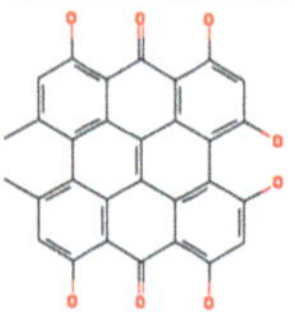	407313	504.44	1.00	7.50	0.13
A6-4722	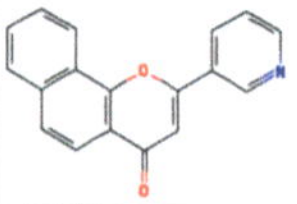	283845	273.0	2.62	30.0	0.09
Pomiferin	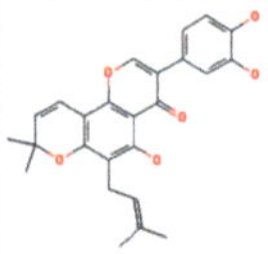	5113	420.45	4.00	63.0	0.06
Nortangeretin	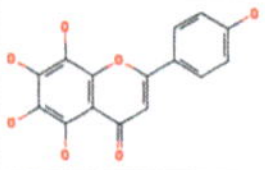	76988	302.0	0.34	15.0	0.02

Data expressed in micromolar were obtained by determining the incorporation of ^{32}P-phosphate into a synthetic peptide substrate in a radioactive-based kinase assay. Columns: (1) shows the compound names. In those cases where no common compound name was available, the DTP plate number is shown; the full chemical name appears in the text. (2) shows the compound structures; (3) NSC no., the NSC number is a numeric identifier for substances submitted to the National Cancer Institute (NCI) for testing and evaluation. It is a registration number for the Developmental Therapeutics Program (DTP) repository. NSC stands for National Service Center. (4) M_w, molecular weight, (5) IC_{50} CK2α (μM), (6) IC_{50} $CK2\alpha_2\beta_2$ (μM), (7) ratio of $CK2\alpha/CK2\alpha_2\beta_2$

compounds are performed with recombinant CK2α, compounds, which specifically target the CK2 holoenzyme, may escape detection.

Although this report is focusing on protein kinase CK2, we would like to emphasize the possible role of off-target molecules.

We believe strongly that compounds that have been identified under a screening against a particular protein kinase are very likely not acting entirely through this particular kinase but also through off-targets.

A literature search for the compounds shown in Table 1 confirms our multi-kinase inhibitor hypothesis. This will also change the current view of how to identify the best targets in cancer therapy (perhaps relevant for all multifunctional diseases).

It is mandatory to verify the efficacy of identified inhibitors using biomarkers to establish them as promising targets for CK2 but also for off-targets which are involved in angiogenesis, proliferation, and programmed cell death.

7 Characterization of the CK2 Inhibiting Compounds

7.1 Compounds with IC50 Ratios of CK2α/CK2 Holoenzyme >1

7.1.1 Coumestrol, NSC 22842

Coumestrol is a known phytohormone which has recently been shown to suppress HIF-1α by inhibiting ROS-mediated sphingosine kinase 1 in hypoxic PC-3 prostate cancer cells [17].

Liu et al. [18] identified coumestrol when screening the NCI's natural product library. It turned out to be a reversible ATP-competitive CK2 inhibitor with an IC50 value of 0.228 μM in good agreement with the IC50 value of 0.19 μM which we obtained for the holoenzyme (Table 1).

Coumestrol inhibited cancer cell growth partially through downregulation of CK2-specific AKT phosphorylation.

Of interest was the observation by Lee et al. [19] that coumestrol induces senescence through protein kinase CK2 inhibition-mediated ROS production in human cancer cells.

We show here that coumestrol has the lowest IC50 for the CK2 holoenzyme, among the compounds shown in Table 1. The IC50 value for the catalytic subunit was 15 times higher, i.e., 2.7 μM. Results from structural studies involving the catalytic CK2α subunit and the CK2 holoenzyme together with coumestrol could perhaps explain the remarkable differences seen between the two IC50 values and lead to more insight about possible allosteric binding sites on CK2 molecules.

7.1.2 Naganol (Suramin/Germanin), NSC 34936

Naganol is a clinically prescribed drug for the treatment of human African trypanosomiasis, cancer, and infection. Despite its importance in clinical and basic research, the biological actions of this molecule are still incompletely understood.

Naganol shows an IC50 value of 0.25 μM for the holoenzyme and 0.75 μM for the catalytic CK2α subunit.

There is one report showing that naganol/suramin inhibits DNA-dependent protein kinase activity with an IC50 of 1.7 μM. Inhibition of DNA-PK sensitizes cells to ionizing radiation, e.g., under cancer treatment. Together with our observations that naganol inhibits the antiapoptotic protein kinase CK2, one may conclude that it is a promising compound also for cancer treatment.

Moreover, with a molecular mass of 1,429 Da, it is the largest molecule that has been identified during our screening [20, 21].

7.1.3 Resorufin, NSC 12097

This compound has been described by [22] as a CK2 inhibitor showing a similar IC50 ratio as observed with naganol, i.e., 2.73. However, the potency is five times less. Klopffleisch et al. [23] showed co-crystallization of CK2α and resorufin. The results obtained suggested that resorufin penetrates relatively deeply into a region of the ATP-binding site that is typically blocked by a conserved water molecule [24]. Obviously, resorufin binding requires no clearly defined hydrogen bonds to the enzyme matrix, in particular not to the hinge/αD region.

A curious feature of the crystal packing of the CK2α-resorufin complex that was not found in any other human or maize CK2α crystal form is an "arginine zipper," i.e., a set of six alternating arginine side chains from two symmetry-equivalent enzyme subunits. To our knowledge, such an extended motif as the arginine zipper has not been described before.

7.1.4 Boldine, NSC 65689

Boldine is the major and most characteristic alkaloidal constituent of the boldo tree (*Peumus boldus* Mol. and *Lindera aggregata*). Boldine is an effective antioxidant in both biological and non-biological systems [25, 26] and protects against free radical-induced lipid peroxidation or enzyme inactivation. In addition, it has alpha-adrenergic antagonist activities in vascular tissue, and it has also been reported to have hepatoprotective, cytoprotective, antipyretic, and anti-inflammatory effects. Only recently, it has been shown that boldine inhibited cell growth and cell cycle arrest, and it appears to be linked to the inactivation of ERK and AKT activity in T24 human bladder cells [27]. The fairly low IC50 values for CK2 inhibition, i.e., 0.7 and 1.7 μM for the holoenzyme and the catalytic α-subunit, respectively, suggest that the observed therapeutic effects of boldine for the treatment of urinary bladder cancer may also partly be explained by CK2 inhibition, an interesting molecule worthwhile to be pursued.

7.1.5 B8-4722, NSC 134674

4,5,6,7-Tetrachloro-benzotriazole

This compound has not been shown so far to be a protein kinase inhibitor. The compound has been shown to have virus-inhibitory activity, e.g., influenza B virus multiplication [28]. We have identified B8-4722 to inhibit CK2 holoenzyme with an IC50 of 2 μM and the CK2α with an IC50 of 4.72 μM.

7.1.6 Galloflavin, NSC 107022

Galloflavin has been shown to interfere with lactate dehydrogenase (LDH) activity [29]. Galloflavin induces death in human breast cancer cells by affecting distinct signaling pathways [30]. Moreover, the suppression of LDH activity was accompanied by MYC downregulation in Burkitt lymphoma cells through the inhibition of Sirtuin-1 (SIRT1) [31].

This observation was very interesting since CK2 has been shown to phosphorylate SIRT1 [32]. Moreover, it was shown that CK2 is the regulator of SIRT1 substrate-binding affinity, deacetylase activity, and cellular response to DNA damage [33]. The inhibition of CK2 induces p53-dependent cell cycle arrest in glioblastoma cells through SIRT1 inhibition [34]. Hence, our finding that galloflavin is a CK2 inhibitor for both CK2α and CK2 holoenzyme with fairly good IC50 values makes it a promising compound with therapeutic potential. The obtained IC50 values for CK2α and the holoenzyme were 0.39 and 0.2 μM, respectively. Obviously, one compound can affect totally different molecules such as LDH and a protein kinase such as CK2.

7.2 Compounds with IC50 Ratios of CK2α/CK2 Holoenzyme <1

7.2.1 Stictic Acid, NSC 13091250

Stictic acid is an aromatic organic compound, a product of secondary metabolism in some species of lichens. Stictic acid is the subject of medical research in cancer. Researchers showed that stictic acid extracted from Antarctic lichens has cytotoxic apoptotic effects on hepatocytes [35]. Subsequent cancer research results revealed that the compound has tumor-suppressing effects via action on mutant p53 proteins in cancer cells [36]. In this context, it is interesting to note that p53 is a well-known substrate for CK2 which is modulating p53 activity.

Other authors report on antioxidant activities associated with stictic acid derivatives from the lichen *Usnea articulata* [37].

7.2.2 G11-4726, NSC 156563-L/2

(4Z)-4-[[4-(carboxymethyl)phenyl]hydrazinylidene]-3-oxonapthalene-2-carboxylic acid

The IC50 values obtained for the CK2α and CK2 holoenzyme were 1.3 and 1.60 μM, respectively. No protein kinase-associated results were found in the literature.

7.2.3 H11-4245, NSC 348402

4-(6-Thioguanine)-7-nitro-2, 3-benzoxadiazole

This compound has relatively high IC50 values of 2.5 and 7.7 μM for CK2α and the holoenzyme, respectively. The ratio is 0.33. So far, it has not been described to be a kinase inhibitor or having effects on cellular signaling.

7.2.4 Fisetin, NSC 407010

Fisetin has been described as a CK2 inhibitor [5]. These authors found an IC50 of 0.35 μM. We have identified fisetin to inhibit CK2α with an IC50 of 0.16 μM and the CK2 holoenzyme with an IC50 of 1.19 μM.

Yet, fisetin has also been shown to target MAPK and NF-κB signaling pathways [38]. Moreover, it was identified as a novel dual inhibitor of PI3K/AKT and mTOR [39, 40]. Fisetin induces apoptosis in tumor cells through ERK1/2-mediated activation of caspase-dependent pathways. There are a number of reports describing the involvement of this flavonoid in inflammation [41], in activation of ROS/AMPK signaling [42], and in regulating obesity by targeting mTORC1 signaling [43]. The many observed effects fisetin has on the different cellular targets make it difficult to assign it a specific function, even if only focusing on one class of molecules such as protein kinases.

7.2.5 Hypericin, NSC 407313

Hypericin and its derivatives have been extensively studied mainly for their antitumor, antiviral, and antidepressant properties. Notably, hypericin is one of the most potent natural photodynamic agents derived from St. John's wort.

Hypericin causes the generation of ROS; it is antiangiogenic and involved in multiple molecular pathways (intrinsic/extrinsic apoptotic pathways) and ERK inhibition (for review [44]).

We found IC50 values for hypericin of 1 and 7.5 μM for CKα and the holoenzyme, respectively. The role of CK2 inhibition by hypericin is difficult to assess since no cellular experiments are available.

However, most of the investigations are associated with its photodamage properties [45–47].

7.2.6 Pomiferin, NSC 5113

Pomiferin is a prenylated isoflavonoid from *Maclura pomifera*.

Isoflavones possess strong antioxidant activities inasmuch as they inhibit the production of ROS (reviewed in [48]). It has also been shown to increase the production of reactive oxygen species in mouse macrophages [49]. Antiproliferative activity of pomiferin was shown in normal and transformed breast epithelial cells [50].

Recently, pomiferin has been shown to be a histone deacetylase inhibitor at low micromolar concentrations (IC50 approx. 1 μM) and to inhibit growth of different human cancer cell lines, e.g., kidney, lung, prostate, breast, or colon cancers without affecting the growth of primary human hepatocytes [51].

Pomiferin has also been identified as a novel mTOR and translation inhibitor [52].

We have identified pomiferin as a CK2 inhibitor. The IC50 for the CK2α subunit was determined with 4 μM and is in the range of the IC50 value found for histone deacetylase (1 μM). To our surprise, the IC50 value for the CK2 holoenzyme was 63 μM, approx., 15 times higher than that for the catalytic subunit (Fig. 4a). Since the holoenzyme is the predominant CK2 molecule in the cell, the effect of pomiferin on endogenous CK2 activity should be only marginable.

However, because of the large differences in the IC50 values, pomiferin is certainly an interesting molecule for structural studies when exploring potential allosteric binding sites on the two CK2 molecules.

7.2.7 Nortangeretin, NSC 76988

Nortangeretin (5,6,7,8,4′-pentahydroxyflavone) is a citrus flavonoid known to exhibit remarkable cytotoxicity to leukemia cells [53]. Other pentahydroxyflavones such as quercetin, morin, and taxifolin have been shown to inhibit protein kinase CK2 holoenzyme [5]. The published IC50 values for these three compounds were determined as 0.55, 10, and 29 μM, respectively. We found IC50 for nortangeretin of 0.34 μM for CK2α and 15 μM for the holoenzyme. Hence, nortangeretin exhibits a high IC50 value such as that determined for the pentahydroxyflavones morin and taxifolin for the holoenzyme [5]. Unfortunately, no IC50 data for the catalytic subunit have been published for morin and taxifolin, so that we cannot assess whether the large difference seen in IC50 values between CK2α and the CK2 holoenzyme is a characteristic of pentahydroxyflavones. In any case, there are many data available on quercetin and its many features such as anti-ROS activity and growth inhibition [54–58].

8 The Role of CK2β

The role of the CK2β dimer within the tetrameric holoenzyme structure has been closely investigated [59]. Mutations in the acidic amino acid stretch 55–59 led to autophosphorylation and phosphorylation of the β-subunit [60]. CK2β mutants lacking the acidic amino acids at position 55–57 showed >25-fold stimulation of the catalytic subunit [61].

CK2 downregulation and activation by polybasic peptides were shown by [62]. Moreover, with the elucidation of the holoenzyme structure [59] and subsequent investigations on the structural interactions between CK2α and CK2β, important information was gathered concerning the special function of CK2β within the tetrameric holoenzyme structure [63, 64].

In the same vein, one could imagine that the β-subunit may influence enzyme activity when a small molecular compound binds to the holoenzyme and it would not come to a surprise if the binding to the catalytic subunit alone could lead to different IC50 values. In our present investigation, the compound with the highest IC50 value for the holoenzyme identified throughout our investigations was pomiferin with an IC50 value which is about 15 times higher than that found for the catalytic subunit alone. Here, the β-subunit obviously "protects" against pomiferin action. On the other side of the ranking, we have identified coumestrol which inhibits the holoenzyme with an IC50 of 0.19 μM in comparison to the catalytic α-subunit, where we determined an IC50 of 2.7 μM. Here, the β-subunit does not "protect" against kinase activity inhibition by coumestrol.

The majority of today's kinase inhibitors are ATP selective (type I inhibitors) which makes most of them not very selective. An alternative type of inhibitors would be allosteric inhibitors. Some of our results are indicative for compound-induced allosteric changes affecting the catalytic activity via binding to the β-subunit similar as reported by [65]. In silico modeling could be helpful in supporting these notions.

9 Compound Selection

As already mentioned previously, the CK2 holoenzyme is the major form in the cell. Hence, inhibitors with lower IC50 values for the CK2 holoenzyme would be preferable for in vitro and in vivo studies. Therefore, coumestrol (0.19 μM), galloflavin (0.2 μM), naganol (0.25 μM), and boldine (0.7 μM) are the compounds of choice because of their specificity for the CK2 holoenzyme (Table 1).

We have also performed further investigations applying the compounds in cell cultures using the two glioblastoma cell lines M059J and M059K, which are isogenic cell lines isolated from a surgical sample obtained from a patient with untreated glioblastoma multiforme (see http://www.lgcstandards-atcc.org). The M059J cell line lacks DNA-PKcs, making the cells severalfold more sensitive to ionizing radiation and radiomimetic drugs than M059K. The aim of the experiment treating both cell lines with resorufin was to see whether the inhibition of protein kinase CK2 is more evident in the M059J cells. The reason for our assumption was based on earlier observations by [66] who showed that CK2α′ is significantly increased both at the protein and mRNA level with a concomitant increase in CK2 activity in DNA-PKcs-deficient M059J cells. Results were described in [67], where it was postulated that CK2α′ associates with CK2β to form CK2α′β2 trimeric complexes which might be less "stable" than the tetrameric CK2α-based holoenzyme. One can speculate that the trimeric complexes are more vulnerable toward CK2 inhibitors targeting the ATP-binding site, e.g., resorufin.

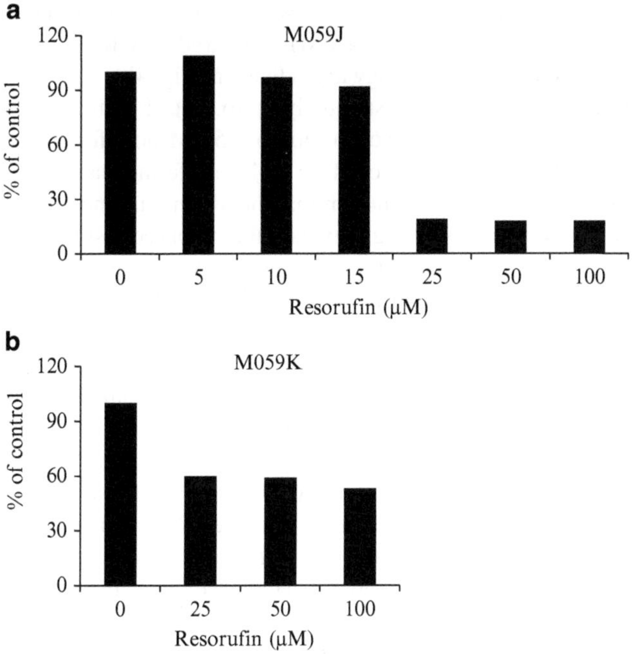

Fig. 3 Kinase activity determination of CK2 from whole cell lysates derived from (**a**) M059J and (**b**) M059K human glioblastoma cell lines in the presence of increasing concentrations of resorufin. Inhibition is expressed in percentage relative to control assays performed in the absence of inhibitor

Figure 3 shows the results with the glioblastoma cell lines M059J and M059K after resorufin treatment.

In Fig. 3a, we show MO59J cells treated with increasing concentrations of resorufin. CK2 activity is decreasing in a dose-dependent manner. There is no distinct drop in activity up to 15 μM. At 25 μM resorufin, the activity drops to 20 % and remains more or less constant up to 100 μM.

The isogenic M059K cell line behaves distinctively different. Here, CK2 activity drops to 60 % of that what was observed in the untreated control and remains at that level up to 100 μM resorufin (Fig. 3b). According to the results obtained by [66], we postulate the presence of a larger number of CK2α′ holoenzyme molecules in M059J cells which would render these cells more vulnerable against CK2 inhibitor treatment in comparison to M059K cells. Indeed, that is what we observed (Fig. 3a). Janeczko et al. [14] who have performed a thorough Ki determination of five different CK2 complexes using different protein substrates found in all cases an up to nine times lower Ki value for the α′-based holoenzyme than for the α-holoenzyme. This is supporting our hypothesis that cell lines with higher CK2α′-based holoenzyme are more sensitive to inhibitor treatment.

We also tested pomiferin, a compound with more than one order of magnitude lower IC50 value for CK2α (IC50 4 μM) compared to CK2 holoenzyme (IC50 63 μM) (Fig. 4a) for cell morphology and adhesion (Fig. 4b).

Both cell lines responded almost equally up to 10 μM pomiferin. A drastic loss of cell adhesion occurred at a concentration of 25 μM pomiferin (Fig. 4b). In the case of the M059K cells, the effect of 25 μM pomiferin treatment was less pronounced (Fig. 4b) supporting the notion that the aforementioned differences in the CK2α′ holoenzyme complexes might explain the observed results.

Cells were treated with increasing concentrations (i.e., up to 25 μM) of pomiferin, hypothesizing that a concentration of 25 μM would be sufficient to inhibit the predominant CK2α′ holoenzyme complexes present in the M095J cells. The results in Fig. 4c show no difference in CK2 activity between the two cell lines.

A possible explanation could be that the 25 μM concentration was too low to inhibit endogenous CK2 complexes or that the binding to the complex is reversible in cellular lysates.

Hence, it is possible that off-targets affected by pomiferin may explain the "adhesion" results seen in Fig. 4b. Pomiferin has also been shown to be a histone deacetylase inhibitor [51]. An overview of the treatment results with respect to CK2 activity in cancer cell lines is shown in Table 2. Here, we show the effect of resorufin, coumestrol, fisetin, and pomiferin on various cancer cell lines. Resorufin inhibits CK2 activity in HTC116 cell by 40 % and in prostate carcinoma cell lines by 70 % inhibition. Cell line-dependent inhibition was already previously shown by [68] where PC-3 cells were less responsive toward cisplatin treatment than other prostate carcinoma cell lines. Similar results were reported by Yde et al. [69] with respect to breast carcinoma cell lines.

10 Conclusion

The sequence of events in order to find potential kinase targets by screening compound libraries involved the following steps: (1) identification of a protein kinase inhibitor from screening suitable compound libraries for example such as the ones provided by DTP/NCI, (2) testing identified positive compounds in various cell lines using established biomarkers, (3) selectivity screen against a kinase panel and (4) in vivo testing in animals.

We believe from our own and others' results that screening for one target usually identifies compounds which additionally target also other molecules and pathways, e.g., involved in angiogenesis and proliferation, and which have antiapoptotic properties. This is what one would like to achieve finally. Our personal view of a "magic bullet" is circumstantial. We believe that it is more desirable to have compounds with defined mechanisms irrespective of the number of molecules involved. Importantly, they should interfere successfully with targets responsible for the disease process.

Finally, the ideal chemopreventive agent should have no/low toxicity, oral availability, known mechanism, and low cost.

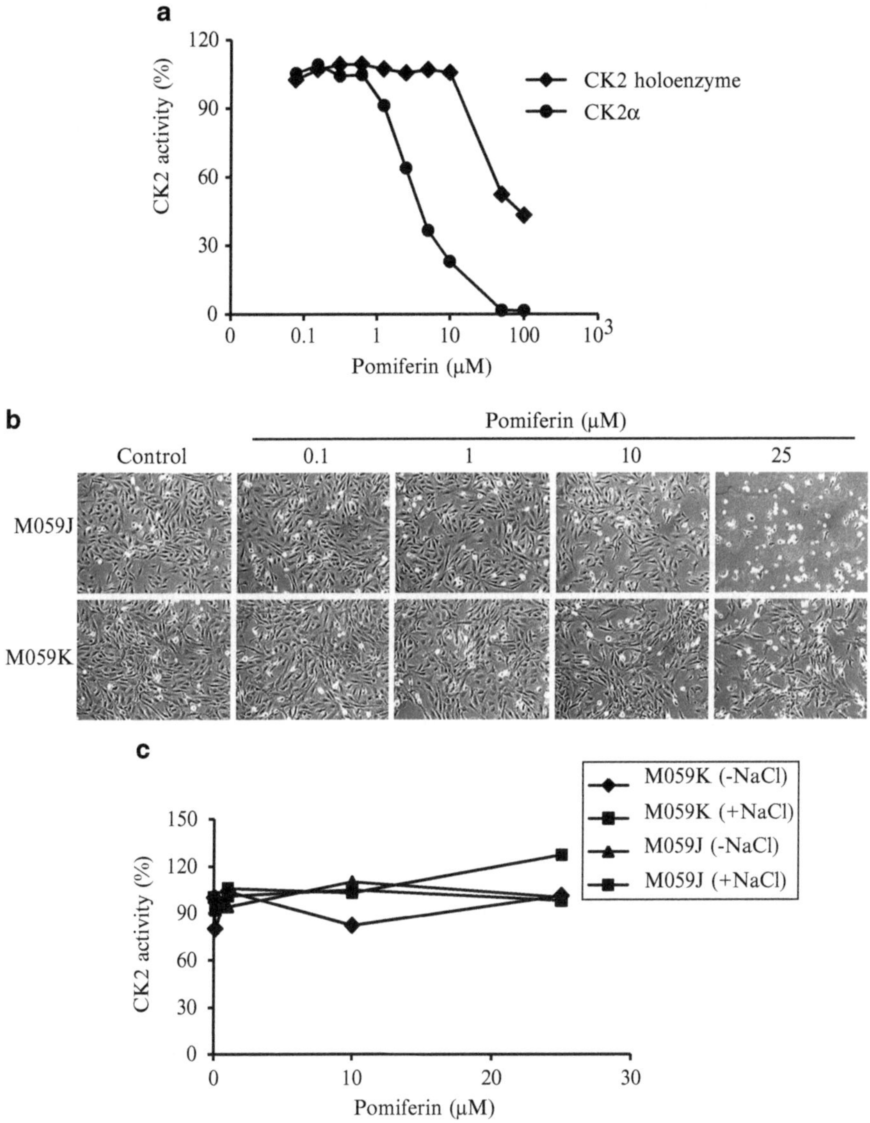

Fig. 4 (**a**) Dose-response study of pomiferin in the presence of CK2 holoenzyme and CK2α, respectively. Radioactive-based kinase assays were performed in the presence of increasing concentrations of pomiferin. Kinase inhibition is expressed in percentage relative to the lowest concentration of compound tested, i.e., 0.1 μM. (**b**) Phase contrast microscopy pictures of M059K and M059J cells left untreated (control) or incubated with increasing concentrations of pomiferin for 24 h. Pictures were taken at 400× magnification. (**c**) Whole protein lysates from cells left untreated or treated as described above were subjected to radioactive-based protein kinase assays in the presence and absence of 150 mM NaCl, respectively. Kinase activity is expressed in percentage relative to the assays performed in the presence of DMSO and whole protein extract from untreated cells

Table 2 Determination of CK2 kinase activity in whole cell lysates from various cell lines treated with the indicated compounds for 24 h

Compound	Concentration (μM)	Cell line	Inhibition (%)
Resorufin	40	HTC116	40
	40	PC-3, DU145, LNCaP	70
Coumestrol	50	M059J	93
	50	M059K	86
	50	MCF-7	43
Fisetin	50	M059J	22
	50	M059K	20
	50	MCF-7	40
Pomiferin	50	M059J	11
	50	M059K	7.6

The percentage of inhibition is expressed relative to the kinase activity of CK2 in protein extracts from untreated cells

Acknowledgments We thank the Drug Synthesis and Chemistry Branch, Developmental Therapeutics Program, Division of Cancer Treatment and Diagnosis, National Cancer Institute, USA, for providing us with plated and vialed samples from the various compound sets. The advice of Dr. David Newman at the start of the investigation is particularly acknowledged. We also thank the many students who have been involved in some of the screenings presented in this report. This work was supported in part by Grosserer M. Brogaard og Hustrus Mindefond and the Danish Council for Independent Research-Natural Sciences (grant 1323-00212A to B. Guerra), the Danish Cancer Society (grant 252-1109-210), and the Danish Research Council (grant 21-04-0517) to O.-G. Issinger.

References

1. Milner JA, Romagnolo DF (2010) Bioactive compounds and cancer. Springer, New York
2. Olsen LF, Issinger O-G, Guerra B (2013) The Yin and Yang of redox regulation. Redox Rep 18:245–252. doi:10.1179/1351000213Y.0000000059
3. Kumamoto T, Fujii M, Hou D-X (2009) Akt is a direct target for myricetin to inhibit cell transformation. Mol Cell Biochem 332:33–41. doi:10.1007/s11010-009-0171-9
4. Walker EH, Pacold ME, Perisic O et al (2000) Structural determinants of phosphoinositide 3-kinase inhibition by wortmannin, LY294002, quercetin, myricetin, and staurosporine. Mol Cell 6:909–919
5. Lolli G, Cozza G, Mazzorana M et al (2012) Inhibition of protein kinase CK2 by flavonoids and tyrphostins. A structural insight. Biochemistry 51:6097–6107. doi:10.1021/bi300531c
6. Golub AG, Bdzhola VG, Kyshenia YV et al (2011) Structure-based discovery of novel flavonol inhibitors of human protein kinase CK2. Mol Cell Biochem 356:107–115. doi:10.1007/s11010-011-0945-8
7. Boldyreff B, Rasmussen TL, Jensen HH et al (2008) Expression and purification of PI3 kinase alpha and development of an ATP depletion and an alphascreen PI3 kinase activity assay. J Biomol Screen 13:1035–1040. doi:10.1177/1087057108326079
8. Guerra B, Issinger O-G (2008) Protein kinase CK2 in human diseases. Curr Med Chem 15:1870–1886

9. Trembley JH, Chen Z, Unger G et al (2010) Emergence of protein kinase CK2 as a key target in cancer therapy. Biofactors 36:187–195. doi:10.1002/biof.96
10. St-Denis NA, Litchfield DW (2009) Protein kinase CK2 in health and disease: from birth to death: the role of protein kinase CK2 in the regulation of cell proliferation and survival. Cell Mol Life Sci 66:1817–1829. doi:10.1007/s00018-009-9150-2
11. Filhol O, Cochet C (2009) Protein kinase CK2 in health and disease. Cell Mol Life Sci 66:1830–1839. doi:10.1007/s00018-009-9151-1
12. Olsen BB, Boldyreff B, Niefind K, Issinger O-G (2006) Purification and characterization of the CK2alpha'-based holoenzyme, an isozyme of CK2alpha: a comparative analysis. Protein Expr Purif 47:651–661. doi:10.1016/j.pep.2005.12.001
13. Yim H, Lee YH, Lee CH, Lee SK (1999) Emodin, an anthraquinone derivative isolated from the rhizomes of Rheum palmatum, selectively inhibits the activity of casein kinase II as a competitive inhibitor. Planta Med 65:9–13. doi:10.1055/s-1999-13953
14. Janeczko M, Masłyk M, Szyszka R, Baier A (2011) Interactions between subunits of protein kinase CK2 and their protein substrates influences its sensitivity to specific inhibitors. Mol Cell Biochem 356:121–126. doi:10.1007/s11010-011-0951-x
15. Stalter G, Siemer S, Becht E et al (1994) Asymmetric Expression of Protein Kinase CK2 Subunits in Human Kidney Tumors. Biochem Biophys Res Commun 202:141–147. doi:10.1006/bbrc.1994.1904
16. Deshiere A, Duchemin-Pelletier E, Spreux E et al (2013) Unbalanced expression of CK2 kinase subunits is sufficient to drive epithelial-to-mesenchymal transition by Snail1 induction. Oncogene 32:1373–1383. doi:10.1038/onc.2012.165
17. Cho S-Y, Cho S, Park E et al (2014) Coumestrol suppresses hypoxia inducible factor 1α by inhibiting ROS mediated sphingosine kinase 1 in hypoxic PC-3 prostate cancer cells. Bioorg Med Chem Lett 24:2560–2564. doi:10.1016/j.bmcl.2014.03.084
18. Liu S, Hsieh D, Yang Y-L et al (2013) Coumestrol from the national cancer Institute's natural product library is a novel inhibitor of protein kinase CK2. BMC Pharmacol Toxicol 14:36. doi:10.1186/2050-6511-14-36
19. Lee Y-H, Yuk HJ, Park K-H, Bae Y-S (2013) Coumestrol induces senescence through protein kinase CKII inhibition-mediated reactive oxygen species production in human breast cancer and colon cancer cells. Food Chem 141:381–388. doi:10.1016/j.foodchem.2013.03.053
20. Hosoi Y, Matsumoto Y, Tomita M et al (2002) Phosphorothioate oligonucleotides, suramin and heparin inhibit DNA-dependent protein kinase activity. Br J Cancer 86:1143–1149. doi:10.1038/sj.bjc.6600191
21. Borges S, Döppler HR, Storz P (2014) A combination treatment with DNA methyltransferase inhibitors and suramin decreases invasiveness of breast cancer cells. Breast Cancer Res Treat 144:79–91. doi:10.1007/s10549-014-2857-2
22. Sandholt IS, Olsen BB, Guerra B, Issinger O-G (2009) Resorufin: a lead for a new protein kinase CK2 inhibitor. Anticancer Drugs 20:238–248. doi:10.1097/CAD.0b013e328326472e
23. Klopffleisch K, Issinger O-G, Niefind K (2012) Low-density crystal packing of human protein kinase CK2 catalytic subunit in complex with resorufin or other ligands: a tool to study the unique hinge-region plasticity of the enzyme without packing bias. Acta Crystallogr D Biol Crystallogr 68:883–892. doi:10.1107/S0907444912016587
24. Battistutta R, Mazzorana M, Cendron L et al (2007) The ATP-binding site of protein kinase CK2 holds a positive electrostatic area and conserved water molecules. Chembiochem 8:1804–1809. doi:10.1002/cbic.200700307
25. Speisky H, Cassels BK (1994) Boldo and boldine: an emerging case of natural drug development. Pharmacol Res 29:1–12
26. O'Brien P, Carrasco-Pozo C, Speisky H (2006) Boldine and its antioxidant or health-promoting properties. Chem Biol Interact 159:1–17. doi:10.1016/j.cbi.2005.09.002
27. Gerhardt D, Bertola G, Dietrich F et al (2014) Boldine induces cell cycle arrest and apoptosis in T24 human bladder cancer cell line via regulation of ERK, AKT, and GSK-3β. Urol Oncol 32:36. doi:10.1016/j.urolonc.2013.02.012

28. Tamm I, Bablanian R, Nemes MM et al (1961) Relationship between structure of benzimidazole derivatives and selective virus inhibitory activity. Inhibition of poliovirus multiplication and cytopathic effects by 2-(alpha-hydroxybenzyl)-benzimidazole, and its 5-chloroderivative. J Exp Med 113:625–656
29. Manerba M, Vettraino M, Fiume L et al (2012) Galloflavin (CAS 568-80-9): a novel inhibitor of lactate dehydrogenase. ChemMedChem 7:311–317. doi:10.1002/cmdc.201100471
30. Farabegoli F, Vettraino M, Manerba M et al (2012) Galloflavin, a new lactate dehydrogenase inhibitor, induces the death of human breast cancer cells with different glycolytic attitude by affecting distinct signaling pathways. Eur J Pharm Sci 47:729–738. doi:10.1016/j.ejps.2012.08.012
31. Vettraino M, Manerba M, Govoni M, Di Stefano G (2013) Galloflavin suppresses lactate dehydrogenase activity and causes MYC downregulation in Burkitt lymphoma cells through NAD/NADH-dependent inhibition of sirtuin-1. Anticancer Drugs 24:862–870. doi:10.1097/CAD.0b013e328363ae50
32. Zschoernig B, Mahlknecht U (2009) Carboxy-terminal phosphorylation of SIRT1 by protein kinase CK2. Biochem Biophys Res Commun 381:372–377. doi:10.1016/j.bbrc.2009.02.085
33. Kang H, Jung J-W, Kim MK, Chung JH (2009) CK2 is the regulator of SIRT1 substrate-binding affinity, deacetylase activity and cellular response to DNA-damage. PLoS One 4:e6611. doi:10.1371/journal.pone.0006611
34. Dixit D, Sharma V, Ghosh S et al (2012) Inhibition of Casein kinase-2 induces p53-dependent cell cycle arrest and sensitizes glioblastoma cells to tumor necrosis factor (TNFα)-induced apoptosis through SIRT1 inhibition. Cell Death Dis 3:e271. doi:10.1038/cddis.2012.10
35. Correché ER, Enriz RD, Piovano M et al (2004) Cytotoxic and apoptotic effects on hepatocytes of secondary metabolites obtained from lichens. Altern Lab Anim 32:605–615
36. Wassman CD, Baronio R, Demir Ö et al (2013) Computational identification of a transiently open L1/S3 pocket for reactivation of mutant p53. Nat Commun 4:1407. doi:10.1038/ncomms2361
37. Lohézic-Le Dévéhat F, Tomasi S, Elix JA et al (2007) Stictic acid derivatives from the lichen Usnea articulata and their antioxidant activities. J Nat Prod 70:1218–1220. doi:10.1021/np070145k
38. Pal HC, Sharma S, Strickland LR et al (2014) Fisetin inhibits human melanoma cell invasion through promotion of mesenchymal to epithelial transition and by targeting MAPK and NFκB signaling pathways. PLoS One 9:e86338. doi:10.1371/journal.pone.0086338
39. Syed DN, Adhami VM, Khan MI, Mukhtar H (2013) Inhibition of Akt/mTOR signaling by the dietary flavonoid fisetin. Anticancer Agents Med Chem 13:995–1001
40. Adhami VM, Syed DN, Khan N, Mukhtar H (2012) Dietary flavonoid fisetin: a novel dual inhibitor of PI3K/Akt and mTOR for prostate cancer management. Biochem Pharmacol 84:1277–1281. doi:10.1016/j.bcp.2012.07.012
41. Currais A, Chiruta C, Goujon-Svrzic M et al (2014) Screening and identification of neuroprotective compounds relevant to Alzheimer's disease from medicinal plants of S. Tomé e Príncipe. J Ethnopharmacol. doi:10.1016/j.jep.2014.06.046
42. Jang KY, Jeong S-J, Kim S-H et al (2012) Activation of reactive oxygen species/AMP activated protein kinase signaling mediates fisetin-induced apoptosis in multiple myeloma U266 cells. Cancer Lett 319:197–202. doi:10.1016/j.canlet.2012.01.008
43. Jung CH, Kim H, Ahn J et al (2013) Fisetin regulates obesity by targeting mTORC1 signaling. J Nutr Biochem 24:1547–1554. doi:10.1016/j.jnutbio.2013.01.003
44. Karioti A, Bilia AR (2010) Hypericins as potential leads for new therapeutics. Int J Mol Sci 11:562–594. doi:10.3390/ijms11020562
45. Gronquist M, Bezzerides A, Attygalle A et al (2001) Attractive and defensive functions of the ultraviolet pigments of a flower (*Hypericum calycinum*). Proc Natl Acad Sci U S A 98:13745–13750. doi:10.1073/pnas.231471698
46. Krammer B, Verwanger T (2012) Molecular response to hypericin-induced photodamage. Curr Med Chem 19:793–798

47. Jocham D, Stepp H, Waidelich R (2008) Photodynamic diagnosis in urology: state-of-the-art. Eur Urol 53:1138–1148. doi:10.1016/j.eururo.2007.11.048
48. Amin A, Gali-Muhtasib H, Ocker M, Schneider-Stock R (2009) Overview of major classes of plant-derived anticancer drugs. Int J Biomed Sci 5:1–11
49. Hošek J, Toniolo A, Neuwirth O, Bolego C (2013) Prenylated and geranylated flavonoids increase production of reactive oxygen species in mouse macrophages but inhibit the inflammatory response. J Nat Prod 76:1586–1591. doi:10.1021/np400242e
50. Yang R, Hanwell H, Zhang J et al (2011) Antiproliferative activity of pomiferin in normal (MCF-10A) and transformed (MCF-7) breast epithelial cells. J Agric Food Chem 59:13328–13336. doi:10.1021/jf202898g
51. Son IH, Chung I-M, Lee SI et al (2007) Pomiferin, histone deacetylase inhibitor isolated from the fruits of *Maclura pomifera*. Bioorg Med Chem Lett 17:4753–4755. doi:10.1016/j.bmcl.2007.06.060
52. Bajer MM, Kunze MM, Blees JS et al (2014) Characterization of pomiferin triacetate as a novel mTOR and translation inhibitor. Biochem Pharmacol 88:313–321. doi:10.1016/j.bcp.2014.01.034
53. Grace MH, Wilsonb GR, Kandil FE et al (2009) Characteristic flavonoids from *Acacia burkittii* and *A. acuminata* heartwoods and their differential cytotoxicity to normal and leukemia cells. Nat Prod Commun 4:69–76
54. Rezaei-Sadabady R, Eidi A, Zarghami N, Barzegar A (2014) Intracellular ROS protection efficiency and free radical-scavenging activity of quercetin and quercetin-encapsulated liposomes. Artif Cells Nanomed Biotechnol 1–7. doi: 10.3109/21691401.2014.926456
55. Rockenbach L, Bavaresco L, Fernandes Farias P et al (2013) Alterations in the extracellular catabolism of nucleotides are involved in the antiproliferative effect of quercetin in human bladder cancer T24 cells. Urol Oncol 31:1204–1211. doi:10.1016/j.urolonc.2011.10.009
56. Kim Y, Kim W-J, Cha E-J (2011) Quercetin-induced growth inhibition in human bladder cancer cells is associated with an increase in Ca-activated K channels. Korean J Physiol Pharmacol 15:279–283. doi:10.4196/kjpp.2011.15.5.279
57. Sak K (2014) Site-specific anticancer effects of dietary flavonoid quercetin. Nutr Cancer 66:177–193. doi:10.1080/01635581.2014.864418
58. Men K, Duan X, Wei XW et al (2014) Nanoparticle-delivered quercetin for cancer therapy. Anticancer Agents Med Chem 14:826–832
59. Niefind K, Guerra B, Ermakowa I, Issinger O-G (2001) Crystal structure of human protein kinase CK2: insights into basic properties of the CK2 holoenzyme. EMBO J 20:5320–5331. doi:10.1093/emboj/20.19.5320
60. Boldyreff B, Meggio F, Pinna LA, Issinger O-G (1994) Efficient autophosphorylation and phosphorylation of the beta-subunit by casein kinase-2 require the integrity of an acidic cluster 50 residues downstream from the phosphoacceptor site. J Biol Chem 269:4827–4831
61. Boldyreff B, Meggio F, Pinna LA, Issinger O-G (1994) Protein kinase CK2 structure-function relationship: effects of the beta subunit on reconstitution and activity. Cell Mol Biol Res 40:391–399
62. Meggio F, Boldyreff B, Issinger O-G, Pinna LA (1994) Casein kinase 2 down-regulation and activation by polybasic peptides are mediated by acidic residues in the 55-64 region of the beta-subunit. A study with calmodulin as phosphorylatable substrate. Biochemistry 33:4336–4342
63. Raaf J, Brunstein E, Issinger O-G, Niefind K (2008) The interaction of CK2alpha and CK2beta, the subunits of protein kinase CK2, requires CK2beta in a preformed conformation and is enthalpically driven. Protein Sci 17:2180–2186. doi:10.1110/ps.037770.108
64. Raaf J, Brunstein E, Issinger O-G, Niefind K (2008) The CK2 alpha/CK2 beta interface of human protein kinase CK2 harbors a binding pocket for small molecules. Chem Biol 15:111–117. doi:10.1016/j.chembiol.2007.12.012
65. Raaf J, Issinger O-G, Niefind K (2008) Insights from soft X-rays: the chlorine and sulfur substructures of a CK2alpha/DRB complex. Mol Cell Biochem 316:15–23. doi:10.1007/s11010-008-9826-1

66. Olsen BB, Fischer U, Rasmussen TL et al (2011) Lack of the catalytic subunit of DNA-dependent protein kinase (DNA-PKcs) is accompanied by increased CK2α' levels. Mol Cell Biochem 356:139–147. doi:10.1007/s11010-011-0954-7
67. Olsen BB, Guerra B (2008) Ability of CK2beta to selectively regulate cellular protein kinases. Mol Cell Biochem 316:115–126. doi:10.1007/s11010-008-9817-2
68. Skjøth IHE, Issinger O-G (2006) Profiling of signaling molecules in four different human prostate carcinoma cell lines before and after induction of apoptosis. Int J Oncol 28:217–229
69. Yde CW, Gyrd-Hansen M, Lykkesfeldt AE et al (2007) Breast cancer cells with acquired antiestrogen resistance are sensitized to cisplatin-induced cell death. Mol Cancer Ther 6: 1869–1876. doi:10.1158/1535-7163.MCT-07-0072

CK2 Inhibitors and the DYRK Family Protein Kinases

Yoshihiko Miyata

Abstract CK2 is a ubiquitous and pleiotropic Ser-/Thr-targeting acidophilic protein kinase. CK2 plays an important role in the aberrant proliferation of malignant cancer cells. Because of constitutive activity of CK2, its inhibitors have been widely used to analyze the physiological function of CK2 in cellular systems. In addition, CK2 inhibitors are regarded as promising cancer chemotherapeutic candidates. Recently, several commonly used CK2 inhibitors have been shown to suppress DYRK (dual-specificity tyrosine-phosphorylation-regulated protein kinase) family protein kinases. Thus, the results obtained with conventional CK2 inhibitors should be carefully interpreted considering their effects on DYRKs. In this chapter, after an introductory section on CK2 and its inhibitors, the structures and activation mechanism of DYRK family protein kinases are portrayed. DYRK1A is one of the pivotal factors encoded in Down's syndrome critical region on human chromosome 21, and dysregulation of DYRK1A may be a molecular basis of various phenotypes observed in Down's syndrome patients. Substrates, physiological function, binding partners, regulatory mechanisms, and CK2 inhibitor sensitivities of DYRK1A are described in detail. Finally, the biological and clinical importance of CK2 and DYRK1A as therapeutic targets will be discussed.

Keywords CK2 • DYRK1A • Phosphorylation • Protein kinase • Inhibitor • Down's syndrome • TBB • Cancer chemotherapeutics • NFAT • Leukemia

1 CK2 and Its Inhibitors

1.1 Protein Kinase CK2

CK2 (previously misleadingly called as "casein kinase 2") is a ubiquitous serine-/threonine-specific protein kinase [1–6] which was discovered many decades ago as a major responsible enzyme for protein phosphorylation. CK2 is a tetrameric

Y. Miyata (✉)
Department of Cell and Developmental Biology, Graduate School of Biostudies, Kyoto University, Kitashirakawa Oiwake-cho, Sakyo-ku, Kyoto 606-8502, Japan
e-mail: ymiyata@lif.kyoto-u.ac.jp

K. Ahmed et al. (eds.), *Protein Kinase CK2 Cellular Function in Normal and Disease States*, Advances in Biochemistry in Health and Disease 12, DOI 10.1007/978-3-319-14544-0_19

enzyme consisting of two α- and/or α′-catalytic subunits and two accessory β-subunits that dimerize to bring two catalytic subunits together. CK2 preferentially phosphorylates Ser/Thr residues that are followed by a cluster of acidic amino acids; thus, CK2 is categorized in the "acidophilic" protein kinase class. This is a sharp contrast to other classical AGC group "basophilic" protein kinases, including PKA and PKC. Many signaling protein kinases belonging to the CMGC group such as mitogen-activated protein kinases (MAP kinases) and cyclin-dependent kinases (CDKs) specifically phosphorylate Ser/Thr residues followed by a Pro residue; thus, they are categorized as "Pro-directed" protein kinases. CK2 shares a common root with CMGC group protein kinases in the phylogenic tree (Fig. 1), but the amino acid sequence of the catalytic domain of CK2 is only distantly related to other CMGC kinases. CK2 plays essential and pivotal roles in many physiological systems by phosphorylating a wide variety of more than 300 cellular proteins [7]. In fact, phospho-proteomic analyses suggest that CK2 alone may be responsible for more than 20 % of the eukaryotic protein phosphorylation sites [8, 9]. However, even after the long history of CK2 studies, it remains unclear how CK2 activity is regulated in cells. CK2 activity is readily detectable in extracts of cells and tissues without any stimulation. In contrast to many other protein kinases that require activating

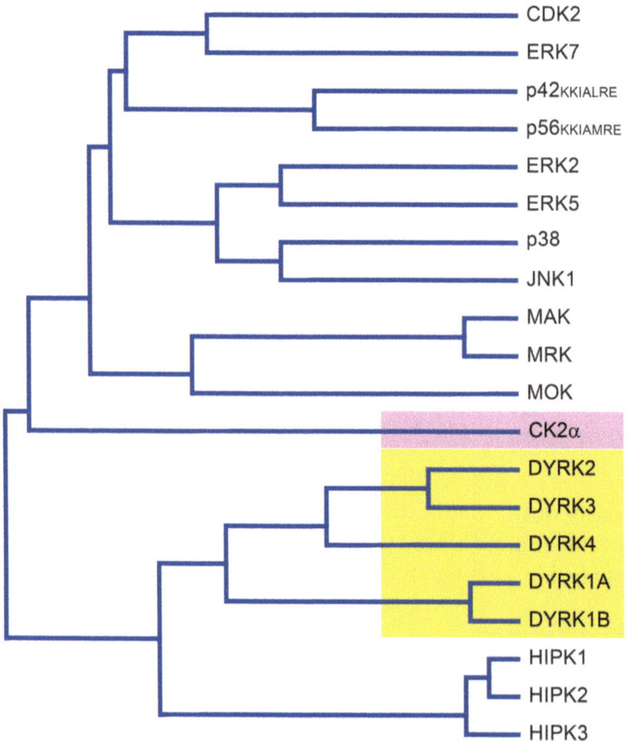

Fig. 1 A phylogenic tree of CMGC group Pro-directed protein kinases and CK2α. Similarity of amino acid sequences of the catalytic domains of indicated kinases was analyzed using the fixed distance scale UMGMA method. CK2 and DYRK family protein kinases are *colored*

low-molecular-weight factors, no pivotal factor that directly activates CK2 in cells is revealed to date. Signaling protein kinases are often positively or negatively controlled by binding to regulatory partner subunits. The β-subunit of CK2 confers substrate specificity in CK2 holoenzyme and CK2β is required for optimal phosphorylation of certain substrates [4, 10, 11]. However, the catalytic CK2α/α′ alone seems to be able to phosphorylate most of CK2 substrates. Many signaling protein kinases are regulated by phosphorylation with upstream "kinase kinases." Several protein kinases, including Cdc2, Src, ERK2, and Akt, have been reported to phosphorylate CK2 [12–15], but none of these kinases seems to be absolutely required for CK2 activity. All the evidence suggests that the regulatory mechanism of CK2 activity is different from that of other well-characterized signaling protein kinases whose activities are responsive to cellular stimulations. Therefore, CK2 has been often suggested to be "constitutively active." On the other hand, CK2 activity is upregulated in highly malignant cancer cells [16, 17]. CK2 activity is intimately related to the cell cycle [18–20] and modulated during the circadian rhythm [21–23]. In addition, CK2 plays an important role in mediating Wnt/β-catenin signaling [24]. These facts altogether indicate that there should be a precise control mechanism for CK2 in cells, despite its constitutive activity when isolated.

1.2 CK2 Inhibitors

CK2 inhibitors have been widely used both in vivo and in vitro to examine the physiological role of CK2. CK2 activity has long been known to be strongly inhibited by polyamines such as heparin, and thus the heparin sensitivity has been regarded as a Merkmal for CK2 [25], but high molecular weights and the highly negatively charged nature of such polyanions limit their applicability to intact cells and bodies. In fact, heparin does not enter into the cells and also it shows many physiological effects on cells other than CK2 inhibition: therefore, it can only be used as a specific CK2 inhibitor in in vitro studies. Several protein kinase inhibitors, including emodin [26], quercetin [27], and apigenin [28, 29] suppress CK2 activity (Table 1); however, they are rather classified as general ATP-competitive ligands that inhibit a large number of protein kinases other than CK2. One of the first clues of a highly specific CK2 inhibitor came from a finding that DRB (5,6-dichloro-1-β-d-ribofuranosylbenzimidazole), an inhibitor of eukaryotic mRNA transcription, inhibits in a parallel manner the activity of CK2 [30]. A screening of derivatives of DRB identified several compounds that showed higher specificity for CK2 among other protein kinases known those days. The most widely used inhibitor for CK2 in the scientific literatures may be 4,5,6,7-tetrabromobenzotriazole (TBB), which was described as a potent and specific inhibitor for CK2 from yeast and mammalian sources in 1995 [31]. As of June 2014, more than 70 publications using TBB as a CK2 inhibitor are included in the PubMed database, and TBB still continues to be used as a specific CK2 inhibitor. TBB strongly inhibits CK2 ($Ki = 120$ nM) but not another ubiquitous acidophilic kinase CK1. This is an advantage of TBB, because some previous CK2 inhibitors such as DRB showed almost comparable inhibition of CK1.

Table 1 Inhibitors for CK2 and DYRK1A. IC_{50} (μM) or % remaining activities collected from literatures are shown

	CK2	DYRK1A
Staurosporine	19.5	0.012
Emodin	0.9–2.0	4.2
Quercetin	19 % at 20 μM	21 % at 10 μM
Apigenin	30	15 % at 10 μM
NBC	0.37	15
dNBC	32	0.6
DRB	6–10	19 % at 10 μM
IQA	0.39	50 % at 10 μM
DMAT	0.15	0.12
DMAT	7 % at 10 μM	5 % at 10 μM
TBB	0.15–1.6	1–4.4
TBB	6 % at 10 μM	4 % at 10 μM
TBBz	10 % at 10 μM	3 % at 10 μM
CX-4945	0.001–0.003	5 % at 0.5 μM for DYRK2
Harmine	>10	0.034
Leucettine L41	0.32	0.012

Note that the data were obtained in different systems with different ATP concentrations and different kinase sources

Recently, because many more kinases are available to examine the effects and specificities of protein kinase inhibitors, it is now apparent that the specificities of some of conventional CK2 inhibitors seem to be inadequate for cell biological and clinical purposes. Notably, many of the commonly used CK2 inhibitors, including DRB, TBB, TBBz (4,5,6,7-tetrabromo-1H-benzimidazole), IQA ([5-oxo-5,6-dihydro-indolo(1,2-a)quinazolin-7-yl]acetic acid), and DMAT (2-dimethylamino-4,5,6,7-tetrabromo-1H-benzimidazole), were revealed to inhibit DYRK family protein kinases as well (Table 1). Therefore, the results obtained with these conventional CK2 inhibitors should be carefully reinterpreted considering their effects on DYRK family protein kinases. In the following sections, a concise description on DYRK family protein kinases will be presented to be of help for those who are interested in and investigating CK2.

2 DYRK Family Protein Kinases

2.1 Structures of DYRK Family Protein Kinases

DYRK (dual-specificity tyrosine-phosphorylation-regulated protein kinase) family protein kinases belong to the CMGC protein kinase group that also includes MAP kinases and CDKs (Fig. 1). In mammalian species, the DYRK family consists of

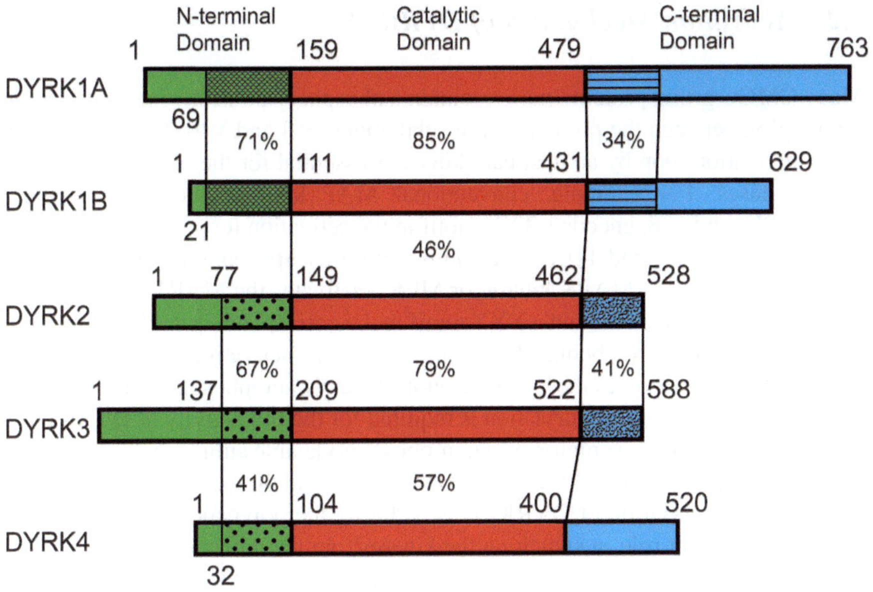

Fig. 2 A schematic illustration of the DYRK family protein kinase structures. Amino acid identities in the N-terminal, kinase catalytic, and C-terminal domains of the five DYRK members with amino acid numbers are shown

five members, including DYRK1A, DYRK1B, DYRK2, DYRK3, and DYRK4. According to the amino acid sequence similarity, DYRK1A and DYRK1B are categorized into "type 1" DYRKs, whereas DYRK2, DYRK3, and DYRK4 are "type 2" DYRKs. All DYRK family protein kinases share a homologous protein kinase catalytic domain in the center of their amino acid sequences (Fig. 2). The amino acid identity in the kinase domain is highest between DYRK1A and DYRK1B (85 %) and then between DYRK2 and DYRK3 (79 %). On the other hand, the amino acid identities between type 1 and type 2 DYRKs in the catalytic domain do not exceed 50 % (Fig. 2). All DYRKs possess characteristic N-terminal and C-terminal domains. DYRK1A and DYRK1B share two regions in their N- and C-terminal domains with amino acid identities of 71 % and 34 %, respectively. On the other hand, all the type 2 DYRKs share in their N-terminal domain a homologous region that is not observed in type 1 DYRKs. Finally, DYRK2 and DYRK3 have in common a similar C-terminal domain with 41 % amino acid identity (Fig. 2). DYRK1A is ubiquitously expressed in many species in multiple tissues and rich in testis and heart [32]. On the other hand, other DYRKs expressed mostly in testis [32]. DYRK1A has been most intensively studied so far; therefore, this chapter focuses mostly on DYRK1A. The detailed description of other members of DYRKs can be obtained in several review articles [33–36].

2.2 Activation Mechanism of DYRK1A

Many CMGC group protein kinases contain a phosphorylation site in the activation loop region between the protein kinase subdomains VII and VIII. Phosphorylation of the activation loop by an upstream kinase is essential for the optimal activity of these kinases. For example, conventional MAP kinases, including ERK1/2, JNK|SAPK, and p38, encode a TXY motif in the activation loop, and phosphorylation of both the Thr and Tyr residues in the motif by upstream activating kinases, MAP kinase kinases (MAPK kinase, or MEK), activates the MAP kinases. DYRK family protein kinases contain a YXY motif in the corresponding region of the activation loop. Instead of being phosphorylated by an activating upstream kinase, DYRK1A phosphorylates the Tyr in the motif by a *cis*-autophosphorylation mechanism [37]. This autophosphorylation is required for the full activity of DYRK1A; a replacement of Tyr in the region by a non-phosphorylatable amino acid significantly obstructs the protein kinase activity of DYRK1A. In addition, it was proposed that the autophosphorylation of DYRKs occurs during the polypeptide translation and plays an important role in switching the kinase from the autophosphorylating Tyr-kinase into a Ser/Thr kinase toward exogenous substrates [38]. A recent analysis modified the scheme and it is proposed that DYRK1A is in a structural equilibrium between the Tyr-autophosphorylating kinase and the Ser-/Thr-targeting kinase and that the Tyr-autophosphorylation stabilizes the conformation suitable for the Ser-/Thr-targeting kinase activity in the equilibrium [39]. The involvement of molecular chaperones has been reported in the maturation process of GSK3β [40]. Similarly, molecular chaperones may play a role in the conformational regulation of DYRKs.

Just like CK2, it seems that DYRK1A does not require a cofactor, upstream activating kinases, or an associating subunit for its activity, and it is "constitutively active" after translation and Tyr-autophosphorylation. A possibility cannot be excluded, however, that phosphorylation of DYRK1A by other kinases modulates its activity. In addition, there should be binding partners of DYRK1A that control the activity, stability, and localization of DYRK1A in cells (see Sect. 4).

2.3 Substrates for DYRK1A

Substrate specificities of DYRKs have been determined using peptide-based in vitro assays [32, 41, 42]. Although there may be slight specificity differences between the members, DYRKs in common preferentially phosphorylate Ser/Thr residues followed by a Pro. Therefore, DYRKs can be categorized into the Pro-directed protein kinases as many other CMGC group kinases. In addition, DYRK1A efficiently phosphorylates Ser/Thr residues with Arg at the −3 position and Pro at the −2 position, resulting in the optimal recognition sequence of RPX[ST]P [41]. This is strikingly different from the phosphorylation consensus sequence of CK2, [ST]XX[DEpSpTpY]; therefore, most of CK2 phosphorylation sites in proteins might not be phosphorylated by DYRK1A, except in rare cases that a sequence fulfills the optimal substrate consensus for both CK2 and DYRK1A.

The number of reported DYRK1A substrates has been rapidly increasing in the last decade. The list of DYRK1A substrates includes eIF2B [43], microtubule-binding protein tau [43–46], splicing factor SF3b1|SAP155 [47], caspase 9 [48, 49], glycogen synthase [50], CRY2 [51], cyclin L2 [52], DSCR1|RCAN1 [53], and several transcription factors, including STAT3 [37, 54], CREB [55], FKHR [56], Gli1 [57], and NFAT [58, 59]. The best characterized DYRK1A substrate, NFAT (nuclear factor of activated T cells), locates in the cytosol when phosphorylated and hence transcriptionally inactive. After dephosphorylation by a Ca^{2+}-activated phosphatase calcineurin, NFAT translocates to the nucleus and cooperates with multiple transcription factors to regulate the target gene expression. Nuclear NFAT protein is phosphorylated by several protein kinases including DYRK1A and translocates back to the cytosol [58, 59]. Therefore, DYRK1A is a negative regulator of NFAT-dependent signaling process.

2.4 Physiological Function of DYRK1A

Among the five members of DYRKs, DYRK1A has attracted most extensive attention, not only because of its ubiquitous expression but also because of its pivotal role in pleiotropic phenotypes observed in Down's syndrome caused by chromosome 21 trisomy. The analysis of partial trisomy cases suggested that Down's syndrome critical region (DSCR) in chromosome 21 between 21q22.1 and 21q22.3 is responsible for Down's syndrome caused by the trisomy. DSCR includes 33 genes, and DYRK1A is one of them [60, 61]. In fact, DYRK1A is overexpressed in Down's syndrome patients [62, 63], and the analyses of mouse models suggested that overexpression of DYRK1A is responsible for at least a part of pleiotropic phenotypes observed in Down's syndrome patients [64–67], although other genes in chromosome 21 DSCR should also play a role. *Minibrain* is a mutant of *Drosophila* that exhibits a marked brain size reduction in the optic lobes and central hemispheres [68]. The responsible gene *mnb* was identified, which encodes a fruit fly homologue of DYRK1A [68]. These results indicate that DYRK1A plays an important role in developmental and functional regulation of neuronal cells from insects to human. In addition, triplication of DYRK1A is necessary and sufficient in model mice to cause the structural and functional retinal alterations that are also observed in Down's syndrome children [69]. Gene knockout mice of the major DYRK1A substrate NFAT share various phenotypic alterations in facial structure, social interaction, motor function, and cardiac morphogenesis, with patients as well as model mice of Down's syndrome [59]. Altogether, DYRK1A is one of the pivotal factors encoded in DSCR and regulates many physiological functions of cells, and dysregulation of DYRK1A leading to NFAT deactivation by hyper-phosphorylation may be a molecular basis of various phenotypes observed in Down's syndrome patients. The functional importance of phosphorylation of other DYRK1A substrates in Down's syndrome awaits further experimental evidence.

3 Effect of CK2 Inhibitors on DYRKs (See Table 1)

The human genome encodes more than 500 protein kinases, and recently the effect of conventional CK2 inhibitors has been tested on many more kinases than before. While most of CK2 inhibitors have no impact on most of protein kinases examined, DYRK family protein kinases are often strongly suppressed by conventional CK2 inhibitors (Table 1). This is rather an unexpected result because the amino acid sequences of DYRKs are only distantly related to that of CK2 (Fig. 1).

IQA is an effective and selective CK2 inhibitor ($Ki = 170$ nM), and the structural details of the binding of IQA to CK2 were described by analyzing the CK2-IQA complex [70]. Examination of the effect of 10 μM IQA on a panel of 44 protein kinases indicated that most of the kinases other than CK2 (90 % inhibition) in the panel were not affected by IQA, except only one kinase, DYRK1A (50 % inhibition). Examination of 34 protein kinases for the effect of TBB and its derivatives resulted in a conclusion that TBB and 2-dimethylamino TBB (with better efficiency in CK2 inhibition) significantly inhibited DYRK1A, but not other kinases [71]. Pagano et al. examined a panel of 70 protein kinases for their sensitivity toward several conventional CK2 inhibitors [72]. While most of protein kinases tested were insensitive to 10 μM DMAT (93 % inhibition of CK2), it strongly inhibited the activity of DYRK1A (95 % inhibition), DYRK2 (94 % inhibition), and DYRK3 (97 % inhibition). Several other protein kinases, including PKD1, PIM1/2/3, and HIPK2/3, were also inhibited by DMAT. The effect of TBB on the same expanded set of 70 protein kinases was also examined, and DYRK and PIM kinases were strongly inhibited [72]. Several other newly developed TBB derivatives also significantly inhibited DYRK1A, whereas there are some compounds that showed only a modest inhibition of DYRK1A remaining sufficient efficacy in CK2 inhibition. Specific inhibition of CK2 without touching DYRK1A activity may be achieved if carefully selected compounds are used with appropriate concentrations.

Leucettines were originally developed as specific CLK and DYRK inhibitors, and leucettine L41 effectively suppressed all the members of DYRKs (IC_{50} for DYRK1A = 10 nM), while many other Pro-directed kinases, including CDKs and ERKs, were not affected [73]. However, the activity of CK2 was strongly suppressed by leucettine L41 (Table 1). Binding proteins for immobilized leucettine L41 were analyzed, and both DYRK1A and CK2 were found to specifically bind to the inhibitor [73]. These results further implicate the significant overlap of the binding and inhibiting spectrum of inhibitors for CK2 and DYRK1A.

All of these analyses suggested that CK2 and DYRKs may share a common structural factor in the ATP binding pocket, which may accept a similar array of ATP-analogous competitive inhibitors. The amino acid sequence alignment of the kinase catalytic domains of CK2α and DYRK1A is shown in Fig. 3. Both of the sequences have in common many characteristic amino acid motifs conserved in most of Ser/Thr kinases. In addition, there are several conserved amino acids between CK2α and DYRK1A, but it remains unclear which of them rule the sensitivity to shared inhibitors for both kinases.

```
                      *     ** *    *              * *            *
CK2α <human>    037:DDYQLVRKLGRGKYSEVFEAINITNNEKVVVKILKPVKKKKIKR--EIKILENLRGGPNI
                    | | +   +|+|    |  |     +| | +||+|  |        |+++|| +
DYRK1A <human>  157:DRYEIDSLIGKGSFGQVVKAYDRVEQEWVAIKIIKNKKAFLNQAQIEVRLLELMNKH-DT

CK2α <human>    095:ITLADIVKDPVS---RTPA-LVFEHVN--------NTDFKQLYQTLTDYDIRFYMYEILK
                         ||+       |    |||| +         || |+ ++  ||       |   |
DYRK1A <human>  216:EMKYYIVHLKRHFMFRNHLCLVFEMLSYNLYDLLRNTNFRGVSLNLTRKFAQQ-MCTALL

                               * ***  **           * * *                *     **
CK2α <human>    143:ALDYCHSMGIMHRDVKPHNVMIDHEHRK-LRLIDWGLAEFYHPGQEYNVRVASRYFKGPE
                     |     + |+| |+|| |+++   +|  ++++|+|       ||     + ||  + ||
DYRK1A <human>  275:FLA-TPELSIIHCDLKPENILLCNPKRSAIKIVDFGSSCQL--GQRIYQYIQSRFYRSPE

                                * ****
CK2α <human>    202:LLVDYQMYDYSLDMWSLGCMLASMIFRKEPFFHGHDNYDQLVRIAKVLG-----------
                    +|+    ||  +|||||||+|  |    || | |    ||+ +|  |||
DYRK1A <human>  332:VLLG-MPYDLAIDMWSLGCILVEMHTG-EPLFSGANEVDQMNKIVEVLGIPPAHILDQAP

CK2α <human>    251:------------TEDL---YDYIDKYNIELDPRFNDILGRHSRKRWERFVHSENQHLVSP
                                |  |    |    |      +   |||  +     |    |  | |
DYRK1A <human>  390:KARKFFENLPDGTWNLKKTKDGKREYKPPGTRKLHNILGVETGGPGGR-RAGESGHTVA-

                                       *
CK2α <human>    296:EALDFLDKLLR---YDHQSRLTAREAMEHPYF  324
                    + | | | +||   ||  +|+    |+ |  |
DYRK1A <human>  448:DYLKFKDLILRMLDYDPKTRIQPYYALQHSFF  479
```

Fig. 3 Alignment of amino acid sequences of the kinase catalytic domains of CK2α and DYRK1A. Identical amino acids are shown by "|" and similar amino acids are shown by "+." Amino acids conserved in most of Ser/Thr kinases are indicated by "*"

4 Functional Regulation of DYRK1A by Cellular Binding Partners

4.1 WDR68

To unveil the physiological function and regulation, the identification of cellular binding partners for DYRK1A should be of potential importance. Several studies revealed that a WD40-repeat protein WDR68 is a major binding partner for DYRK1A and DYRK1B, but not for class 2 DYRKs [50, 74–76]. The amino acid sequence of WDR68 is extremely conserved among species from plant to human, and all known mammalian WDR68 sequences encode exactly the same 342 amino acids. WDR68 is essential for cell proliferation and cell survival in mammalian cultured cells [75]. WDR68 is one of the components of a ubiquitin-conjugating enzyme DDB1-CUL4 complex [77], thus also called as DDB1-CUL4-associated factor 7 (DCAF7). The analysis of its amino acid sequence revealed that WDR68 contains five WD40-repeats; however, our computational structural analysis indicates that WDR68 forms a seven-bladed β-propeller ring [78]. All of these facts suggest that WDR68 plays a fundamental biological role in cells possibly by facilitating protein–protein interactions. Cellular localization of WDR68 is ubiquitous both in the cytoplasm and nucleus [75]. The cellular localization of WDR68 seems to be critical for its proper function, since mislocalization of WDR68 resulted in the developmental malformation of craniofacial structure in zebrafish [79]. Overexpression of DYRK1A induced nuclear accumulation of WDR68 [75]. Moreover, the molecular chaperone TRiC|CCT was essential for the DYRK1A-binding and nuclear accumulation of WDR68 [78]. The balance between cytoplasmic and nuclear WDR68 distribution may be precisely controlled, and DYRK1A should be a pivotal factor for the normal distribution and function of WDR68 in cells. The dysregulation of cellular localization of WDR68 by overexpressed DYRK1A might be a part of the molecular mechanism underlining the pleiotropic pathological alterations observed in Down's syndrome patients.

4.2 14-3-3

The 14-3-3 proteins are a family of regulatory molecules and participate in a wide range of cellular processes through binding to hundreds of structurally and functionally diverse proteins [80, 81]. 14-3-3 proteins recognize a sequence motif containing phospho-Ser/Thr in target proteins. 14-3-3 proteins were identified as binding partners of DYRK1A by a yeast two-hybrid screening [82, 83]. In yeast, a DYRK1A homologue Yak1p and a 14-3-3 homologue Bmh1/2p have been also shown to be associated together [84, 85]. 14-3-3 binds to either the N-terminal region of DYRK1A in a phosphorylation-independent manner or to the PEST domain near the C-terminal region after autophosphorylation of DYRK1A [82, 83]. The binding of 14-3-3 modestly increased DYRK1A kinase activity in vitro, and the inhibition of the 14-3-3 binding to DYRK1A by a small peptide decreased DYRK1A kinase activity. These results indicate that the 14-3-3 binding facilitates the protein kinase activity of DYRK1A.

4.3 REST|NRSF

REST (RE1-silencing transcription factor)|NRSF (neuron-restrictive silencer factor) plays a pivotal role in neuronal differentiation process by modulating transcription of its target genes by binding to a specific DNA element, the repressor element 1 (RE1)|neuron-restrictive silencer element (NRSE) [86, 87]. Target proteins for REST, including ion channels, synaptic proteins, and neurotransmitter receptors, have fundamental functions in neuronal cells. REST acts as a repressor of neuronal differentiation and activates proliferation. REST is induced during normal aging in human neurons, but is lost in Alzheimer's disease patients [88]. REST levels are closely correlated with cognitive preservation during aging, suggesting an important role of REST in neuroprotection [88]. DYRK1A overexpression reduces REST protein levels by facilitating its ubiquitination and degradation [89]. DYRK1A interacts with a SWI/SNF complex that is known to bind to REST [89]. REST stability is regulated by phosphorylation-dependent ubiquitination by an E3 ligase; however, it remains unclear if DYRK1A directly phosphorylates REST. Altogether, the DYRK1A function in neural cell differentiation may in part be ascribed to its REST level regulation. On the other hand, REST can activate DYRK1A transcription via a RE1|NRSE site in the human DYRK1A promoter [90], suggesting a negative feedback loop mechanism that precisely controls the expression levels of REST and DYRK1A. Dysregulation of the DYRK1A–REST combination may result in developmental as well as functional defects of neural system observed in Down's syndrome patients.

5 CK2 and DYRK1A as Therapeutic Targets

5.1 CK2 Inhibitors as Cancer Chemotherapeutic Agents

Early observations suggested that CK2 could be activated by growth factors such as epidermal growth factor and insulin-like growth factor [91–93]; however, recent analyses indicated that this might not generally be the case [94, 95]. On the other hand, CK2 activity is recognized to be higher in rapidly proliferating cells [16, 17], and exogenous overexpression of CK2 in transgenic mice is tumorigenic [96]. CK2 is thus implicated to play an important role in supporting the malignant growth of cancer cells and tumors. There are many known CK2 substrates that are involved in cell growth and proliferation; however, the detailed molecular mechanism of CK2-mediated cell proliferation is not yet fully revealed. CK2-dependent phosphorylation of Cdc37, a kinase-targeting co-chaperone for Hsp90, is essential for the folding and function of many Cdc37/Hsp90 client signaling kinases, including Cdk4 and Raf1 [97–100]. Therefore, phosphorylation of Cdc37 alone might significantly contribute to the important role of CK2 in cell growth and proliferation. CK2 is regarded as a promising molecular target for cancer chemotherapy, and many new-generation CK2 inhibitors have been recently developed. CX-4945 [Silmitasertib®] (5-((3-chlorophenyl)amino)benzo[c][2,6]naphthyridine-8-carboxylic acid) is a potent (Ki < 1 nM) and orally available ATP-competitive inhibitor with unprecedented specificity for CK2α and CK2α′ catalytic subunits [101, 102]. CX-4945 has antiproliferative activity in multiple cancer cell lines, shows antitumor efficacy in mouse models, and is under clinical trials for cancer chemotherapy [103, 104]. In phase I clinical trials, CX-4945 induced stable disease in 20 % of patients with different solid tumors, having promising pharmacodynamic and safety profiles. A combinatorial treatment with CK2 inhibitors and other chemotherapeutic agents might be also a valid therapeutic option.

5.2 Down's Syndrome, Leukemia, and DYRK1A Inhibitors

Involvement of DYRK1A in tumorigenesis is rather complex. Adult Down's syndrome individuals overexpressing DYRK1A show reduced tendency of most of malignant solid tumors of epithelial origin [105, 106]. This suggests that DYRK1A may have a tumor-suppressive function, but the molecular basis for this observation is not yet completely understood. It should be pointed out that DYRK family protein kinases have been proposed to play a role in promoting cell apoptosis [36], which is intimately involved in the exclusion of cancer cells in the body. On the contrary, children with Down's syndrome have a markedly increased risk of developing both acute megakaryoblastic leukemia and acute lymphoblastic leukemia as compared with children who do not have Down's syndrome [105, 106]. Therefore, DYRK1A could be proleukemic in children and antitumorigenic in adults.

The molecular mechanism behind this paradox is only partially figured out. In general, activation of the NFAT pathway is considered cancer promoting through several mechanisms [107]. The NFAT family proteins were originally discovered in T cells and shown to facilitate T cell activation and proliferation. In addition, NFAT pathway can enhance angiogenesis by activating transcription of VEGF (vascular endothelial growth factor). As describe in Sects. 2.3 and 2.4, increased expression of DYRK1A suppresses NFAT function by the phosphorylation-dependent nuclear export of NFAT. DYRK1A, in cooperation with another protein DSCR1|RCAN1 encoded in DSCR, suppresses VEGF-dependent endothelial cell proliferation [108]. The suppression of tumor angiogenesis by the DYRK1A-provoked NFAT inhibition may explain the lower rate of epithelial cancers in Down's syndrome adults. Angiogenesis is critical for the growth and expansion of cancer cells in solid tumor, whereas most of leukemic cells in blood do not rely on angiogenesis for proliferation. This could be one of the reasons why leukemia is increased, while solid tumors are decreased in Down's syndrome patients. Obviously, contributions of other DYRK1A substrates and other proteins encoded in chromosome 21 DSCR cannot be ruled out in the trisomy-related tumor suppression.

In any case, cell biological analyses with specific DYRK1A inhibitors may shed new light on the molecular basis of Down's syndrome and tumorigenesis as well. In addition, DYRK1A inhibition may have a therapeutic benefit. Recently, several specific DYRK1A inhibitors have been identified [109]. Harmine (7-methoxy-1-methyl-9H-pyrido[3,4-b]indole), originally isolated from South American vine, was reported to suppress the activity of DYRK1A [110], though it also shows a potent inhibitory effect on monoamine oxidase A [111]. Malinge et al. suggested that DYRK1A inhibitors may be clinically useful in the context of Down's syndrome-related acute megakaryoblastic leukemia by demonstrating that harmine inhibited the growth of megakaryoblastic leukemic cell lines with trisomy 21 [112]. Since some of conventional CK2 inhibitors have potent inhibitory activity on DYRK1A (see Sect. 3), one may consider the CK2 inhibitors as lead compounds for the development of DYRK1A-specific inhibitors that have no impact on CK2 activity.

Down's syndrome is caused by congenital alteration by the chromosome trisomy, and the therapeutic aim should be the modulation of DYRK1A from 1.5-fold increase back to the normal level [111]. This should be quite challenging having only a narrow therapeutic range for the DYRK1A inhibitors, but certainly worth the further investigation.

6 Conclusion

CK2 is a ubiquitous and constitutively active protein kinase implicated in the malignant proliferation of cancer cells. Low-molecular-weight specific inhibitors for CK2 have been developed both for biological and clinical applications. Recent studies revealed that many conventional CK2 inhibitors including DMAT and TBB also suppress the activity of DYRK family protein kinases; therefore, the alterations in tissues and cells observed with these drugs should not be ascribed solely to CK2. New-generation highly specific CK2 inhibitors such as CX-4945 have been recently developed and tested for clinical applications as cancer chemotherapeutic agents. DYRK1A plays a pivotal role in Down's syndrome, and DYRK1A-dependent phosphorylation of NFAT is a key event that causes various phenotypes and also low incidence of solid tumors in Down's syndrome patients. Specific inhibitors for DYRK1A may also have biological and clinical importance.

References

1. Olsen BB, Guerra B, Niefind K, Issinger OG (2010) Structural basis of the constitutive activity of protein kinase CK2. Methods Enzymol 484:515–529
2. Niefind K, Issinger OG (2010) Conformational plasticity of the catalytic subunit of protein kinase CK2 and its consequences for regulation and drug design. Biochim Biophys Acta 1804:484–492
3. Olsten ME, Litchfield DW (2004) Order or chaos? An evaluation of the regulation of protein kinase CK2. Biochem Cell Biol 82:681–693
4. Filhol O, Martiel J-L, Cochet C (2004) Protein kinase CK2: a new view of an old molecular complex. EMBO Rep 5:351–355
5. Pinna LA (2003) The raison d'être of constitutively active protein kinases: the lesson of CK2. Acc Chem Res 36:378–384
6. Litchfield DW (2003) Protein kinase CK2: structure, regulation and role in cellular decisions of life and death. Biochem J 369:1–15
7. Meggio F, Pinna LA (2003) One-thousand-and-one substrates of protein kinase CK2? FASEB J 17:349–368
8. Salvi M, Sarno S, Cesaro L, Nakamura H, Pinna LA (2009) Extraordinary pleiotropy of protein kinase CK2 revealed by weblogo phosphoproteome analysis. Biochim Biophys Acta 1793:847–859
9. Pagano MA, Cesaro L, Meggio F, Pinna LA (2006) Protein kinase CK2: a newcomer in the "druggable kinome". Biochem Soc Trans 34:1303–1306
10. Bolanos-Garcia VM, Fernandez-Recio J, Allende JE, Blundell TL (2006) Identifying interaction motifs in CK2β—a ubiquitous kinase regulatory subunit. Trends Biochem Sci 31:654–661

11. Bibby AC, Litchfield DW (2005) The multiple personalities of the regulatory subunit of protein kinase CK2: CK2 dependent and CK2 independent roles reveal a secret identity for CK2β. Int J Biol Sci 1:67–79
12. Bosc DG, Slominski E, Sichler C, Litchfield DW (1995) Phosphorylation of casein kinase II by p34*cdc2*. Identification of phosphorylation sites using phosphorylation site mutants in vitro. J Biol Chem 270:25872–25878
13. Donella-Deana A, Cesaro L, Sarno S, Ruzzene M, Brunati AM, Marin O, Vilk G, Doherty-Kirby A, Lajoie G, Litchfield DW, Pinna LA (2003) Tyrosine phosphorylation of protein kinase CK2 by Src-related tyrosine kinases correlates with increased catalytic activity. Biochem J 372:841–849
14. Ji H, Wang J, Nika H, Hawke D, Keezer S, Ge Q, Fang B, Fang X, Fang D, Litchfield DW, Aldape K, Lu Z (2009) EGF-induced ERK activation promotes CK2-mediated disassociation of α-catenin from β-catenin and transactivation of β-catenin. Mol Cell 36:547–559
15. Nguyen le XT, Mitchell BS (2013) Akt activation enhances ribosomal RNA synthesis through casein kinase II and TIF-IA. Proc Natl Acad Sci U S A 110:20681–20686
16. Ahmad KA, Wang G, Slaton J, Unger G, Ahmed K (2005) Targeting CK2 for cancer therapy. Anticancer Drugs 16:1037–1043
17. Xu X, Landesman-Bollag E, Channavajhala PL, Seldin DC (1999) Murine protein kinase CK2: gene and oncogene. Mol Cell Biochem 191:65–74
18. Marshak DR, Russo GL (1994) Regulation of protein kinase CKII during the cell division cycle. Cell Mol Biol Res 40:513–517
19. Lorenz P, Pepperkok R, Pyerin W (1994) Requirement of casein kinase 2 for entry into and progression through early phases of the cell cycle. Cell Mol Biol Res 40:519–527
20. Homma MK, Homma Y (2008) Cell cycle and activation of CK2. Mol Cell Biochem 316:49–55
21. Allada R, Meissner RA (2005) Casein kinase 2, circadian clocks, and the flight from mutagenic light. Mol Cell Biochem 274:141–149
22. Tsuchiya Y, Akashi M, Matsuda M, Goto K, Miyata Y, Node K, Nishida E (2009) Involvement of the protein kinase CK2 in the regulation of mammalian circadian rhythms. Sci Signal 2:ra26
23. Lin JM, Kilman VL, Keegan K, Paddock B, Emery-Le M, Rosbash M, Allada R (2002) A role for casein kinase 2α in the *Drosophila* circadian clock. Nature 420:816–820
24. Seldin DC, Landesman-Bollag E, Farago M, Currier N, Lou D, Dominguez I (2005) CK2 as a positive regulator of Wnt signalling and tumourigenesis. Mol Cell Biochem 274:63–67
25. Tuazon PT, Traugh JA (1990) Casein kinase I and II—multipotential serine protein kinases: structure, function and regulation. Adv Second Messenger Phosphoprotein Res 23:123–164
26. Yim H, Lee YH, Lee CH, Lee SK (1999) Emodin, an anthraquinone derivative isolated from the rhizomes of *Rheum palmatum*, selectively inhibits the activity of casein kinase II as a competitive inhibitor. Planta Med 65:9–13
27. Davies SP, Reddy H, Caivano M, Cohen P (2000) Specificity and mechanism of action of some commonly used protein kinase inhibitors. Biochem J 351:95–105
28. Hessenauer A, Montenarh M, Götz C (2003) Inhibition of CK2 activity provokes different responses in hormone-sensitive and hormone-refractory prostate cancer cells. Int J Oncol 22:1263–1270
29. Zhao M, Ma J, Zhu HY, Zhang XH, Du ZY, Xu YJ, Yu XD (2011) Apigenin inhibits proliferation and induces apoptosis in human multiple myeloma cells through targeting the trinity of CK2, Cdc37 and Hsp90. Mol Cancer 10:104
30. Zandomeni R, Zandomeni MC, Shugar D, Weinmann R (1986) Casein kinase type II is involved in the inhibition by 5,6-dichloro-1-β-D-ribofuranosylbenzimidazole of specific RNA polymerase II transcription. J Biol Chem 261:3414–3419
31. Szyszka R, Grankowski N, Felczak K, Shugar D (1995) Halogenated benzimidazoles and benzotriazoles as selective inhibitors of protein kinases CK I and CK II from *Saccharomyces cerevisiae* and other sources. Biochem Biophys Res Commun 208:418–424
32. Becker W, Weber Y, Wetzel K, Eirmbter K, Tejedor FJ, Joost H-G (1998) Sequence characteristics, subcellular localization, and substrate specificity of DYRK-related kinases, a novel family of dual specificity protein kinases. J Biol Chem 273:25893–25902

33. Mercer SE, Friedman E (2006) Mirk/Dyrk1B: a multifunctional dual-specificity kinase involved in growth arrest, differentiation, and cell survival. Cell Biochem Biophys 45:303–315
34. Aranda S, Laguna A, de la Luna S (2011) DYRK family of protein kinases: evolutionary relationships, biochemical properties, and functional roles. FASEB J 25:449–462
35. Becker W (2012) Emerging role of DYRK family protein kinases as regulators of protein stability in cell cycle control. Cell Cycle 11:3389–3394
36. Yoshida K (2008) Role for DYRK family kinases on regulation of apoptosis. Biochem Pharmacol 76:1389–1394
37. Wiechmann S, Czajkowska H, de Graaf K, Grötzinger J, Joost HG, Becker W (2003) Unusual function of the activation loop in the protein kinase DYRK1A. Biochem Biophys Res Commun 302:403–408
38. Lochhead PA, Sibbet G, Morrice N, Cleghon V (2005) Activation-loop autophosphorylation is mediated by a novel transitional intermediate form of DYRKs. Cell 121:925–936
39. Walte A, Rüben K, Birner-Gruenberger R, Preisinger C, Bamberg-Lemper S, Hilz N, Bracher F, Becker W (2013) Mechanism of dual specificity kinase activity of DYRK1A. FEBS J 280:4495–4511
40. Lochhead PA, Kinstrie R, Sibbet G, Rawjee T, Morrice N, Cleghon V (2006) A chaperone-dependent GSK3β transitional intermediate mediates activation-loop autophosphorylation. Mol Cell 24:627–633
41. Himpel S, Tegge W, Frank R, Leder S, Joost HG, Becker W (2000) Specificity determinants of substrate recognition by the protein kinase DYRK1A. J Biol Chem 275:2431–2438
42. Campbell LE, Proud CG (2002) Differing substrate specificities of members of the DYRK family of arginine-directed protein kinases. FEBS Lett 510:31–36
43. Woods YL, Cohen P, Becker W, Jakes R, Goedert M, Wang X, Proud CG (2001) The kinase DYRK phosphorylates protein-synthesis initiation factor eIF2Bepsilon at Ser539 and the microtubule-associated protein tau at Thr212: potential role for DYRK as a glycogen synthase kinase 3-priming kinase. Biochem J 355:609–615
44. Ryoo SR, Jeong HK, Radnaabazar C, Yoo JJ, Cho HJ, Lee HW, Kim IS, Cheon YH, Ahn YS, Chung SH, Song WJ (2007) DYRK1A-mediated hyperphosphorylation of Tau. A functional link between Down syndrome and Alzheimer disease. J Biol Chem 282:34850–34857
45. Liu F, Li B, Tung EJ, Grundke-Iqbal I, Iqbal K, Gong CX (2007) Site-specific effects of tau phosphorylation on its microtubule assembly activity and self-aggregation. Eur J Neurosci 26:3429–3436
46. Kimura R, Kamino K, Yamamoto M, Nuripa A, Kida T, Kazui H, Hashimoto R, Tanaka T, Kudo T, Yamagata H, Tabara Y, Miki T, Akatsu H, Kosaka K, Funakoshi E, Nishitomi K, Sakaguchi G, Kato A, Hattori H, Uema T, Takeda M (2007) The DYRK1A gene, encoded in chromosome 21 Down syndrome critical region, bridges between beta-amyloid production and tau phosphorylation in Alzheimer disease. Hum Mol Genet 16:15–23
47. de Graaf K, Czajkowska H, Rottmann S, Packman LC, Lilischkis R, Lüscher B, Becker W (2006) The protein kinase DYRK1A phosphorylates the splicing factor SF3b1/SAP155 at Thr434, a novel *in vivo* phosphorylation site. BMC Biochem 7:7
48. Laguna A, Aranda S, Barallobre MJ, Barhoum R, Fernández E, Fotaki V, Delabar JM, de la Luna S, de la Villa P, Arbonés ML (2008) The protein kinase DYRK1A regulates caspase-9-mediated apoptosis during retina development. Dev Cell 15:841–853
49. Seifert A, Allan LA, Clarke PR (2008) DYRK1A phosphorylates caspase 9 at an inhibitory site and is potently inhibited in human cells by harmine. FEBS J 275:6268–6280
50. Skurat AV, Dietrich AD (2004) Phosphorylation of Ser640 in muscle glycogen synthase by DYRK family protein kinases. J Biol Chem 279:2490–2498
51. Kurabayashi N, Hirota T, Sakai M, Sanada K, Fukada Y (2010) DYRK1A and glycogen synthase kinase 3β, a dual-kinase mechanism directing proteasomal degradation of CRY2 for circadian timekeeping. Mol Cell Biol 30:1757–1768
52. de Graaf K, Hekerman P, Spelten O, Herrmann A, Packman LC, Büssow K, Müller-Newen G, Becker W (2004) Characterization of cyclin L2, a novel cyclin with an arginine/serine-rich domain: phosphorylation by DYRK1A and colocalization with splicing factors. J Biol Chem 279:4612–4624

53. Jung MS, Park JH, Ryu YS, Choi SH, Yoon SH, Kwen MY, Oh JY, Song WJ, Chung SH (2011) Regulation of RCAN1 protein activity by Dyrk1A protein-mediated phosphorylation. J Biol Chem 286:40401–40412
54. Matsuo R, Ochiai W, Nakashima K, Taga T (2001) A new expression cloning strategy for isolation of substrate-specific kinases by using phosphorylation site-specific antibody. J Immunol Methods 247:141–151
55. Yang EJ, Ahn YS, Chung KC (2001) Protein kinase Dyrk1 activates cAMP response element-binding protein during neuronal differentiation in hippocampal progenitor cells. J Biol Chem 276:39819–39824
56. Woods YL, Rena G, Morrice N, Barthel A, Becker W, Guo S, Unterman TG, Cohen P (2001) The kinase DYRK1A phosphorylates the transcription factor FKHR at Ser329 in vitro, a novel in vivo phosphorylation site. Biochem J 355:597–607
57. Mao J, Maye P, Kogerman P, Tejedor FJ, Toftgard R, Xie W, Wu G, Wu D (2002) Regulation of Gli1 transcriptional activity in the nucleus by Dyrk1. J Biol Chem 277:35156–35161
58. Gwack Y, Sharma S, Nardone J, Tanasa B, Iuga A, Srikanth S, Okamura H, Bolton D, Feske S, Hogan PG, Rao A (2006) A genome-wide *Drosophila* RNAi screen identifies DYRK-family kinases as regulators of NFAT. Nature 441:646–650
59. Arron JR, Winslow MM, Polleri A, Chang CP, Wu H, Gao X, Neilson JR, Chen L, Heit JJ, Kim SK, Yamasaki N, Miyakawa T, Francke U, Graef IA, Crabtree GR (2006) NFAT dysregulation by increased dosage of DSCR1 and DYRK1A on chromosome 21. Nature 441:595–600
60. Galceran J, de Graaf K, Tejedor FJ, Becker W (2003) The Mnb/Dyrk1A protein kinase: genetic and biochemical properties. J Neural Transm Suppl 67:139–148
61. Hämmerle B, Elizalde C, Galceran J, Becker W, Tejedor FJ (2003) The Mnb/Dyrk1A protein kinase: neurobiological functions and Down syndrome implications. J Neural Transm Suppl 67:129–137
62. Guimerá J, Casas C, Estivill X, Pritchard M (1999) Human *minibrain* homologue (*MNBH/DYRK1*): characterization, alternative splicing, differential tissue expression, and overexpression in Down syndrome. Genomics 57:407–418
63. Dowjat WK, Adayev T, Kuchna I, Nowicki K, Palminiello S, Hwang YW, Wegiel J (2007) Trisomy-driven overexpression of DYRK1A kinase in the brain of subjects with Down syndrome. Neurosci Lett 413:77–81
64. Altafaj X, Dierssen M, Baamonde C, Martí E, Visa J, Guimerà J, Oset M, González JR, Flórez J, Fillat C, Estivill X (2001) Neurodevelopmental delay, motor abnormalities and cognitive deficits in transgenic mice overexpressing *Dyrk1A* (*minibrain*), a murine model of Down's syndrome. Hum Mol Genet 10:1915–1923
65. Ahn KJ, Jeong HK, Choi HS, Ryoo SR, Kim YJ, Goo JS, Choi SY, Han JS, Ha I, Song WJ (2006) DYRK1A BAC transgenic mice show altered synaptic plasticity with learning and memory defects. Neurobiol Dis 22:463–472
66. Martínez de Lagrán M, Altafaj X, Gallego X, Martí E, Estivill X, Sahún I, Fillat C, Dierssen M (2004) Motor phenotypic alterations in TgDyrk1a transgenic mice implicate DYRK1A in Down syndrome motor dysfunction. Neurobiol Dis 15:132–142
67. Galdzicki Z, Siarey R, Pearce R, Stoll J, Rapoport SI (2001) On the cause of mental retardation in Down syndrome: extrapolation from full and segmental trisomy 16 mouse models. Brain Res Rev 35:115–145
68. Tejedor F, Zhu XR, Kaltenbach E, Ackermann A, Baumann A, Canal I, Heisenberg M, Fischbach KF, Pongs O (1995) *Minibrain*: a new protein kinase family involved in postembryonic neurogenesis in Drosophila. Cell 14:287–301
69. Laguna A, Barallobre MJ, Marchena MA, Mateus C, Ramírez E, Martínez-Cue C, Delabar JM, Castelo-Branco M, de la Villa P, Arbonés ML (2013) Triplication of *DYRK1A* causes retinal structural and functional alterations in Down syndrome. Hum Mol Genet 22:2775–2784
70. Sarno S, de Moliner E, Ruzzene M, Pagano MA, Battistutta R, Bain J, Fabbro D, Schoepfer J, Elliott M, Furet P, Meggio F, Zanotti G, Pinna LA (2003) Biochemical and three-dimensional-structural study of the specific inhibition of protein kinase CK2 by [5-oxo-5,6-dihydroindolo-(1,2-a)quinazolin-7-yl]acetic acid (IQA). Biochem J 374:639–646

71. Pagano MA, Andrzejewska M, Ruzzene M, Sarno S, Cesaro L, Bain J, Elliott M, Meggio F, Kazimierczuk Z, Pinna LA (2004) Optimization of protein kinase CK2 inhibitors derived from 4,5,6,7-tetrabromobenzimidazole. J Med Chem 47:6239–6247
72. Pagano MA, Bain J, Kazimierczuk Z, Sarno S, Ruzzene M, Di Maira G, Elliott M, Orzeszko A, Cozza G, Meggio F, Pinna LA (2008) The selectivity of inhibitors of protein kinase CK2: an update. Biochem J 415:353–365
73. Tahtouh T, Elkins JM, Filippakopoulos P, Soundararajan M, Burgy G, Durieu E, Cochet C, Schmid RS, Lo DC, Delhommel F, Oberholzer AE, Pearl LH, Carreaux F, Bazureau JP, Knapp S, Meijer L (2012) Selectivity, cocrystal structures, and neuroprotective properties of leucettines, a family of protein kinase inhibitors derived from the marine sponge alkaloid leucettamine B. J Med Chem 55:9312–9330
74. Mazmanian G, Kovshilovsky M, Yen D, Mohanty A, Mohanty S, Nee A, Nissen RM (2010) The zebrafish *dyrk1b* gene is important for endoderm formation. Genesis 48:20–30
75. Miyata Y, Nishida E (2011) DYRK1A binds to an evolutionarily conserved WD40-repeat protein WDR68 and induces its nuclear translocation. Biochim Biophys Acta 1813: 1728–1739
76. Morita K, Lo Celso C, Spencer-Dene B, Zouboulis CC, Watt FM (2006) HAN11 binds mDia1 and controls GLI1 transcriptional activity. J Dermatol Sci 44:11–20
77. Jin J, Arias EE, Chen J, Harper JW, Walter JC (2006) A family of diverse Cul4-Ddb1-interacting proteins includes Cdt2, which is required for S phase destruction of the replication factor Cdt1. Mol Cell 23:709–721
78. Miyata Y, Shibata T, Aoshima M, Tsubata T, Nishida E (2014) The molecular chaperone TRiC/CCT binds to the Trp-Asp 40 (WD40) repeat protein WDR68 and promotes its folding, protein kinase DYRK1A binding, and nuclear accumulation. J Biol Chem 289:33320–33332
79. Wang B, Doan D, Roman Petersen Y, Alvarado E, Alvarado G, Bhandari A, Mohanty A, Mohanty S, Nissen RM (2013) Wdr68 requires nuclear access for craniofacial development. PLoS One 8:e54363
80. Darling DL, Yingling J, Wynshaw-Boris A (2005) Role of 14-3-3 proteins in eukaryotic signaling and development. Curr Top Dev Biol 68:281–315
81. Morrison DK (2009) The 14-3-3 proteins: integrators of diverse signaling cues that impact cell fate and cancer development. Trends Cell Biol 19:16–23
82. Alvarez M, Altafaj X, Aranda S, de la Luna S (2007) DYRK1A autophosphorylation on serine residue 520 modulates its kinase activity via 14-3-3 binding. Mol Biol Cell 18:1167–1178
83. Kim D, Won J, Shin DW, Kang J, Kim YJ, Choi SY, Hwang MK, Jeong BW, Kim GS, Joe CO, Chung SH, Song WJ (2004) Regulation of Dyrk1A kinase activity by 14-3-3. Biochem Biophys Res Commun 323:499–504
84. Moriya H, Shimizu-Yoshida Y, Omori A, Iwashita S, Katoh M, Sakai A (2001) Yak1p, a DYRK family kinase, translocates to the nucleus and phosphorylates yeast Pop2p in response to a glucose signal. Genes Dev 15:1217–1228
85. Lee P, Paik SM, Shin CS, Huh WK, Hahn JS (2011) Regulation of yeast Yak1 kinase by PKA and autophosphorylation-dependent 14-3-3 binding. Mol Microbiol 79:633–646
86. Negrini S, Prada I, D'Alessandro R, Meldolesi J (2013) REST: an oncogene or a tumor suppressor? Trends Cell Biol 23:289–295
87. Qureshi IA, Gokhan S, Mehler MF (2010) REST and CoREST are transcriptional and epigenetic regulators of seminal neural fate decisions. Cell Cycle 9:4477–4486
88. Lu T, Aron L, Zullo J, Pan Y, Kim H, Chen Y, Yang TH, Kim HM, Drake D, Liu XS, Bennett DA, Colaiácovo MP, Yankner BA (2014) REST and stress resistance in ageing and Alzheimer's disease. Nature 507:448–454
89. Lepagnol-Bestel AM, Zvara A, Maussion G, Quignon F, Ngimbous B, Ramoz N, Imbeaud S, Loe-Mie Y, Benihoud K, Agier N, Salin PA, Cardona A, Khung-Savatovsky S, Kallunki P, Delabar JM, Puskas LG, Delacroix H, Aggerbeck L, Delezoide AL, Delattre O, Gorwood P, Moalic JM, Simonneau M (2009) DYRK1A interacts with the REST/NRSF-SWI/SNF chromatin remodelling complex to deregulate gene clusters involved in the neuronal phenotypic traits of Down syndrome. Hum Mol Genet 18:1405–1414

90. Lu M, Zheng L, Han B, Wang L, Wang P, Liu H, Sun X (2011) REST regulates *DYRK1A* transcription in a negative feedback loop. J Biol Chem 286:10755–10763
91. Sommercorn J, Mulligan JA, Lozeman FJ, Krebs EG (1987) Activation of casein kinase II in response to insulin and to epidermal growth factor. Proc Natl Acad Sci U S A 84:8834–8838
92. Ackerman P, Osheroff N (1989) Regulation of casein kinase II activity by epidermal growth factor in human A-431 carcinoma cells. J Biol Chem 264:11958–11965
93. Klarlund JK, Czech MP (1988) Insulin-like growth factor I and insulin rapidly increase casein kinase II activity in BALB/c 3T3 fibroblasts. J Biol Chem 263:15872–15875
94. Litchfield DW, Dobrowolska G, Krebs EG (1994) Regulation of casein kinase II by growth factors: a reevaluation. Cell Mol Biol Res 40:373–381
95. Miyata Y, Nishida E (2007) Analysis of the CK2-dependent phosphorylation of serine 13 in Cdc37 using a phospho-specific antibody and phospho-affinity gel electrophoresis. FEBS J 274:5690–5703
96. Seldin DC, Leder P (1995) Casein kinase IIα transgene-induced murine lymphoma: relation to theileriosis in cattle. Science 267:894–897
97. Miyata Y (2013) The pivotal role of CK2 in the kinome-targeting Hsp90 chaperone machinery. The Wiley-IUBMB series on biochemistry and molecular biology: protein kinase CK2. John Wiley & Sons, Inc. (Hoboken, NJ, U.S.A.)
98. Miyata Y (2009) CK2: the kinase controlling the Hsp90 chaperone machinery. Cell Mol Life Sci 66:1840–1849
99. Miyata Y, Nishida E (2004) CK2 controls multiple protein kinases by phosphorylating a kinase-targeting molecular chaperone Cdc37. Mol Cell Biol 24:4065–4074
100. Miyata Y, Nishida E (2004) Supervision of multiple signaling protein kinases by the CK2-Cdc37 couple, a possible novel cancer therapeutic target. Ann N Y Acad Sci 1030:150–157
101. Siddiqui-Jain A, Drygin D, Streiner N, Chua P, Pierre F, O'Brien SE, Bliesath J, Omori M, Huser N, Ho C, Proffitt C, Schwaebe MK, Ryckman DM, Rice WG, Anderes K (2010) CX-4945, an orally bioavailable selective inhibitor of protein kinase CK2, inhibits prosurvival and angiogenic signaling and exhibits antitumor efficacy. Cancer Res 70: 10288–10298
102. Battistutta R, Cozza G, Pierre F, Papinutto E, Lolli G, Sarno S, O'Brien SE, Siddiqui-Jain A, Haddach M, Anderes K, Ryckman DM, Meggio F, Pinna LA (2011) Unprecedented selectivity and structural determinants of a new class of protein kinase CK2 inhibitors in clinical trials for the treatment of cancer. Biochemistry 50:8478–8488
103. Pierre F, Chua PC, O'Brien SE, Siddiqui-Jain A, Bourbon P, Haddach M, Michaux J, Nagasawa J, Schwaebe MK, Stefan E, Vialettes A, Whitten JP, Chen TK, Darjania L, Stansfield R, Bliesath J, Drygin D, Ho C, Omori M, Proffitt C, Streiner N, Rice WG, Ryckman DM, Anderes K (2011) Pre-clinical characterization of CX-4945, a potent and selective small molecule inhibitor of CK2 for the treatment of cancer. Mol Cell Biochem 356:37–43
104. Kim J, Kim SH (2012) Druggability of the CK2 inhibitor CX-4945 as an anticancer drug and beyond. Arch Pharm Res 35:1293–1296
105. Hasle H, Clemmensen IH, Mikkelsen M (2000) Risks of leukaemia and solid tumours in individuals with Down's syndrome. Lancet 355:165–169
106. Nižetić D, Groet J (2012) Tumorigenesis in Down's syndrome: big lessons from a small chromosome. Nat Rev Cancer 12:721–732
107. Müller MR, Rao A (2010) NFAT, immunity and cancer: a transcription factor comes of age. Nat Rev Immunol 10:645–656
108. Baek KH, Zaslavsky A, Lynch RC, Britt C, Okada Y, Siarey RJ, Lensch MW, Park IH, Yoon SS, Minami T, Korenberg JR, Folkman J, Daley GQ, Aird WC, Galdzicki Z, Ryeom S (2009) Down's syndrome suppression of tumour growth and the role of the calcineurin inhibitor DSCR1. Nature 459(7250):1126–1130
109. Becker W, Sippl W (2011) Activation, regulation, and inhibition of DYRK1A. FEBS J 278:246–256

110. Göckler N, Jofre G, Papadopoulos C, Soppa U, Tejedor FJ, Becker W (2009) Harmine specifically inhibits protein kinase DYRK1A and interferes with neurite formation. FEBS J 276:6324–6337
111. Becker W, Soppa U, Tejedor FJ (2014) DYRK1A: a potential drug target for multiple Down syndrome neuropathologies. CNS Neurol Disord Drug Targets 13:26–33
112. Malinge S, Bliss-Moreau M, Kirsammer G, Diebold L, Chlon T, Gurbuxani S, Crispino JD (2012) Increased dosage of the chromosome 21 ortholog *Dyrk1a* promotes megakaryoblastic leukemia in a murine model of Down syndrome. J Clin Invest 122:948–962

Sensitivity of Protein Kinase CK2 to ATP/GTP and Specific Inhibitors Depends on Complexity of Interacting Proteins

Andrea Baier and Ryszard Szyszka

Abstract In yeast cells five isoforms of protein kinase CK2 may simultaneously exist composed of $\alpha\alpha'\beta\beta'$, $\alpha_2\beta\beta'$, $\alpha'_2\beta\beta'$, and free catalytic α, α'. Each isoform exhibits properties typical for CK2, but they differ in their substrate specificity and sensitivity to specific modulators. Enzymes tested with protein substrates differently interacting with CK2 subunits were tested for their ATP/GTP binding capacity as well as with two commonly used ATP-competitive inhibitors TBB and TBI. Obtained results show that both ATP/GTP binding capacity and sensitivity to specific inhibitors are determined by the composition of CK2/substrate complexes. Both inhibitors, TBB and TBI, decrease cell growth to extend devoting interactions with different CK2 isoforms present in the cell; however, the presence of the regulatory $\beta\beta'$ dimer has a high importance toward sensitivity. Conceivably, only selected CK2-mediated processes in the cell can be inhibited by a given inhibitor concentration.

Protein kinase CK2 denotes one of the most pleiotropic serine/threonine protein kinases with hundreds of substrates already known. As the list of targets for CK2 continues to grow, it becomes evident that CK2 has the potential to participate in the regulation of fundamental cellular processes, such as gene expression, protein synthesis, proliferation, apoptosis, and differentiation/transformation. Protein kinase CK2 is distributed ubiquitously in eukaryotic organisms, where it appears as tetrameric complex composed of two catalytic subunits (α/α') associated with a dimer of regulatory β-subunits. Tetrameric CK2 complexes may contain identical (i.e., $\alpha_2\beta_2$ or $\alpha'_2\beta_2$) or nonidentical (i.e., $\alpha\alpha'\beta_2$) catalytic subunits, and the composition of the holoenzyme may have influence on CK2 properties, namely, nucleotide and protein substrate specificity and sensitivity to effectors. In contrast with the majority of other protein kinases, the catalytic subunit of CK2 has never been found in an inactive

A. Baier
Department of Molecular Biology, Faculty of Biotechnology and Environmental Sciences, Institute of Biotechnology, The John Paul II Catholic University of Lublin, Lublin, Poland
e-mail: szyszkar@kul.pl

R. Szyszka (✉)
Department of Molecular Biology, The John Paul II Catholic University of Lublin Institute of Biotechnology, Lublin, Poland
e-mail: szyszkar@kul.pl

K. Ahmed et al. (eds.), *Protein Kinase CK2 Cellular Function in Normal and Disease States*, Advances in Biochemistry in Health and Disease 12, DOI 10.1007/978-3-319-14544-0_20

conformation because of unique intramolecular constraints, a feature which is commonly referred to by saying that CK2 is "constitutively active." Such a lack of evident down-regulatory devices is likely to reflect the pleiotropy of this enzyme but is also probably responsible for the pathogenic potential of this kinase, whose aberrantly high activity has been associated with numerous diseases.

Keywords CK2 • Phosphorylation • ATP-competitive inhibitors • Protein–protein interactions

1 Properties of CK2

Protein kinase CK2 has a tetrameric structure which consists of two catalytic subunits (~42 kDa α and ~38 kDa α′) and a dimer of regulatory subunits (~28 kDa β) existing as configurations $\alpha\alpha'\beta_2$, $\alpha_2\beta_2$, or $\alpha'_2\beta_2$; however, the subunits may exist and function individually in cells. The two catalytic subunits are linked through a homodimer of the β-subunits in animal cells [1] or heterodimer of ββ′-subunits in yeasts [2].

While the catalytic subunits exhibit 90% amino acid sequence homology in the N-terminal region, the regulatory subunit does not demonstrate any sequence similarity. The β-subunit plays an important role in the assembly and stability of the holoenzyme and directs the recruitment of substrates, thus modulating the enzyme's substrate selectivity and specificity. The holoenzyme formation significantly amplifies CK2 catalytic activity [3–5].

There are some peculiar properties associated with protein kinase CK2 which are not found in any other protein phosphotransferase: (1) the enzyme is constitutively active, (2) it can use ATP as well as GTP, and (3) it is found elevated in most investigated rapidly proliferating tissues and tumors [3–6].

The importance of the protein kinase CK2 is evidenced by studies demonstrating its essential involvement in development, proliferation, and survival [3]. Mice with knockout of the catalytic subunit CK2α′ are viable but produce sterile offspring, indicating its involvement during spermatogenesis [7], whereas mice with knockout of the CK2β induce embryonic lethality [8]. This underscores the high degree of functional redundancy between both catalytic subunits CK2α and CK2α′ with non-absolute compensation.

Analogous results were obtained in yeast *Saccharomyces cerevisiae*, which also harbors two catalytic isoenzymes CK2α and CK2α′, designated CKA1 and CKA2 [9–11]. After disruption of either CKA1 or CKA2 gene, cells remain viable, while disruption of both of them is synthetically lethal. This indicates that the two isoforms of catalytic subunits can compensate for each other in the context of viability as it was observed in mice. Differently from mammalian holoenzymes, yeast CK2 contains two different regulatory subunits, designated CKB1 and CKB2 [9]. It was shown that in yeast haploid and diploid strains lacking either one or both beta subunit genes, cells are viable, demonstrating that the regulatory subunit of CK2 is dispensable. Such strains exhibit wild-type behavior with regard to growth on both fermentable and nonfermentable carbon sources, mating, sporulation, spore germination, and resistance to heat shock and nitrogen starvation but are salt-sensitive, which suggests a role for CK2 in ion homeostasis [12].

2 Biological Function of CK2

CK2 is a protein serine/threonine kinase which is a highly conserved and ubiquitous protein kinase. It is localized in the cytoplasmic and nuclear compartments, which accords with its multiple functional activities in the cell.

A number of protein kinases have an influence on apoptotic pathways by regulating caspases or events downstream of caspases. Of particular note are those proteins where phosphorylation by CK2 proximal to their cleavage sites has been shown to protect from caspases cleavage which results in apoptosis inhibition (e.g., Bid, Max, PTEN) [13]. In addition, phosphorylation of caspase-2 by CK2 inhibits its dimerization and activation [14]. Similarly, CK2 modification of Ser^{348} on caspase-9 appears to render the protease refractory to cleavage by active caspase-8 [15]. In case of ARC protein modification, CK2 increases its ability to localize to the mitochondria and inhibit caspase-8 activity [16].

Adequate to this is also the fact that a large number of putative substrates for CK2 have been identified in various compartments of the cell. New evidence from several laboratories has further reinforced the involvement of CK2 in signal transduction related to many cellular functions, thus underscoring the significance of its functional role in normal and abnormal cell growth and proliferation. Additionally, the shuttling of the kinase to various compartments in response to physiological and stress stimuli appears to be a key feature of the functional regulation of its activity in the cell. An important feature of CK2 biology is the recognition that the enzyme localizes dynamically to various compartments of the cell under different conditions [17].

Over 300 CK2 substrates have been identified, and more are being rapidly added to the list (reviewed in [18]). Three biggest functional groups of CK2 substrates are:

- Signaling proteins (e.g., β-catenin, androgen receptor Bid, calmodulin, DARPP-32, NLS receptor, RII subunit of PKA, inhibitor 2 of PP1, PKC IIβ, NIPP-1, PTEN)
- Transcription factors (e.g., c-MYB, c-MYC, CREB, CREM, IκBβ, p53, Max, TFIIIA)
- Proteins affecting functions of DNA/RNA and protein synthesis (e.g., DNA ligase, DNA topoisomerases I and IIα, eIF2, nucleolin, RNA polymerases I and III, Nopp140, TBP, ribosomal proteins P0, P1, and P2) [18]

CK2 phosphorylates protein substrates with minimal consensus sequence of Ser/Thr-X-X-Asp/Glu/pSer (where X denotes any nonbasic amino acid), which is distinct from other known intracellular kinases. CK2 phosphorylates its substrates by utilizing either ATP or GTP as a phosphate donor. This unique property of CK2 is due to the substitution of Val^{66} for Ala in the kinase-conserved region. This provides its catalytic core at the nucleotide-binding site with enough space for a hydrogen-bonding frameshift, allowing water molecules to utilize either adenine or guanine to employ the full hydrogen-bonding potential of this cleft [4].

Elevated levels of CK2 have long been associated with increased cell growth and proliferation, in normal as well as in cancer cells. It was shown that CK2 expression and activity are elevated in many cancer cells, including cancers of the head and neck, lung, prostate, and immune system [19–21]. The overexpression of CK2 attenuates apoptosis induced by chemotherapeutic drugs. This antiapoptotic effect is due to the ability of CK2 to phosphorylate in response to apoptotic stimuli pro-apoptotic proteins such as Max, Bid, connexin 45.6, and HS1 [3, 22]. In all cases, phosphorylation by CK2 protects these proteins from caspase-mediated degradation. On the other hand, downregulation of CK2 expression or inhibition of its activity facilitates cell death triggered upon drug exposure, ligation of death receptors, and ionizing radiation [16, 23]. In addition to this it has been shown that inhibition of CK2 with antisense oligodeoxynucleotides is a potent inducer of apoptosis in cancer but not in normal cells. It has been shown that inhibition of CK2 activity induces apoptotic cell death by activation of caspase-3, fragmentation of DNA, and externalization of phosphatidylserine. As a result of CK2 inhibition, a significant increase of hydrogen peroxide—a critical mediator of apoptosis—was observed in leukemia cells [16]. In a related study, pharmacological downregulation of CK2 activity by a panel of specific inhibitors, or knockdown of CK2α expression by RNA interference, was able to induce cell death in R-CEM line. It was shown that inhibitors of CK2 could promote an increased uptake of chemotherapeutic drugs inside the cells and sensitize them to drug-induced apoptosis in a cooperative manner [24]. Furthermore, it was shown that inhibition of CK2 resulted in induction of reactive oxygen species (ROS)-mediated apoptosis in leukemia cells [16]. Taken together, presented data suggest the involvement of CK2 in carcinogenesis by endowing cells the ability to evade apoptosis and as such present a novel target for potential exploitation for cancer management. Therefore considerable efforts have been made in recent years to develop specific and cell-permeable CK2 inhibitors.

3 Substrates Can Regulate CK2 Viability to Bind ATP/GTP

Protein kinase CK2 differs from other phosphotransferases in utilizing as phosphate donor ATP as well as GTP, being constitutively active and lacking factors that activate or inhibit the enzyme in response to stimuli [3, 25, 26]. The catalytic subunits of protein kinase CK2 have always active conformation and capability to phosphorylate substrate proteins both as part of a holoenzyme and in the absence of regulatory β-subunits. Many of CK2 protein substrates can serve at the same time as regulators of phosphotransferase activity. Studies in yeast and in human cells demonstrate that different isoforms of CK2 interact with a large number of cellular proteins [27]. Some of these, e.g., Nopp140 or nucleolin interactions with CK2, simply reflect enzyme–substrate interactions, but some other proteins, e.g., Cdc37,

Table 1 Yeast protein CK2 substrates and their characteristic

Protein substrate	Modified sequence	Short characteristic	References
RPP2B	EEEAKEE**S**[100]DDDMGF	Yeast 60S acidic ribosomal protein P2-beta. No interaction with CK2 subunits	[31–34]
FIP1	YSDS**S**[73]DDD**S**[77]DSDLE	Pre-mRNA polyadenylation factor, component of the cleavage and polyadenylation factor (CPF) complex. Interact with CK2α. FIP1 phosphorylation by α but not α′-subunit is inhibited by ββ′	[2, 35, 36]
ELF1	DDGDEG**S**[95]DSDYESD DGEID**S**[117]DEEEVDSD	A conserved transcription elongation factor from RNA polymerase promoter. Interact with all subunits of CK2. ELF1 phosphorylation by both α and α′ subunits is enhanced by ββ′	[37, 38]
SVF1[a]	DEEE**S**[210]**S**[211]ADEDD GD**S**[227]EEE**S**[231]G**S**[233]EEE EEE**S**[237]DSEEVE	Required for survival in response to oxidative stress. Involved in diauxic shift. Interact with CK2β. Regulatory ββ′ inhibit SVF1 phosphorylation by α but not by α′ catalytic subunit of CK2	[39–41]

[a]Putative phosphorylation sites

Hsp90, and Pin1, may alter or stabilize enzyme catalytic activity at the same time [3, 26–28]. In many case CK2 substrates interact with catalytic and/or regulatory subunits of protein kinase, as shown for FGF-1, FGF-2, Hsp90, and eIF2β [3, 26–30]. In addition, there are reports which demonstrate that CK2 subunits mediate effects other than phosphorylation by interacting as adaptor/scaffold/targeting proteins [3, 26, 27].

We examined all yeast isoenzymic forms of CK2, namely, enzymes with subunit composition αα′ββ′, $\alpha'_2\beta\beta'$, $\alpha_2\beta\beta'$ and free catalytic subunits α and α′ for their capacity to bind both phosphate donors with different protein substrates listed in Table 1. Proteins used as protein substrates are:

- Ribosomal protein P2B (RPP2B)—A component of the ribosomal stalk formed by five acidic proteins organized as a pentameric complex (P0–[P1A–P2B]/[P1B–P2A]), with P0 acting as an anchor linking the stalk to the ribosome and directly interacting with the large rRNA GTPase-associated domain [31] involved in the interaction between translational elongation factors EF2 and the ribosome [32]. Free (cytoplasmic) P2 stimulates the phosphorylation of the eIF2α subunit (Sui2) by protein kinase Gcn2 [33]. P2 also regulates the accumulation of P1 (RPP1A and RPP1B) in the cytoplasm. In all P-proteins forming ribosomal stalk, the conservative C-terminal end is phosphorylated by CK2 at serine residues [34].
- FIP1—Subunit of cleavage polyadenylation factor (CPF) interacting directly with poly(A) polymerase (Pap1) to regulate its activity. Amino acids 80–105 of

the FIP1 polypeptide are responsible for binding and for the inhibition of Pap1 activity as well as essential for cell viability [35]. CK2 phosphorylates FIP1 proteins at sites in close neighborhood of this region (Ser[73] and Ser[77]), which probably influences Pap1 activity and inhibits polyadenylation process [36]. Co-immunoprecipitation experiments show that FIP1 interact with CK2α subunit. Phosphorylation of FIP1 polypeptide by α but not α′ subunit is inhibited by ββ′ [2].

- ELF1—Factor maintaining proper chromatin structure in regions of active transcription elongation [37]. ELF1 is phosphorylated by CK2 at serine residues (Ser[95] and Ser[117]) [38], and probably this modification destabilizes interaction with Pol II and proteins Spt4 and Spt6 [37]. The N-terminal part of ELF1 contains the sequence K[3]RKKSTRK[10] with high homology to sequences K[75]KKIKRE[82] and K[85]MKKIYRE[92] present in CK2α and CK2α′, respectively. Co-immunoprecipitation experiments show that ELF1 interact with catalytic as well as regulatory subunits of yeast CK2 [38].
- SVF1—Protein with a potential role in cell survival pathways, required for the diauxic growth shift, and for yeast survival under conditions of oxidative stress, including cold stress [39]. Expression in mammalian cells increases survival under conditions inducing apoptosis. In yeast, SVF1 may be functionally complemented by exogenous expression of the human antiapoptotic Bcl-x(L) gene [40]. SVF1 protein is multiphosphorylated in vitro by CK2 (up to six Ser residues). Co-immunoprecipitation experiments show that SVF1 interacts only with regulatory β-subunits of yeast CK2. CK2 reconstitution experiments show that the dimer of regulatory subunits ββ′ inhibits SVF1 phosphorylation only in case of $\alpha_2\beta\beta'$ holoenzyme [41].

As shown in Tables 2 and 3, the efficiency of phosphate donor binding is different and depends on the composition of CK2 as well as from the protein substrate used in the experiment, suggesting that protein substrates together with CK2 regulatory subunits may have influence on utilization of ATP as well as GTP as phosphate donor. All isoforms of CK2 possess lower affinity for GTP with exception of free catalytic α-subunit and ribosomal P2B protein (respectively 39 μM and 42 μM). They bind ATP (Table 2) with *Km* values ranging from 2.4 μM for the complex of CK2α′/ELF1 until 54 μM for the complex of CK2α/SVF1. GTP utilization is much lower and has *Km* values from 23 to 24 μM for complexes of αα′ββ′ holoenzyme with ELF1 and SVF1 until much over 100 μM for many complexes especially those containing the $\alpha_2\beta\beta'$ holoenzyme.

Those results are in particular interest especially in relation to results presented for human CK2, which is able to utilize both phosphate donors at much lower concentration (ATP with *Km* values of 12, 15, 5, and 6 μM and GTP with *Km* values of 20, 21, 10, and 9 μM for α, α′, $\alpha_2\beta_2$, and $\alpha'_2\beta_2$, respectively, assayed with RRRDDDSDDD peptide substrate) [42]. In the report of Jensen et al. is shown that, unlike most CK2s, catalytic CK2α subunit of the protozoan parasite *Trypanosoma brucei* discriminates highly between ATP and GTP and only weakly accept guanosine triphosphate [43].

Table 2 Values of *Km* for ATP of different isoforms of yeast CK2

	Composition of yeast CK2				
	α	α'	$\alpha\alpha'\beta\beta'$	$\alpha_2\beta\beta'$	$\alpha'_2\beta\beta'$
Protein substrate	*Km* (μM)				
RPP2B	39.0	25.0	28.2	31.6	49.0
FIP1	46.0	6.6	13.7	26.2	25.0
ELF1	7.2	2.4	4.4	10.4	43.8
SVF1	53.0	28.0	7.5	16.2	18.0

Table 3 Values of *Km* for GTP of different isoforms of yeast CK2

	Composition of yeast CK2				
	α	α'	$\alpha\alpha'\beta\beta'$	$\alpha_2\beta\beta'$	$\alpha'_2\beta\beta'$
Protein substrate	*Km* (μM)				
RPP2B	42.0	80.0	55.0	>100	>100
FIP1	72.0	53.0	29.0	53.0	>100
ELF1	30.0	37.0	23.0	32.0	>100
SVF1	>100	82.0	24.2	42.0	75.0

4 ATP-Competitive Inhibitors of CK2

The degree of homology across the family of protein kinases is relatively high, especially within their catalytic sites. Nevertheless, the majority of CK2 inhibitors interact with the conserved ATP-binding site. Several classes of ATP-competitive inhibitors have been identified, showing variable effectiveness (Fig. 1) [44–49]. This class of inhibitors represents small chemical compounds acting as adenine mimics, such as halogenated benzimidazole derivatives, condensed polyphenolic derivatives, and indoloquinazoline-based compounds displaying high specificity for CK2 and efficiency in nanomolar concentrations. However, the molecular architecture of this multisubunit enzyme could offer alternative strategies to inhibit CK2 functions. As a result, recently, many laboratories show an increasing interest in identifying more specific CK2 inhibitors that do not directly compete with ATP and exhibit different mechanisms of action, e.g., targeting substrate-binding site, targeting the CK2β subunits, or disrupting subunit interaction [50–54].

In the case of protein kinase ATP-competitive inhibitors, the potency of the inhibitor in the cell depends on the K_m for ATP of its target, which means that such inhibitors will inhibit in cells those kinases that have a higher K_m [55]. Protein kinase CK2 has a relatively low K_m for ATP, ~10–20 μM, and therefore can be affected by this problem more than other kinases.

Nevertheless, during the last years a number of different classes of chemical compounds that target the CK2 active site have been characterized as ATP-competitive inhibitors, such as halogenated compounds, 4,5,6,7-tetrabromo-*1H*-benzimidazole

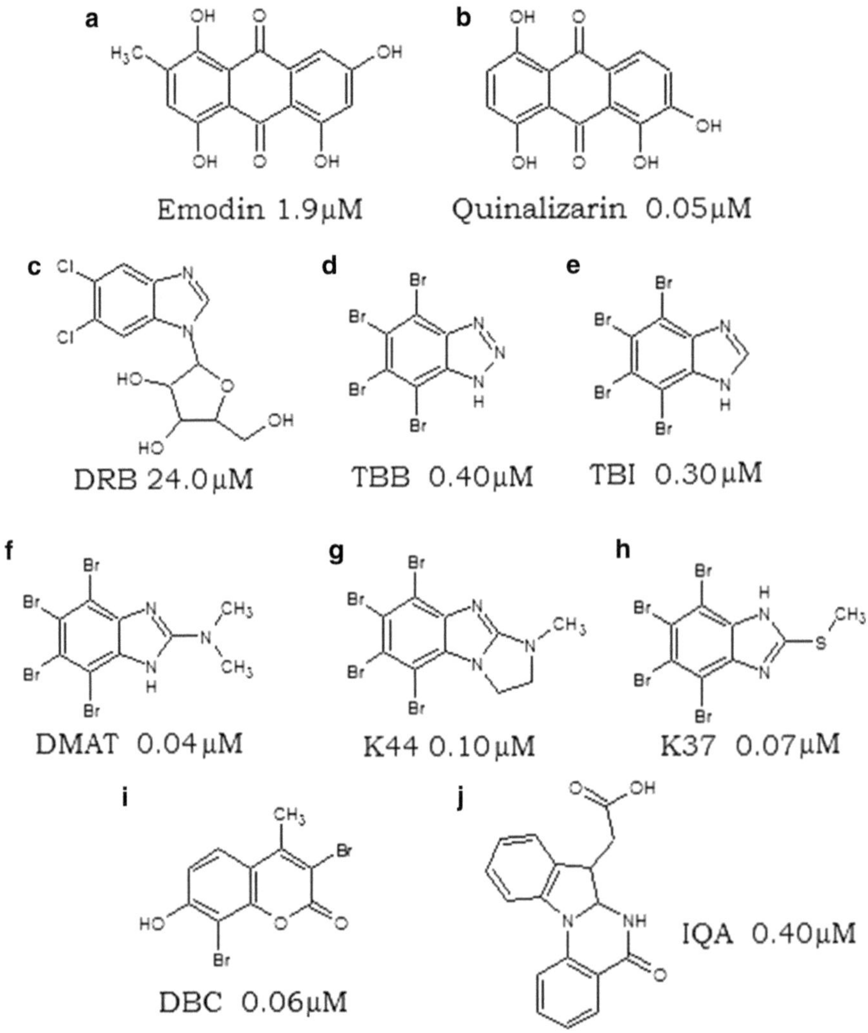

Fig. 1 Chemical formula of selected ATP-competitive CK2 inhibitors, with the value of the reported *Ki*

(TBB) derivatives [46], indoloquinazoline-based compounds [45, 56], and condensed polyphenolic derivatives [57]. They display high specificity for CK2 and high efficiency in the low micromolar range. Some of them show good cell permeability but differ in their physiological effects. This physiological difference probably reflects their off-target activity toward other ATP-binding proteins with similar nucleotide-binding sites [45, 47].

5 Interactions Between Protein Kinase CK2 and Substrates Influence Sensitivity to ATP-Competitive Inhibitors

A key role in cellular signal transduction pathways plays interactions between proteins. The regulatory CK2β subunit by bringing CK2α and/or CK2α′ into proximity with its partners and substrates enhances CK2 signaling [58].

TBB and TBI (Fig. 1d, e) have capacity to inhibit CK2 from various sources [59, 60]. In addition, it was shown that TBI does not influence the activity of free CK2α′ toward acidic ribosomal P-proteins in vitro and in vivo [48, 60]. Phosphorylation of yeast ribosomal P-proteins is known as one of the mechanisms regulating translational activity of the ribosomes. Ribosomal P-proteins can be phosphorylated in vitro by all CK2 isoforms. In our experiments translational activity of the yeast cells was monitored by comparison of the proportion between polysomes, 80S monosomes, and free ribosomal subunits (the most sensitive indicators of the physiological state of the cell and cell growth). It was shown that only TBB at 5 μM concentration induces significant increase of monosomes and free ribosomal subunit level correlated with the decrease of polysomes. The observed effect was correlated with inhibition of P-proteins phosphorylation [48] and indicates very high specificity of TBB in vivo just with complex P-proteins/CK2α′.

Both inhibitors (TBB and TBI) were tested with all five molecular isoforms of CK2 and different substrates. At 5 μM concentration of TBI, just few reactions were strongly inhibited (RPP2B with α, α′, and αα′ββ′; ELF1/$\alpha_2\beta\beta'$; SVF1/αα′ββ′), while the same TBB concentration significantly inhibits all CK2 reactions with substrates listed in Table 1.

As is shown in Tables 4 and 5, the efficiency of both investigated inhibitors of CK2 depends on phosphorylation complex composition. The *Ki* values vary from 0.2 μM for complexes RPP2B/$\alpha'_2\beta\beta'$, ELF1/α, and ELF1/αα′ββ′ with TBB until values over 50 μM for SVF1/$\alpha_2\beta\beta'$ and SVF1/$\alpha'_2\beta\beta'$ with TBI and ATP used as phosphate donor. It should be noted that in many cases, *Ki* is up to twice higher when GTP was used as a phosphate donor.

Obtained results show that besides enzyme–substrate interactions, there are other ones which have influence on the 3D structure of the ATP-binding pocket. Those interactions in protein complexes can change CK2 affinity for ATP/GTP as well as their sensitivity to ATP-competitive inhibitors.

Additionally when both inhibitors were used for their effect on the growth on yeast cells (a wild type and having single deletions of the genes CKA1, CKA2, CKB1, or CKB2), they show decreases of cell growth to extend devoting interaction with different CK2 isoforms present in the cell. Of high importance in sensitivity to both inhibitors is the presence of regulatory subunits. Only strains with deletion of genes coding for one of the regulatory subunits (CKB1 or CKB2) resulting in only free catalytic subunits (CK2α and CK2α′) present in the cell [2] show increased sensitivity to both investigated inhibitors.

Table 4 *Ki* values of different isoforms of yeast CK2 with ATP used as phosphate donor

	Composition of yeast CK2									
	α		α′		αα′ββ′		$\alpha_2\beta\beta'$		$\alpha'_2\beta\beta'$	
	Ki (μM)									
Protein substrate	TBB	TBI	TBB	TBI	TBB	TBI	TBB	TBI	TBB	TBI
RPP2B	0.6	0.4	1.4	0.9	0.4	1.4	1.8	21.0	0.2	4.3
FIP1	1.8	12.4	1.2	17.6	0.5	5.0	3.8	6.7	0.4	14.2
ELF1	0.2	5.3	1.0	19.0	0.2	13.9	2.5	1.8	0.8	7.2
SVF1	1.9	7.2	0.5	32.7	1.1	3.1	4.3	>50	2.8	>50

Table 5 *Ki* values of different isoforms of yeast CK2 with GTP used as phosphate donor

	Composition of yeast CK2									
	α		α′		αα′ββ′		$\alpha_2\beta\beta'$		$\alpha'_2\beta\beta'$	
	Ki (μM)									
Protein substrate	TBB	TBI	TBB	TBI	TBB	TBI	TBB	TBI	TBB	TBI
RPP2B	1.3	0.3	1.8	3.8	1.6	2.8	3.5	25.2	3.4	8.2
FIP1	2.4	5.1	1.4	6.3	2.0	2.4	4.2	3.4	4.3	9.8
ELF1	0.6	1.2	1.6	7.2	0.8	0.6	3.7	1.2	1.0	3.5
SVF1	1.1	15.0	2.4	3.2	2.6	0.6	2.2	25.8	6.7	28.2

Similar differences as described above were obtained when the sensitivity of human CK2α and CK2α′ was tested with 4,5,6,7-tetrabromo- and 4,5,6,7-tetraiodo-1H-benzimidazole derivatives and two different substrates: RPP2B and synthetic CK2 substrate (RRRDDDSDDD). In this case investigated iodo derivatives showed, in most cases, stronger inhibitory properties than respective brominated congeners, but differences showed considerable dependence on the substrate used [49]. The α′-subunit was mostly inhibited to a higher extent than the α-subunit.

6 Conclusions

Protein kinase CK2 is one of the most pleiotropic, ubiquitous, and constitutively active phosphotransferases, with cytosolic and nuclear localization in mammalian cells. It has many cellular targets and forms different signaling complexes which reflect the multifunctional nature of this enzyme. It is believed that CK2 promotes tumorigenesis because its protein content and/or activity are enhanced in many human cancers and rapidly proliferating tissues [3, 22]. Moreover, CK2 may play an important role in other human disorders and conditions such as Alzheimer's disease, ischemia, chronic alcohol exposure, or HIV infection [61]. During the last years, different classes of CK2-specific inhibitors were developed, and a few of them present an appropriate drug-like profile.

Conceivably (bearing in mind all presented results in vitro and in vivo), a given inhibitor concentration can inhibit only selected processes in the cell mediated by protein kinase CK2. All results together show that each of the five CK2 isoforms may regulate different processes and the way of its regulation, and sensitivity of each protein complex to a specific inhibitor depends on their components—between them are also protein substrates of CK2. The presented findings reveal the great challenge that is faced in order to understand the regulation of protein kinase CK2. An understanding of how these protein substrates/regulators interact with CK2 may also help to develop new CK2-specific inhibitors, inhibitors that may be important for treatment of cancer and other diseases.

References

1. Graham KC, Litchfield DW (2000) The regulatory β subunit of protein kinase CK2 mediates formation of tetrameric CK2 complexes. J Biol Chem 275:5003–5010
2. Kubiński K, Domańska K, Sajnaga E, Mazur E, Zieliński R, Szyszka R (2007) Yeast holoenzyme of protein kinase CK2 requires both β and β′ regulatory subunits for its activity. Mol Cell Biochem 295:229–235
3. Litchfield DW (2003) Protein kinase CK2: structure, regulation and role in cellular decisions of life and death. Biochem J 369:1–15
4. Niefind K, Guerra B, Ermakowa I, Issinger O-G (2001) Crystal structure of human protein kinase CK2: insights into basic properties of the CK2 holoenzyme. EMBO J 20:5320–5331
5. Pinna LA (2002) Protein kinase CK2: a challenge to canons. J Cell Sci 115:3873–3878
6. Guerra B, Issinger O-G (2008) Protein kinase CK2 in human diseases. Curr Med Chem 15:1870–1886
7. Escalier D, Silvius D, Xu X (2003) Spermatogenesis of mice lacking CK2alpha′: failure of germ cell survival and characteristic modifications of the spermatid nucleus. Mol Reprod Dev 66:190–201
8. Buchou T, Vernet M, Blond O, Jensen HH, Pointu H, Olsen BB, Cochet C, Issinger O-G, Boldyreff B (2003) Disruption of the regulatory beta subunit of protein kinase CK2 in mice leads to a cell-autonomous defect and early embryonic lethality. Mol Cell Biol 23:908–915
9. Glover CW (1998) On the physiological role of casein kinase II in *Saccharomyces cerevisiae*. Prog Nucleic Acid Res Mol Biol 59:95–133
10. Rethiswamy A, Birnbaum MJ, Glover CV (1998) Temperature sensitive mutations of the CKA1 gene reveal a role for casein kinase II in maintenance of cell polarity in *Saccharomyces cerevisiae*. J Biol Chem 273:5869–5877
11. Poole A, Poore T, Bandhakavi S, McCann RO, Hanna DE, Glover CV (2005) A global view a CK2 function and regulation. Mol Cell Biochem 274:163–170
12. Bidwai AP, Reed JC, Glover CV (1995) Cloning and disruption of CKB1, the gene encoding the 38-kDa beta subunit of *Saccharomyces cerevisiae* casein kinase II (CKII). Deletion of CKII regulatory subunits elicits a salt-sensitive phenotype. J Biol Chem 270:10395–10404
13. Duncan JS, Turowec JP, Vilk G, Li SSC, Gloor GB, Litchfield DW (2010) Regulation of cell proliferation and survival: convergence of protein kinases and caspases. Biochim Biophys Acta 1804:505–510
14. Shin S, Lee Y, Kim W, Ko H, Choi H, Kim K (2005) Caspase-2 primes cancer cells for TRAIL-mediated apoptosis by processing procaspase-8. EMBO J 24:3532–3542
15. McDonnell MA, Abedin MJ, Melendez M, Platikanova TN, Ecklund JR, Ahmed K, Kelekar A (2008) Phosphorylation of murine caspase-9 by the protein kinase casein kinase 2 regulates its cleavage by caspase-8. J Biol Chem 283:20149–20158

16. Hanif IM, Ahmad KA, Ahmed K, Pervaiz S (2009) Involvement of reactive oxygen species in apoptosis induced by pharmacological inhibition of protein kinase CK2. Ann N Y Acad Sci 1171:591–599
17. Faust M, Montenarh M (2000) Subcellular localization of protein kinase CK2—a key to its function? Cell Tissue Res 301:329–340
18. Meggio F, Pinna LA (2003) One-thousand-and-one substrates of protein kinase CK2? FASEB J 17:349–368
19. Ahmad KA, Harris NH, Johnson AD, Lindvall HC, Wang G, Ahmed K (2007) Protein kinase CK2 modulates apoptosis induced by resveratrol and epigallocatechin-3-gallate in prostate cancer cells. Mol Cancer Ther 6:1006–1012
20. Kim JS, Eom JI, Cheong JW, Choi AJ, Lee JK, Yang WI, Min YH (2007) Protein kinase CK2alpha as an unfavorable prognostic marker and novel therapeutic target in acute myeloid leukemia. Clin Cancer Res 13:1019–1028
21. Trembley JH, Wang G, Unger G, Slaton J, Ahmed K (2009) CK2: a key player in cancer biology. Cell Mol Life Sci 66:1858–1867
22. Duncan JS, Litchfield DW (2008) Too much of a good thing: the role of protein kinase CK2 in tumorigenesis and prospects for therapeutic inhibition of CK2. Biochim Biophys Acta 1784:33–47
23. Yamane K, Kinsella TJ (2005) CK2 inhibits apoptosis and changes its cellular localization following ionizing radiation. Cancer Res 65:4362–4367
24. Di Maira G, Brustolon F, Bertacchini J, Tosoni K, Marmiroli S, Pinna LA, Ruzzene M (2007) Pharmacological inhibition of protein kinase CK2 reverts the multidrug resistance phenotype of a CEM cell line characterized by high CK2 level. Oncogene 26:6915–6926
25. Pinna LA (1990) Casein kinase 2: an "eminence grise" in cellular regulation? Biochim Biophys Acta 1054:267–284
26. Olsten ME, Litchfield DW (2004) Order or chaos? An evaluation of the regulation of protein kinase CK2. Biochem Cell Biol 82:681–693
27. Gyenis L, Litchfield DW (2008) The emerging CK2 interactome: insights into the regulation and functions of CK2. Mol Cell Biochem 316:5–14
28. Miyata Y (2009) Protein kinase CK2 in health and disease: CK2: the kinase controlling the Hsp90 chaperone machinery. Cell Mol Life Sci 66:1840–1849
29. Skjerpen CS, Nilsen T, Wesche J, Olsnes S (2002) Binding of FGF-1 variants to protein kinase CK2 correlates with mitogenicity. EMBO J 21:4058–4069
30. Llorens F, Roher N, Miro FA, Sarno S, Ruiz FX, Meggio F, Plana M, Pinna LA, Itarte E (2003) Eukaryotic translation-initiation factor eIF2beta binds to protein kinase CK2: effects on CK2alpha activity. Biochem J 375:623–631
31. Krokowski D, Tchórzewski M, Boguszewska A, Grankowski N (2005) Acquisition of a stable structure by yeast ribosomal P0 protein requires binding of P1A-P2B complex: *in vitro* formation of the stalk structure. Biochim Biophys Acta 1724:59–70
32. Santos C, Ballesta JP (2005) Characterization of the 26S rRNA-binding domain in *Saccharomyces cerevisiae* ribosomal stalk phosphoprotein P0. Mol Microbiol 58:217–226
33. Jiménez-Díaz A, Remacha M, Ballesta JP, Berlanga JJ (2013) Phosphorylation of initiation factor eIF2 in response to stress conditions is mediated by acidic ribosomal P1/P2 proteins in *Saccharomyces cerevisiae*. PLoS One 8:e84219
34. Szyszka R (1999) Protein kinases phosphorylating acidic ribosomal proteins from yeast cells. Folia Microbiol 44:142–152
35. Helmling S, Zhelkowsky A, More CL (2001) Fip1 regulates the activity of poly(A) polymerase through multiple interactions. Mol Cell Biol 21:2026–2037
36. Zieliński R, Hellman U, Kubiński K, Szyszka R (2006) Fip1—an essential component of the *Saccharomyces cerevisiae* polyadenylation machinery is phosphorylated by protein kinase CK2. Mol Cell Biochem 286:191–197
37. Prather D, Krogan NJ, Emili NJ, Greenblat JF, Winston F (2005) Identification and characterization of Elf1, a conserved transcription elongation factor in *Saccharomyces cerevisiae*. Mol Cell Biol 25:10122–10135

38. Kubiński K, Zieliński R, Hellman U, Mazur E, Szyszka R (2006) Yeast Elf1 factor is phosphorylated and interacts with protein kinase CK2. J Biochem Mol Biol 39:311–318
39. Vander Heiden MG, Choy JS, Vander Weele DJ, Brace JL, Harris MH, Bauer DE, Prange B, Kron SJ, Thompson CB, Rudin CM (2002) Bcl-x(L) complements *Saccharomyces cerevisiae* genes that facilitate the switch from glycolytic to oxidative metabolism. J Biol Chem 277:44870–44876
40. Brace JL, Vander Welle DJ, Rudin CM (2005) Svf1 inhibits reactive oxygen species generation and promotes survival under conditions of oxidative stress in *Saccharomyces cerevisiae*. Yeast 22:641–652
41. Masłyk M, Kochanowicz E, Zieliński R, Kubiński K, Hellman U, Szyszka R (2008) Yeast surviving factor Svf1 as a new interacting partner, regulator and in vitro substrate of protein kinase CK2. Mol Cell Biochem 312:61–69
42. Dobrowolska G, Lozeman FJ, Li D, Krebs EG (1999) CK2, a protein kinase of the next millennium. Mol Cell Biochem 191:3–12
43. Jensen BC, Kifer CT, Brekken DL, Randall AC, Wang Q, Drees BL, Parsons M (2007) Characterization of protein kinase CK2 from *Trypanosoma brucei*. Mol Biochem Parasitol 157:28–40
44. Battistutta R (2009) Structural bases of protein kinase CK2 inhibition. Cell Mol Life Sci 66:1868–1889
45. Sarno S, Moro S, Meggio F, Zagotto G, Dal BD, Ghisellini P, Battistutta R, Zanotti G, Pinna LA (2002) Toward the rational design of protein kinase casein kinase-2 inhibitors. Pharmacol Ther 93:159–168
46. Pagano MA, Andrzejewska M, Ruzzene M, Sarno S, Cesaro L, Bain J, Elliott M, Meggio F, Kazimierczuk Z, Pinna LA (2004) Optimization of protein kinase CK2 inhibitors derived from 4,5,6,7-tetrabromobenzimidazole. J Med Chem 47:6239–6247
47. Duncan JS, Gyenis L, Lenehan J, Bretner M, Graves LM, Haystead TA, Litchfield DW (2008) An unbiased evaluation of CK2 inhibitors by chemoproteomics: characterization of inhibitor effects on CK2 and identification of novel inhibitor targets. Mol Cell Proteomics 7:1077–1088
48. Zień P, Abramczyk O, Domańska K, Bretner M, Szyszka R (2003) TBBz but not TBBt discriminates between two molecular forms of CK2 *in vivo* and its implications. Biochem Biophys Res Commun 312:623–628
49. Janeczko M, Orzeszko A, Kazimierczuk Z, Szyszka R, Baier A (2012) CK2α and CK2α′ subunits differ in their sensitivity to 4,5,6,7-tetrabromo- and 4,5,6,7-tetraiodo-1H-benzimidazole derivatives. Eur J Med Chem 47:345–350
50. Prudent R, Cochet C (2009) New protein kinase CK2 inhibitors: jumping out of the catalytic box. Chem Biol 16:112–120
51. Prudent R, Moucadel V, Laudet B, Barette C, Lafanechere L, Hasenknopf B, Li J, Bareyt S, Lacoste E, Thorimbert S, Malacria M, Gouzerh P, Cochet C (2008) Identification of polyoxometalates as nanomolar noncompetitive inhibitors of protein kinase CK2. Chem Biol 15:683–692
52. Laudet B, Moucadel V, Prudent R, Filhol O, Wong YS, Royer D, Cochet C (2008) Identification of chemical inhibitors of protein-kinase CK2 subunit interaction. Mol Cell Biochem 316:63–69
53. Perea SE, Reyes O, Baladron I, Perera Y, Farina H, Gil J, Rodriguez A, Bacardi D, Marcelo JL, Cosme K, Cruz M, Valenzuela C, López-Saura PA, Puchades Y, Serrano JM, Mendoza O, Castellanos L, Sanchez A, Betancourt L, Besada V, Silva R, López E, Falcón V, Hernández I, Solares M, Santana A, Díaz A, Ramos T, López C, Ariosa J, González LJ, Garay H, Gómez D, Gómez R, Alonso DF, Sigman H, Herrera L, Acevedo B (2008) CIGB-300, a novel proapoptotic peptide that impairs the CK2 phosphorylation and exhibits anticancer properties both *in vitro* and *in vivo*. Mol Cell Biochem 316:163–167
54. Farina HG, Benavent AF, Perera Y, Rodríguez A, Perea SE, Castro BA, Gomez R, Alonso DF, Gomez DE (2011) CIGB-300, a proapoptotic peptide, inhibits angiogenesis *in vitro* and *in vivo*. Exp Cell Res 317:1677–1688
55. Knight ZA, Shokat KM (2005) Features and selective kinase inhibitors. Chem Biol 12:621–637

56. Vangrevelinghe E, Zimmermann K, Schoepfer J, Portmann R, Fabbro D, Furet P (2003) Discovery of a potent and selective protein kinase CK2 inhibitor by high-throughput docking. J Med Chem 46:2656–2662
57. Meggio F, Pagano MA, Moro S, Zagotto G, Ruzzene M, Sarno S, Cozza G, Bain J, Elliott M, Deana-Deana A, Brunati AM, Pinna LA (2004) Inhibition of protein kinase CK2 by condensed polyphenolic derivatives. An in vitro and in vivo study. Biochemistry 43:12931–12936
58. Bibby AC, Litchfield DW (2005) The multiple personalities of the regulatory subunit of protein kinase CK2: CK2 dependent and CK2 independent roles reveal a secret identity for CK2beta. Int J Biol Sci 1:67–79
59. Domańska K, Zieliński R, Kubiński K, Sajnaga E, Masłyk M, Bretner M, Szyszka R (2005) Different properties of four molecular forms of protein kinase CK2 from *Saccharomyces cerevisiae*. Acta Biochim Pol 52:947–951
60. Zień P, Bretner M, Zastąpiło K, Szyszka R, Shugar D (2003) Selectivity of 4,5,6,7-tetrabromobenzimidazole as an ATP-competitive potent inhibitor of protein kinase CK2 from various sources. Biochem Biophys Res Commun 306:129–133
61. Cozza G, Bortolato A, Moro S (2010) How druggable is protein kinase CK2? Med Res Rev 30:419–462

Index

K. Ahmed et al. (eds.), *Protein Kinase CK2 Cellular Function in Normal and Disease States*, Advances in Biochemistry in Health and Disease 12, DOI 10.1007/978-3-319-14544-0

MIX
Papier aus verantwortungsvollen Quellen
Paper from responsible sources
FSC® C105338
www.fsc.org

If you have any concerns about our products,
you can contact us on
ProductSafety@springernature.com

In case Publisher is established outside the EU,
the EU authorized representative is:
Springer Nature Customer Service Center GmbH
Europaplatz 3, 69115 Heidelberg, Germany

Printed by Libri Plureos GmbH
in Hamburg, Germany